SOLAR SYSTEM

RADIO ASTRONOMY

SOLAR SYSTEM RADIO ASTRONOMY

*Lectures presented at
the NATO Advanced Study Institute
of the National Observatory of Athens:
Cape Sounion August 2-15, 1964*

Edited by
Jules Aarons

*Air Force Cambridge Research Laboratories
Bedford, Massachusetts*

Distributed by

PLENUM PRESS
NEW YORK
1965

ISBN-13: 978-1-4615-8605-0   e-ISBN-13: 978-1-4615-8603-6
DOI: 10.1007/978-1-4615-8603-6

Library of Congress Catalog Card Number 65-14086

©1965 Ionospheric Institute of the National Observatory
of Athens, Greece
Softcover reprint of the hardcover 1st edition 1965

Contents

THE INTERPLANETARY MEDIUM

THE MOON

THE PLANETS

Preface

The Ionospheric Institute of the National Observatory of Athens has had two interests in recent years: the study of the ionosphere and the study of the sun. In our previous Advanced Study Institutes in 1960, 1961, and 1962, we have emphasized the ionosphere. For the Advanced Study Institute of 1964, however, we invited Dr. Jules Aarons of the Air Force Cambridge Research Laboratories to collaborate in preparing and directing a program of studies of the sun, the moon, the planets, and the interplanetary medium.

The lectures of this Advanced Study Institute form essentially an advanced course in radio astronomy. Without being a textbook on the matter, we feel that the present book can be considered as an excellent reference for those students starting their research work in the field of solar system radio astronomy. All lecturers tried to present their subjects in a simple form based upon their extensive personal experience, but without emphasizing their personal research. We must recognize that it was an excellent achievement for them to keep their text exactly at the level indicated by the Program Director, and outlined by the general program of Advanced Study Institutes of NATO. We are deeply grateful to all the invited scientists for their outstanding contributions in lecturing on their subjects in a clear and authoritative manner.

The Scientific Affairs Division of NATO, in its aid to basic research, sponsors various programs. Among them is the Program of Advanced Study Institutes. As a Research Director in a country undergoing scientific growth, I can state that this Summer School is one of the most successful programs of aid. It provides a new form of meeting, different from the well-known conferences, symposia, or colloquia. Scientists from developed countries stimulate and aid research undertaken by scientists in countries under development. Living together in one pleasant place during the period of the Summer Schools provides the Institute members with many opportunities for worthwhile discussions not only in the conference rooms, but during leisure time as well. I feel that this informal and amicable exchange of ideas on scientific matters is of great value and most fruitful. Each individual is given the greatest possible opportunity for an exchange of creative ideas through the organization of Advanced Study Institutes.

In behalf of the participants, the lecturers, and especially the young team of radio astronomers of the National Observatory of Athens, I express my gratitude to NATO's Scientific Affairs Division for the foresight shown in sponsoring the Cape Sounion Advanced Study Institute on Solar System Radio Astronomy.

Michael Anastassiades

Professor, University of Athens
Director of the Advanced Study Institute of NATO
on Solar System Radio Astronomy

vii

Introduction

In the latter part of the summer of 1964, the National Observatory of Athens directed and hosted an Advance Study Institute on Solar System Radio Astronomy. The aim was to satisfy the dual need of supplying a written technical review of radio and radar astronomy as it related to the physics of the solar system and of conducting a school plus a conference. Invitations were extended to astronomers who had contributed to the field; they were asked to deliver lectures on the subjects proposed to them as well as write papers for a comprehensive volume on Solar System Radio Astronomy. The result is this volume.

In planning the publication, stress was placed on areas which had not been adequately covered by the plethora of review articles which have appeared recently. If reviews had appeared in the literature, a quite different approach to the subject was sought. Two examples among many are Weaver's review of passive lunar observations and Righini's advanced introduction to physical characteristics of the sun as obtained by optical techniques.

At the end of the Institute, both the lecturers and the participants are saturated with knowledge and discussion. One has the feeling that a great body of knowledge has been gathered by the tools of radio and radar astronomy, but it also was apparent that there are nagging and recurring questions that must be answered.

The final meeting of the Institute, a round table discussion, brought to light the ideas of the lecturers as to what might constitute future studies in solar system radio and radar astronomy.

The observations of the sun are varied both in type and in number, but there is open season on ideas as to the means by which flares generate radio frequency emission. Even within this volume, there are several theories as to the generation and radiation mechanisms of various burst types. Some observations of the sun have as yet inadequate materiel. The studies of magnetic field changes near sunspots during times before and after flare emission have limited data; new equipment in solar observatories is being set up to make these measurements. New observations of solar X-rays have just become available; these must be added to the background of observations that provide the foundation for theoretical studies. Millimeter observations of both the solar slowly varying component and solar bursts will become important in the future.

Absolute flux measurements are now the subjects of many discussions. The use of radio measurements from the solar centers of activity has increased with the correlation of solar flux values with atmospheric density, E-layer critical frequency, and other atmospheric parameters. The need is growing for both knowing the absolute value of the flux from the sun at a particular wavelength and comparing these measurements at various phases of the sunspot cycle. Both these solar data and others must be integrated from a worldwide point of view; observatories with radio telescopes on the same frequency report greatly varying values for a particular day. The integration of data on a worldwide basis is necessary in order to understand the evolution of a series of flares from a particular sunspot region.

The addition of polarization observations, absolute flux measurements, and spectral index recordings (of bursts) will aid in the construction of models of solar flare emission. Solar physicists were in agreement as to the need for both theoretical studies in magnetohydrodynamics, laboratory plasma experiments as well as additional measurements to understand the mechanism of the solar flare and its emissions.

The Cavendish laboratories, using their large interferometer, have recently observed scintillation of radio stars which they attribute to the motion and structure of the interplanetary medium. However, their derived velocities are somewhat low for the solar wind. Possibilities exist using this technique to reconstruct three-dimensional spatial variations and, of course, to study the outflow of large solar features and the solar corpuscular wind.

The participants agreed that the field of planetary studies was just beginning, both from the viewpoint of ground observations as well as deep space vehicles. Weaver has stressed the importance of good infrared studies of both the moon and the planets. The possibility of microwave spectroscopy yielding results was discussed; molecular spectra may be obtainable from ground measurements. Radar astronomy, of course, has been making great strides in determining the astronomical unit and the rotational period of the planets as well as the surface reflectivity.

Lunar studies, in spite of the possibility of an astronaut scooping up the surface material, will increase in the future from the earth. Infrared observations, polarization measurements (both by passive radio measurements and active radar), and long wavelength radar reflectivity studies (7 – 10 m) as well as the resolution of small regions, were areas where the participants felt research could and would be extended.

One comes away from a meeting of this type with two apparently diametrically opposed points of view; on one hand, a vast amount of observation and theoretical material on solar, lunar, and planetary processes has been gathered, but opposing this a vast amount needs to be done to answer the simplest of questions about solar flares and general solar activity, lunar surfaces, and planetary surfaces and atmospheres.

It is hoped that this book will serve as an introduction to solar system radio astronomy, just as the work done to date on solar physics will merely be an introduction to future understanding of solar processes. I would like to thank John Castelli for his assistance and advice in the preparation of this volume.

Jules Aarons

Editor
Air Force Cambridge Research Laboratories
U.S.A.

Opening Address

It is a great privilege and pleasure for me to attend, as a representative of the NATO Science Committee, the opening session of the Advanced Study Institute on Solar System Radio Astronomy. This institute is one of forty-three schools sponsored this year by the Science Committee within the framework of its Advanced Study Institute Program.

It is worth remembering that the Advanced Study Institute Program was started with five schools in 1959, and has been gradually expanded to the present level. Accordingly, the budget allocated by NATO for this purpose has been increased from 100,000 to 650,000 dollars over the same period.

The Advanced Study Institute Program is not intended for the support of regular scientific meetings such as conferences, symposia, or colloquia, but for courses or seminars at an advanced level, which last from two weeks to two months. There is no special form of organization, but it is expected that the schools provide for concentrated studies in a restricted scientific field.

Study institutes are held at any time of the year, although most of them take advantage of holiday periods during the summer months, so as to be able to bring together faculty staff and students from various nations. The schools are geographically situated in a country which is a member of the Atlantic Alliance; however, participation is not limited to citizens of NATO nations.

The subject of a summer school is chosen from any of the natural sciences and technology, including mathematics. No special form of application for a NATO grant is required if a scientist wishes to organize an institute. Eligible applications are considered by the staff of the Division for Scientific Affairs, and a small panel of three members of the Science Committee makes the selection of recipients and determines the amount of money. Decisions are taken on the basis of the anticipated quality of the study, the proposed degree of international participation concerning both faculty and students, and the desire to achieve a reasonable distribution of grants among different sciences and different countries.

The NATO grants are made completely unrestricted so as not to influence the management of the school. As a matter of policy, the director of the institute bears full responsibility and enjoys the liberty to establish the program and select teaching staff and students. Incidentally, many schools simultaneously receive money from other sources.

Since the introduction of the Advanced Study Institute Program in 1959 until today, NATO has provided financial assistance to about a hundred and fifty summer schools. Approximately fifty of these have published their courses in the form of brochures and books which are available in the literature. The Program is now well established and well known in the Western scientific community and will continue in the years to come.

Rudolf Schrader

Deputy Assistant Secretary General for Scientific Affairs
North Atlantic Treaty Organization
Paris, France

Solar Characteristics from Optical Observations

G. Righini

Osservatorio Astrofisico di Arcetri
Florence, Italy

INTRODUCTION

The sun is a dwarf star, type G 2 in the Harvard spectral classification, whose absolute visual magnitude is $M_v = +4.79$. It is therefore a typical main sequence star in the H-R diagram; in addition, its place in the Galaxy is rather peripheral since its distance from the center is about 27,000 light years, i.e., two-thirds of the radius of the disk. The sun revolves around the galactic center with a velocity of approximately 200 km/sec; the period of this motion is of the order of 10^8 years, i.e., one order of magnitude shorter than the admitted life of our solar system.

Since the average distance from the earth is 150×10^6 km, the sun is the only star which shows a visible radius; this, at a mean distance subtends an angle of $960'' = 0.005$ rad. Since the radius R_0 is 7×10^5 km, it follows that $1'' = 725$ km. Other physical parameters of the sun — the mass $M = 2 \times 10^{33}$ g, the mean density $\bar{\rho} = 1.41$ g/cm^3, the surface temperature T, and the luminosity L — will be discussed in the next section.

It is customary to distinguish three zones in the whole solar body: the photosphere, the chromosphere, and the corona. The prominences are features which have their "roots" in the chromosphere but they develop in the low corona.

THE PHOTOSPHERE

Solar Constant, Luminosity, and Temperature

The solar constant S is the flux received by a 1-cm^2 surface placed outside our atmosphere. Its value at the mean sun–earth distance is

$$S = 1.39 \times 10^6 \text{ erg/sec-cm}^2$$

The luminosity L is the total flux emitted by the sun, which is readily obtained by multiplication, with the surface of the sphere having a radius equal to the sun–earth distance. Its value is

$$L = 3.90 \times 10^{33} \text{ erg/sec}$$

1

TABLE I

Some Data about the Sun

Spectral type....................	G 2
Absolute visual magnitude........	+4.79
Radius $R_\odot$....................	7×10^5 km
Mass M........................	2×10^{33} g
Mean density $\bar{\rho}$................	1.41 g/cm^3
Effective temperature T_e	5800°K
Luminosity L	3.90×10^{33} erg/sec
Mean distance sun–earth........	150×10^6 km
Angular radius	0.005 rad = 960"
1" of arc	725 km
Solid angle	6.8×10^{-5} sr

and divided by the mass it gives the radiation per unit mass as 1.96 erg/sec-g. At the surface of the sun the radiation emitted is

$$F = 6.41 \times 10^{10} \text{ erg/cm}^2\text{-sec}$$

Thus by applying the Stefan law $F = \sigma T^4$ for blackbody emission, one obtains the so-called effective temperature

$$T_e = 5800°K$$

This temperature is only a useful parameter, which once introduced into the Stefan law reproduces the observed emission of the sun per cm-sec which allows a comparison to be made among the stars.

Nuclear Processes Inside the Sun

It is well known today, thanks to the works of Bethe *et al.*, that the energy radiated by the stars is produced by nuclear reactions. In the case of the sun, the main process is the association of four protons to give a nucleus of helium. Since each proton has a mass 1.0084 (referred to oxygen-16) and the resulting nucleus has a total mass of 4.004, the difference between this and the sum of the four pro-

TABLE II

Total Energy Produced by Some Phenomena

	Energy (ergs)
Violent earthquake	10^{25}
H-bomb (100 megaton)...............	4×10^{24}
Volcanic eruption	10^{24}
A-bomb (20 kton)..................	8×10^{20}
Lightning	2×10^7
Solar emission per second..........	3.90×10^{33}

tons, which amounts to 0.028 (equal to 0.7% of the total mass) is transformed into energy according to

$$E = mc^2$$

This process also called $p \longrightarrow p$ (or proton – proton) reaction can be represented as follows:

(a) ^{1}H (p, β^+, ν) ^{2}D

(b) ^{2}D (p, γ) ^{3}He

(c) ^{3}He (^{3}He, 2p) ^{4}He

where p is the proton, β^+ the positive electron, γ the quantum γ, and ν the neutrino. It is evidently necessary to have two reactions of the type (a) and (b) in order to produce two nuclei ^{3}He, which, interacting with two protons can produce one ^{4}He. The neutrinos are lost from the energy balance output, and on the whole the energy of the γ quanta produced by reaction (b) and those produced by annihilation of β^+ colliding with negative electrons amounts to 26.207 MeV.

Another process of less importance in the sun is the so-called carbon cycle, which leads to the production of five γ quanta with an energy output of 25.026 MeV.

Equilibrium Conditions of the Sun

The nuclear reactions require that the energy of the interacting particles be very high in order for them to penetrate the potential barriers of the nuclei. Inside a star such energy can only be thermal, which indicates that the temperature must be very high. Simple considerations about the mechanical equilibrium of a gaseous body lead to the same conclusion. In fact, the sun is a body in a state of gravitational hydrostatic equilibrium, which is described by the simple differential equation

$$dP = -g\rho\, dr$$

where P is the pressure, ρ is the local density, r is the local radius, and

$$g = G\frac{M(r)}{r^2}$$

is the local acceleration of gravity. Combining the two equations, we obtain

$$\frac{dP}{dr} = -\frac{G\,M(r)}{r^2}\,\rho(r)$$

Also

$$dM(r) = 4\pi r^2 \rho\, dr$$

and from the gas law

$$P = (k/H)(\rho/\mu) T$$

where k is the Boltzmann constant, H is the mass of the hydrogen atom, μ is the atomic weight (based on oxygen = 16), and T is the temperature. The last three equations allow us to deduce simply that in the center of the sun the pressure and the temperature must satisfy the relations

$$P_c \geq 1.326 \times 10^{15} \text{ dyn/cm}^2$$
$$T_c \geq 7.4 \times 10^6 \, ^\circ\text{K}$$

if $\mu = 1$ (pure hydrogen).

Solar Models

The procedure outlined in the preceding paragraph is a very crude one, and was only intended to illustrate that the physical theory of the equilibrium of a gas body requires the existence of a high temperature in the center. The more advanced theory of internal constitution of the stars permits us to compute refined solar models which give pressure, density, and temperature as a function of radius. The average of the best models gives the following conditions at the center of the sun:

$$T_c = 14 \times 10^6 \, ^\circ\text{K}$$
$$\rho_c = 98 \text{ g/cm}^3$$
$$P_c = 2.0 \times 10^{17} \text{ dyn/cm}^2$$

Table III gives the kinetic energy of a Maxwellian particle at a temperature comparable to that of the sun's center.

The probability of penetration into a nucleus is expressed by the exponential "Gamow factor." The product of the Gamow factor and the Maxwellian distribution shows a maximum for an energy that, at the temperature of the sun's center,

TABLE III

Percentage of Particles with Kinetic Energy Greater Than $E_0(T = 14 \times 10^6 \, ^\circ\text{K})$

E_0, eV	Percentage
0.7×10^3	88
2.1×10^3	39
4.2×10^3	11
8.2×10^3	0.7
14.0×10^3	2×10^{-4}
140.0×10^3	10^{-40}

is about 6.2×10^3 eV. For this energy, the number of processes is large enough to maintain the energy output of the sun for at least 75×10^9 years.

The Convection Zone

All solar models define a "core" where the temperature is high enough to maintain the nuclear processes. The energy produced there flows through the solar material to the outer surface. Except for the core, which is probably in a state of convective equilibrium, the major part of the solar mass is in radiative equilibrium, i.e., energy is transmitted without displacement of material; near the surface, the situation again changes, and convective currents set in.

That the sun is not a mixed star is evident from the abundance of the isotopes of hydrogen and carbon, which can be predicted for the inner core and observed in the atmosphere. From the cross sections observed in the laboratory, Fowler computed the following abundance ratios for the core of the sun:

$$^2H/^1H = 2.6 \times 10^{-17} \quad \text{and} \quad ^{13}C/^{12}C = 4.3 \pm 1.6$$

assuming

$$T_c = 13 \times 10^6 \,^\circ K \quad \text{and} \quad \rho_c = 150 \text{ g/cm}^3$$

The corresponding values observed in the solar atmosphere are instead

$$^2H/^1H = 4 \times 10^{-5} \quad \text{and} \quad ^{13}C/^{12}C = 10^{-4}$$

which in turn are quite different from the terrestrial values

$$^2H/^1H = 1.4 \times 10^{-4} \quad \text{and} \quad ^{13}C/^{12}C = 10^{-2}$$

The difference between the solar and terrestrial values is rather peculiar since with the exception of the lightest gases the relative abundance of the elements is almost the same as in the earth. Table IV contains the most recent data on the number abundance of the elements in the sun.

The predominant elements are hydrogen and helium, which have an overwhelming abundance. Because He lines do not occur in the Fraunhofer spectrum,

TABLE IV

Element	$\log N$	Element	$\log N$
H	12	N	8.15
He	11.3	O	8.68
Li	0.83	Na	6.14
Be	2.43	Mg	7.27
C	8.28	Fe	6.83

the ratio He/H remains uncertain and has to be derived from the spectra of prominences.

Solar matter maintains its state of radiative equilibrium as long as the actual temperature gradient is smaller than the temperature gradient which would hold for convective equilibrium. This is the well-known "stability criterion" of K. Schwarzschild:

$$\left|\frac{dT}{dh}\right|_{rad} < \left|\frac{dT}{dh}\right|_{ad}$$

which can also be written as

$$\left(\frac{d \ln T}{d \ln P}\right)_{rad} < \left(\frac{d \ln T}{d \ln P}\right)_{ad}$$

The theory shows that outside the core (which is in convective equilibrium) the adiabatic gradient is much larger than the radiative one. However, where the degree of hydrogen ionization tends to diminish this is no longer true because, when the hydrogen is only partially ionized, the value of $\gamma = c_p/c_v$ is small and the adiabatic gradient, $(\gamma - 1)/\gamma$, becomes smaller than the radiative gradient. For such layers we have the formation of convective currents which transport almost all of the energy. The convection zone extends deep inside the sun, up to 1/10 of the solar radius, reaching layers where the pressure is 10^{11} dyn/cm^2 (66,000 km below the solar limb).

The currents are turbulent since the Reynolds number, Re $\approx vl/c_s\delta$, is large, where v is the velocity of the currents, l is the mixing length, δ the mean free path, and c_s the velocity of sound. Because $v \approx c_s$ we find, using $l \approx 100$ km $= 10^7$ cm and $\delta = 2 \times 10^{-2}$ cm, that Re $= 10^9$. At the upper boundary of the convection currents the conditions for radiative equilibrium are again fulfilled.

The Granulation

Direct observation of the solar surface in white light shows the granulation which is in fact composed of the convective current viewed end on. Our atmosphere interferes severely with this kind of observation because of scattering and scintillation. New techniques have been employed to obtain pictures of the solar granulation from a balloon at a height of 25 km (Schwarzschild [1]), but comparably good results have been achieved by Rösch [2] photographing the sun from the Pic du Midi.

The size of the granules is of the order of $1'' = 700$ km, but some fine details of the order of $0.''3$ are sometimes visible inside the granules. The average lifetime of the granules obtained either from time correlation (Schwarzschild) or from the evolution of individual granules is of the order of 9 min, independent of whether or not they are in a magnetic region; larger granules seem to have longer lifetimes.

Granules are bright on a dark background. Their average temperature is difficult to obtain because of the perturbing effect of the atmosphere and other instru-

TABLE V

Limb Darkening of the Solar Disk

λ (Å)	$\sin \theta = 0.00$	$\sin \theta = 0.95$
3230	100	38.2
4330	100	45.0
5340	100	54.8
6700	100	62.9
8600	100	69.9
10310	100	73.0
12250	100	75.6
16550	100	81.5
20970	100	83.8

mental effects. Nevertheless Bahng and Schwarzschild were able to correct the observed data using photographs of the solar limb. They obtained a root-mean-square (rms) fluctuation of $\pm 92°K$ between granules and intergranular material.

The theory of the hydrogen convection zone of Biermann, Unsöld, and Vitense (Unsöld [3]) explains many aspects of the granulation. In this theory it is assumed that the mixing length l (or characteristic length) is of the order of the scale height $H = RT/\mu g$. The predicted size of the granules is of the same order as the scale height. The observations give a larger size (1.5 to 2 times greater). The mean turbulent velocity $\bar{v}$ computed following this theory is 1.7 km/sec.

Disk Radiation and Limb Darkening

The visible radiation of the whole sun has a spectrum whose distribution is not very different from that of a blackbody of temperature $T_e = 5800°K$. Its maximum falls around $\lambda = 4700$ Å and amounts to 30×10^{13} erg/cm^2-sec for a band of $\Delta \lambda = 1$ cm. Fairly large deviations from a blackbody curve are observed around the maximum and in the UV region; this confirms the concept that the effective temperature is only a useful parameter but has not a thermodynamical meaning, since in the sun and in the stars the fundamental conditions of thermodynamical equilibrium are not fulfilled. In fact, the spectral distribution of the radiation emitted by the center of the disk is quite different; it has a maximum at $\lambda = 4560$ Å which amounts to 39×10^{13} erg/cm^2-sec for a band $\Delta \lambda = 1$ cm.

The solar disk therefore does not have uniform brightness but shows "limb darkening," with an amplitude which increases with decreasing wavelength.

The data of Table V were derived using the heliocentric angle θ as defined in Fig. 4 and $\sin \theta = r/R_\odot$. The observations are reliable only up to a point which is at a distance of $0.97R_\odot$ or 0.5 from the limb. Beyond this point, because of the atmospheric scintillation and the light scattered into the apparatus, the measurements are inaccurate. However, during the last phase of a total eclipse it is possible to study the structure of the outer limb of the solar disk whose photometric curve is given in Fig. 2.

The abscissas are distances in kilometers from a reference level which will be discussed later on. It is clear from the figure that the brightness drops to almost

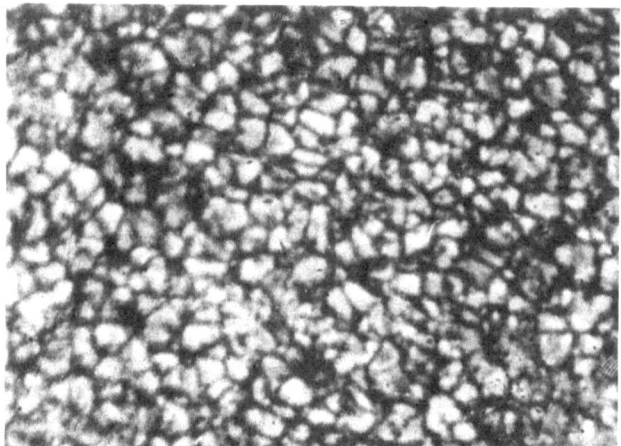

Fig. 1. Photospheric granulation. Photography taken by
Stratoscope I at an altitude of 24,000 m (Ap. J. 130, 345
(1959).

zero in about 700 km ≈ 1″. For this reason, the solar disk appears to be bounded
by a sharp edge.

The Solar Atmosphere — The Reference Level

We now encounter the problem of locating the photosphere or, in other words,
establishing a "reference level" from which to measure the depths. For this we define the "optical depth," which is expressed by

$$\tau = \int_0^t \kappa \rho \, dt$$

where κ is the absorption coefficient per gram of solar matter, ρ is the density, and
t is the length measured inward from the reference level. The absorption coeffi-

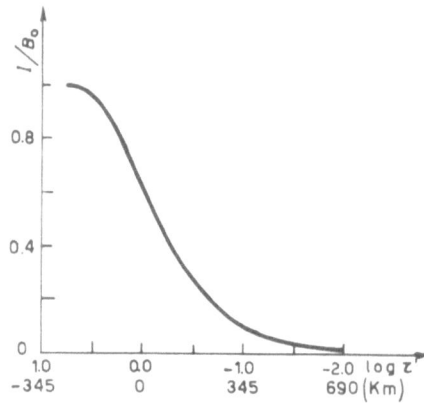

Fig. 2. Photometric profile of the solar
limb.

cient is wavelength-dependent and can assume a very high value at wavelengths corresponding to the center of strong absorption lines.

It is convenient to consider separately the continuum absorption coefficient κ and the monochromatic one κ'; the first varies slowly with λ as shown in Fig. 3; the second is generally zero except inside a Fraunhofer line, where it can assume a very high value.

The absorption properties of solar matter are regulated chiefly by the formation of the negative hydrogen ion H^-, whose ionization potential is 0.75 eV. In the infrared the free–free transitions play an important role, while in the far ultraviolet the photoionization of hydrogen and of the metals dominates. Sometimes it is convenient to use the mean absorption coefficient $\bar{\kappa}$, which is a weighted mean over frequency. This gives of course a mean optical depth $\bar{\tau}$ for the whole spectrum.

The intensity I (erg/sec-cm^2-sr) of the solar radiation depends upon the emissivity of the different layers and the opacity of the solar matter as well as the heliocentric angle θ.

$$I(0, \theta) = \int_0^\infty S(\tau) \exp(-\tau \sec \theta) \sec \theta \, d\tau$$

where $S(\tau)$, the source function, is in fact the Planck function for the temperature corresponding to the depth τ.

Varying θ from 0 to 90°, we have the "limb darkening" of the solar disk. If $\theta = 90°$, we have $I(0, \pi/2)$, the limb intensity, which is equal to the source function $S(0)$ of the upper layer of the photosphere.

The optical depth τ and the geometrical depth t are reckoned along the radius of the sun; the optical depth τ', which we now introduce (Fig. 4), is measured along the line of sight. For a point outside the solar limb we have

$$\tau' = \int_{-\infty}^{+\infty} \kappa \rho \, dy$$

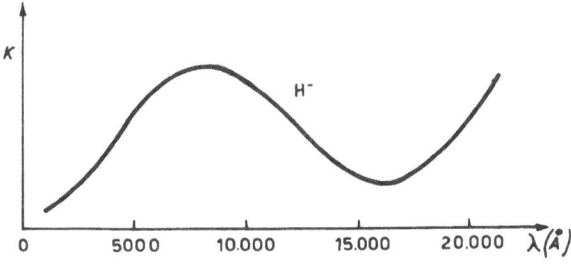

Fig. 3. Absorption coefficient in the solar atmosphere.

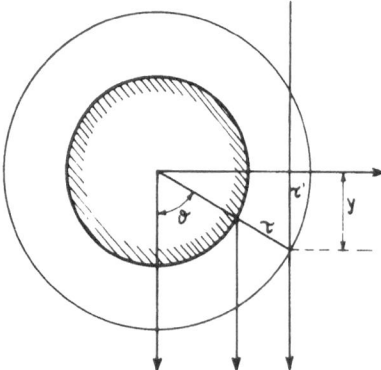

Fig. 4. Heliocentric angle θ, radial optical depth τ, and tangential optical depth τ'.

"Solar atmosphere" refers to the outer layer of the photosphere, where the temperature decreases from the admitted value of about 6000 to 4000°K. This solar atmosphere viewed at the edge of the disk gives the structure of the limb shown in Fig. 2; being in a state of hydrostatic equilibrium, the pressure and the density distribution can be represented by a barometric law

$$P = P_0 e^{-ah}$$

where h is the geometrical height and $a = g\mu/kT$ is the reciprocal of the "scale height" H for an isothermal atmosphere. In the solar atmosphere, which can be considered to a first approximation isothermal with a temperature $T = 4000°K$, we have to take into account the "turbulence term," which tends to increase the scale height H:

$$H = \frac{1}{g}\left(\frac{RT}{\mu} + \frac{v_{turb}^2}{2}\right)$$

where v_{turb} is the velocity of turbulence.

The curve of Fig. 2 can be represented by

$$I = B_0(1 - e^{-\tau'})$$

where B_0 is the Planck function which corresponds to the limb temperature and τ' is the optical depth measured along the line of sight. The reference level is located conventionally at $\tau' = 1$, i.e., at a point where the brightness has a value of 66% of that of the limb. By means of a simple consideration, it is possible to deduce the relation between τ' and the scale, in kilometers

$$x = -345 \log \tau'$$

TABLE VI

Physical Conditions

	τ_{5000}	$T, °K$	$\log P_g$	$\log P_e$
Reference level	0.003	3980	3.70	-0.47
Photosphere	0.67	6045	5.00	1.52

As an example, the physical conditions at the reference level and at the photosphere are given in Table VI. The optical depth has been computed for a wavelength of 5000 Å; the pressures are in dyn/cm^2.

The Spectrum of the Photosphere

The spectrum of the photosphere is a typical Fraunhofer spectrum or a continuum emission spectrum crossed by absorption lines (Fig. 5a). The total energy of the continuum absorbed by the lines amounts to 9.4%.

Each absorption line has a measurable width which is a compound of (a) the natural width of the line, (b) the Doppler width due to the random thermal motions of the absorbing atoms, and (c) the turbulent motion of the absorbing matter. Strong lines show a profile with a core and, more or less, extended wings. For lines of weak or medium strength, the profile is determined chiefly by the instrument. The area of the profile expressed as the width of a band of the neighboring continuum is called the equivalent width W, and is expressed in angstroms or milliangstroms. The monochromatic absorption coefficient increases rapidly and reaches a maximum for the central wavelength of the line; it follows that each point of the profile is formed at a different depth in the solar atmosphere, the wings mainly in the very deep layers, and the core of the lines in the upper layers.

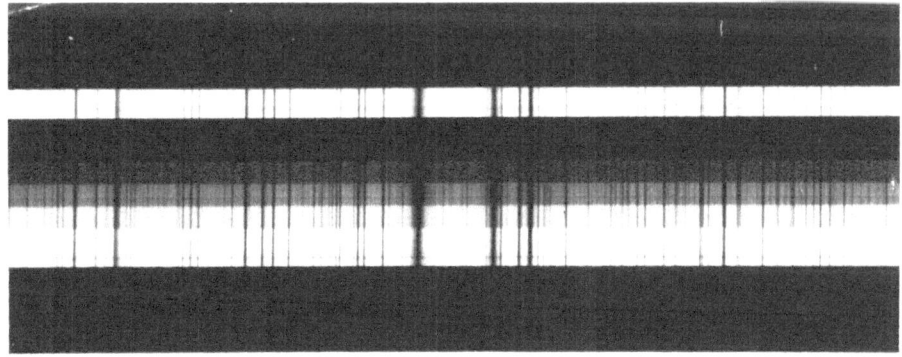

Fig. 5a. Spectrum of the photosphere around 5.180 Å, photographed at the Arcetri observatory. (The strong lines are the Mg b-lines.)

Turbulence in the Solar Atmosphere

The convective currents which are at the origin of the granulation produce some effects in the line profiles. We have to distinguish two extreme cases:

a. The "turbulent elements" are optically thin ($\kappa l \ll 1$), so that the light received from one point of the solar surface is produced by many elements. This case is called microturbulence and its effect is similar to the thermal motion of the gas. The total velocity can be written as

$$v_{tot}^2 = 2RT/\mu + v_{turb}^2$$

This effect produces a line broadening and an increase of the equivalent width of strong and medium strong lines.

b. If the elements are optically thick ($\kappa l \gg 1$) we have macroturbulence plus a part of the microturbulent field. The elements with $\kappa l \gg 1$ produce Doppler displacements on the line.

In the solar atmosphere an intermediate case exists in which both effects act at the same time. The various parts of a line profile, being formed at different depths, are widened and shifted according to the local conditions of the micro and macroturbulence.

Waddell [4], using high resolving power, has obtained the best results from the study of the line profiles. They refer to $\tau_0 = 0.3$ and are $(\overline{v})_{rad} = 1.8$ km/sec and $(\overline{v})_{tang} = 3.0$ km/sec.

Macroturbulence due to granular Doppler shift appears in Fraunhofer lines when observed with high resolving power. Lines which present this effect are called "wriggled" or "zigzag" lines. The Doppler shift gives velocities between 0.3 to 0.9 km/sec for mean optical depths around 0.2 to 0.3. All the results, though fairly scattered, point to a steady increase of the turbulence with height. Roughly speaking, the turbulent velocity ranges from 0.5 km/sec for a depth of 200 km to 3 km/sec for the 0 depth (reference level). The observations show that this velocity increases in the chromosphere.

The convective elements may ascend to $\tau_0 = 0.3$; the upper limit of the photosphere is about $\tau = 0.7$ and the lower limit $\tau_0 = 2$. Therefore, the layer which emits the solar radiation is very thin.

The fact that the nonconvective (radiative) part of the solar atmosphere is also turbulent can be explained by the formation of acoustic noise originating in the

TABLE VII

Equivalent Width W of the Strongest Fraunhofer Lines

Name	Atom	W, $\overset{\circ}{A}$
K	Ca II	19.1
H	Ca II	14.4
$H\beta$	H I	3.75
D_2	Na I	0.83
D_1	Na I	0.60
$H\alpha$	H I	4.1

granulation zone. If $\Delta \bar{v}$ is the velocity amplitude of a wave which travels at a speed c_s in a medium of density ρ, then the acoustical energy flux is $\rho(\Delta \bar{v})^2 c_s$. This flux is constant if the energy dissipation is negligible. Therefore, the mean turbulent velocity should increase with height according to $(\rho \, c_s)^{-\frac{1}{2}}$. The observations qualitatively confirm this law.

The Curve of Growth

The equivalent width W of the line is proportional to the number of absorbing atoms per square centimeter of the surface of the sun only for the weak lines; the complete relationship between the equivalent width and the number of absorbing atoms is known as the "curve of growth," which is shown in Fig. 6. On the ordinate, $\log_{10}$ of the equivalent width W divided by $2\Delta\lambda_D$ is plotted, where

$$\Delta\lambda_D = (2RT/\mu)^{\frac{1}{2}}$$

is the Doppler width of the line. At the solar temperature $\Delta\lambda_D = 0.017$ Å for the violet iron lines and 0.158 Å for the hydrogen line H_β.

On the abscissas $\log(NHf/\Delta w_D)$ — in essence the number of absorbing atoms per square centimeter of the solar surface divided by the Doppler width expressed in angular frequency — is plotted.

Three parts of the curve of growth can be distinguished: (a) weak absorption, $W/2\Delta\lambda_D < 0.1$; (b) strong absorption, $W/2\Delta\lambda_D \approx 2$; (c) very strong absorption, $W/2\Delta\lambda_D > 8$.

If the atmosphere is turbulent, the component of the velocity along the line of sight v_{turb} tends to increase the Doppler width; in fact

$$\overline{\Delta\lambda_D}^2 = 2RT/\mu + v_{\text{turb}}^2$$

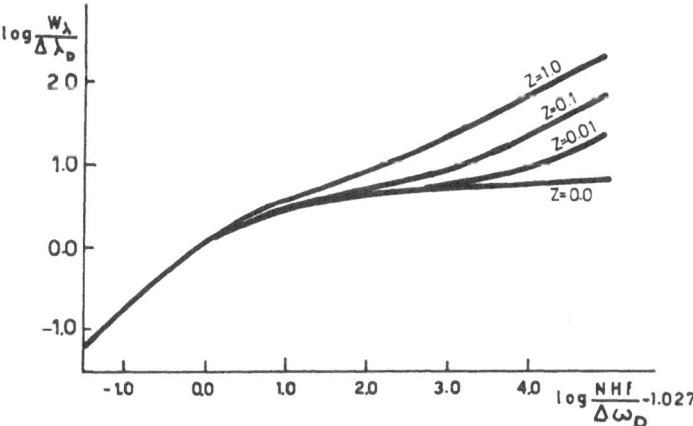

Fig. 6. The curve of growth; the relationship between the equivalent width of the line and the number of absorbing atoms (see text).

With a suitable technique employing the "curve of growth," it is possible to deduce not only the temperature T but also the velocity v_{turb} of the microturbulent field of the atmosphere.

For the sun, the best accepted value is

$$v_{turb} = 1.0 - 1.3 \text{ km/sec}$$

which is smaller than the value found by Waddell from the studies of line profiles. This is quite understandable since the curve of growth gives only a value of v averaged over the different layers of the atmosphere.

The Sunspots

The most prominent phenomena belonging to the photosphere are the spots. A sunspot is a photospheric region cooler than its surroundings and perhaps less transparent. It is composed of a nucleus or umbra surrounded by a brighter penumbra. The mean umbral area is on the average only 16% that of the whole spot. The penumbra is not uniform, but shows a filamentary structure of elongated granules whose mean life is 30 min. Medium and large spots may show a bright irregular structure around the penumbra (bright ring) 2 to 3% brighter than the surrounding photosphere. The umbra sometimes shows very small bright points, perhaps less than 350 km in diameter.

The number and the areas A of the spots follows, from the long-term point of view, a cycle with a period of 11 yr. This is called the solar cycle and is common to many other solar phenomena.

Since the measurements of the area A (which is commonly expressed in 10^{-6} of the area of the hemisphere of the sun) are not easy, an empirical index has been adopted in order to characterize the solar activity as far as the spots are concerned. This is the so-called "wolf number" or "relative number" R, expressed as

$$R = k(10 g + s)$$

where k is a constant depending on the instruments, g is the number of groups, and s is the total number of spots. Although the relative numbers reflect a certain amount of subjective judgment, they are quite convenient for practical purposes and are of course statistically correlated with the total area of the spots present on the disk.

The spots show a typical drift in heliographic latitude which repeats every cycle. At the very beginning of a new cycle, the average latitude is rather high — 30 to 35° (the exceptional case is 40°) — in both hemispheres. At the end of the cycle the latitude is about 5°.

The Magnetic Field of the Sunspots

Around the year 1908, G. E. Hale discovered that the sunspots possess a magnetic field with a rather high intensity.

The Zeeman splitting of the solar line is not at all easy to detect because of the structure of the line itself; the Zeeman splitting $\Delta\lambda$ is given by

$$\Delta\lambda = g(\lambda^2/4\pi mc^2)H$$

in angstroms:

$$\Delta\lambda = g \times 1.17 \times 10^{-5}(\lambda/\lambda_{5000})^2 \times H$$

where H is expressed in Gauss and g is the well-known Landé factor. For example, a singlet line ($g = 1$) having $\lambda = 5000$ Å in a magnetic field of 2800 G gives a splitting $\Delta\lambda = 0.033$ Å which is more or less equal to the Doppler width of an iron line. Since the magnetic field of a spot is mainly perpendicular to the surface, only a longitudinal effect can be measured.

In bipolar spots, the leading spot always has a polarity opposite that of the following one; in addition, if in the northern hemisphere the pattern is polarity north leading and south following, the reverse is the case in the southern hemisphere. At the beginning of a new cycle, the situation is reversed and is restored after two cycles. Therefore, as far as the magnetic polarity of the spots is concerned, the solar cycle lasts 22 years.

The maximum field intensity is statistically related to the area A of the sunspots by the formula

$$H = 3700(A + 66)/A \quad (A \text{ in units of } 10^{-6} \text{ of the visible hemisphere})$$

In and around the spot, the field is distributed according to the experimental law deduced by Mattig [5]

$$H = H_M(1 - r^4) \exp(-2r^2)$$

where r is the distance from the center of the spot, expressed in units of the penumbral radius.

On the average spots have a common magnetic life history. The area develops to its maximum in 10 days and the magnetic field intensity follows this development, but while afterward the area decreases, the field remains practically constant until the end of the spot's life (50 to 60 days). The highest field that has been measured in a spot is 4500 G.

The brightness of the umbral part of the spot is about 0.2 to 0.3 that of the

TABLE VIII

Hale's Law of the Magnetic Polarity in a Bipolar Group

Bipolar group	Following	Leading	Following	Leading	Following	Leading
Northern hemisphere	S*	N	N	S	S	N
Southern hemisphere	N	S	S	N	N	S
Cycle number		n		$n + 1$		$n + 2$

*S = south; N = north.

photosphere; that of the penumbra is 0.75 that of the photosphere. The effective temperature of the spots is about 4600°K, compared with 5785°K for the undisturbed sun. The excitation temperature is lower in the spot, and the electron pressure is much smaller than in the photosphere. Typical values are

$$T = 3900°K \qquad \Delta \log P_g = -0.84 \qquad \Delta \log P_c = -1.40$$

Because of this low temperature, the contribution of the ion H^- to the opacity is much smaller than in the photosphere. In a spot the opacity comes more from the free – free and free – bound transitions in the field of H^+. Furthermore, it has been found from the line intensity that the microturbulent velocity is high (from 1.9 to 3.0 km/sec) compared with 1.2 km/sec for the photosphere at the same level.

The Evershed Effect

The Evershed effect is a radial mass motion with velocity v_E found mainly in the penumbra of the spots. Individual values of v_E may vary from 2 to 6 km/sec and are clearly related to the height in the photosphere. Faint metallic lines show an outward streaming motion, stronger lines no motion at all, and very strong lines an inward streaming motion of up to 3 km/sec.

The velocity is zero near the border of the umbra and attains its maximum value near the outer border of the penumbra. The maximum value of v is linearly related to the radius of the umbra (v is larger for larger spots) and appears to increase with the intensity of the magnetic field.

The Coriolis force tends to give curved paths to the streams, thus resembling cyclonic or anticyclonic motions around the spots. That v_E is not purely radial but has a small tangential component is due to the Coriolis force.

For faint metal lines, the sign of rotation is reversed as compared with hydrogen and calcium because of the outstreaming motion.

Examples for a typical spot are

Total area .... A = 350 × 10^{-6} of the visible hemisphere ($2\pi R_0^2$)
Umbral diameter .... U = 17,500 km; relative umbral intensity 0.3
Penumbral diameter .... P = 37,000 km; relative intensity penumbra ... 0.8
Bright ring diameter.... R = 50,000 km; relative intensity.......... 1.03
Maximum magnetic field.H_M = 3100 G
Evershed effect........v_E = +3 km/sec .... in the photosphere
                       v_E =   0 .... h = 500 km
                       v_E = -4 .... h = 2000 km

Energy of the order of 10^8 erg/g-sec is absorbed in the umbra and in about half of the penumbra, and is emitted outside this region in the bright ring, which is faintly brighter than the photosphere because it has a large area.

Spots occur mainly in groups and are transient phenomena. Spot groups are classified according to their magnetic polarity as unipolar α, bipolar β, multipolar γ, and according to their stage of development from A to J. A is the beginning of a spot group (1st day), F is the phase of largest development (10th to 12th day), and J is the dying phase of a spot group. Not all groups show the whole cycle of development. The lifetime depends on the maximum group area; sometimes lifetimes of 40 days and more have been observed.

Spot groups exhibit systematic internal motions. In general, spots are diverging, but the leading spots move westward more rapidly than the trailing ones move eastward. The latitude motions are generally very small, of the order of $0°.03$ per day. If the heliographic latitude is smaller than $16°$ the spots tend to move toward the equator, while spots at greater latitude have a poleward motion.

Before 1940, a spot was believed to be a big convective region where the low temperature was produced from the adiabatic expansion of the ascending gas. The picture today is somewhat different. In fact, since a strong magnetic field suppresses convective motions by the effect of Foucault currents, no convection will occur above the level where the magnetic energy density is of the same order as the energy of turbulent motion. For $H = 2000$ G, this level is well below the photosphere.

A contribution to the low spot temperature can be attributed to the process of evaporation of hot ions and electrons in a diverging magnetic field. For this to occur, it is sufficient that the energy of the ion + electron be 3 eV.

Solar Rotation

Observing the sunspots on the solar disk, Galileo at the end of the year 1611 was able to detect the rotation of the sun. We know today that the rotation axis is not perpendicular to the ecliptic, but is inclined at $7°$. This, combined with the inclination of the rotation axis of the earth ($23.5°$), has as a consequence that the axis of the sun coincides with the north–south direction only at the beginning of the months January and July. The position angle P reckoned positive eastward is given below.

The sun does not rotate as a rigid body, but its period, instead, varies with heliographic latitude. For a latitude of $16°$, a "sidereal rotation" of 25.38 days is adopted; the corresponding "synodic period," 27.27 days, is obtained taking into account the daily angular motion of the earth around the sun. Inside the sunspot zone the daily sidereal rotation of the sun is well represented by the formula

TABLE IX

Position Angle P of the Axis of the Sun

Date	P		Date	P
January 6	$0°.0$		July 7	$0°.0$
February 5	-13.7		August 7	$+13.0$
March 6	-22.7		September 8	$+22.7$
April 7	-26.4		October 10	$+26.4$
May 7	-23.1		November 9	$+23.0$
June 6	-13.7		December 8	$+13.5$

$$14°38 - 2°7 \sin^2 \phi$$

where ϕ is the heliographic latitude. The solar rotation can also be measured spectroscopically by comparing the spectra of the east and west limbs and measuring the Doppler displacement. Average values ranging from 1.91 to 2.02 km/sec have been obtained by many investigators.

The General Magnetic Field

The accuracy of the photographic Zeeman splitting measurements is ±50 G. For detection of smaller field intensities, Babcock has developed a more sensitive photoelectric method employing a large dispersion spectrograph (11 mm/Å). The sensitivity of the method is better than 1 G. Two slits are placed on the steep sides of the profile of a magnetically field-sensitive line; each slit is followed by a photomultiplier. In front of the slit is an electro-optical retardation plate followed by a nicol prism. An alternating voltage applied to the $\lambda/4$ plate varies the retardation between $\pm\lambda/4$ with a frequency of 120 c/s. This produces a very small periodic displacement of the line and therefore a modulated signal. The Babcock magnetograph is able to measure fields down to 0.3 G. A weak "poloidal" (dipolelike) general field is restricted to latitudes 30° from the poles.

Between 1953 and 1956, this field was consistently positive in the north, negative in the south. Its intensity varied from place to place and its boundaries were irregular. The field component along the line of sight was about 1 G. The angle between the rotational and the magnetic axes of the sun was smaller than 0°7. During the sunspot minimum of 1954–55, poloidal fields were found near the equator. About the middle of 1957 the polarity near the south heliographic pole was reversed; reversal of the field near the north pole was observed in November 1958. At present, the sun's polar field is parallel to that of the earth.

The total magnetic flux of the poloidal field is 8×10^{21} Mx; the summation of all the magnetic fluxes of the sunspots yields 0.5 to 8×10^{21} Mx (if the signs of the magnetic fluxes are included in the summation).

THE CHROMOSPHERE

The Chromosphere at the Limb and on the Disk

The reference level where the transversal optical depth $\tau' = 1$ is the base of the chromosphere. This is therefore the extension of the solar atmosphere toward the corona.

The optical thickness in the continuum is very small; along the line of sight it is less than 1, radially it is as low as 0.004 and therefore the chromosphere is not visible in white light. However, the monochromatic absorption coefficient of the chromospheric matter can be rather high, for instance the chromosphere is visible when observed at the wavelength 6562.808 Å because here the monochromatic absorption coefficient is very large due to the presence of the hydrogen line H_α. In such cases, the chromosphere can be observed not only at the limb (which means that τ' is large) but also on the disk (τ also very large).

The Chromosphere at the Limb

The observations in this case can be performed simply with a spectroscope. If the slit is set radially, it is seen that the H_α absorption line ends in a bright lance. The height of the bright emissions gives the height of the chromosphere. Measurements performed for a long period at Arcetri (Fracastoro [6]) have shown that the chromosphere is high at the poles during the minima of solar activity and vice versa.

With a Lyot filter it is quite easy to see the chromosphere in H_α: it is also possible to obtain good spectra of the low chromosphere in good seeing conditions with a spectrograph attached to a solar tower. But the more complete and developed spectra are obtained at solar eclipses. Before the second contact of a solar eclipse the sun's disk is reduced to a very thin crescent, the Fraunhofer lines disappear, and a bright emission spectrum takes their place. The emission spectrum lasts only a few seconds and hence it is called the "flash" spectrum.

The flash spectrum is the spectrum of the chromosphere. The lines due to neutral elements have more or less the same relative intensity as in the absorption spectrum; the lines of ionized elements are enhanced. The interpretation of the observed line intensities is difficult since we observe a surface intensity which is the product of a double integration: along the line of sight and, radially to the sun, perpendicular to the line of sight. Complications arise from the fact that self-absorption tends to diminish the line intensity.

The Chromosphere on the Disk

In the central part of a strong Fraunhofer line the selective absorption coefficient $k(\Delta\lambda)$ is so large that the chromosphere becomes opaque. It is therefore possible to observe or photograph the chromosphere on the disk; this can be accomplished using filters of the Lyot type or spectroheliographs.

The average level from which the radiation that can be observed originates is the one which has an optical depth of 1. Lower levels contribute very little because of the absorption, higher levels too contribute very little because of the lower temperature. Therefore, the geometrical height h^* at which the chromospheric features that we observe on a spectroheliogram are located can be computed using the equation

TABLE X
Values of h^* for Different Absorption Lines

Line	h^*, km	Line	λ	h^*, km
H_α	5000	Mg	5183	300
H_β	3000	Fe	4348	400
H_γ	2000	Ca	4227	800
H_δ	1000	Na	5890	300
K(Ca II)	4000	Sr	4087	900
K(wings)	1600			

$$\int_{h*}^{\infty} k(\Delta\lambda)N(\lambda)\,dh = 1$$

where k is the selective absorption coefficient, $N(\lambda)$ is the density of atoms capable of absorbing the radiation, and h is the geometrical height.

Since $k(\Delta\lambda)$ has a maximum for $\lambda = 0$ (center of the line) and decreases rapidly with increasing $\Delta\lambda$, it is possible to obtain different layers of the chromosphere by observing the same line in the core or in the wings.

Phenomena Observed in a Quiescent Chromosphere

Viewed at the limb, the chromosphere is not a homogeneous layer but rather has a spiked structure. This was observed for the first time by Father Secchi [7], who, referring to the chromosphere, used the term "prateria ardente" or "burning prairie." Spicules are small, short-lived structures rising and descending with velocities of the order of 10 km/sec and are only observed above 5000 km. Below that height the spicules do not exist or are so numerous that they cannot be individuated. Chromospheric spicules may penetrate into the corona up to heights of some ten thousands of kilometers (Fig. 7).

On the disk in spectroheliograms taken with a high resolving power, three types of elements can be observed:

a. Small granules ("fine mottling") whose diameters range from 600 to 1600 km. These elements are observable only under very good conditions.
b. Clusters of granules, called "coarse mottles," with diameters between 2000 and 8000 km.
c. Coarse mottles are often arranged in a "coarse network" with a characteristic length of 25,000 to 50,000 km.

Fig. 7. Spicules observed at the edge of the sun. Drawing made by P. Tacchini at Palermo, March 22, 1873 (Memorie della Societa degli Spettroscopisti Italiani — Vol. II).

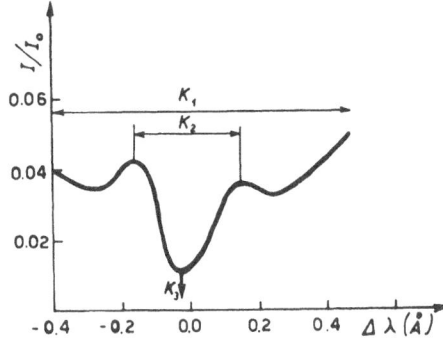

Fig. 8. Structure of the Ca II (K) line in the solar atmosphere.

These structures occur in all kinds of spectroheliograms but with different relative importance.

Low chromosphere spectroheliograms (h = 1000 km), such as those obtained employing metal lines (with the exception of Ca II lines), display dark mottles in a brighter field. The visibility of this structure is better in the steep part of the line and can be attributed to a Doppler effect in the mottles. The mean velocity is 0.4 km/sec.

Medium chromosphere spectroheliograms $1000 < h < 4000$ km can be obtained in the case of the Ca II (K) line, in H_γ, H_β, and in the wings of H_α. The K line is well suited because of its structure. In fact, the profile of this broad line is conventionally divided into three regions labeled 1, 2, and 3 (see Fig. 8). K_1 spectroheliograms refer to the base of the chromosphere and do not show details other than those visible in metal line spectroheliograms. Coarse mottles are visible as bright elements; the network is also visible. In K_2, the mottles and the network are both well-pronounced. In K_3 spectroheliograms the coarse mottling is resolved (if the atmospheric conditions are very good) in small granules (fine mottles) of the order of 700 km (Fig. 9). The lifetime of the coarse mottles of the undisturbed sun is between 6 and 14 hr; the network has a greater lifetime.

High chromosphere spectroheliograms ($h > 3000$ km) are the ones obtained in the core of H_α (Fig. 10). Fine mottles are the dominant feature of these pictures; spectroheliograms taken in the wings of H_α show instead coarse mottles as a predominant structure.

Recently, Leighton [8] has tried to solve the complicated problem of the interpretation of spectroheliograms by studying the velocity field at chromospheric level. He takes two spectroheliograms with the slits placed on opposite sides of the line core, then takes a contact print of one plate and superposes it on the other. If there is no Doppler shift between the two plates, a uniform gray is produced, otherwise, lighter regions indicate velocity of recession and vice versa. The method suitably calibrated permits measurement of the velocity field of the chromosphere.

Leighton's results may be summarized as follows:

a. Large cells with diameters of about 16,000 km spaced on the average of 30,000 km and with a lifetime of about 1 day show a horizontal outflow of matter with a typical velocity 0.5 km/sec. These cells can be identified with the coarse mottles.

b. Bright elements tend to move upward in the lower chromosphere and to move downward in the middle chromosphere.

c. The rms velocity appears to be constant over height, but the "cell size" of the vertical velocity field increases with height from 1700 to 3500 km.

d. Vertical velocities show a period in time of the order of 5 min.

e. At high levels (center of H_α), the Doppler field changes its appearance very rapidly, with a lifetime of about 30 sec.

The granular structure at high levels has a mean size of 3600 km. At lower levels (wings of H_α), the solar surface is relatively quiescent; motions are predominantly downward in a pattern of structure similar to the Ca II network.

Physical Conditions in the Chromosphere

Inside a star the conditions required by thermodynamic equilibrium are almost completely fulfilled. In fact, the stellar plasma is nearly opaque and we can consider the radiation in each point of the star mass as blackbody radiation. Therefore, the physical conditions at every point can be described with only one local temperature; briefly it can be said that the star mass is in LTE (Local Thermodynamic Equilibrium). The situation is a little different in the solar atmosphere where, although LTE can be admitted, it is found that the physical conditions are better described by additional temperatures such as the electron temperature T_{el},

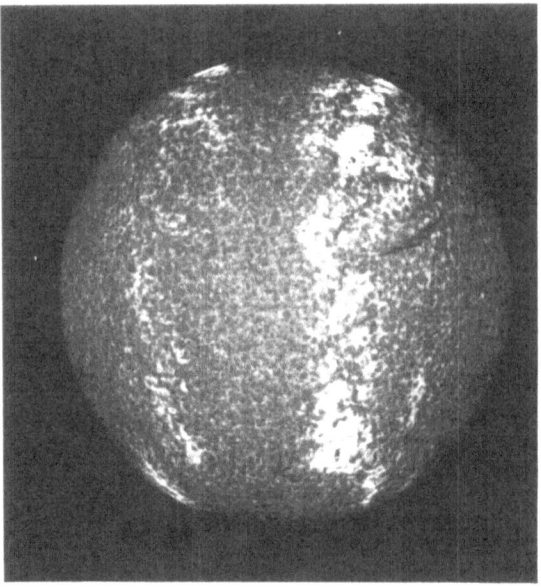

Fig. 9. K_{23} spectroheliogram taken August 9, 1956, photographed at the Arcetri Observatory by M. C. Ballario.

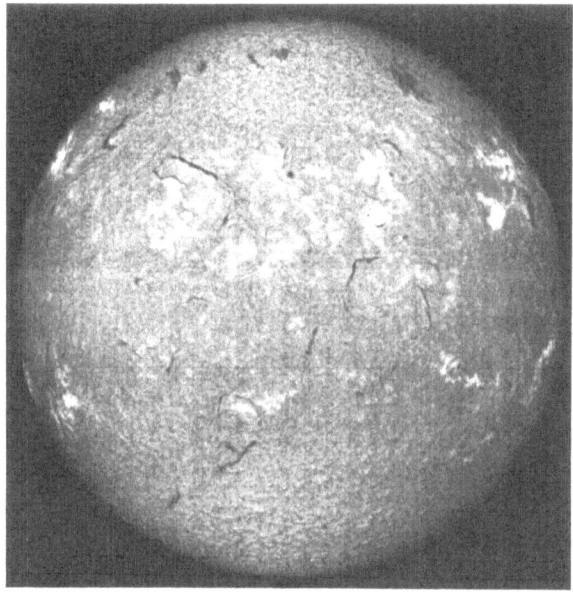

Fig. 10. H_α spectroheliogram, June 4, 1958, photographed at the Arcetri Observatory by M. C. Ballario.

the kinetic temperature T_{kin}, the excitation and ionization temperatures T_{ex} and
T_{ion}, and so on. The deviations from LTE are more and more evident in the chromosphere, where the electron temperature is quite different from the temperature
of the incoming radiation; therefore it is not possible to compute the population
of the atomic levels by the Boltzmann law, nor to identify T_{el} with T_{ion} or the T_{ex}.
However, the results are up to now rather uncertain and although a difference
among the temperatures exists, it is preferable to give only average values here.
Temperatures and number of atoms are shown in Fig. 11.

The interpretation of chromospheric spectra yields information about turbulence along the line of sight (rms velocity η) which is obtained from the width of
the emission line after the elimination of the thermal component, and about turbulence perpendicular to the sun's surface (rms velocity ξ) from the chromospheric
gradients. The line width is determined by the total velocity

$$\eta^2 + \frac{2RT}{\mu}$$

and the scale height H by

$$H = \frac{1}{2g}\left(\frac{RT}{\mu} + \xi^2\right)$$

The average results are that ξ increases from the base up to 3000 km; there is no
indication of anisotropy in the turbulence field.

TABLE XI

Velocity of Turbulence in the Chromosphere

Kilometers	0	1000	2000	3000
Kilometer/second	2	7	12	16

In a more refined analysis of the chromospheric structure, one has to take into account the spicular nature of the chromosphere and the behavior of the spiculae in comparison with the surrounding matter. In this case, there are two kinds of elements — "cold" and "hot." Spiculae are hot elements in the low chromosphere which become cold elements in the high chromosphere and vice versa for the inter-spicular matter.

In addition, there is a slight difference in density between spicular and inter-spicular matter; the latter is less dense but hotter, at least above 6000 km.

The Faculae

The faculae are bright structures on the disk which are quite visible in spectroheliograms of the strong Fraunhofer lines and are also faintly visible in the continuous spectrum close to the limb. They have a close relationship to the spots, and examination of the faculae shows a structure similar to that observed in the spectroheliograms. The faculae are brighter than the photosphere, the excess of brightness in total radiation of 2% increasing to 6% near the limb. In monochro-

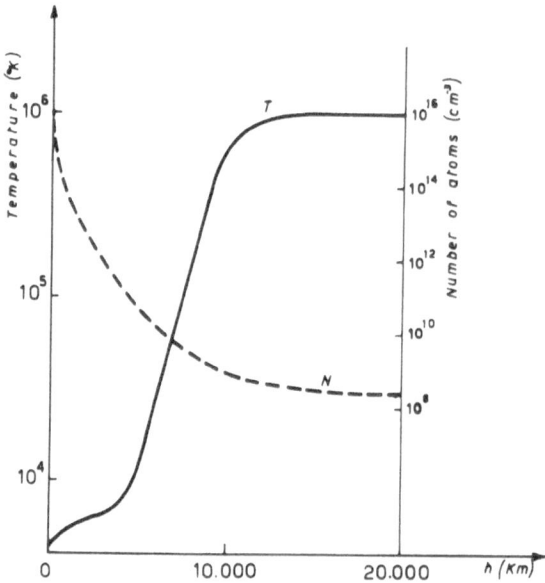

Fig. 11. Temperature and number of atoms in the chromosphere.

matic light this relative brightening toward the limb is also observed.

In the low chromosphere, the facular temperature exceeds that of the surrounding matter, but it increases with depth more slowly than that of the undisturbed photosphere. Equality of temperature occurs around $\tau = 0.6$. Below this, the facular temperature is lower than that of its surroundings.

Babcock [9] has found that apart from the poloidal magnetic field of about 1 G observable only at high latitudes, there are two other kinds of fields at lower latitudes. The stronger type appears as areas of opposite magnetic polarity and are the BM regions; sometimes, but rarely, one finds MM regions; still more rare and weaker are the UM regions. Spots occur in the BM region when it is young; faculae are always associated with BM regions, where the field intensity is greater than 2 G.

Filaments often mark the border line between regions of opposite polarity; BM regions are correlated with enhanced coronal activity. In addition, very short-lived magnetic regions sometimes occur, with lifetimes of a few hours. These regions do not contain sunspots. Large numbers of very small areas with fields up to 100 G are sometimes observed; small visible spots develop in such areas one or two days later, and thus the occurrence of a visual spot is preceded by the birth of a local magnetic field.

The Flares

Monochromatic Observations

A flare is a short-lived and sudden brightening of a local region of the solar disk, always occurring in a center of activity, and rarely in faculae without spots.

Flares are sometimes observed in white light, but only in H_α and Ca II are they really a very prominent phenomenon (Fig. 12). They are divided according to their area and intensity into classes of importance 1, 2, 3, and 3^+; class 1^- contains the subflares. Large flares are irregular and display a filamentary structure; small flares tend to circular shapes.

The flare is a phenomenon which originates in the chromosphere but develops in the lower corona. Statistical studies of flare areas show that the projection areas do not follow a $\cos \theta$ law because flares are not two-dimensional features. From the area variation between the center and limb, it can be concluded that flares are extended in height but are not perpendicular to the surface. Their preferred inclination is $\pm 50°$ to the radius of the sun. The average height of a flare of importance 2 is 14,000 to 16,000 km and at the center of the disk a flare covers an area of 10^8 km^2.

Flares occur almost simultaneously in different centers of activity with frequencies which are certainly higher than could be attributed to chance. The observations can be explained by assuming that every flare is a center of a perturbation which is able to trigger other flares at appropriate places. Sometimes when a flare appears in a center, a filament disintegrates in another center. Photographs in the wings of H_α, taken at the Lockheed Solar Observatory, also show shadows starting from the flares similar to waves moving away from the perturbation center.

TABLE XII

Typical Values for Flares of the Different Classes Are Listed Below

Importance	Duration, min	Mean area	H: Intensity	Frequency
1^-	—	72×10^{-6}	$0.6\ I_{cont}$	—
1	20	160	0.8	0.72
2	30	349	1.2	0.25
3	60	973×10^{-6}	1.4	0.03
3^+	180	—	2	—

The Type II radio bursts which are often observed at metric wavelengths may be associated with the traveling perturbation which triggers the flares.

Flares have a recurrence tendency; they reoccur in the same place even after days have passed, but it has also been observed that flares which occur nearby seem to be quite independent events. The light curve of a flare typically has a sudden rise with a slow decrease; sometimes the "flare" part lasts only a few seconds and is therefore very difficult to observe.

Spectroscopic Observations

A typical profile of H_α for different phenomena is given in Fig. 13. The points X and Y, for which the flare disappears into the background, limit the width of the emission line. This width is very useful because it can easily be measured with the spectrohelioscope and the Lyot filter, and is correlated with the intensity of H_α. The emission lines of a flare often show a depression on the violet side which can be ascribed to an absorption of H_α gas expelled by the flare at a velocity of 100 km/sec. The density of atoms in the cloud is estimated at 5×10^{12} atoms/cm^2

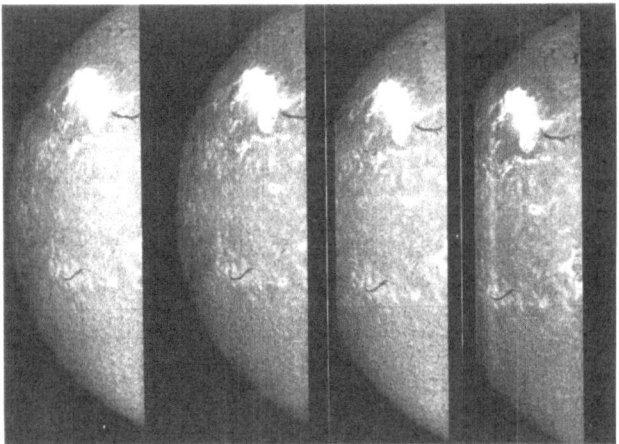

Fig. 12. Flare of importance $3\pm$ spectroheliograms from 9:15 to 9:25 taken June 1, 1960, by M. C. Ballario. Arcetri Observatory photograph.

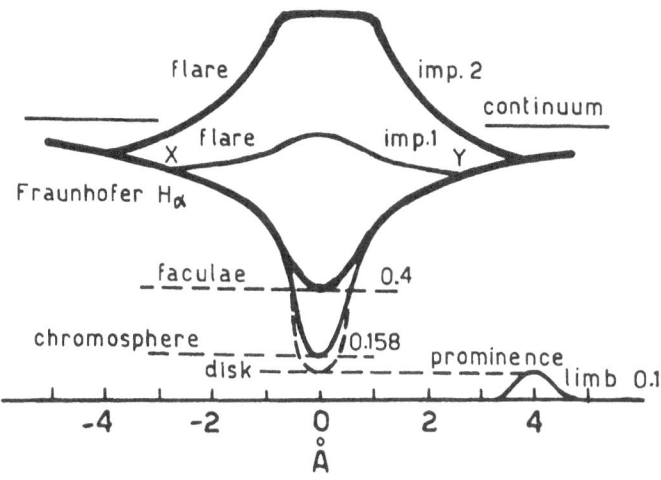

Fig. 13. Profile of the line H_α for different phenomena.

The H_α line as well as the higher Balmer lines are very broad because of the Stark effect produced by free charges. An electron density of 10^{13} electrons/cm^3 is deduced from the extension of the line wings. The intensity of the flare at various distances from the center of the line gives the source function and the optical depth τ, since

$$I(\Delta\lambda) = I_0(\Delta\lambda)\exp(-\tau\sec\theta) + E[1 - \exp(-\tau\sec\theta)]$$

It follows that in the center of the line ($\Delta\lambda = 0$), τ is very large and the source function E gives a temperature of about $8000°K$. The number of atoms in the second quantum state turns out to be 10^{16} above 1 cm^2 of the photosphere for large flares.

Strong flares often emit a continuous spectrum and many other lines; sometimes more than 100 lines have been observed and the Balmer lines extend as far as H_{15}. Further lines are probably concealed by the absorption spectrum. The classical observation of Carrington [10] (on September 1, 1859), who observed a flare in white light, was repeated only by Unno and Shimizu at the Tokyo Observatory (Natuki et al. [11]) on February 23, 1956.

Other phenomena connected with flares are the "bombs" and the "moustaches." Both are emissions over a broad region (30 Å or more) which are probably concentrated in short-lived granules with diameters of the order of 400 km. Bombs and moustaches probably differ only in their lifetimes; moustaches are short-lived events.

These phenomena show up in the wings but not in the center of H_α; this means that they are subchromospheric or photospheric phenomena. The amount of energy liberated is about 5×10^2 erg/cm^3-sec, about 1000 times the thermal photospheric energy density. Severny thinks that these phenomena are produced by nuclear reaction because the depression in the spectrum at 6560.8 Å, tentatively attributed by him to D_α, seems reinforced in the moustaches.

Along this line it should be remarked that Goldberg and his collaborators tried to interpret the asymmetry of the H_α line in limb flares by the presence of a D_α line of fairly high intensity. They claimed that deuterium could be produced in flares in large amounts.

Rocket measurements of X-rays at the time of flares have shown that the normal solar spectrum is enhanced in a range of wavelengths around 8 Å. No increase of L_α radiation has been found.

Associated Phenomena

A flare may produce effects on quiescent filaments which are far away from the flare. Sometimes the filament is "activated," i.e., becomes darker and its size increases; sometimes the filament dissolves. This latter case occurs if the filament was unstable.

The most striking effect of a flare is the ejection of a "surge," i.e., a filament or a streamer of knots of luminous gas at the limb but dark if projected on the disk. The surge is often quasi-oscillatory since it may rise many times in the same place; it is possible that the same surge material is involved. Surges are ejected after the first phase of the flare. The delay is of the order of 4 min. The lifetime of a surge ranges from a few minutes to hours. Flare surges may reach a length of 10^5 km and they do not occur over a flare but at an average distance of 15,000 km.

Sometimes surges are also observed in white light as dark structures for a few minutes. If this absorption is produced by the scattering of free electrons, the density of a surge should be of the order of $10^{13}/cm^3$. However, the average density of a surge is probably only 10^{10} to 10^{11} H atoms/cm^3.

The flare theories will not be discussed here, but it should be noted that existing theories attempt to explain a flare as a local heating of the chromosphere. Because a flare has a height of 16,000 to 20,000 km and therefore occurs in the corona, the origin of the flare should be attributed to a cooling of the coronal material with a subsequent increase of density.

Severny [12] has observed that flares occur where the magnetic structure presents a neutral line or a neutral point. He found that after the flare the magnetic structure is completely destroyed. How this is possible is not clear since the magnetic measurements are made at a lower level than the flare, which occurs in the low corona.

The Prominences

The prominences are masses of solar matter which are sustained above the chromosphere, embedded in the low corona. Their average dimensions show that they are "blades" of solar gas; in fact, an average prominence has a thickness of 6600 km, a height of 42,000 km, and a length of 200,000 km. At the limb viewed in H_α, a prominence is a bright feature which sticks out from the edge of the sun (Fig. 14); on the disk prominences appear as dark filaments.

Fig. 14. Inner corona and prominences at the eclipse of June 19, 1936. Photographed by the Arcetri Observatory expedition to Sara (Orenburg), U.S.S.R.

There are many types of prominences, but we can distinguish two classes: quiescent prominences and moving prominences. Quiescent prominences are a long-lived feature, some originating in the spot group at the age of about 1 rotation. The prominence, seen on the disk as a filament, is on the average inclined 38° to the meridian. Filaments originating outside spots show no differences. During further development, the filament increases in length at a rate of 10^5 km/ rotation. It reaches its greatest length after three solar rotations. Its inclination grows until the filament is more or less directed along the parallel. At the end of its life, the filament breaks up and dissolves. This is also true for low-latitude filaments.

High-latitude filaments drift toward the pole; they are oriented nearly parallel to the equator. The poleward drift is interpreted as a movement of the excitation region.

Moving prominences are also divided into subclasses such as active, eruptive, spot prominences, surges, and spiculae: the names also explain the characteristics.

Bright prominences show a distinct continuous spectrum; the prominence is optically thin for continuous radiation and therefore is not visible in white light. There is a 15% polarization of the continuum probably due to Thomson scattering by free electrons. Observations of the continuous spectrum allow us to determine N_e and T_e if the thickness is known. A value of T_e of $14,000°K$ is probably the closest to reality.

The class of moving prominences contains different types of unstable features which are connected with one or another of the phases of activity of a center. They can have a different name (loop, knot, funnel, etc.) according to their characteristics. Upward velocities of eruptive prominences may be very great (up to 1130 km/sec). The greatest height reached is 1.5×10^6 km.

The theory of prominences is not yet completely satisfactory but some progress has been made in the last years.

It is required first that pressure equality $P_{cor} = P_{prom}$ be realized. This means

$$(NT_e)_{cor} = (NT_e)_{prom}$$

At $h = 50,000$ km, $(NT_e)_{cor} = 1.4 \times 10^{14}$ and $(NT_e)_{prom} = 3.0 \times 10^{14}$. In the vertical direction, a prominence and the corona will not be in hydrodynamical equilibrium since the density of the former is one hundred times greater than that of the latter.

The existence of a prominence depends on its thickness. A prominence is continuously bombarded by coronal matter and the radiation of the prominence should exceed the absorption of coronal energy. For instance, a prominence absorbs from the corona 4×10^5 erg/cm^2-sec and has to emit the same quantity in order to be in equilibrium. This leads once again to $T = 15,000°K$, $N_e = 10^{10}/cm^3$ for heights greater than 10,000 km.

The prominences have an emission spectrum where the lines of H, He, and Ca II dominate. Many lines of other metals are also present, but outside of total eclipses, it is possible to study only the strongest lines. The spectra obtained during the eclipses show a structure very similar to that of the chromosphere. The average physical conditions of a prominence are listed in Table XIII.

THE CORONA

Generals

The corona is a bright aureole which is observed around the sun during the total phase of a solar eclipse. Its brightest inner part is only some 10^{-6} of the

TABLE XIII

Physical Conditions of a Prominence

Kinetic temperature	$T_{kin} = 8000°\,K$
Electron density	$\log N_e = 10.5\ (cm^{-3})$
Number of hydrogen atoms	$\log N_H = 11\ (cm^{-3})$
Neutral hydrogen	0.2%
Neutral helium	50%
Abundance ratio	$H/He = 5$

solar brightness $B_\odot$ and is therefore unobservable in full light due to the brightness of the sky which is on the average $10^{-4}B_\odot$. Only with special precautions in sites where the light scattered by the atmosphere has a low level $(10^{-6}B_\odot)$ is it currently possible to observe the green emission line of the corona and to measure the distribution around the disk of the monochromatic green corona. Today it is accepted that the radiation of the corona is the resultant of three components called respectively K, F, and L. The K component shows a continuous spectrum with no absorption lines except perhaps a very shallow depression in the region of the strong H and K lines of the solar spectrum. However, the F component possesses an absorption spectrum of the same type as the Fraunhofer solar spectrum except for one peculiarity: the lines are shallower than in the solar spectrum and the central absorption decreases going toward the sun's limb. These two components merge and are separable only with spectroscopic methods. The L corona presents an emission spectrum of about 30 lines, which are produced by highly ionized elements.

During totality, when the sky has a brightness around $10^{-9}B_\odot$ the sum of the components $K + F$ can be observed as far as four radii from the limb, and the L component usually goes no further than $0.5R_\odot$ from the limb.

The instrument invented by Lyot and called the Coronagraph permits observations of the L corona between distances of 0.07 and $0.4R_\odot$ from the limb without an eclipse; the new photoelectric coronagraph, also invented by Lyot, which measures the polarization of the white corona, can detect coronal streamers up to a limb distance of 1 to $1.5R_\odot$.

The Spectrum of the Corona

If during a total eclipse one takes a spectrum putting, for instance, the slit of the spectrograph along the equatorial radius, the spectrum obtained will show the following characteristics:

a. From the limb up to an altitude of some minutes the spectrum is a continuum (K component).

b. On this continuum, the emission lines of the L component appear; of these, the green (5303 Å) and the red (6374 Å) are very strong.

c. The outer part is an absorption spectrum in which the lines gradually become sharper moving radially outward (F component).

The general brightness of the spectrum decreases rapidly so that often the inner part is overexposed if proper measures are not taken to lower its brightness. The spectral distribution is very similar to that of the solar spectrum, at least for the K component, and is slightly redder for the F component.

The current interpretation is that the K and F components are photospheric radiation scattered by particles which constitute the solar corona: for the K corona the particles would be high-temperature electrons, for the F component small solid particles. In fact, it is easy to show that if the scattering particles are high-velocity thermal electrons, the Fraunhofer lines disappear in the K corona; from this it can be inferred that the temperature has to be of the order of 10^6 $^\circ$K.

On the other hand, there is a continuous transition of brightness between the F corona and the zodiacal light, as Blackwell and Saito have shown; since the zodiacal light is solar light scattered by interplanetary particles which seem to constitute a sort of lenticular disk around the ecliptic, the interpretation of the F corona seems to be quite reasonable.

It is quite clear that electrons and particles are mixed in all parts of the solar corona, and only their ratio is different.

Brightness and Electron Content of the K and F Corona

Brightness *vs.* distance is not symmetrical around the disk. Along the equator the brightness is higher and the gradient smaller than along the polar axis; there are also some variations in the brightness and the gradient during the solar cycle. At sunspot maximum the total flux emitted by the corona is 1.3×10^{-6} of the total solar flux. At sunspot minimum this value is only 0.8×10^{-6}. In any case, the radiation of the L corona is quite negligible.

It is possible to separate the two components K and F with a simple technique since the lines in the F spectrum appear shallower because of the amount of the K continuum. By comparing the central depth of the F lines with the central depth of the photospheric lines, one can deduce the amount of K corona radiation at each point. In addition, if it is assumed that the K corona has a symmetrical distribution around the sun, it is easy to deduce the electron density N_e. Since the radiation scattered by electrons is polarized, taking into account the geometrical conditions of the scattering process in the corona, the measurement of the degree of polarization offers another way to disentangle the K from the F corona. The maximum values of the brightness of the K and F components and of N_e are given in Fig. 15.

The L Corona

Some thirty emission lines have been found in the coronal spectrum of which only the strongest can be observed with the coronagraph. The lines can be divided into four classes according to their behavior; the lines in the fourth group appear only in active regions and therefore are lines of high excitation. The equivalent width of the coronal lines range from 0.3 to 48 Å; they are usually expressed in terms of the intensity of the coronal continuum.

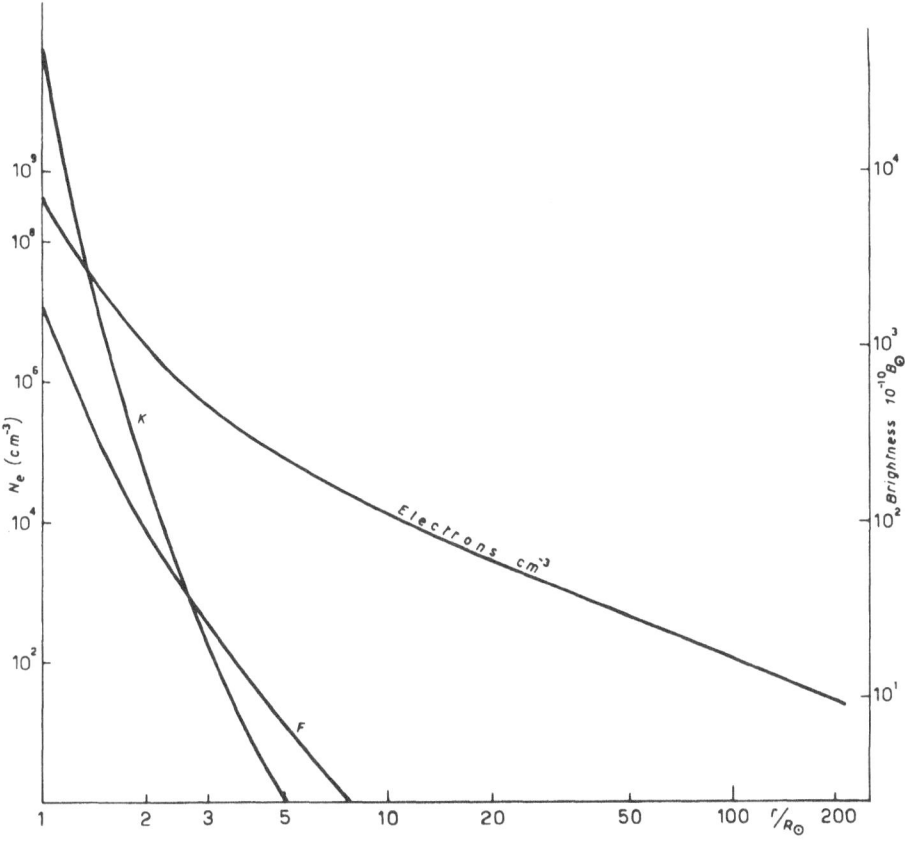

Fig. 15. Brightness and electron density of the K and F corona.

The lines are due to forbidden transitions between the odd or even levels; this shows that the lifetime of the upper states is long and that the probability of a collisional downward transition is very small.

Examples of strong lines are

$$3,388 \ldots\ldots\ldots \text{Fe XIII} \ldots\ldots\ldots W = 10 \text{ Å}$$
$$5,303 \ldots\ldots\ldots \text{Fe XIV} \ldots\ldots\ldots W = 20 \text{ Å}$$
$$10,746 \ldots\ldots\ldots \text{Fe XIII} \ldots\ldots\ldots W = 48 \text{ Å}$$
$$10,798 \ldots\ldots\ldots \text{Fe XIII} \ldots\ldots\ldots W = 30 \text{ Å}$$

Their ionization potential is between 325 and 375 eV.

High excitation lines

$$4086 \ldots\ldots\ldots \text{Ca XIII} \ldots\ldots\ldots \text{Ionization potential} = 655 \text{ eV}$$
$$4412 \ldots\ldots\ldots \text{A XIV} \ldots\ldots\ldots \text{Ionization potential} = 682 \text{ eV}$$
$$5694 \ldots\ldots\ldots \text{Ca XV} \ldots\ldots\ldots \text{Ionization potential} = 814 \text{ eV}$$

are rather faint. Their equivalent width is around 0.3 Å. The excitation of the coronal lines can occur either by absorption of radiation or by electron collision.

In the former case, if one admits that the corona has the same temperature everywhere, the equivalent width has to be independent of the height; in the latter case, W should decrease with increasing height. Observations show that W decreases up to a distance of one radius from the limb, then remains constant.

The Temperature of the Corona

We have already mentioned that the absence of Fraunhofer lines in the K corona indicates an electron temperature higher than 10^5 °K. The L corona offers another way to obtain the temperature. In fact, in its spectrum the ions Fe X, Fe XI, Fe XIII, and Fe XIV are present and it is therefore possible to deduce an electron temperature taking into account the relative abundances of the iron ions. The values which have been obtained by Shklovski [13] range from 600,000 to 1,200,000°.

The law of the hydrostatic equilibrium applied to the coronal electron distribution with the additional hypothesis of a constant temperature, again yields a kinetic temperature of 1.6 to 1.15×10^6 °K. Finally, the width of the emission line leads to a kinetic temperature of 2×10^6 °K.

There are some discrepancies between the electron temperature, which is around 10^5 °K, and the kinetic temperature, which is larger than 10^6 °K — discrepancies which can be partly due to the fact that the electron temperature has been mainly measured in the quiet minimum corona while the emission lines belong mainly to the active corona.

The Corona and the Solar Cycle

The aspect of the corona varies from one eclipse to the other because it is related to the phase of the solar cycle. The corona can be described as a more or less symmetrical aureole from which streamers stick out; these streamers can be traced some solar radii away from the disk. During the maxima of solar activity, the streamers are more or less uniformly distributed around the disk; at the minima, they are concentrated along the equator. In addition, at the poles of the sun one observes short brushes which are reminiscent of the structure of the lines of force around a magnetized sphere.

The shape of the corona is more closely related to the prominences than to the spots, and particularly with the number of prominences which have a heliographic latitude greater than 40°.

Apart from the gross aspects of the solar corona, during the period of high solar activity the monochromatic corona shows a great deal of correlation with some chromospheric and photospheric phenomena. For instance, some emission lines are intensified over active spots and quite often a brightening of the line 5694 Å is observed at the time of a flare. Typical are the "coronal condensations," which can be observed in the K and L corona. These are regions where the corona is denser and the temperature higher.

There is some evidence that at least during the minima the corona shows some "cold" regions (Deutsch and Righini [14]). In fact, low temperature emission lines of Ca II have often been observed with the coronagraph; in addition, during the eclipse of July 20, 1963, the Ca II (*K*) line was obtained in a spectrum of the outer corona.

The Activity Centers

It is now clear that photospheric, chromospheric, and coronal phenomena are tightly connected; when the sun is active, spots, chromospheric flares, and coronal condensation are found more or less in the same region.

A "center of activity" is the totality of all phenomena accompanying the birth of a sunspot. The primary event is probably the arrival of a magnetic disturbance at the sun's surface. The life of an activity center may be more than 200 days and therefore it sometimes covers as many as 10 rotations of the sun.

The sequence of the events is briefly as follows:

Day   0 — Magnetic field
Day   1 — Faculae and later spots appear
Days   2 to 5 — Flare activity
Day   11 — Maximum of flare activity; spots, maximum development of coronal condensations, coronal line 5694 observed
Day   27 — Few flares, large faculae, spots almost disappeared, long filaments
Day   54 — Filaments very long, spots disappear completely
Day 108 — Faculae dissolved, filaments at maximum length
Day 135 — Prominence seems to migrate toward the poles
Day 270 — Dissolution of filaments, development of coronal streamers

The Heating of the Corona

The mechanical energy produced by the turbulence of the photospheric convection zone is partly dissipated in the chromosphere and in the upper photosphere, but the portion which arrives at the level of the low corona is still very important. In fact, the loss by radiation of the corona is very small, and it is easy to show that the mechanical energy dissipated equalizes the loss by radiation at a level where the electron density is still 4×10^9 electron/cm^3 or in the high chromosphere.

In the corona, the incoming mechanical energy is more or less 100 times the energy lost by radiation. This excess of energy is in part transported back to the high chromosphere, since the corona is a good heat conductor, where it heats the transition zone. This mechanism leads to a computed temperature of about $10^6\,°$K, which is quite acceptable. Another part of the mechanical energy is lost by evaporation of the coronal matter since the particles in the corona at a temperature of $10^6\,°$K have velocities which are near to the "escape" velocity.

From the study of the cometary tails it can be concluded that the sun emits at least 10^{36} protons/sec, which means a mass loss of 10^{12} g/sec; near the earth this "solar wind" yields a particle density from 2 to 100 particles/cm^3, assuming that the velocity is of the order of 500 km/sec.

The Sun Viewed from Outside the Atmosphere

The recent progress of space research has produced an extension of our knowledge of the sun. Thanks to the excellent work of groups of research workers, who installed spectrographs, ionization chambers, and photoelectric spectrometers on board rockets and satellites, we now have a considerable knowledge of the UV and X-ray spectra of the sun.

The spectrum of the sun has been extended down to a wavelength of 13 Å and many strong lines of highly ionized elements have been discovered. Spectroheliograms of the sun in the lines He II 304, Fe XVI 335, and Fe XVI 361 have shown the existence of monochromatic plages which are comparable with those observed in Ca II (K) line.

Spectroheliograms in Ly_α have shown structures very similar to those observed in H_α. Pictures of the sun in X-rays have been obtained with a pinhole camera and with grazing incidence mirrors; the features observed are similar to the enhanced spots observed in the microwave radio domain. The X-ray monitoring satellites have pointed out that the flux is highly variable and is enhanced at the occasion of solar flares; no enhancement has been observed in Ly_α. However, enhancements of the X-ray flux have been observed many times without the presence of a flare.

The X-ray radiation of the quiet corona is emitted mainly at wavelengths above 20 Å; radiation of shorter wavelengths is emitted in highly excited regions, such as coronal condensations, where due to the high temperature ($T \approx 10^7$ °K) the spectrum may extend down to 1.5 Å. The free–free and the free–bound transitions of the electron in the fields of the highly ionized atoms are mainly responsible for the emission of wavelengths under 100 Å. Longer wavelengths can be emitted by bound–free transition of hydrogen $\lambda < 912$ Å and by recombination in the helium continuum for $\lambda < 227$ and $\lambda < 504$ Å.

The Solar Terrestrial Relationship

The terrestrial ionosphere is maintained by solar radiation, which supplies the energy required for the ionization of the atmospheric gases. Conventionally, the ionosphere is divided into layers at various heights. It is evident that the study of the ionospheric behavior can give some ideas about the emission of the sun in the range indicated.

The situation shown in the table corresponds to the quiet ionosphere; during solar events some types of ionospheric disturbances are observed:

a. Mögel–Dellinger effect or short-wave fade-out (SWF).
b. Fade-out or disappearance of ionospheric reflections from all layers.
c. Increase of absorption of cosmic noise at $\lambda \approx 10$ to 15 m (SCNA).

d. Magnetic crochets which are displacements of small amplitude in the magnetic components.
e. Sudden enhancements of atmospherics (SEA) or reinforcements of signals
   at long waves.
f. Sudden phase anomaly (SPA) — difference of phase observed between ground
   and sky waves.
g. Sudden field strength anomaly (SFA) due to interference between sky and
   ground waves.

All these effects are due to the electromagnetic radiation of the flare, or even
more, to the increase of ionization in the D layer. Recent rocket and satellite experiments have shown that hard X-rays are responsible for this effect since they
are very efficient in producing a large number of electrons in low strata.

In addition, there are three types of particle emissions which have to be considered in the solar–terrestrial relationship:

1. Cosmic ray particles emitted by great flares, whose rigidity is between 3
   and 10 GeV.
2. Slower particles emitted together with flares or after the flares, which
   produce magnetic storms.
3. Recurrent particle streams which produce small recurrent magnetic storms
   with a period of 27 days.

The events described in (1) are rather rare and are shown by using the techniques for the observation of cosmic rays. More frequent are the events of type (2)
which are observed with a magnetometer. Usually the arrival of the solar particles
in the magnetosphere is shown by the onset of the storm. (For 3 to 6 hr the magnetic
field intensity increases more than $10\,\gamma$ and then it decreases to -20 or $-30\,\gamma$ with
a very slow recovery.) These storms are sporadic, while the storms of type (3) are
regular since they repeat with a period which is equal to the solar rotation. In general, the recurrent storms are not related to flares. Recurrent storms show a long
life since they repeat for 20 rotations or more.

Since the magnetic character of each day can be represented by a figure called
the international character figure, which ranges from 0 (quiet day) to 2 (disturbed
day), these recurrent storms are easy to detect simply by plotting the character
figure for a sequence of days belonging to the same solar rotation. The duration of
this perturbation is of the order of one day, which can indicate that the beam of solar
particles is rather sharply defined.

TABLE XIV

Ionospheric Layers and Ionizing Radiation

h, km	Region	Source of principal ionization	Radiation, Å
80	D	NO	Ly_α (1215)
110	E	O_2, O, (N_2)	≤ 80
160	F_1	N_2, O	≤ 200
260	F_2	N_2	200 to 900

REFERENCES

1. M. Schwarzschild, *Ap. J.* **130**, 345 (1959).
2. J. Rösch, *Intern. Astron. Union, Symp.*, 12th (1960), p. 313.
3. A. Unsöld, *Physik der Sternatmospharen* (Springer, Berlin, 1955), p. 234.
4. J. Waddell, *Ap. J.* **127**, 284 (1958).
5. W. Mattig, *Z. Astrophys.* **31**, 273 (1953).
6. M. G. Fracastoro, *Mem. Soc. Astron. Ital.* **19**, 45 (1948).
7. A. Secchi, *Le Soleil* (Gauthier-Villars, Paris, 1870).
8. R. B. Leighton, *Ap. J.* **135**, 474 (1962).
9. H. W. Babcock and H. D. Babcock, *Ap. J.* **121**, 349 (1955).
10. R. C. Carrington, *Monthly Notices Roy. Astron. Soc.* **20**, 13 (1859).
11. M. Natuki, T. Hatanaka, and W. Unno, *Publ. Astron. Soc. Japan* **8**, 52 (1956).
12. A. B. Severny, *Astron. Zh.* **35**, 335 (1958).
13. I. S. Shklovski, *Izv. Krymsk. Astrofiz. Observ.* **6**, 105 (1950).
14. A. Deutsch and G. Righini, *Ap. J.* **140**, 313 (1964).

This review has been extracted mainly from the following books:

G. Abetti, *The Sun* (Faber and Faber, London, 1955).

C. W. Allen, *Astrophysical Quantities* (The Athlone Press, 1963).

C. De Jager, "Structure and Dynamics of the Solar Atmosphere," in: *Handbuch der Physik*, Vol. 52 (Berlin, 1959).

K. O. Kiepenheuer, *The Sun* (Ann Arbor, The University of Michigan Press, 1959).

G. P. Kuiper, ed., *The Sun* (Chicago, The University of Chicago Press, 1953).

D. H. Menzel, *Our Sun* (Harvard University Press, 1959).

J. C. Pecker and E. Schatzman, *Astrophysique Générale* (Masson & Cie. Editeurs, 1959).

H. J. Smith and E.v.P. Smith, *Solar Flares* (Macmillan, New York, 1963).

R. N. Thomas and R. G. Athay, *Physics of the Solar Chromosphere* (Interscience Publishers, New York, 1961).

A. Unsöld, *Physik der Sternatmospharen*, second ed. (Berlin, 1955).

M. Waldmeier, *Die Sonnenkorona I and II* (Birkhäuser, Basel, 1951).

Chapter 2

Introduction to the Study of Solar Radio Emission*

J. F. Denisse

Observatoire de Paris
Paris, France

OBSERVATIONAL TECHNIQUES

As in the optical domain, radio frequency radiation which comes from the sun is characterized by its intensity and its polarization which, in general, vary as a function of time and frequency. With antennas of large dimensions, the solar regions from which these emissions come and eventually the diameter or even the structure of these regions, can also be determined.

If the receiving system works with a unique and monochromatic antenna, the system is called a radiometer, and its resolving power is measured generally in degrees (in other words greater than the apparent solar diameter), and only the total radiation from the sun is received.

The simplest interferometer is obtained with two antennas, separated by a distance D; the radiations received by the two antennas are combined and the reception pattern obtained is a sinusoid with an angular period $\alpha \sim \lambda/D$. This interferometer works in a manner similar to a two-slit interferometric unit. With two antennas sufficiently far apart one can locate precisely the position of emission sources on the sun and measure their diameters if these sources are not too complex. In receiving with two antennas with different polarizations one can also measure the polarization of the wave by combining radiation received by the two antennas with suitable phase delays.

If instead of two antennas, several aligned antennas are used, a radiation pattern is obtained which is a combination of the sinusoidal pattern of the antennas considered in pairs. The multiple antenna interferometer, first perfected by W. N. Christiansen, consists of a group of antennas regularly spaced, which acts as a grating. The radiation pattern of this instrument is composed of lobes with an opening of $\sim \lambda/D$ (where D is the distance between the outer antennas) and separation of the lobes of λ/d (where d is the distance between two consecutive antennas). This apparatus is particularly well adapted to solar studies due to the fact that the angle between lobes is greater than the angular diameter of the sun.

---

*For a more detailed discussion of the various questions which are mentioned in this introduction, one should read the excellent recent article of J. P. Wild, S. F. Smerd, and A. A. Weiss, "Solar Bursts," which appeared in 1963 in the *Ann. Rev. Astron. Astrophys.* [1].

In phasing the antennas of a multiple interferometer relative to one another, it is possible to displace the receiving antenna pattern and thus to explore different regions in space either consecutively or simultaneously. This complication is unavoidable, particularly for the interferometer with a north–south orientation for which the diurnal motion is inadequate to resolve the emission sources observed by the individual interferometer fringes; it is also necessary for the observation of rapid phenomena. Interferometers of this type with variable patterns have been successfully operated, in particular, by P. Wild, in Australia, and E. J. Blum, in France.

The total radiation received from the sun during a period of activity varies very rapidly with time and with frequency. The dynamic spectrograph of P. Wild, which allows for the analysis of the spectrum of solar radiation as a function of time, is a radiometer with rapid frequency sweeping. The result of the observation is presented in the form of intensity modulation of motion picture film which progresses as a function of time; this equipment has allowed the identification of very important types of radio emission.

THE RADIO FREQUENCY CHARACTERISTICS
OF THE SOLAR ATMOSPHERE

From the radio point of view the solar atmosphere can be considered completely defined at each point by the electron density (which is also that of the ions), by the magnetic field which exists in the region, and by the temperature.

It is convenient to express the electron density by means of the critical frequency of the medium defined by $f_p = 9 \times 10^{-3} N^{1/2}$, and the intensity of the magnetic field H by the gyrofrequency of the electrons $f_H = 2.8 H$ (in these formulas, f_p and f_H are expressed in Mc/s, N in cm^{-3}, and H in G). At all levels of the solar atmosphere f_p and f_H decrease with increasing altitude.

In the corona, where the density is always below 10^9 electrons/cm^3, the critical frequency is less than 300 Mc/s and reaches 30 Mc/s at the altitude of one solar radius. The gyro frequency f_H can be much greater than f_p above sunspots, where the magnetic field can reach values of several thousand gauss, but it is probably always lower than f_p for altitudes greater than one-half of a solar radius above the photosphere; there the coronal magnetic fields do not exceed tens of gauss.

In the absence of a magnetic field, and neglecting collisions, one can calculate that the propagation index μ of a radio wave of frequency f in a plasma with a critical frequency f_p is given by the basic formula (Denisse and Delcroix [2] and Ratcliffe [3]).

$$\mu^2 = 1 - \frac{f_p^2}{f^2}$$

This formula implies that when a wave of frequency f penetrates an increasingly dense solar atmosphere the index of propagation decreases monotonically, and is equal to zero at a particular altitude for which the critical frequency f_p

equals the frequency f. For all altitudes below this critical altitude, $f < f_p$, the index of propagation μ is purely imaginary; no propagation can exist here. It follows that all solar emission received on a given frequency f necessarily originates at an altitude greater than the critical altitude where $f_p = f$.

As the index of refraction of the corona differs from unity, the ray trajectories are not generally rectilinear and must be calculated, especially in the case of long waves, which can propagate with low absorption in the outer corona, where the density gradients are very weak.

In the presence of a magnetic field, the solar plasma becomes birefringent; any radiation is divided into two waves with different polarizations, one of which is the ordinary wave, the other the extraordinary wave. For example, when the propagation is parallel to the magnetic field (longitudinal propagation), the indices of the two waves are

$$\mu_{oL}^2 = 1 - \frac{f_p^2}{f(f + f_H)} \quad \text{and} \quad \mu_{eL} = 1 - \frac{f_p^2}{f(f - f_H)}$$

In this case, the two waves are circularly polarized in an inverse sense, the extraordinary wave rotates in the same sense as the electrons under the effect of the magnetic field.

When the propagation is perpendicular to the magnetic field (transverse propagation), one has two linearly polarized waves with indices

$$\mu_{oT} = \mu \quad \text{and} \quad \mu_{eT} = \mu_{eL}$$

The above equations are valid only in the case of homogeneous solar atmospheres. In fact, there certainly exist in the corona small-scale density variations; these can be clearly seen when the radiation from a radio source is observed passing through the corona, giving rise to the diffusion phenomenon, which was extensively used for the study of the outer corona (Hewish [4]).

Radio waves are absorbed in the solar atmosphere due to electron–ion collisions. The optical thickness

$$\tau = \int^X \frac{x}{\mu} \rho \, ds$$

varies as $N^2/\mu f^2 T^{3/2}$ and depends on the ray path under consideration (Wild et al. [1]). The absorption calculated for long waves ($f < 30$ Mc/s) is small; by contrast the absorption can be very important for shorter wavelengths which originate near the critical frequency, especially for the density gradients which correspond to the undisturbed solar atmosphere. However, many emissions are observed which appear to be emitted from regions close to the critical altitude of the mean corona – without appearing to undergo noticeable absorption; this is perhaps due to the fact that these emissions are produced in regions where abnormally high gradients exist.

THE DIFFERENT TYPES OF SOLAR RADIO EMISSIONS

The Quiet-Sun Radiation

The emission mechanism of the quiet sun does not differ fundamentally from emission observed in the optical domain. Both have their origin in thermal processes operating in the solar atmosphere; in this case, the emission is due to the radiation produced by electron – ion collisions. It is the same process by which absorption is produced and which has been mentioned above. To calculate completely the radiation emitted by the quiet sun, it is sufficient to know the variation of the electron density $N_e(r)$ and the temperature $T(r)$ of the solar atmosphere as a function of distance r from the center of the sun (Wild *et al.* [1] and Newkirk [5]).

Radio frequency observations of the quiet sun have permitted the precise calculation of the temperature and the density of the solar atmosphere.

Above the centers of solar activity one observes an increase of electron density – the condensations. These condensations produce an increase in optical thickness in the regions of higher temperature and consequently an increase of thermal emission; this radiation due to the condensation is called the slowly varying component (Newkirk [6]).

Radio Emissions Associated With Flares

When an important center of solar activity appears on the disk, it is the seat of numerous chromospheric flares. In general, these flares are followed by a sequence of radio and other events whose study has revealed a wealth of details and has fundamentally changed our conceptions of the flare mechanism.

The Preliminary Phase

As we have already indicated, a center of solar activity, even moderately active, is found associated with plages which are more or less radio emissive and which originate in the relatively dense condensations lying above the chromospheric faculae. In general, the eruptive activity of a sunspot group coincides with an increased brightness on centimetric waves (Pick [7]) of the centers involved. These bright regions are strongly localized in the immediate vicinity of the spots (narrow condensations) and reveal by the polarized character of their radiation their association with the magnetic fields of the spots (Kundu [8]). The duration of this preliminary phase, which manifests itself as a localized "heating" of the active region, is essentially variable from one event to another and can last several days.

The Explosive Phase

The start of the flare proper is marked by an explosive phase in the course of which many types of remarkable events are produced. A fast increase of emission of centimetric waves is observed which lasts for a few minutes; this is the centimetric burst. The source of this emission is stable and appears to be localized high in the chromosphere; it certainly must have its origin in radiation from high-energy electrons which manifest themselves simultaneously by also emitting hard X-rays.

At the same time there appears in the corona the burst called Type III, whose duration does not exceed a few seconds and which is radiated by a perturbation which rises in the corona with an extremely high velocity, of the order of a fraction of the speed of light. It is believed that the Type III burst is also produced by high-energy electrons escaping from the eruptive center. The very existence of this burst, its brevity, the fact that it often recurs many times are important properties for understanding the flare phenomenon. They show actually that in the lapse of a fraction of a second an activity center is able to accelerate a considerable number of particles up to the energy of the order of a fraction of 1 MeV. The origin of this acceleration is unknown and to date is not even partially understood.

Frequently, the Type III burst leaves in the corona a fleeting trace of its passage; this is the Type V burst. The source of this burst is stable and fades after only a few minutes (Wild et al. [9]). This explosive phase of the flare has also been observed optically (Giovanelli [10]) but the rapid increase in H_α brightness gives no indication of the mechanism through which the high-energy particles are created.

The Expansion Phase

When the explosive phase is particularly intense, some remarkable effects result which extend more slowly into the outer corona and reach even into interplanetary space. It is in the course of this expansion phase that several minutes after the start of the flare two quite characteristic bursts appear: the bursts of Types II and IV. The Type II burst, as the Type III burst, corresponds to an ascending perturbation but the speed of the Type II is of the order of thousands of kilometers per second, which is the speed of the hydrodynamic supersonic shock wave produced by the initial explosion.

The radiation of this burst is very complex. It fluctuates rapidly and presents all sorts of peculiarities [splitting, existence of harmonics, polarization (Roberts [11])] which have not as yet been adequately explained; the origin of the radiation itself is not clear. However, the shock wave itself is rather well known since its structure has been analyzed quite recently, in some detail, by deep-space probes. The shock wave produced in the vicinity of the earth correlates with sudden commencement geomagnetic storms and with decreases of interstellar cosmic rays of the Forbush type.

The Type IV burst finally corresponds also to an ascending disturbance, but quite different from the preceding. This disturbance does not appear as localized as does the shock wave described earlier. The radiation is a rather stable continuum whose source has been followed as far out as five to six solar radii above the photosphere. This source appears as an immense cloud, whose dimensions can reach tens of solar diameters before finally disappearing. Its duration is variable from a 10-min period to several hours (Boischot [12]).

It is thought that the radio emission of this cloud is due to the radiation from relativistic electrons which stay embedded in coronal magnetic fields, which are relatively intense (tens of gauss).

The Type IV burst, discovered by A. Boischot, owes much of its importance to the fact that the various geophysical effects of flares appear to be directly related to the properties of the burst, in particular geomagnetic storms and their velocity of propagation across interplanetary space (Caroubalos [13]) as well as the production of solar cosmic rays (Avignon and Pick [14]).

The Final Phase

When all traces of the optical flare have disappeared, there still remains in the corona (Pick [7]) a permanent and stable radio emission, the continuum storm, whose radiation can persist several days. The source of this radiation is located relatively low in the corona, near the critical altitude, while the Type IV burst takes place at very much higher altitudes. At first the storm radiation is remarkably stable, but little by little very short-lived bursts (Type I) are observed superimposed on the continuum, becoming more and more numerous, so that this emission can no longer be distinguished from ordinary radio noise storms, whose fluctuating radiation constitutes the most common source of solar radio emission. Most frequently, these storms originate shortly after an eruption, possibly of little importance, under conditions still poorly defined (Le Squeren [15]), but without the complete sequence of events which have just been described.

It was proposed to link the existence of these storms to the presence in the corona of high-energy particles which are generated in the course of the flare, and which for many days remain trapped near the sun in suitable magnetic configurations (Denisse [16]).

It is important to note that the preceding description of the different phases of the flare must be considered as extremely schematic and only a few quite rare flares are accompanied by the complete sequence of the radio emissions. For instance, the Type III bursts and the radio noise storms are common events which often accompany quite minor flares. By contrast, bursts of Type II, Type IV, and continuum storms are relatively rare and constitute probably the most significant characteristics of important flares. Finally, a large number of other emissions exist which are relatively rare and complex; these we have not described, for their study is only beginning and their physical significance unknown.

THE EMISSION PROCESSES

Optical observation of the solar corona is quite difficult, for the only atomic lines existing in the visible portion of the spectrum arise from highly ionized atoms and are relatively rare. In the radio spectrum, on the other hand, the ionized plasma which forms the corona possesses definite electron resonance frequencies (Denisse and Delcroix [2]) which are directly related either to the critical frequency of the plasma f_p or to the gyro frequency of the electrons f_H;

for the solar atmosphere we have seen that these frequencies are located in the radio spectrum accessible to radio-astronomical observation. Furthermore, in the corona the mean free path of the particles is often very large compared with the scale of variation of the corona itself. The result is that particles of very different energies can coexist; in particular, one can expect to observe in the corona electron distributions quite far removed from those which correspond to thermal equilibrium. These conditions are favorable to setting into motion amplification processes analogous to those which are produced in masers, for example, and these phenomena certainly play an important role in the generation of solar radiowaves.

The radiation from the quiet sun and that from the less active condensations are explained quite well as thermal emission from the solar atmosphere and reflect as we have stated earlier the electron temperature of the medium (Newkirk [5]).

By contrast, all other types of radio emission are too intense or possess too distinctive characteristics to be interpreted as resulting from a thermal process. It is certain that their interpretation must call upon nonthermal processes which are both more efficient and often more complex. As we have seen, these processes are tied to either the gyro frequency of the electron f_H or to the frequency f_p of coherent oscillations of the coronal plasma.

In their gyrating movement around the lines of force of the magnetic field, low-energy electrons of the coronal plasma radiate a monochromatic wave of frequency f_H, which is obviously an extraordinary wave; the formulas outlined earlier show that near the frequency f_H (and for $f > f_H$) the extraordinary wave index is always imaginary, that is to say that this emission cannot propagate in the plasma. When higher-energy electrons are present, then, in addition to the frequency f_H, these electrons also radiate harmonic frequencies (Wild *et al.* [1]), which can propagate normally. This radiation, called gyrofrequency radiation, is probably the origin of some observed emissions.

When the superthermal electrons become relativistic, then the spectrum radiated attains higher and higher harmonics. For electrons with high enough energy, continuous spectrum called synchrotron radiation can even be reached, for which the maximum energy is radiated in the vicinity of the frequency

$$f_c \sim f_H \left(\frac{\epsilon}{m_0 c^2} \right)^2$$

which can be greater than f_H if the energy of the electrons ϵ is above their energy of mass $m_0 c^2$ (Le Roux [17]).

On the other hand the ionized plasma of the corona can be the seat of coherent oscillations of thermal electrons at the frequency f_p; these oscillations, which are weakly excited in a plasma in thermal equilibrium by the fluctuations due to the particle movement, can be strongly amplified in the presence of high-energy particles with which they are often strongly coupled (Bohm and Gross [18]).

Two cases can be distinguished: when each energetic particle acts independently to excite the plasma waves, the radiation obtained is the classic Čerenkov effect (Cohen [19]); on the other hand when the particles act coherently, as in a beam to excite plasma, plasma oscillations which can reach much higher amplitudes are obtained (Bohm and Gross [18]). Both these processes probably take place in the corona, but it should be noted that plasma oscillations generated in both processes are waves which exist only in the plasma and can only radiate beyond the corona, in the interplanetary medium, if they can be transformed into transverse electromagnetic waves. The conditions for this transformation have been studied by many authors (Bohm and Gross [18] and Ginzburg and Zheleznyakov [20]), and require nonlinear properties of the conducting medium, particularly for the emission of radiation at the harmonic frequency $2f_p$.

It is interesting to note that all of the processes which have just been described are based on the existence of superthermal particles which play a fundamental role and can only be produced during the occurrence of flares; therefore, it is not surprising that all the radio emissions observed in the sun appear as more or less related to eruptive phenomena.

To decide the role which these different processes play in the interpretation of solar radio emissions, it is necessary to note that emissions of the plasma oscillation type must be produced in a relatively narrow bandwidth near the critical altitude $f \approx f_p$, which depends only on the local electron density N. Furthermore, although the Čerenkov emission generated by a multitude of independent particles can be relatively stable in time, by contrast plasma oscillations, amplified by a bunching of coherent particles, must exist for only a brief period.

On the other hand, it has been seen that synchrotron radiation is not related in a definite way to the frequency f_H, and that it can be generated at any altitude; moreover, this radiation must be stable since it is emitted by a large number of independent particles.

If the preceding considerations are used as a guide, it is generally admitted that the Type III burst is the result of plasma oscillations generated by a flux of electrons in the hundreds of KeV range, emitted mostly at the very start of the eruption (Wild et al. [21]). By contrast, the Type IV burst is interpreted as synchrotron radiation of relativistic electrons trapped in coronal magnetic fields (Boischot and Denisse [22]).

Interpretation of other sources of emission is much less certain; it is possible that gyro radiation plays a role in the emission of the most active condensations (Zheleznyakov [23]), and that synchrotron radiation is the origin of centimetric bursts (Takakura [24]). It has been proposed that synchrotron radiation is equally responsible for Type V bursts (Wild et al. [1]), but it seems that Čerenkov radiation is better fitted to explain the ensemble of the properties of this burst type. No really satisfactory interpretation has yet been advanced to explain the Type II burst, which is very complex and is probably a combination of several simultaneous processes, one of which is probably the Čerenkov effect (Wild et al. [1]). Finally, one proposal has interpreted the Type I bursts as a transitory amplification of gyro emissions (Twiss and Roberts [25]). Nevertheless, these proposed mechanisms are still quite uncertain and do not rest on sufficiently precise experimental data.

To conclude this brief review, it can be stated that observation of solar radio emissions has revitalized our knowledge of solar activity in a very general sense. A few decades ago, the major problem of solar studies was research into the origin of solar magnetic fields, especially those associated with sunspots. Today, although this problem has not been completely resolved, it has been superseded by another one quite as important, which is to know how these magnetic fields are associated with the creation of highly energetic particles. Of course, this vital problem is of interest for many astronomical researchers (as well as in geophysics), but it seems that the sun, thanks to radio astronomy is the best source of information and the object through which progress can best be made on this important question.

REFERENCES

1. J. P. Wild, S. F. Smerd, and A. A. Weiss, *Ann. Rev. Astron. Astrophys.* 1, 291, 1963.
2. J. F. Denisse and J. L. Delcroix, *Théorie des ondes dans les plasmas*, Dunod, ed., Paris (1961).
3. J. A. Ratcliffe, *The Magnetoionic Theory and Its Applications to the Ionosphere*, Cambridge (1959).
4. A. Hewish, *Proc. Roy. Soc. Ser. A.* 228, 238 (1955).
5. G. Newkirk, *Paris Symposium on Radio Astronomy*, Stanford University Press, Stanford, California (1959), p. 149.
6. G. Newkirk, *Astrophys. J.* 133, 983 (1961).
7. M. Pick, *Ann. Astrophys.* 24, 183 (1961).
8. M. R. Kundu, *Ann. Astrophys.* 22, 1 (1959).
9. J. P. Wild, K. V. Sheridan, and A. A. Neylan, *Australian J. Phys.* 12, 369 (1959).
10. R. G. Giovanelli, *Australian J. Phys.* 11, 350 (1958).
11. J. A. Roberts, *Australian J. Phys.* 12, 327 (1959).
12. A. Boischot, *Ann. Astrophys.* 21, 273 (1958).
13. C. Caroubalos, *Ann. Astrophys.* 27, 1 1964).
14. Y. Avignon and M. Pick, *Compt. rend. acad. sci.* 249, 2276 (1959).
15. A. M. Le Squeren, *Ann. Astrophys.* 26, 97 (1963).
16. J. F. Denisse, *Inform. Bull. S.R.O.E.*, No. 4, 3 (1960).
17. E. Le Roux, *Ann. Astrophys.* 24, No. 1, 71 (1961).
18. D. Bohm and E. P. Gross, *Phys. Rev.* 75, 1864 (1949).
19. M. H. Cohen, *Phys. Rev.* 123, 711 (1961).
20. V. L. Ginzburg and V. V. Zheleznyakov, *Astron. Zh.* 36, 233 (1959).
21. J. P. Wild, J. A. Roberts, and J. D. Murray, *Nature* 173, 532 (1954).
22. A. Boischot and J. F. Denisse, *Compt. rend. acad. sci.* 253, 1539 (1961).
23. V. V. Zheleznyakov, *Soviet Astron. – AJ* 7, No. 5, 630 (1964).
24. T. Takakura, *J. Phys. Soc. Japan* 17, Suppl. A II, 243 (1962).
25. R. Q. Twiss and J. A. Roberts, *Australian J. Phys.* 11, 424 (1958).

Chapter 3

A Survey of Radio Observations of Solar Eclipses

John P. Castelli and Jules Aarons

Air Force Cambridge Research Laboratories
Bedford, Massachusetts, United States of America

INTRODUCTION

From earliest recorded time man has been fascinated by eclipses of the sun. For the astronomer, the eclipse served to reveal features of the chromosphere and the corona. More recently, development in perfecting and utilizing the coronograph has provided a continuous monitoring of coronal activity but the eclipse still serves in many optical solar studies. For the radio astronomer, the eclipsing process gave resolution to his solar mapping, revealing the temperature and angular diameters of enhanced regions of solar activity associated with sunspots, plages, and prominences. The eclipse gave the radio astronomer data on coronal temperatures and on the extent of the corona at many wavelengths. The development of high-resolution radio interferometers has lessened the need for eclipse measurements, but many eclipse measurements continue to add to our understanding of solar physics.

Mitchell [1] states that, according to Oppolzer's *Canon of Eclipses*, during each century there is an average of 237 solar eclipses, 84 of which are partial, 77 annular, 10 annular and total (in which the vertex of the moon's shadow just reaches the earth), and 66 total. Because half of the total eclipses are observable only in inaccessible places, the average of all the usable total eclipses is approximately one every third year.

The volume from which the above data are drawn is the famous *Canon der Finsternisse* of Professor Oppolzer, who, in the last century, tabulated 8000 solar eclipses and 5200 lunar eclipses (without computers). The translation of this work in 1962 by Gingerich was accompanied by an introduction by Menzel and Gingerich [2] in which they pointed out that working back to the eclipse in the year 700 B.C. cumulative errors of as much as 20 min exist. Even for the July 20, 1963, total solar eclipse, Oppolzer does not show the path of totality falling within the U.S., as indeed it did.

Calculations of modern eclipses, as found in the annual American Ephemeris and Nautical Almanac issued by the U. S. Naval Observatory, are computed by the method of F. W. Bessel. The details are shown in *Explanatory Supplement to the Astronomical Ephemeris* and the *American Ephemeris and Nautical Almanac*,

London, 1960, prepared jointly by the Nautical Almanac office of the United States and the United Kingdom [3].

In accordance with requests from the Solar Eclipse Commission of the International Astronomical Union, the principal circumstances of all solar eclipses are calculated and made available to astronomers several years in advance of the Ephemeris. Advance data are given in the U. S. Naval Observatory Circular and in "Astronomical Phenomena," an annual publication of the U. S. Naval Observatory.

SIMPLIFIED ECLIPSE GEOMETRY

There are numerous thorough treatments of solar eclipse geometry and predictions which are available. Briefly, a partial or total solar eclipse takes place when the moon is new, and the earth moves into the shadow cast by the moon. If the distance L in Fig. 1 is such that the shadow lies below the earth, we have a total eclipse with the width of the totality region, AB. The width may be as much as 240 km at certain locations. Within a diameter of 6500 km, the eclipse will be partial with varying percentages.

If the distance from the moon to the earth is greater than the length of the shadow, the moon does not eclipse the entire sun and we have an annular eclipse with a bright ring of sunlight appearing. The path of the annular eclipse is somewhat wider than that of the total eclipse. Beyond the path of totality or annularity, the eclipse is observed as partial; it also is extremely useful in radio astronomy. It should be noted that even though calculations are made for a particular eclipse, optical observations of the event may yield a small time correc-

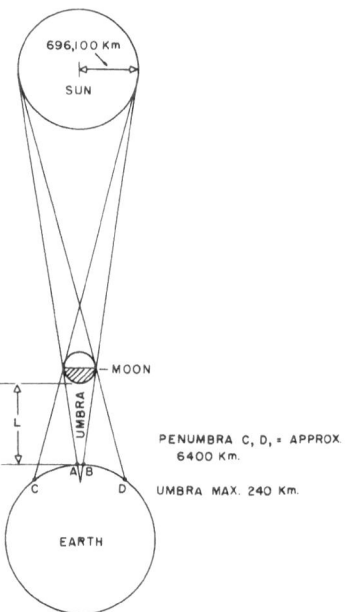

Fig. 1. Simplified solar eclipse geometry.

tion between observed and calculated positions, with the moon's position being the difficult one to determine.

Duration at totality may be as long as 7 min (a rarity) or less than 1 min; the path width of totality is typically 150 km.

THE ECLIPSING PROCESS IN RADIO ASTRONOMY

If optical observations of a solar eclipse are made through a telescope equipped with neutral density or narrow-band filters, the picture is quite simple. As the moon moves in front of the sun, it occults solar features such as sunspots, plages, and prominences, with a knife-edge sharpness. At totality, between second and third contact, only the solar corona is visible. After totality the moon moves on uneclipsing the same solar features as during the eclipsing period with a different egress geometry compared with the period between first and second contact.

Basically, information can be gained from solar eclipses relating to the following:

1. The emissivity of centers of radio activity on the sun. The radio diameter and brightness of radio emissive regions associated with sunspots, plages, and prominences can be obtained with a definition not possible with any known interferometer. The technique used is to find sudden changes in the solar flux level between first and second contact or between third and fourth contact. The polarized component of an enhanced region can be found, as can its height and spectrum; a comparison with optical features can be made. Records of the radio eclipsing of the sun are shown in Figs. 2a and 2b; the former is of the sun with enhanced regions on it, and the latter is an eclipse record of the quiet sun. Sudden changes in the slope mean that emission areas have been covered or uncovered. The techniques will be covered in greater detail in a subsequent portion of this paper.

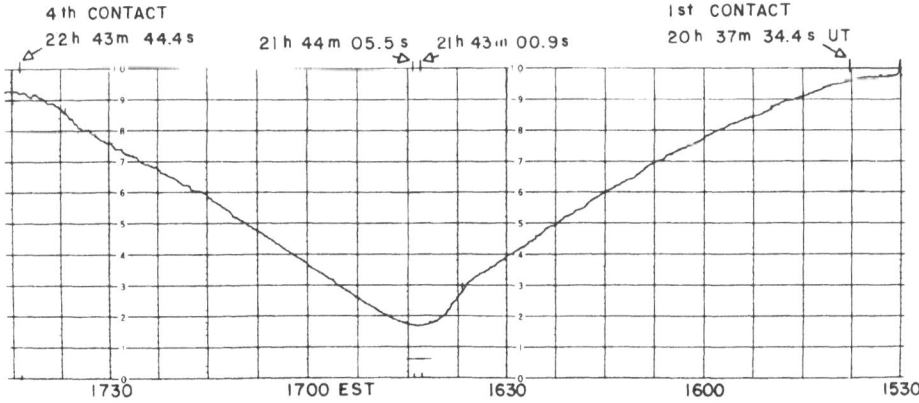

Fig. 2a. Unretouched eclipse curve July 20, 1963, at Hermon, Maine, at $\lambda = 10.1$ cm showing covering of east limb region at about 16:37 and showing an apparent cold region 17:17 − 17:20 (AFCRL 1964).

2. Total eclipses of the sun are used to find the radio diameter of the solar corona at various wavelengths and at different periods of the sunspot cycle. At totality at meter wavelengths, a residual, for example, of 25% or more of the solar energy will show that the solar corona has not been completely eclipsed.

3. A model of solar brightness distribution can be derived from eclipse records. The observed brightness distribution across the disk of the sun depends on the variation in intensity of an emergent ray at various positions from the center of the solar disk to the limb. At optical wavelengths there is a marked limb-darkening effect. As temperature increases with depth in the photosphere, central direct rays originating at greater depths are hotter than oblique rays at the limb. At centimeter radio wavelengths, the reverse situation applies. The corona is thin but not transparent and there is an increase in temperature from the chromosphere out to the corona. Therefore, the direct central rays emerging from the deeper regions in the chromosphere are not as intense as the oblique rays at the limb which originate in higher and hotter regions; limb brightening is the result. The situation changes at the meter wavelengths. Although temperature increases from the inner to the outer corona, the corona is more opaque to meter wavelengths. The increased absorption due to the greater length of the ray path toward the limb causes limb-darkening at these very long wavelengths. The shape of the eclipse curve attests to the presence of limb brightening or darkening. If there is limb brightening, a sharp decrease will be seen at the time of first contact as if an enhanced region has been eclipsed. The same effect will be seen near fourth contact.

DATA INTERPRETATION

In order to reduce the curves shown in Figs. 2a and 2b to solar temperatures and diameters, several problems of data interpretation must be overcome.

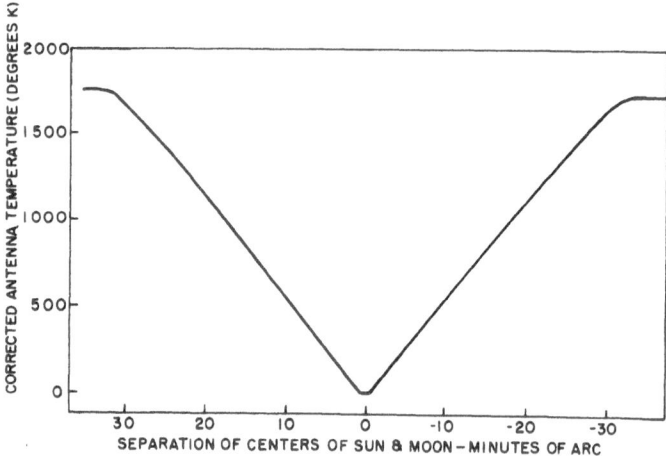

Fig. 2b. Eclipse curve corrected for gain and atmospheric variations, λ = 8.65 mm June 30, 1954, Oskarshamn (Coates [21]).

1. The basic eclipse curve may reflect effects of atmospheric scintillation, particularly when low-angle observations are made of the active sun with enhanced regions.
2. Attenuation of the atmosphere varies with water vapor pressure and content, and oxygen content in the centimeter and millimeter ranges.
3. The lunar temperature must be carefully measured to note its contribution to the residual signal at totality. At millimeter wavelengths, it may reach 2 to 2½% of the sun's temperature.

DIAMETERS OF ENHANCED REGIONS FROM ECLIPSE CURVE SLOPE ANALYSIS

An eclipse curve $P(t)$ is a record of power as a function of time. The derivative of this curve, $P'(t)$ is the change in power as a function of time. The slope is then proportional to the brightness distribution of circular arcs across the face of the sun. Therefore, the slope curve can be used to distinguish irregularities in solar brightness which show up much more prominently here than on the eclipse curve itself. Knowing the total flux and the residual energy allows one to normalize any slope curve in arbitrary units (e.g., displacement per minute) to the true slope curve in percentage of total flux per minute, for each point, after correcting for atmospheric attenuation and other factors.

The actual slope is obtained by sampling the eclipse curve at fixed time intervals of, for example, 0.1 to 4 min, and plotting rate of change as a function of the angular distance between the center of the moon and the center of the sun.

Although the eclipse mechanism affords great angular resolution, it must be kept in mind that the ultimate limit of detectability of small changes in the eclipse curve is fixed by the output fluctuations of the radiometer as seen on the recorder. Since the solar antenna temperature contribution is very high, it may determine the minimum fluctuation level. The well-known basic expression for peak to peak fluctuations is

$$\Delta T \cong K \frac{(F - 1)\ 290 + Ta}{\sqrt{BT}}$$

Low-noise receivers, therefore, are of little use except perhaps during the central portion of the eclipse when Ta is greatly diminished. Nor can equipment time constant be increased if time resolution is to be preserved. It would seem that even by greatly increasing equipment bandwidth within practical limits, a sensitivity commensurate with eclipse resolution is incapable of realization.

THE RADIO DIAMETER OF THE SUN

The comparison of times of first and fourth radio contact *vs.* first and fourth optical contact during an eclipse is a good indication of the radio diameter of the sun at a given frequency. Geometrical considerations show that if the moon begins the occulting of the radio energy from the solar corona 2 min before first contact (approximately $1'$ of arc) then the radio diameter is the optical diameter plus approximately $2'$. During the total solar eclipse of July 20, 1963, the 10-cm radio diameter of the sun, enlarged by this amount, yielded a value of $1.06R_\odot$. This assumes nothing about the energy distribution — only that there is a significant amount of radio energy at a height of 44,000 km above the photosphere.

A second approach to the evaluation of the radio diameter of the sun at a particular wavelength for a single eclipse can be derived from the residual energy measured at totality. In this case, the energy in a ring-shaped area is considered rather than the level of energy at points in equatorial and polar planes. Recording times of first and last radio contact helps to establish area limits.

If there is an enhanced region on the limb, both of these methods may be in error. A residual measurement of 20% at 25 cm was made during the eclipse of October 2, 1959 (Castelli et al. [48]). However, there was a bright region on the limb extending to a average height of 80,000 km above the photosphere. It is clear that the rotation of this area would have produced pronounced differences in the residual measurement if the eclipse had occurred a day or two earlier or later. The optical features associated with the enhanced region in the radio spectrum may not be visible; the photospheric component may be on the nonvisible portion of the sun.

In regard to this, an attempt to use eclipses historically to show the extension of the corona during low and high sunspot years should be done only after noting the eclipse magnitude at the given observation site and after observing caution concerning the nonvisible enhanced solar regions.

The following eclipses to be noted in the text were of listed magnitude:

> June 30, 1954.................1.035
> October 2, 1959..............1.02
> February 15, 1961............1.03
> July 20, 1963................1.01

At wavelengths of 5 cm and shorter, the residual at totality is strongly affected by the eclipse magnitude since the radiation at these wavelengths is emitted by the chromosphere and the lower corona. This can be seen by referring to Fig. 3. Here the geometry is based on a uniform disk.

The task of normalizing data from one eclipse to the next is very difficult. The slopes are frequently analyzed in different fashions. The positions, angular diameter, and brightness temperatures of enhanced regions reflect a certain amount of subjectivity on the part of the analyst. Perhaps the biggest problem of all is to determine accurately the absolute temperature of the sun and its power flux density. If these are in error by $\pm 30\%$, and this seems to be often the case where we compare the data of various observers, then the assigned temperature of enhanced regions will be very much in error. It is indeed remarkable that the results from some eclipses hold up so well when the data are assembled.

THE HISTORY OF SOLAR RADIO ECLIPSE MEASUREMENTS

In the history of radio astronomy, eclipses have played an important role. With the realization in 1945 that the sun was a source of radio energy, radio astronomy observatories began to instrument seriously for eclipse observations. The early years were characterized by a great deal of controversy. Astronomers developed models of the solar atmosphere that predicted what radio distribution could be expected. The radio astronomers came up with chromospheric and coronal temperatures that necessitated revision of the thinking. With the combination of optical and radio observations, new values of solar coronal and chromospheric parameters emerged with solar eclipse measurements assisting in determining many of these.

Arbitrarily, we might divide radio eclipse observations into three periods. The first, 1945–1952, saw many early discoveries such as the coronal contribution and the enhanced region contribution to the total solar flux. The second period was characterized by a study of limb brightening or darkening during the low-sunspot year of 1954. Observations made in the third period during and after the I.G.Y. concentrated on temperature, polarization, and angular diameter of enhanced regions.

The first radio observation of a partial eclipse was made by Dicke and Beringer [4] in Boston, Massachusetts, on July 9, 1945, at 1.25 cm. During the November 23, 1946, partial eclipse, A. E. Covington [5] in Ottawa, Canada, observing on 10.7 cm, determined that he had detected several bright regions, one with an excess of 1.5×10^6 °K above the average surface temperature of

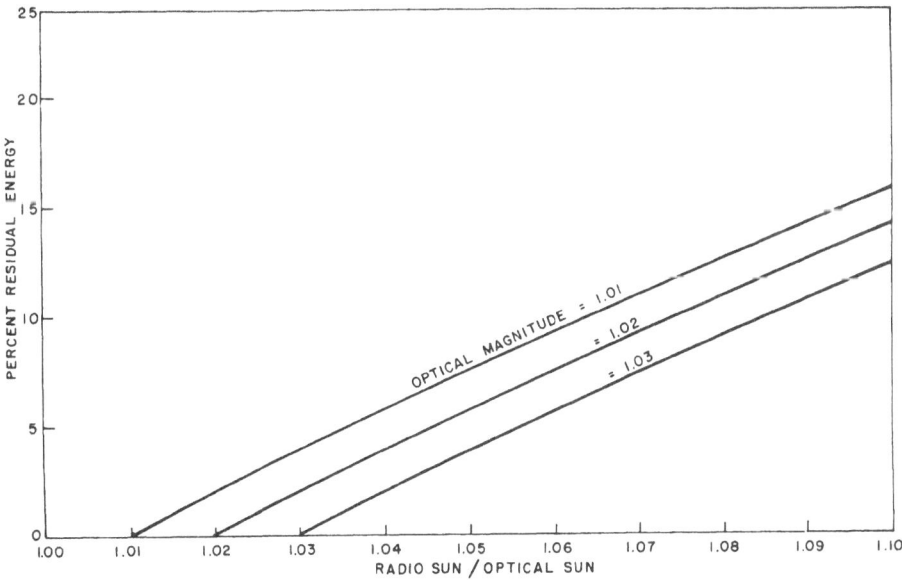

Fig. 3. Residual energy (based on constant brightness distribution, circular sun).

5.6×10^4 °K. This enhanced region had an area equal to 2.2% of the solar disk. It was the first observation of the contribution of small areas of solar activity to the total solar temperature. Another bright region, apparently located at the solar limb, caused the signal to decrease 3 min before first contact. A study of optical records several years after the eclipse indicated that a prominence was present in the region eclipsed. The area had an effective temperature of 3×10^6 °K.

The total eclipse of May 20, 1947, was observed by two groups — one Russian and one American. Taking observations at sea in the South Atlantic at a wavelength of 1.5 m, Khaykin and Tchikhachev [6] found a residual energy at totality of 40%. While they could not draw conclusions about the enhanced regions because of the generally disturbed condition of the sun, the fact that a large percentage of solar energy was observed during totality was evidence that the radiating area at meter wavelengths was the solar corona with a large portion of the corona uneclipsed even during totality. The same event was observed by Hagen [7] at 3.2 cm with a 3% residual. The conclusion of the American group was that limb brightening could be observed.

On November 1, 1948, a partial eclipse was observed in Australia at several different wavelengths and at one wavelength from several different sites. The reason for the latter is as follows: The moon's eclipsing action yields a one-dimensional picture of the variation of solar power with time of passage of the moon across the solar disk. No indication is given where along the moon's arc temperature differences exist. With a group of two or more stations it is possible to create a geometry of intersecting arcs from which the location of enhanced regions can be found. The localizing of bright regions by three stations used in the Australian observations is shown in Fig. 4.

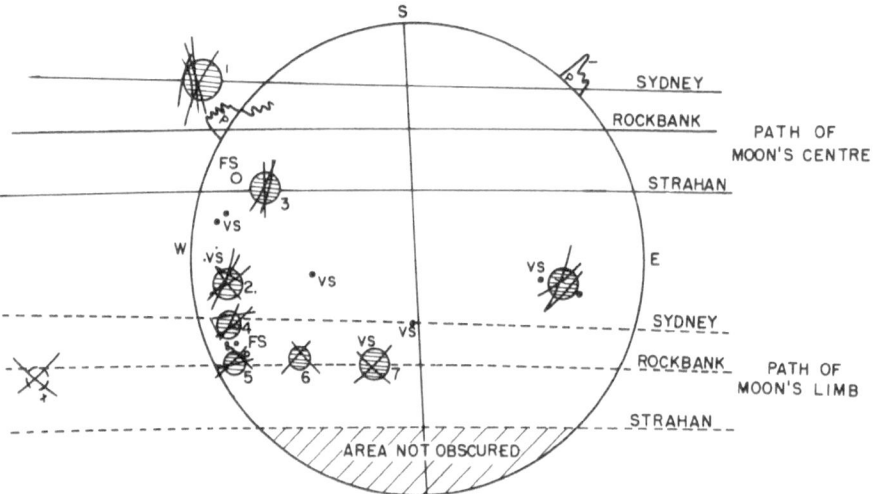

Fig. 4. The position of small bright areas on the solar disk, and their relation to optical features during the eclipse of November 1, 1948. Bright areas are indicated by numerals (wavelength 50 cm). VS, visible sunspot groups; FS, positions of sunspots 27 days before the eclipse; P, prominence (Christiansen, Yabsley, and Mills [10]).

The technique first used by Christiansen, Yabsley, and Mills [10] at $\lambda = 50$ cm was adopted for eclipse measurements from this point forward and was used a number of times with good results. By plotting the path of the moon across the sun for each of the three stations and relating the times of maximum slope of each of the eclipse curves to geometrical conditions, it was possible to pinpoint the location of bright areas to specific solar regions. Three of the radio centers of activity so detected were identified with sunspot groups. Three other sunspot groups that could be observed by optical means showed no radio brightness. Three radio centers of activity were identified with solar latitudes and longitudes where sunspot groups had appeared twenty-seven days earlier. One region which showed activity was not even on the solar disk but 0.25 solar radii beyond; it was identified with the position of a solar prominence. The observations showed, therefore, that some radio centers of activity were associated with sunspots and others were not. The temperature of the centers of activity at 50 cm was 0.5×10^6 °K as an average; individual areas varied by 10 : 1.

The indications were then clear that a new index of solar activity was available — those solar regions which were active in the radio frequency sense and those which were quiescent. The characteristics of the regions, size, and identification with plages and prominences would have to be explored by interferometric techniques over long periods of time but the separation of active and quiescent spot regions was made.

Polarization measurements made during the same eclipse at 50 cm were inconclusive. The objective was to determine the general solar magnetic field. The conclusion was that if any existed its value must be less than 8 G. Thus began the slow whittling down of the assumed value of the general solar magnetic field, then thought to be about 50 G.

At the higher frequencies, the results, however, did not yield all that was hoped for. At 3.18 cm no sources were detected by Minnett and Labrum [8]. The eclipse curve was consistent with a uniform sun of $1.1R_\odot$ or one with a 74%, 26% disk bright ring distribution. At 10 cm one small area yielded a temperature of 10^6 °K (Piddington and Hindman [9]). It was felt that the data required a model with 68% of the energy from a uniform disk and with 32% of the radiation concentrated near the limb. Attempts to measure a general magnetic field were negative. When the eclipse curves were compared with two models, neither seemed to fit the situation. Figure 5 demonstrates the model and the observations. The finding that 10,000 km is the highest level from which intense radiation was received was acceptable.

1949 – 1952

During the period of 1949 – 1952 an increasing number of eclipse expeditions was launched. Although equipment was often marginal, it had improved to the point where reliable flux measurements could be taken. The most pressing question was whether or not there was limb brightening or darkening in each wavelength region. Stanier [11] found no limb brightening at 60 cm; Laffineur and

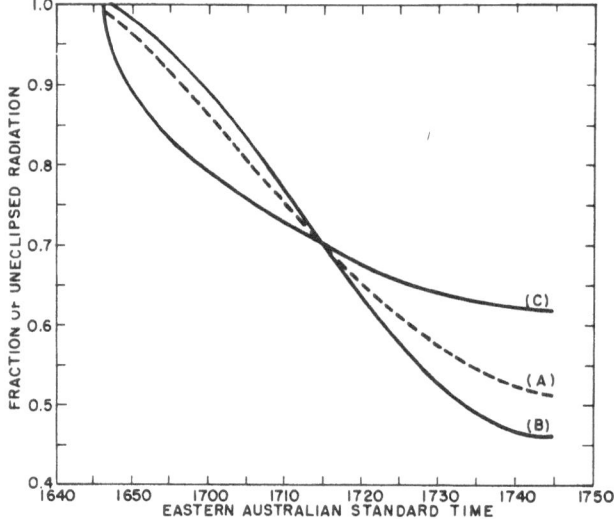

Fig. 5. Eclipse curve (A) taken November 1, 1948, at
λ = 10 cm compared with (B) theoretical curve for uniform
disk and (C) theoretical curve for circumferential ring
(Piddington and Hindman [9].

Michard (Laffineur *et al.* [12]), and Servajean and Steinberg (Laffineur *et al.*
[13]) could support no limb brightening at either 25 or 54.4 cm. Hagen [14] in
1950 observing in Alaska found limb brightening at 3.2 cm with residuals at
totality and the temperature of an enhanced region as follows:

Wavelength	Residual	T of Enhanced Region
3 cm	8%	2×10^6 °K
10 cm	17%	4×10^7 °K
65 cm	26%	

The annular eclipse of September 1, 1951, was interesting in that Blum,
Denisse, and Steinberg [15] did not detect an excess temperature at 3.2 cm for a
region above one sunspot area (neither had the Australians in 1948). However,
another spot region showed a temperature of 6×10^5 °K. The curve was consistent with a $1.07R_\odot$ uniform sun. At 1.78 m the mean radio diameter was 35%
greater than the optical diameter. A residual of 47% was observed with a departure from circular symmetry noted (Fig. 6). This seems to be the first time it
was suggested that the equatorial regions were more extended than the polar. A
uniformly bright ellipsoid would preserve the observed distribution. Hewish [16]
confirmed this in 1954.

Data at 1.78 m from three widely separated sites in Africa were analyzed by
Coutrez *et al.* [17]. Eight bright regions were located. The ellipticity of the

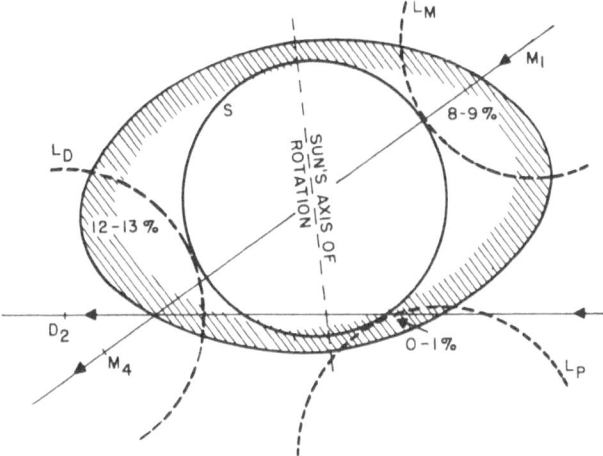

Fig. 6. Model of corona deduced by Blum, Denisse, Stein-
berg [15] at 1.78 m from eclipse observations of 1951, 1952.
Figures indicate intensity reductions observed at Paris (L$_P$),
Dakar (L$_D$), and Markala (L$_M$).

radio sun was apparent but the radio axis did not seem to coincide with the optical axis. Radio bright regions at an altitude of $0.4R_\odot$ coincided with an enhanced region observed at the green coronal line of 5303 Å. This was located at
the base of a large coronal streamer. Other bright regions were related to the
5303 Å emission (Fig. 7).

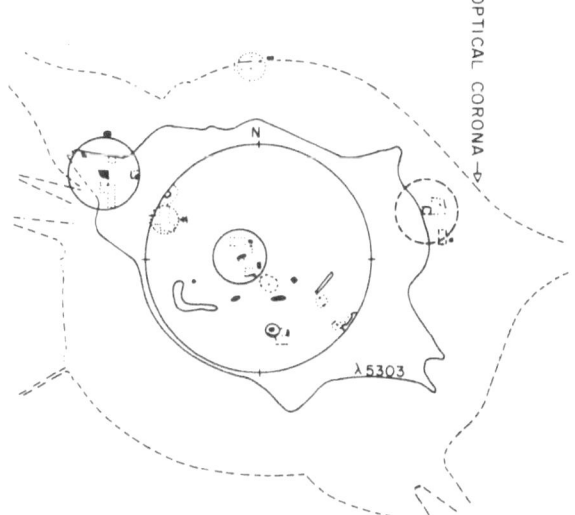

Fig. 7. Sources of radio
energy at 178 m in the solar
corona during the eclipse of
February 25, 1952. Agree-
ment between sources of
maximum coronal emission
at $\lambda = 5303$ Å, coronal
streamers and details on the
disk are clearly seen. Solid
circles are areas of maximum
1.78 m radiation (Coutrez
et al. [17]).

The total eclipse of February 25, 1952, was observed by many groups. The results of this and a partial eclipse on February 14, 1953, indicated that for centimeter and millimeter wavelengths there was limb brightening and probably an elevated temperature at the center of the disk.

THE ECLIPSE OF JUNE 30, 1954

Perhaps the greatest number of radio measurements ever made for any eclipse were made on June 30, 1954. Observations of partial eclipsing were made in Canada, U. S., England, France, Finland, Germany, and Holland. The eclipse was observed as total in Norway and Sweden and in parts of the Soviet Union. Optical observations were poor in general, but radio records were excellent. Totality was observable near noon in Europe. Residual signals noted were all lower than values placed in the literature before or since.

Solar activity was at a minimum on that date. No plages or spots were found on the disk; prominences were not reported. The principal aims of the observers were (1) to determine the sun's radial brightness distribution, (2) to determine the residual flux from the solar corona during minimum sunspot activity, and (3) to contrast the coronal form at the equator and at the poles.

Interpretation of observational data on these points prior to the eclipse was not clear or at best only qualitative. Christiansen and Warburton [18] had found that the quiet sun at 21 cm had limb brightening at the equator and limb darkening at the poles. Alon, Arsac, and Steinberg [19] had found limb brightening at 3.2 cm using both interferometers and eclipse measurements; at wavelengths longer than 21 cm, the questions of limb brightening or darkening and of equatorial vs. polar intensity were not clear.

Observations made at 8 mm in 1954 by Hagen [20] were the closest that radio astronomy came to optical wavelengths. The longest wavelength was 7.9 m observed by Hewish. The spectrum between these points was well covered.

Table I lists data reported by various observers for this event. Data listed in this table and in references from which the data were drawn are displayed graphically in Fig. 8. In many cases radial brightness distribution curves were not presented, or were presented in different ways. We have attempted to present all these data in Fig. 8. Limb brightening is apparent in the shorter wavelength region up to 60 cm. Above this wavelength no limb brightening was detected. At the shorter wavelengths, polar vs. equatorial distributions are generally not reported. Some observers have reported ellipticity at 10 cm; others have not. At the millimeter wavelengths the brightness distribution is essentially circular. At wavelengths greater than 60 cm observers had generally found coronal extension greater at the equator than at the poles.

1954 – 1959

After 1954, there was a sharp decrease in eclipse observations until the eclipse of February 15, 1961. In data reduction and instrumentation subsequent to June 1954, there was a more positive interest in plages and sunspot regions.

TABLE I (June 30, 1954)

λ	Observer	$T_{D_\odot}$, °K	Flux Wb/m²-c/s × 10^{-22}	Radio dia. from time of contacts	Hgt. of equiv. layer above photosphere based on $R/R_\odot$ at 1st contact, km	Residual % at max. phase	% Eclipse	Remarks
8.6 mm	Coates et al. [21]	8,530	2160	1.0068 $R_\odot$	4,740	<0.5	100	Limb brightening over 10%, center to limb taper
3 cm	Hachenberg et al. [22]	22,000	485	1.04 $R_\odot$	28,000	17	85.5	L.B.
3.2 cm	Tchikhatchev [23]					0.98	100	
3.4 cm	Troitzky					0.71	100	
3.4 cm	Fokker et al. [24]					30	78	Max. radio later than max. optical
9.4 cm	Haddock Mayer et al. [25]	33,000	73			7	100	L.B.
10 cm	Fokker et al.					28	78	
10 cm	Tchikhatchev					6.9	100	
10 cm	Troitzky					5.9	100	
10.5 cm	Hey [26]	42,000	75			28	71	L.B. slight ellipticity
10.7 cm	Covington et al. [27]	33,000	69	1.06 – 1.14 $R_\odot$	42,000	19.2	86	L.B. equatorial
20 cm	Hachenberg et al.			1.2 Equatorial / 0.9 Polar	70,000	24	91.9 / 85.5	Limb intensity 2.5 times central
23 cm	Tchikhatchev					9.9	100	Sweden
55 cm	Fokker et al.					30	78	Meudon
55 cm	Laffineur [28]			~ 1.14 $R_\odot$	98,000	14.7 / 42		
60 cm	Eriksen et al. [29]	250,000	13	1.4 $R_\odot$ Equatorial	280,000	18	100	
60 cm	Hewish [16]			1.5 $R_\odot$ Equatorial	350,000	34		L.B.
1 m	Tchikhatchev					20	92	
1.4 m	Hewish	500,000	5	1.8 $R_\odot$ Equatorial	560,000	22		
1.5 m	Priester and Dröge [30]	810,000	7	1.9 $R_\odot$ Equatorial	630,000	27	90.7	
1.5 m	Tchikhatchev					25	92	
1.5 m	Fokker et al.			1.2 $R_\odot$ Equatorial	140,000	27	78	
1.5 m	Eriksen et al.			1.8 $R_\odot$ Equatorial	560,000	27	100	No L.B.
3.5 m	Tchikhatchev					43	92	
3.7 m	Hewish	750,000	1.4	2.5 $R_\odot$ Equatorial / 1.4 Polar	1,750,000 / 280,000	40		
3.7 m	Tuominen [31]			>2 $R_\odot$ Equatorial	>700,000	52	90	

NOTE: Generally, observers at wavelengths greater than 60 cm found coronal extension greater at equator than poles.

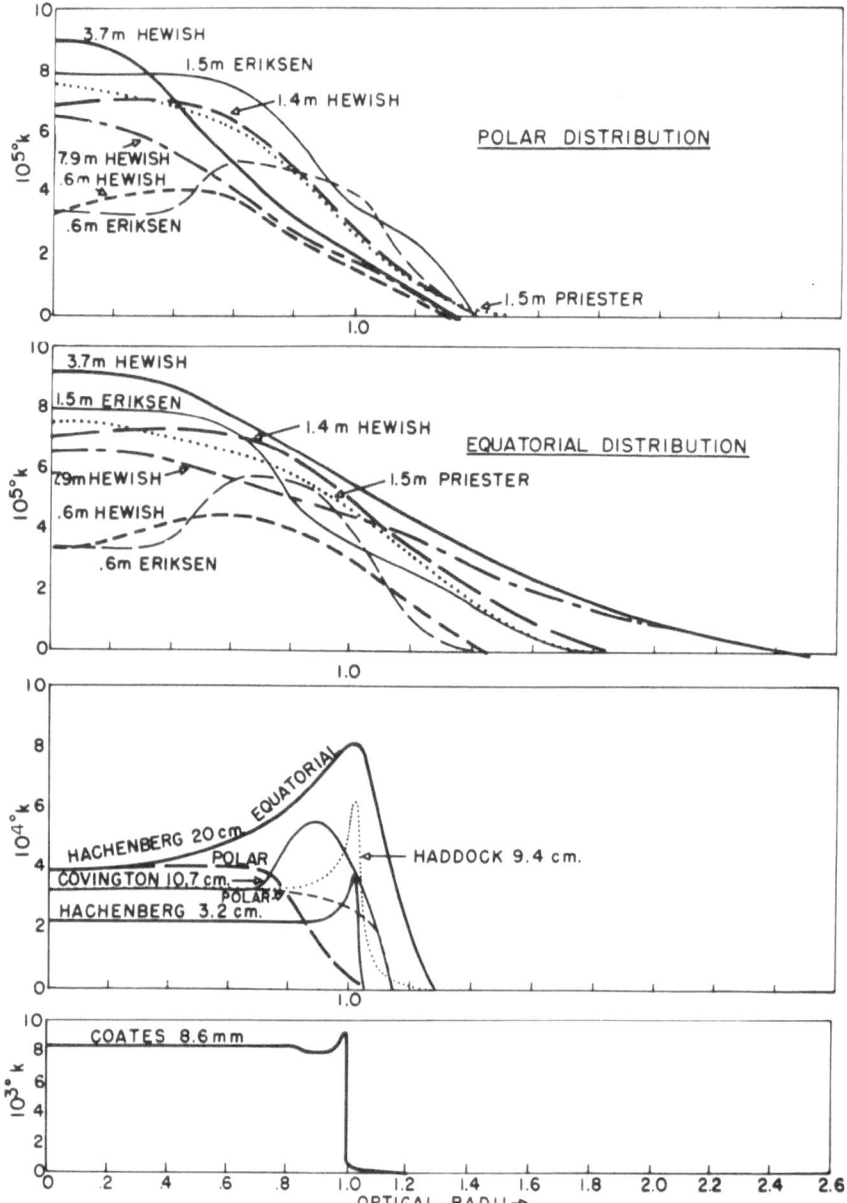

Fig. 8. Composite of solar distribution for the June 30, 1954, eclipse based
on data from Hewish, Haddock, Eriksen, etc.

Multifrequency recordings were made with the intent to determine spectra of enhanced centers of activity. Concentration on one solar area during a specific
eclipse could yield observations on the angular diameter of the center, the temperature of the area, and perhaps the height of the region above the photosphere.

Each of these parameters could be studied as a function of frequency.

The partial eclipse of June 20, 1955, was observed at 10 and 8 cm from three sites in Japan. The occultation path was useful in locating radio bright regions and correlating the slope variations of the eclipse curve with both sunspots and calcium plages (Hatanaka *et al.* [32]).

In Fig. 9, area 1 corresponds to a temperature of 1×10^6 °K, area 2 to 2×10^6 °K, and area 3 to 3×10^6 °K. For the first time detailed agreement was noted between radio spot and calcium plage regions.

The annular eclipse of April 19, 1958, marked the beginning of serious polarization studies of radio bright regions. As early as 1948, eclipse polarization studies were made with rather inconclusive results. For the 1958 event, which was especially interesting since it marked the year of maximum solar activity, the Pulkovo Observatory in cooperation with Chinese scientists made preliminary measurements on Hainan Island at 5.1-, 3.3-, and 2.0-cm wavelengths. Their primary purpose was to study emission regions above sunspots, size, and brightness temperature of polarized and unpolarized components. Once the frequency dependence of the brightness temperature of a coronal condensation for the ordinary and extraordinary wave is known, the magnetic field and, hopefully, the emission mechanism can be determined. Gelfreich [33] describes how the kinetic temperature and electron density for the region can be found.

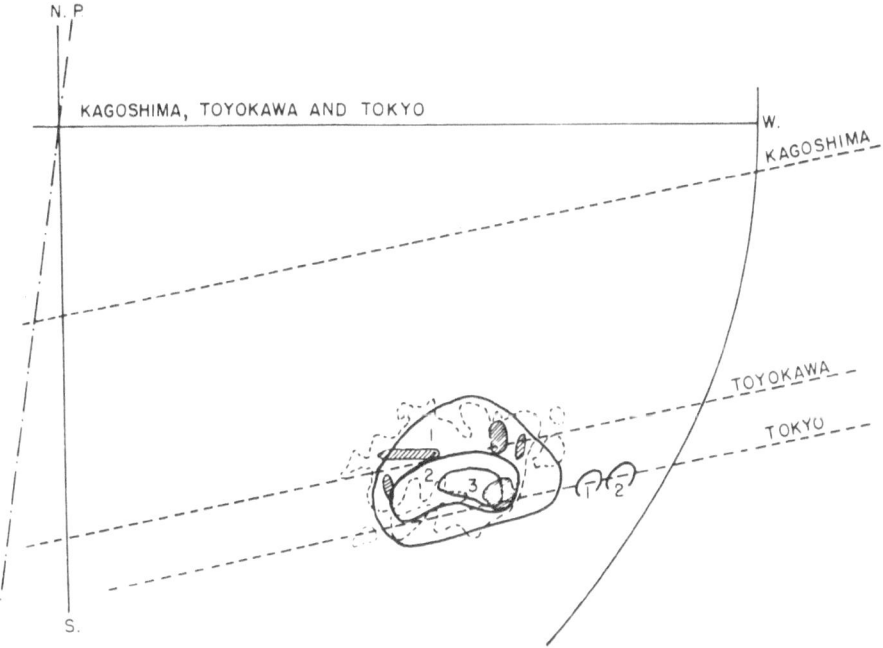

Fig. 9. Comparison between radio and optical observations. Dotted lines indicate the paths of the northern limb of the moon at three stations. Solid lines marked 1, 2, and 3 refer to relative brightness. Shaded area gives optical location of optical sunspot. Dotted line encloses calcium plage region. (Hatanaka [32].)

Based on eclipse observations, the following data were obtained relative to the polarized and unpolarized temperature of a particular region:

	Unpolarized	Polarized	Percentage
5.1 cm	3.6×10^6 °K	0.775×10^6 °K	17.7%
3.3 cm	1.65×10^5 °K	0.350×10^6 °K	17.5%
2.0 cm	200×10^3 °K	120×10^3 °K	37.5%

(The 2-cm data may be unreliable since the author felt the emission was below equipment sensitivity.)

During the same eclipse Tanaka (Tanaka *et al.* [34]), following a year's work at 3.19, 8, 15, and 30 cm, observing the polarization of bursts and bright regions, was able to show from eclipse data that the source of polarization was about 1′ over the spot and that these regions are often bipolar in form, reversing polarity on either side of the solar equator. Some of these data are shown in Fig. 10. Tanaka and Steinberg [35] have recently extended this work. The existence of the extraordinary wave and a high magnetic field was confirmed.

An important point in all this is that the eclipse mechanism provided the necessary resolution, and through calculation of the slope from the basic curve the angular extent of the region as well as the polarized and unpolarized apparent temperatures.

A basic point in polarization studies is that in the presence of a magnetic field, radiation in an ionized medium can propagate in two modes. That is, the medium within the magnetic field divides the radio wave into two components. These are conventionally called the ordinary mode, which is the same as left-hand polarization, and the extraordinary mode or right-hand polarization.

There has been some confusion on the percentage of polarization from enhanced regions measured at certain frequencies. Swarup [47] points out that

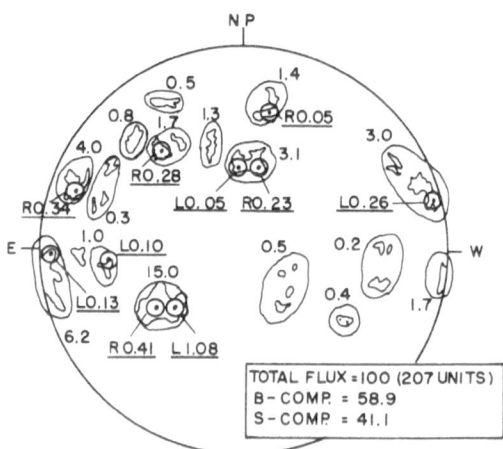

Fig. 10. An 8-cm map of the S-component on April 19, 1958, derived from eclipse observations. (Tanaka [34].)

based on interferometer and eclipse data generally, the degree of polarization is greater than 30% at 3 cm, approximately 10% at 10 cm, and less than 2% at 21 cm. The size of the source is only slightly smaller at 3 cm than at 21 cm, with diameters varying between 2' and 4'.

The observations and analysis of Molchanov [36] and his colleagues led to an important result, i.e., the determination of the spectra of centers of solar activity. One solar region was studied in detail by using multifrequency observations during the eclipse of April 19, 1958. The spectrum obtained is shown in Fig. 11 with data obtained from 0.8 to 30 cm. The spectrum peaked between 5 and 6 cm. Residuals of 6, 11.5, and 16% for 3.2, 4.5, and 5.1 cm, respectively, were found. The importance of this work is twofold: it yields the spectral maximum of an enhanced region and, from the exponent of the spectral index, the magnetic field surrounding a spot area can be determined.

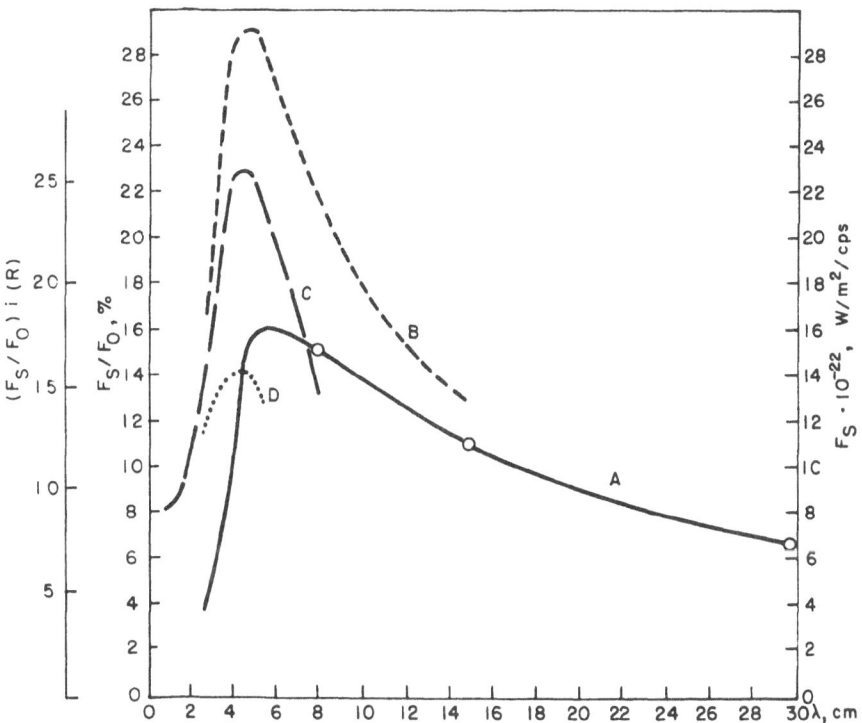

Fig. 11. Ratio of enhanced center of activity flux to background solar radiation from sunspot group #188, solar eclipse of April 19, 1958. Note peak at 5–6 cm (Molchanov [36]). An F_S/F_O, is relative spectrum of bright region; for the same corrected to equivalent flux at sun's center use left scale. B is absolute spectrum of region based on background flux measurements on eclipse day. C is absolute spectrum of region based on background flux averaged over several years. D is absolute spectrum of region based on Rayleigh–Jeans approximation. B, C, and D use right-hand scale.

THE RADIO SPECTROHELIOGRAM

A development which reduced the urgency of eclipse observations was the radio spectroheliograph. In principle, the grating interferometer can give a detailed contour map of the sun; in practice, interferometric resolution during the 1950's and early 1960's was limited to $2'$ to $4'$; eclipse measurements have shown detail to better than $1'$. An inherent weakness of the interferometer is the difficulty involved in locating the optical center of the sun on the solar heliogram; eclipses have at times helped to remove this uncertainty.

Although in several analyses comparison between spectroheliograms and eclipse slopes have yielded only fair results, the spectroheliograms serve to eliminate confusion between centers of radio activity and limb brightness. As the results of the Australian observations of the 1949 eclipse have shown, radio centers can be identified with areas where sunspots were observed on earlier solar rotations; the spectroheliograms show only radio centers of activity and, therefore, a truer display of solar activity at radio wavelengths. Secondly, the spectroheliogram has removed certain ambiguities from the eclipse curve since the occulting action is one-dimensional. In turn, the eclipse data improve the spectrograms since the antenna pattern of the grating interferometer is not as uniform as the designers would like.

A spectroheliogram was used during the partial eclipse observed in Japan on June 20, 1955; the interferometer pattern was used to deduce the variation of flux from an intense radio source. On the basis of three-station observations in the 8 to 10 cm range, the brightness temperature of a spot 10^6 $^\circ$K was established. A model with a uniform distribution of $1.1R_O$ and a bright zone along the equator best fitted the observations.

In a more elaborate analysis of a partial eclipse on April 18, 1959, Krishnan and Labrum [37] found that at 21 cm the eclipse record could be explained by assuming a quiet-sun distribution similar to that of Christiansen and Warburton for sunspot minimum but with the temperature gradient of the limb component increased. In addition, they found excellent correspondence with faculaires as seen in the K line of calcium but with the radio centers located radially above the plages at a height of about 70,000 km (Fig. 12).

OCTOBER 2, 1959; FEBRUARY 15, 1961; AND JULY 20, 1963

An increasing amount of interest was exhibited in the total eclipses of 1959, 1961, and 1963. Instrumentation and analysis concentrated on (1) comparing spectroheliograms with eclipse curves; (2) calculating the height of enhanced regions as a function of wavelength; and (3) increasing instrumentation in the millimeter region.

The U. S. Air Force Cambridge Research Laboratories recorded the three eclipses above, concentrating on accuracy of absolute measurements and on readability of slopes of enhanced regions. Slope curves were carefully constructed to determine the intensity, height, and angular extent of bright regions in the 3-, 10-, 25-, and 135-cm regions. Tables 2a, b, and c give the

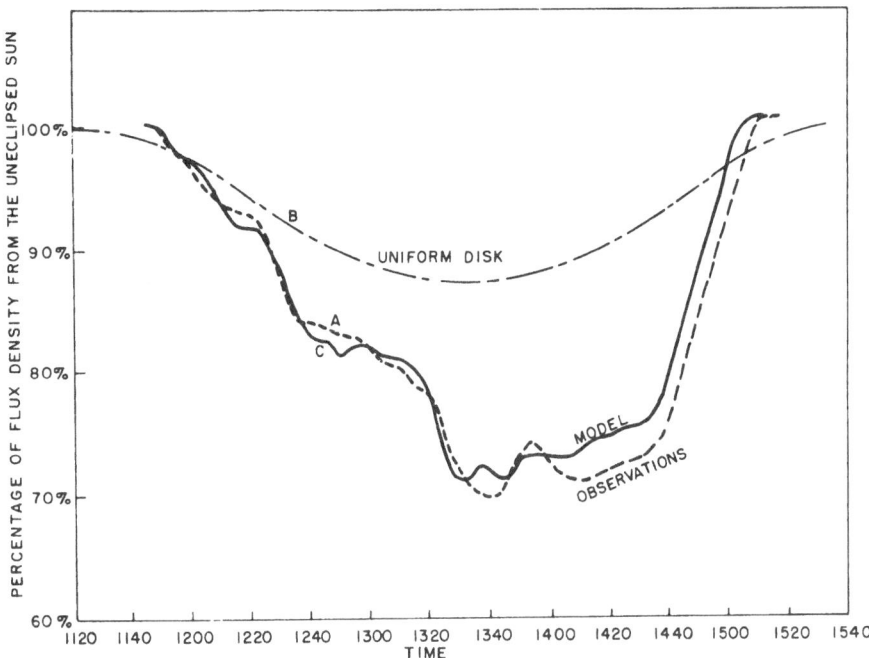

Fig. 12. Correspondence of eclipse observations and solar model adopted (uniform disk with limb brightening); bright regions are included in model curve (Krishnan and Labrum [37]).

pertinent data from measurements made during these eclipses. It should be noted that the 1959 event came close to a period of high sunspot activity, whereas the 1963 event occurred near minimum sunspot number. However, it should be pointed out that the 1963 eclipse showed centers of solar activity while the 1954 eclipse (in the midst of quiet sun conditions) did not.

The spectroheliogram constructed for October 1, 1959, by Christiansen at CSIRO is shown in Fig. 13. During the eclipse on October 2, 1959, it was possible to measure the height of an emitting region just coming around the east limb (south of the solar equator). It was found that the centers of emission heights lay between 13,000 to 26,000 km above the solar disk at 10 cm; 26,000 to 35,000 km at 25 cm; and 105,000 to 140,000 at 1.3 m. These centers of activity, thought to be overdense regions centering about a plage or a sunspot group, extend radially outward as a column into the corona to great heights. The usefulness of an eclipse to determine the height of the region is clear. When the region is near the limb, accurate height determinations can be made. The region referred to in Fig. 13 was found to extend to a height of 80,000 km at 25 cm by U. S. Air Force scientists. During the April 1959 eclipse, Krishnan and Labrum had measured emission heights at 70,000 km at 21 cm.

Figure 14 shows slope data at 10 cm reduced from the October 2, 1959, AFCRL eclipse records (Straka [39]). It is compared with an artificially

TABLE IIa (October 2, 1959)

Wavelength	Observer	Radio diameter	Residual percentage	Eclipse type	Height of emission region, km	Remarks
10 cm	Straka [39]	1.02 – 1.04	17.5	T	13,000 – 26,000	
23 cm	Castelli et al. [48]	1.04 – 1.05	20	T	26,000 – 35,000	
1.34 m	Kidd [39]	1.17	35	T	105,000 – 140,000	

TABLE IIb (February 15, 1961)

Wavelength	Observer	Radio diameter	Residual percentage	Eclipse type	Height of emission region, km	Remarks
4.3 mm	Tolbert et al. [44]			T		200°K Solar flare at totality
1.6 cm	Su-Shih-Wen [40]	1.0	4.3	T		
3.2 cm	Su-Shih-Wen	1.05	5.7	0.937	35,000	
3.1 cm	Castelli et al. [48]		7.6	T		
3.1 cm	Arcetri [43]		5.5 circ. 4.5 lin.	T		
3.2 cm	Weissig [49]	1.07 – 1.08	4.2	T	50,000	
20 cm	Hachenberg [60]	1.3	21.5	T	~200,000	
21 cm	Gosachinsky [50]	1.09	42.0	0.83	65,000	No circular polarization found in spot
23.6 cm	Castelli et al. [48]	1.34	25.3	T	>100,000	
1.45 m	Alekseev [42]	1.28	34	T	196,000	
2.0 m	Alekseev	1.35	43.5	T	245,000	
3.3 m	Alekseev	1.52	53	T	364,000	
4.0 m	Alekseev	1.78	66	T	545,000	

TABLE IIc (July 20, 1963)

Wavelength		Observer	Radio diameter	Residual percentage	Eclipse type	Height of emission region, km	Remarks
3.2	mm	Tolbert [45]	1.0	0	T		Prominence detected on S.E. limb during totality
4.3	mm	Chisholm, J., MIT [51]	1.0	0	T		
8.6	mm	Chisholm, J., MIT	1.0	0	T		
8.6	mm	AFCRL, Martin, E. [52]		7.5	0.94		
3.2	cm	Castelli et al. [53]	1.03	5.4	T	22,000	
4.6	cm	Higgs and Broten [54]	1.08				Strong polar limb darkening
7.3	cm	Bell Labs. [55]					2 Polarizations
10.1	cm	Castelli et al. [53]	1.06	11	T	44,000	
11.1	cm	Covington [56]	1.06	14.3	T	44,000	
25	cm	Castelli et al. [53]	1.06	13.2	0.95	44,000	
33	cm	Miner [57]	1.15	14.4	0.95	105,000	
75	cm	Webb [58]		25			

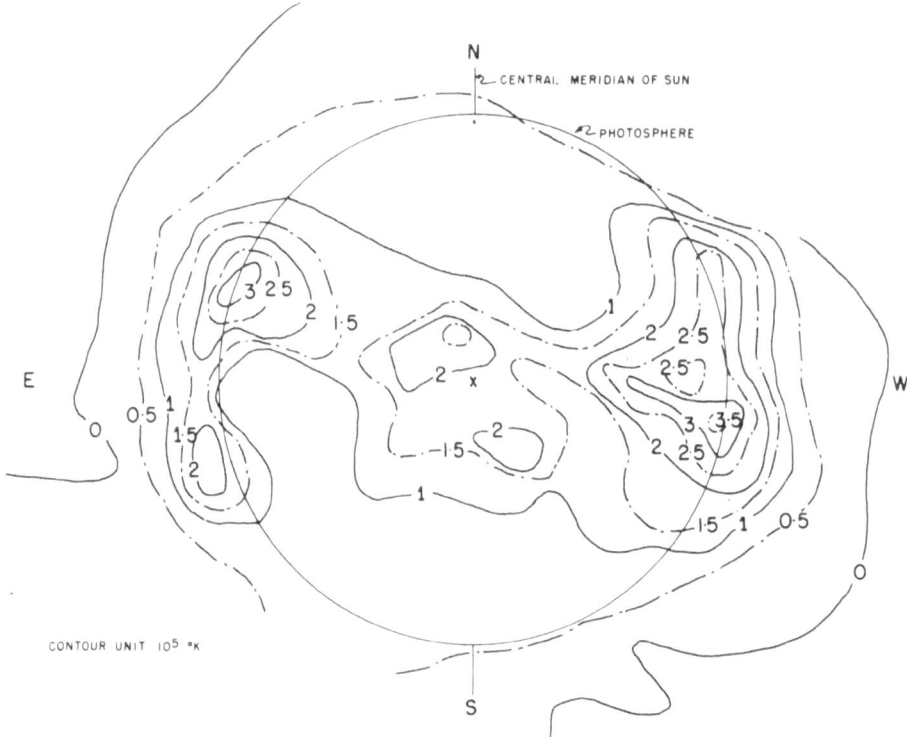

Fig. 13. Interferometric map of the sun at 1420 Mc on October 1, 1959, at 0200 U. T.
(Courtesy of W. N. Christiansen.)

eclipsed heliogram. Of particular interest is the width of the slope curve at points "D" and "G" which indicates the extent of active regions to be 2.5′ and 1.2′ of arc.

An excellent example of combining eclipse with interferometer data was accomplished by Asper *et al.* [38] during the July 20, 1963, eclipse (observed as a partial eclipse at Stanford University). The solar heliogram taken with a 2-min resolution was combined with an eclipse curve taken with a wide-beam antenna on the same day. An east limb region could be located with great accuracy (Fig. 15).

One of the most accurate means of determining the radial extent of the solar radio emission is by eclipse records of the time of first or last radio contact compared with first and last optical contacts. On this basis, the maximum height of radio emission can be determined. In Figs. 16a and b the height parameter is plotted *vs.* wavelength for different eclipses. The eclipse data is also compared with interferometer (noneclipse) observations. It is interesting to note that the height of radio regions was apparently greater for low sunspot years than for more active periods.

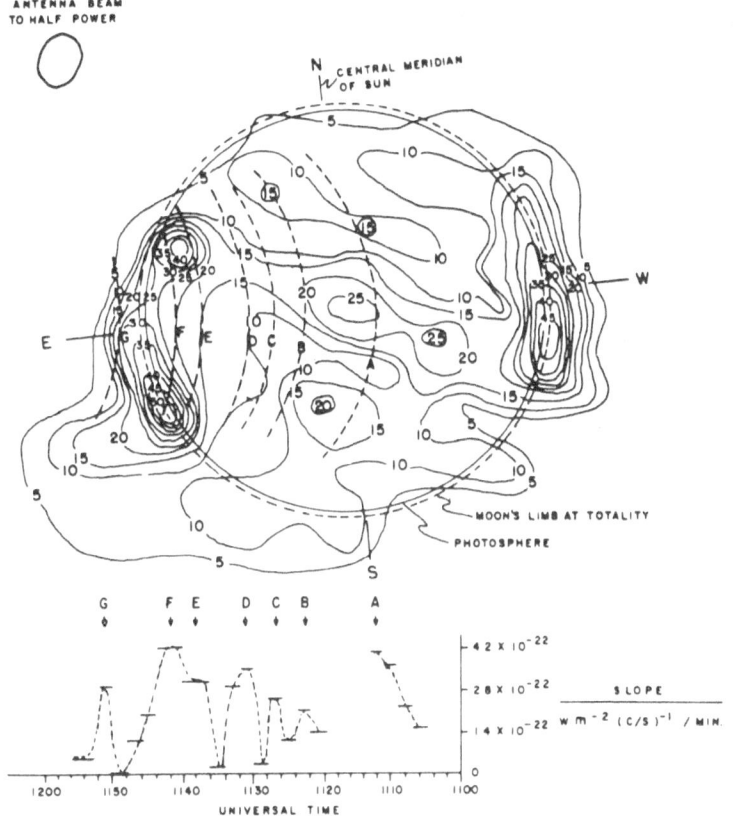

Fig. 14. Radiation contours at 9.1 cm observed at Stanford at about 2100 U. T. on Oct. 1, 1959. Lower curve is that of slopes of corrected 10-cm data taken at AFCRL.

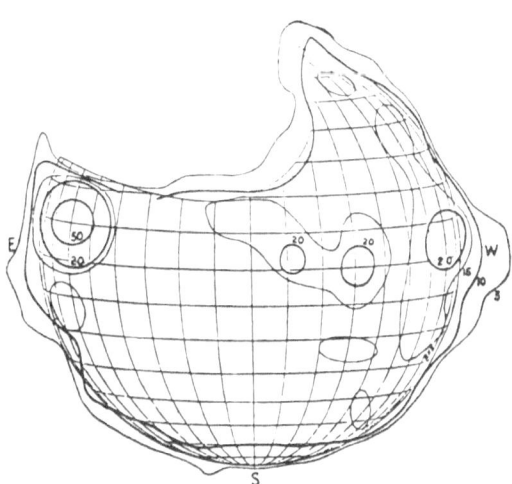

Fig. 15. Stanford 9.1-cm spectro-heliogram, July 20, 1963, 20 hr 11 min−21 hr 16 min U. T. Brightness unit = 2600°K. East limb spot location determined from 1963 eclipse location 58° ± 2° E, 10° ± 2° N.

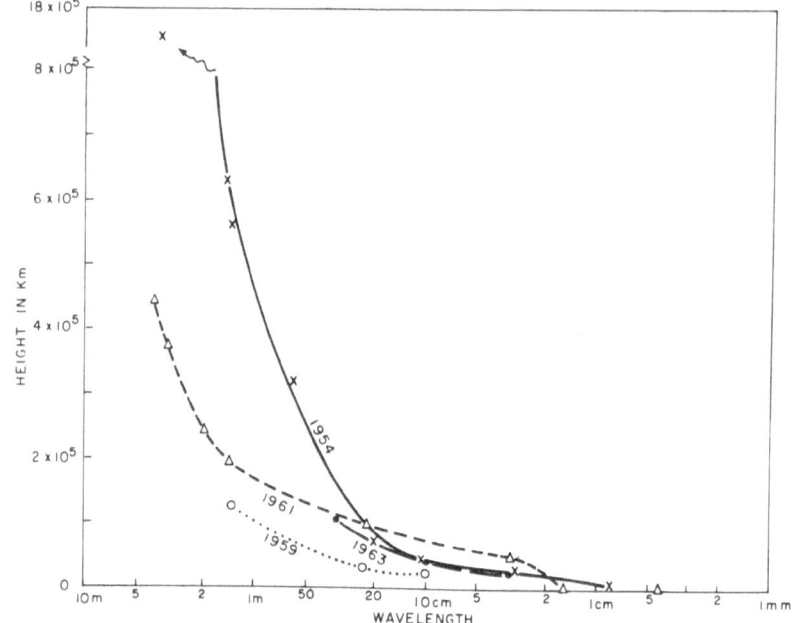

Fig. 16a. Maximum height of solar emission as determined from times of first
and last contact for several eclipses — based on Tables 1 and 2a, b, and c.

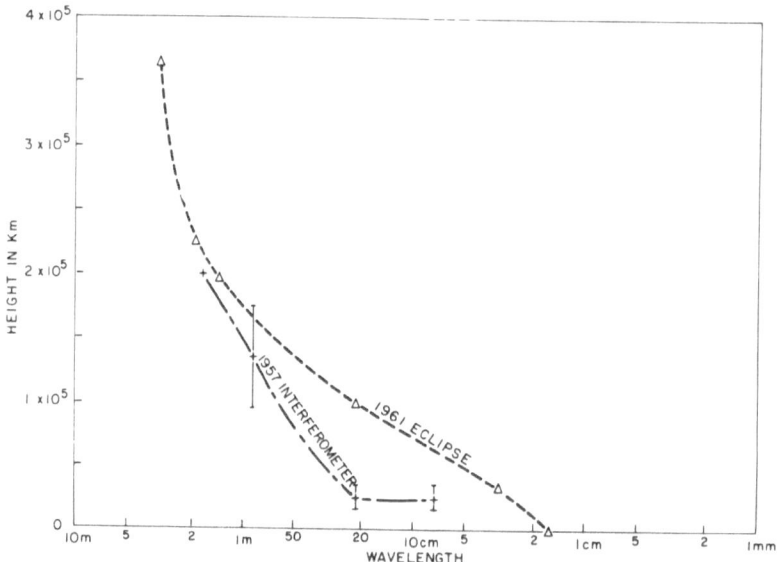

Fig. 16a. Maximum height of solar emission as determined from times of first
and last contact for several eclipses — based on Tables 1 and 2a, b, and c.

A summary of Soviet centimeter observations over nine years is presented in Table III. Of special significance is the 1952–1954 data. During the eclipses in these years the solar brightness was practically the same although the radio diameter decreased from 1952 to 1954. Su Shih-Wen [40] explains this by indicating that the chromospheric contraction is apparently much weaker than the development of solar activity. This height–diameter data is somewhat contrary to that presented in Fig. 16a.

Peterova and his co-workers [41], however, making observations of the July 31, 1962, annular eclipse and the April 19, 1958, annular eclipse at 4.5 cm, concluded also that the dimension of the 4.5-cm radio disk decreased during periods of increased activity. When they compared the residual in 1962 with that in 1958 they found them to be about equal although the open ring was twice as large in 1958 as in 1962.

RESIDUALS

An important part of the eclipse observations is the comparison of eclipse data taken in different years. When the eclipse magnitude varies, it is obvious that for the same solar distribution the residual at totality will change greatly. For an eclipse magnitude of 1.01, the moon covers a height 7000 km above the

TABLE III*

Year and type of eclipse	Wavelength, cm	Residual percentage	Effective radio diameter[†]	Average solar-disk temperature, °K
1952 (total)	3.2	5.2	1.057	14,000
	10.0	15.0	1.12	49,000
1954 (total)	3.2	0.71	1.038	14,000
	10.0	5.94	1.067	55,000
1956 (partial)	3.2		1.1	19,000
	10.0		1.1	70,000
1958 (annular)	1.6		1.03	9,000
	3.2		1.05	21,000
	10.0		1.05	100,000
1961 (total)	1.6	4.3	1.048	11,000
	3.2	5.7	1.05	15,000

*Shih-Wen et al. [40].
[†]Radio diameter at 3.2 cm averaged over 9 years is 1.049
      1954 minimum = 1.038
      1956 maximum = 1.1 unreliable small phase

photosphere; for a magnitude of 1.02, the moon covers 14,000 km, and so forth. This difference is quite important for wavelengths from 1 to 5 cm. For longer wavelengths it is less important.

Figure 17 is a plot of residual energy as a function of frequency for different eclipses. It is based on data in Tables I and II. At meter wavelengths the various curves are not very sensitive to the total eclipse magnitude. During the 1954 and 1961 eclipses the magnitude was very similar (1.03), yet the 1954 residual was much lower. Again, comparing 1959 and 1963, the residual for 1963 with a 1.01 magnitude is lower than 1959 with a higher magnitude. Clearly then, residuals are lower for low solar activity years.

In the meter wavelength region, Alekseev, observing the 1961 eclipse, made important residual, total intensity, and effective radio diameter measurements (Alekseev et al. [42]). These are shown in Fig. 18. It should be noted that the 1952 residuals compare very well with the 1961 residuals.

The measurement of polarization residuals has continued during these eclipses. The 1961 event found the Arcetri Observatory (Noci et al. [43]) making both linear and circular polarization measurements. At totality at 3.1 cm, the

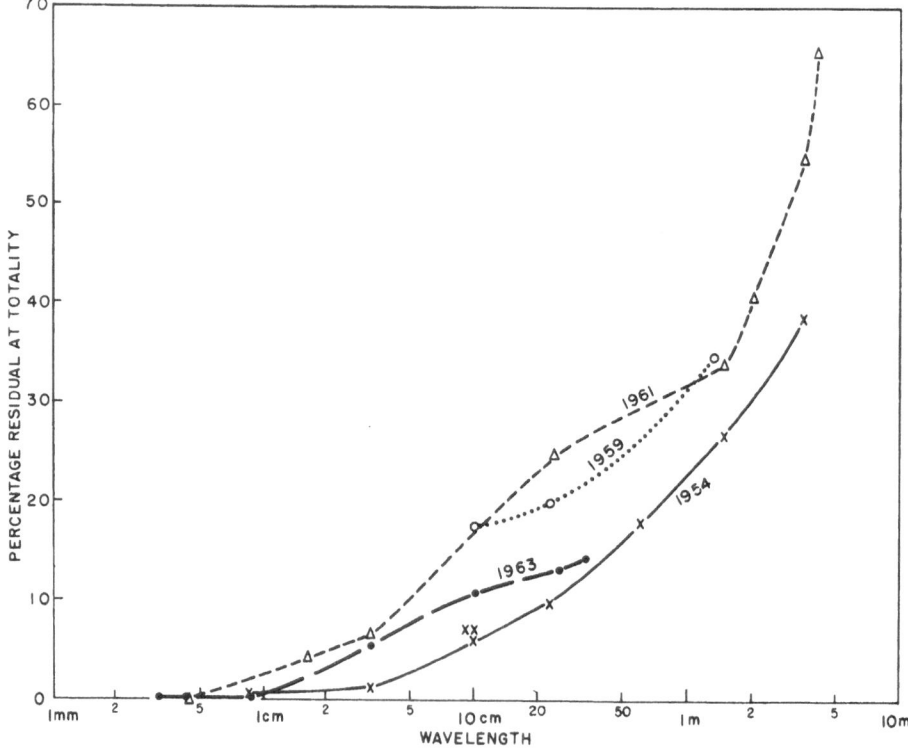

Fig. 17. Residual solar energy at totality vs. λ for four total eclipses based on Tables I, IIa, b, and c.

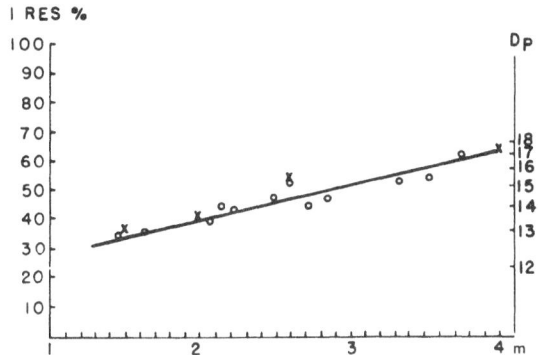

FIG. 18a RESIDUAL INTENSITIES AT THE TIME OF THE TOTAL
PHASE OF THE ECLIPSE; THE SCALE OF EFFECTIVE
RADIO DIAMETERS IS SHOWN ON THE RIGHT. CROSSES
ARE USED TO DENOTE THE RESULTS OBTAINED DUR-
ING THE ECLIPSE OF 25 FEB. 1952; CIRCLES, 15 FEB. 1961

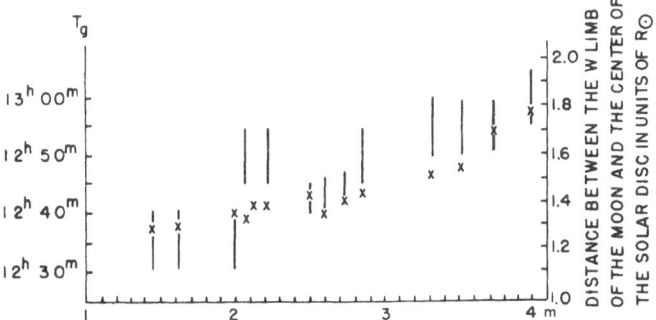

FIG. 18b THE TIME OF THE FOURTH RADIO CONTACT AS A
FUNCTION OF THE WAVELENGTH. THE DISTANCE BE-
TWEEN THE WEST LIMB OF THE MOON AND THE
CENTER OF THE SOLAR DISC IS SHOWN ON THE
RIGHT; CROSSES ARE USED TO DENOTE THE VALUES
DERIVED FROM THE RESIDUAL INTENSITIES. (ALEKSEEV ET AL.)

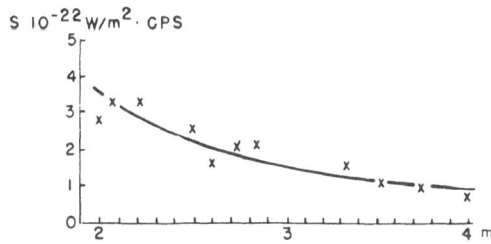

FIG. 18c VARIATION OF THE INTENSITY OF SOLAR
RADIO EMISSION WITH WAVELENGTH (ALEKSEEV ET AL. [42])
ECLIPSE OF 15 FEB. 1961

Fig. 18. Meter wavelength characteristics February 15, 1961.

circular polarization residual measured 5.5% while the linear polarization residual measured 4.5%.

During recent years, millimeter instrumentation has advanced to shorter wavelengths. In particular, the University of Texas has been operating at 4.3 and 3.2 mm at lower atmospheric absorption windows. When a scanning technique with a pencil-beam antenna was employed, solar emission variations as great as 2:1 between adjacent regions were found during the 1961 eclipse (Tolbert and Straiton [44]); regions were equal in size to $\frac{1}{10}$ of the solar disk.

In 1963 shortly after second contact a brightness temperature equal to one-half of the antenna temperature received from an area equal to $\frac{1}{75}$ of the solar disk was recorded. This large signal was attributed to a stationary prominence located on the southeast limb of the sun (Tolbert et al. [45]). It should be noted that Covington [5] in his first observation in 1946 interpreted an abrupt change in his eclipse curve as being associated with an optical prominence. During the November 1948 event the Australians also noted the apparent effect of a prominence. In February 4 – 5, 1962, Covington [46], doing 10.7 strip scans in Canada while Waldmeier was optically observing totality in New Guinea, presented evidence that radio energy is associated with prominences.

The radio association with solar prominences is particularly important since prominences must, therefore, contribute an important part of the slowly varying component. Until now most correlations of optical and radio enhanced regions have stressed plage regions and sunspot groups.

The July 20, 1963, total eclipse saw a large increase in the number of observations. It may be that this heralds a renewed interest in utilizing the eclipse as a tool in solar research. Results of some of these are reported in this review in Table IIc. The material is useful for establishing the spectrum of an east limb enhanced region. The region was not well isolated since there was a second weaker region at about the same longitude in the southern hemisphere. Incomplete, and in some cases preliminary, eclipse data reveal the following characteristics of the region:

Reference	λ		Size	Flux (10^{-22})	T_B
Castelli et al. [53]	25	cm	4.2'	1.87	3.6×10^5
Covington [56]	11.1	cm	1.5'	4.5	$(1.35 \times 10^6)*$
Castelli et al. [53]	10.1	cm	2.2'	4.9	6.5×10^5
Asper et al. [38]	9.1	cm	1.5'	5.0	1×10^6
Higgs and Broten [54]	4.6	cm	1.25'	$(3.74)*$	2.05×10^5
Castelli et al. [53]	3.27 cm		1.4'	4	1.6×10^5

*Computed by authors.

Additional data are available at 4 mm, 8 mm, 2 cm, 3 cm, 75 cm, and other wavelengths.

One puzzle which has not as yet been solved is that of apparent "cold" regions. In principle the eclipse slope from first to second contact should always be negative and the slope from third to fourth contact positive. In practice it has been found by several groups during earlier eclipses (1961) that there are short periods of time when the slope does not change perceptibly. Essentially the regions are colder than the solar background. This has been found by AFCRL scientists at 10, 24, and 33 cm and by the University of Texas at 4.3 mm. Both ingress and egress records show the same regions.

During the 1963 eclipse evidence of what is interpreted to be the existence of a "cold" region across the solar disk, covering an area of the eastern half of the sun, was indicated. The record of Fig. 2a at times between roughly 17:17 EST and 17:20 EST seems to suggest this conclusion. It is noteworthy that the same change in slope characteristic is present also on the 33, 25, and 3.27 cm data. A careful evaluation of this region has been undertaken.

CONCLUSION

In almost twenty years of radio solar eclipse measurements, our knowledge of the sun has increased rapidly. Optical data have played an important part in radio data analysis. The quiet sun of 1954 quite accurately established the solar radio distribution as a function of wavelength for that period. Earlier and subsequent eclipses have established the radio distribution with a fair degree of accuracy. Enhanced regions of activity have been correlated with optical features, and their height, polarization, and intensity have been studied. Progress in interferometer design has greatly aided the study of the sun. As in the past, multifrequency observations are most important to determine the spectrum and mechanism producing enhanced regions of activity.

Data must be carefully processed and all significant information presented. One need above all others seems to stand out; that is, the need for greater absolute accuracy of measurements by all observers.

The eclipse has played an important role in solar physics. It will play a less important role in the future but it is believed that eclipse observations will be important in pointing out directions for long-term, systematic observations.

ACKNOWLEDGMENT

The authors are deeply grateful to John Clemens and Paul Lyons of Wentworth Institute for their assistance in many aspects of this paper. They are likewise grateful to Doreen Ushakoff and Carl Ferioli for biographical and observational research, and to many others who have in many ways contributed help and information.

REFERENCES

1. Samuel Alfred Mitchell, *Eclipses of the Sun* (Columbia Univ. Press, New York, 1951), 5th ed.
2. Prof. T. R. von Oppolzer, *Canon of Eclipses* (*Canon der Finsternisse*). Owen Gingerich (Dover Publications, New York, 1962).

3. *Explanatory Supplement to the Astronomical Ephemeris* and the *American Ephemeris and Nautical Almanac*, issued by H. M. Nautical Almanac Office, 1961.
4. R. H. Dicke and R. Beringer, *Ap. J.* **103**, 375 (1946).
5. A. E. Covington, *Nature (London)* **159**, 405 (1947).
6. S. E. Khaykin and B. M. Tchikhachev, *Izv. Akad. Nauk. SSSR* **12**, 38 (1948).
7. J. P. Hagen, NRL, Washington, Report 3504 (1949).
8. H. C. Minnett and N. R. Labrum, *Australian J. Sci. Res. A* **3**, 60 (1959).
9. J. H. Piddington and J. V. Hindman, *Australian J. Sci. Res. A* **2**, 524 (1949).
10. W. N. Christiansen, D. E. Yabsley, and B. Y. Mills, *Australian J. Sci. Res. A* **2**, 506 (1949).
11. H. M. Stanier, *Nature (London)* **165**, 354 (1950).
12. M. Laffineur, R. Michard, R. Servajean, and J. L. Steinberg, *Ann. Astrophys.* **13**, 337 (1950).
13. M. Laffineur, R. Michard, J. L. Steinberg, and S. Zisler, *Compt. Rend. Acad. Sci. (Paris)* **228**, 1636 (1949).
14. J. P. Hagen, *Astronom. J.* **56**, 39 (1951).
15. E. J. Blum, J. F. Denisse, and J. L. Steinberg, *Ann. Astrophys.* **15**, 184 (1952).
16. A. Hewish, *Fourth I.A.U. Symposium on Radio Astronomy* (Cambridge University Press, 1957), p. 298.
17. R. Coutrez, A. Koeckelenbergh, and E. Pourbaix, "Commun de' L'Observatoire de Belgique," No. 60 (1953).
18. W. N. Christiansen and J. A. Warburton, *Australian J. Phys. A* **6**, 190 (1950).
19. I. Alon, J. Arsac, and J. L. Steinberg, *Compt. Rend. Acad. Sci. (Paris)* **237**, 300 (1953).
20. J. P. Hagen, *Solar Eclipses and the Ionosphere* (Pergamon Press, London, 1956), p. 253.
21. R. J. Coates and J. E. Gibson, NRL, Washington, Report 5114 (1958).
22. O. Hachenberg, F. Fürstenberg, and H. Prinzler, *Z. Astrophys.* **39**(4), 232 (1956).
23. B. M. Tchikhatchev, *Fourth I.A.U. Symposium on Radio Astronomy* Paper 57 (Cambridge University Press, 1957), p. 311.
24. A. D. Fokker, J. C. DeMunck, and L. D. DeFeiter, *Solar Eclipses and the Ionosphere* (Pergamon Press, London, 1956), p. 272.
25. C. H. Mayer, R. M. Sloanaker, J. P. Hagen, and F. T. Haddock, I.A.U. Symposium, 4th, Papers 47 and 48 (1957), p. 269.
26. J. S. Hey and U. A. Hughes, *Observatory* **76**, 226 (1956).
27. A. E. Covington, W. J. Medd, G. A. Harvey, and N. W. Broten, *J. Roy. Astron. Soc. Can.* **49**, 235 (1955).
28. M. Laffineur, "I.A.U. Symposium, 4th, Paper 55 (1955), p. 304.
29. G. Eriksen, O. Hauge, and E. Tandberg-Hansen, *Astrophysica Norvegica Oslo* **5**, 131 (1955).
30. W. Priester and F. Dröge, *Z. Astrophys.* **37**, 132 (1955).
31. J. Tuominen, J. Riihimaa, and K. Tuori, *Ann. Astrophys.* **3**, 147 (1955).
32. T. Hatanaka, K. Akabane, F. Moriyana, H. Tanaka, and T. Kakinuma, *Solar Eclipses and the Ionosphere* (Pergamon Press, London, 1956), p. 264.
33. G. Gelfreich, D. Kotrol'kov, N. Rishkov, and N. Soboleva, *Paris Symposium on Radio Astronomy* (Stanford Univ. Press, 1959), p. 125.
34. H. Tanaka and T. Kakinuma, *Paris Symposium on Radio Astronomy* Paper 41 (Stanford Univ. Press, 1959), p. 215.
35. H. Tanaka and J. L. Steinberg, *Ann. Astrophys.* **27**, 29 (1964).
36. A. P. Molchanov, *Soviet Astron. – AJ* **5**, 651 (1962).
37. T. Krishnan and N. R. Labrum, *Australian J. Phys.* **14**, 403 (1961).
38. H. K. Asper, R. N. Bracewell, J. Deuter, T. Krishnan, J. Picken, and S. Rose, *J. Geophys. Res.* **69**, 1805 (1964).
39. R. M. Straka, (AFCRL #158, Bedford, Mass., April 1961).
40. Su Shih-Wen, Hsiao Kuang-Chia, Wu Hauai-Wei, Tung-Wu, Wu Chin-Ch'v, V. S. Troitskiy, V. L. Rakhlin, K. M. Strezhneva, and M. R. Zelinskaya, *Izv. Vysshikh Uchebn. Zavedenii Radiofiz.*, **5**, 807 (1962).
41. N. G. Peterova, A. P. Molchanov, and V. G. Nagnibeda, "Solar Data" 8th (Akademiia Nauk, 1963).
42. Yu, I. Alekseev, V. I. Babii, V. V. Vitkevich, M. V. Gorelova, and A. G. Sukhovei, *Soviet Astron. – AJ* **6**, 504 (1962).

43. G. Noci, M. Piattelli, G. Righini, and G. Tagliaferri, "Final Report, Contract AF61(052)-430" (Arcetri-Florence, Italy).

44. C. W. Tolbert and A. W. Straiton, *Astrophys. J.* **135**, 826 (1962).

45. C. W. Tolbert, L. C. Krause, and A. W. Straiton, *Astrophys. J.* July 1964 (in press).

46. A. E. Covington, *J. Roy. Astron. Soc. Can.* **57**, 253 (1963).

47. G. Swarup, T. Kakinuma, A. E. Covington, Gladys A. Harvey, R. F. Mullaly, and J. Rome, *Astrophys. J.* **137**, 1251 (1963).

48. J. P. Castelli, H. W. Cohen, R. M. Straka, and J. Aarons, *Icarus* **2**, 317 (1963).

49. G. C. Weissig and U. N. Borovik, "Bulletin of Solar Data," June 1961, pp. 61-63.

50. I. V. Gosachinsky, T. M. Yegorova, and N. F. Rizhkov, "Bulletin of Solar Data," June 1961, pp. 70-73.

51. J. H. Chisholm, (MIT) private communication, 1964.

52. E. Martin, (AFCRL) private communication, 1963.

53. J. P. Castelli, AFCRL Report, to be published.

54. L. A. Higgs and N. W. Broten, (NRC Ottawa) private communication, 1964.

55. J. Hett, (Bell Telephone Laboratories) private communication, 1964.

56. A. E. Covington, (NRC Ottawa) private communication, 1964. (Report to be published in *J. Roy. Astron. Soc. Can.*)

57. R. J. Miner, AFCRL Report #63-796 (AFCRL, Bedford, Mass., October 1963).

58. H. Webb, (Univ. ILL.) private communication, 1964.

59. W. N. Christiansen *et al.*, *Ann. Astrophys.* **23**, 1 (1960).

60. O. Hachenberg, M. Popowa, and H. Prinzler, *Z. Astrophys.* **58**, 36 (1963).

Chapter 4

The Slowly Varying Component of Solar Radiation

Monique Pick

Observatoire de Paris-Meudon
Paris, France

ABSTRACT

The slowly varying component is most important on centimetric and decimetric wavelengths. A brief description of its properties for this domain of frequencies has been given. The slowly varying component at 169 Mc/s has been detected with the Nançay interferometer. Finally, the different hypotheses proposed to explain the emission of the slowly varying component are discussed succinctly.

INTRODUCTION

It has been known for some fifteen years that, superimposed on thermal emission from the quiet sun, there is emission at radio wavelengths associated with the presence of centers of activity on the sun (Covington [1], Lehany and Yabsley [2], and Denisse [3]).

The development of instruments with great resolving power has revealed that in fact this part of the solar radio emission originates in localized sources associated with the hot dense regions lying above the centers of activity (Christiansen *et al.* [4]). These centers of radio emission are called "radioelectric condensations." When they exist, these sources rotate with the rest of the sun and give rise to the slowly varying component. They are mostly observed in the region of centimetric and decimetric wavelengths and cover a wide spectrum of frequencies. Their radiation is relatively stable. Their dimensions, which vary with the frequency of observation, do not exceed those of the faculae. The lifetimes of these condensations lie between that of sunsunspots and faculae.

It has also been observed that there exists an emission on meter wavelengths which is also related to the optical faculae. The characteristics of this radiation do not necessarily correspond to those observed at higher frequencies (Moutot and Boischot [5]).

The study of the properties and emission mechanisms of the slowly varying component should make it possible to obtain a better understanding of the physical conditions, temperature, and electron density in the solar atmosphere over the centers of activity. Moreover, the localized sources of emission are always

situated in regions of strong magnetic fields and may provide the means of obtaining information about their structure.

Numerous investigations carried out in recent years have shown that the
centers of activity which are associated with the strongest radio condensations
are the seats of the type of chromospheric flare which is accompanied by radio
outbursts and sometimes gives rise to corpuscular emission (Kundu [6], Pick
[7], and Caroubalos [8]). Therefore, an understanding of these privileged
centers should certainly be a very useful pointer for the forecasting of certain
very interesting solar and geophysical phenomena such as Type IV solar radio
bursts, sudden commencement geomagnetic storms, polar cap ionospheric absorption, as well as exceptional variations in the intensity of cosmic radiation.

At present interferometers exist which give one- or two-dimensional images
of the sun with a resolving power which at centimeter wavelengths may reach
about 1' of arc. The interferometer at Ottawa (Covington et al. [9]) operates at
a 10-cm wavelength with a resolving power of 1.2' and the one at Toyokowa [10]
operates at 3.2 cm. The latter is the combination of a 16-aerial array (with a resolving power of 2.2') and a two-aerial interferometer, and it achieves a resolving
power of 0.7'. Instruments which give a two-dimensional image of the sun are
the Stanford interferometer (Swarup [11]), which has a lobe $3'1 \times 3'1$ at 9.1 cm
and the Sydney unit (Christiansen et al. [12]) $3' \times 3'$ at 21 cm.

Finally, at meter wavelengths, we shall mention the Nançay interferometer:
the resolving power is 3.5' in the east–west direction and varies with the declination observed between 8 and 12' in the north–south direction.

The above represents only a small sample typical of existing instruments
(Blum [13]).

THE SLOWLY VARYING COMPONENT ON CENTIMETRIC AND DECIMETRIC WAVELENGTHS

Height

Because of the difference in height in the solar atmosphere, the radio-
condensation moves across the disk faster than the corresponding optical
center. Hence, a measurement of the relative motion of these two centers gives
a measure of the height of the radio source (Fig. 1). It should be noted that the
proper motion of a source is certainly less than 1'. On the other hand, at decimetric wavelengths the refraction of waves in the corona increases appreciably
toward the solar limb and so one should not use measurements made more than 5
days before or after central meridian passage.

It should be noted that the altitudes found vary greatly according to the frequency of observation. Thus, at the 21-cm wavelength, the heights vary between
20,000 and 100,000 km, the mean value being about 40,000 km. The precision of
the measurements is estimated as approximately 10,000 km (Christiansen and
Mathewson [14]). At 10 cm and centimetric wavelengths, the heights are generally less (Kakinuma and Swarup [15], [10]). On the average they are between
10,000 and 20,000 km.

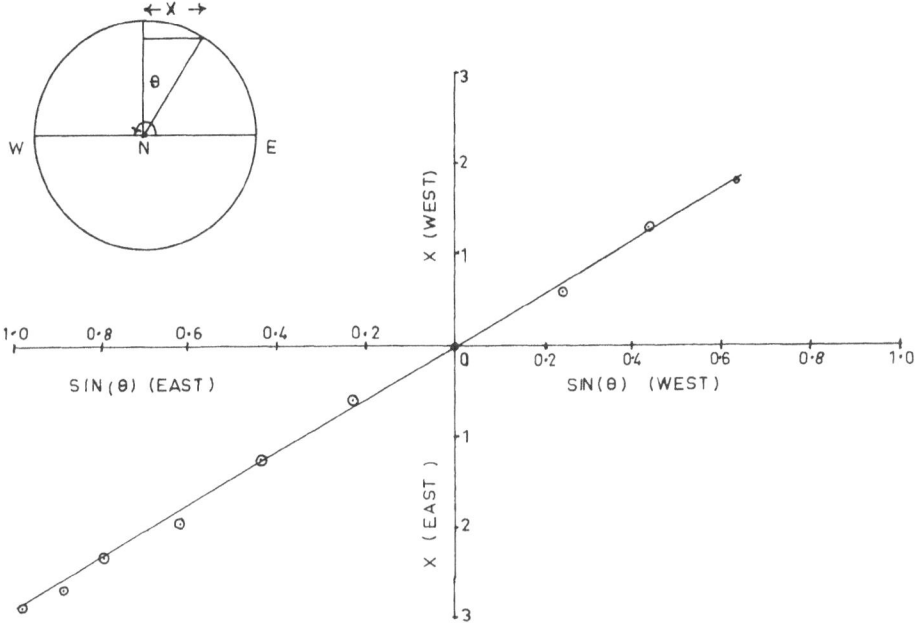

Fig. 1. Determination of the height of radio condensations from their movement across the solar disk.

In a detailed study, Swarup et al. [16] have shown that, as a general rule, the height for 3 cm is less than for 10 cm. Nevertheless, it appears that this situation can be reversed in a few exceptional cases.

Structure of Localized Sources

The radio condensations observed at 21 cm show great similarity to the optical faculae in their form, extent, and brilliance, and in the position of their geometrical centers (Christiansen et al. [17] and Christiansen and Mathewson [14]). On the other hand, it appears that at wavelengths shorter than 10 cm the sources of the slowly varying component are better correlated with the sunspots than with the faculae — indicating the role of the magnetic field in this frequency range (Swarup [18]). It seems that one observes the superposition of two emissive regions, attached to a single center of activity, whose relative intensities vary according to the frequency at which they are observed.

The first region, which dominates at decimetric wavelengths, corresponds to sources whose diameters are comparable to those of the flocculi (plages and faculaires) and whose radiation is weakly polarized (Fig. 2).

The second region is bound to narrow and intense condensations. These condensations, whose predominance indicates periods of strong solar activity, are observed particularly at centimetric wavelengths. Their diameters are comparable to those of spots, and their radiation is strongly polarized.

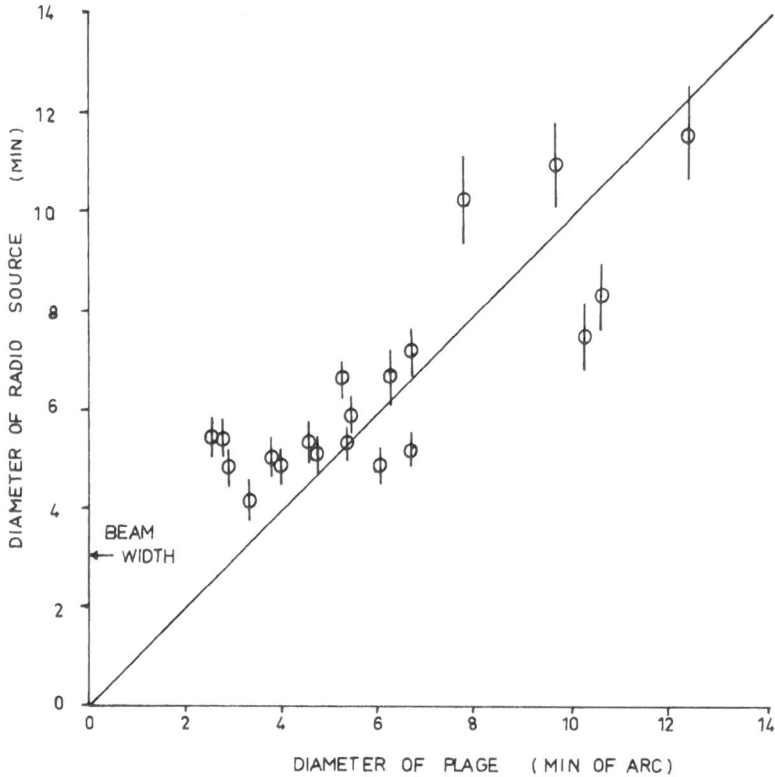

Fig. 2. Relation between the diameters of radio condensations measured at
21 cm and those of the faculae plages.

The existence of these two regions was first demonstrated at a wavelength
of 3.2 cm by Kundu [6] and has been confirmed at 9.1 cm (Kakinuma and Swarup
[15]) and 8 mm (Salomonovich [19]).

At these wavelengths, a typical source consists of an intense region of very
small diameter (smaller than or of the order of $1'.5$) and of a less intense region
of larger diameter ($\sim 5'$). Nevertheless, at 10 cm it seems that, depending on the
cases, a predominance of one or the other of the two components can exist. The
most intense sources seem to have the narrowest diameters.

Brightness Temperature

The flux density S of a source of uniform brightness T_b, which subtends a
solid angle Ω at the point of observation, is given, in 10^{-22} W/m^2-c/s, by the
formula

$$S = 0.27(T\Omega/\lambda^2) \tag{1}$$

The dimensions of the most intense sources of the slowly varying component are comparable to the resolving powers of the instruments and sometimes even smaller. Therefore, the brightness temperature determined from equation (1) represents a lower limit.

At 21 cm, the brightness temperatures determined are on the average of the order of 5×10^5 °K but they can be as much as 1.5×10^6 °K (Christiansen and Mathewson [14]). At 10 cm, the value usually found is of the order of 4.5×10^5 °K. The most intense sources have temperatures in the range of 1.6 to 3.8×10^6 °K (Swarup et al. [16]). At 3 cm, Kundu indicates brightness temperatures of the order of 5×10^5 °K, for the narrowest regions (Moutot and Boischot [36]).

Polarization

Measurements of the polarization of the slowly varying component can yield valuable information about both the emission mechanism and the structure of the magnetic fields of the centers of activity.

The high-frequency radiation is circularly polarized. The degree of polarization decreases as the frequency diminishes. At 3 cm, the degree of circular polarization is found to be of the order of 20 to 30% (Tanaka and Kakinuma [20]). At 10 cm, it is approximately 10%, and at 20 cm it falls to 2% (Christiansen and Mathewson [14]).

The source of the polarized radiation is confined to a very small area above the spots [10]. Measurements made at 4000 Mc/s show that, for a bipolar spot, two sources exist, circularly polarized in opposite directions, the sense of polarization corresponding in each case to emission in the extraordinary mode. It must be noted that the direction of polarization, measured over the source as a whole, is found to change sign near the CMP (Tanaka and Kakinuma [21]).

Variation With Longitude of the Flux Radiated

The variation of the intensity of a source with its longitude makes it possible to obtain an indication of the form of the emissive region. Two effects contribute to the variation of the flux: (1) the systematic variation with the longitude of the center, and (2) important day-to-day changes which are related to the associated optical center's flare activity.

In order to find the law which governs the first effect, a statistical study is necessary. At 21 cm one finds a cosine law, as shown in Fig. 3 (Christiansen and Mathewson [14]). This is characteristic of an emissive region having the form of a plane surface of negligible thickness compared to the diameter.

The measurements at 3 cm show that the variation with longitude is less rapid (Pick and Steinberg [22]). On the other hand, it seems that the departure from the cosine law is more important for the weak sources than for the intense sources (Swarup [37]). Unlike the strong sources, the weak sources would seem to be optically thin, in which case the form of the source is no longer relevant.

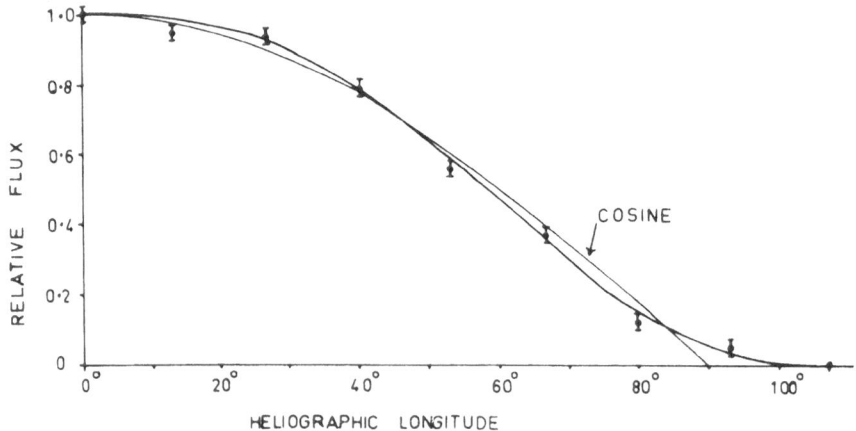

Fig. 3. Directivity of the emission from radio condensations at 21 cm.

Spectrum

Eclipse observations (Tanaka and Kakinuma [20] and Castelli and Aarons [38]) statistical analyses of observations of the total solar flux, and, also studies of individual centers carried out using interferometric techniques (Swarup *et al.* [16]) have shown that many sources have higher flux densities at 10 cm than at 3 cm and 21 cm.

The mean variation of the radio flux of localized sources with wavelength shows a maximum in the region of $\lambda \approx 5$ to 8 cm. In the case of radio condensation associated with optical plages which are devoid of spots, it seems that the value of the flux is independent of wavelength (Molchanov [23]).

This seems indicative of the role played by the magnetic field in the formation of the emission spectrum of the condensations. Figure 4 represents the average spectrum of condensations which are strong at 10 cm, as deduced from interferometric observations. The form of this spectrum is in agreement with that deduced from eclipse observations. One can ask whether the form of this spectrum is general, and, in particular, if it is retained in the case of the privileged centers which give rise to Type IV bursts.

It has indeed been reported that the centers which are the seat of strong solar activity are among the most intense at 3.2 cm. They are the centers which give rise to many Type IV bursts (Kundu [6] and Pick [7]). This property may be a peculiarity of the radiation at 3.2 cm; at 21 cm the most intense condensations are not necessarily centers of solar activity (Caroubalos [8], [24]). Therefore, it seems probable that this category of centers has a different spectrum. Another argument in favor of this distinction is given by the study of the relation between flux density of the condensations and the area of the sunspots in the associated optical centers.

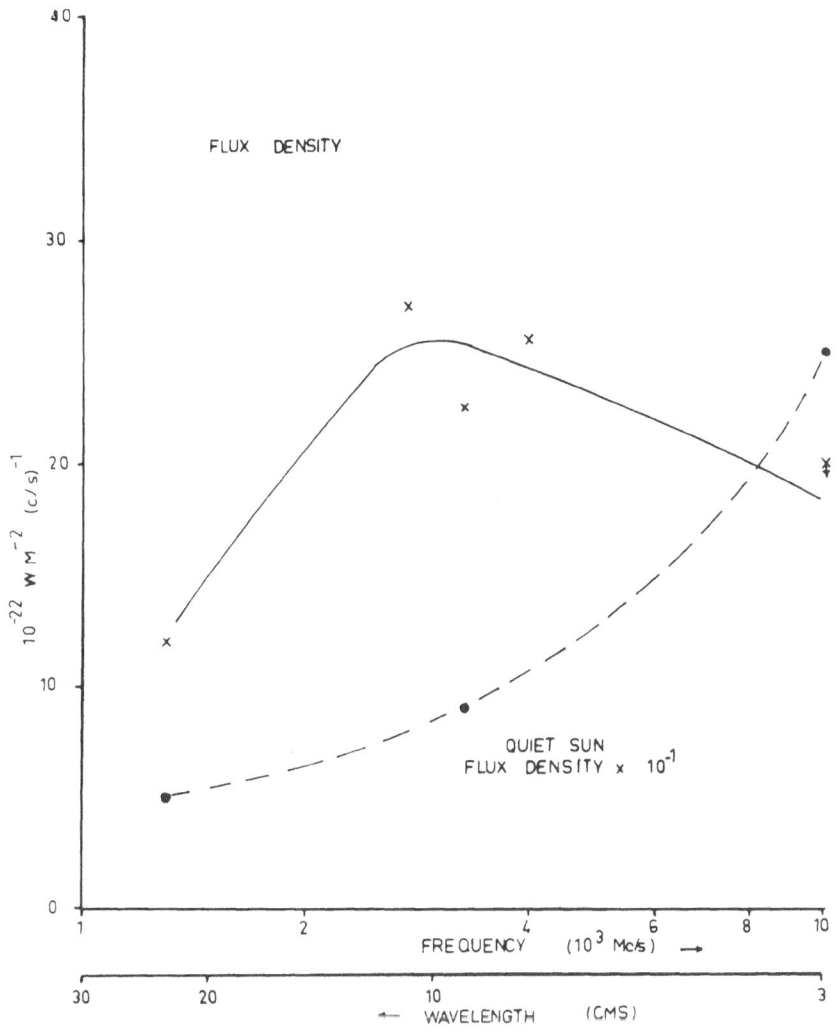

Fig. 4. Spectrum of the slowly varying component.

Relation Between Flux Density and Optical Area

It is well established that the flux density of radio condensations is very closely related to the total area of the sunspots of the corresponding group. However, this relation, which is very close at 21 and 9.1 cm, is much less close at 3.2 cm (Kakinuma and Swarup [15]). It appears, therefore, that the area of the spots is not the only parameter determining the flux density of a condensation at 3.2 cm.

Caroubalos [8, 24] has repeated this study, distinguishing those centimetric condensations which are responsible for one or more Type IV bursts from the rest

of the coronal condensations. Figure 5 shows the relation between the flux density (at 3.2 cm) of a center and the optical importance of the corresponding center of activity. In this figure, one sees that for the same optical importance, it
is the centers related to the most intense condensations which are accompanied
by Type IV bursts. Hence, the great dispersion of Fig. 5 can be attributed to
the existence of two branches corresponding to the ''active'' centers and the
''inactive'' centers, respectively. Finally, let us point out that the morphology
of the optical centers which are accompanied by Type IV bursts indicates that
they do constitute a well-differentiated family of centers (Caroubalos and
Martres-Tropé [25] and Avignon *et al.* [26]).

SLOWLY VARYING COMPONENT AT METRIC WAVELENGTHS

For wavelengths greater than 20 cm, the slowly varying component decreases
gradually. Nevertheless, it has been clearly observed at 60 and 88 cm (Swarup
and Parthasarathy [34] and Firor [35]).

At metric wavelengths it becomes difficult to detect the thermal radiation of
centers of activity. This is partly because, at these wavelengths, the brightness

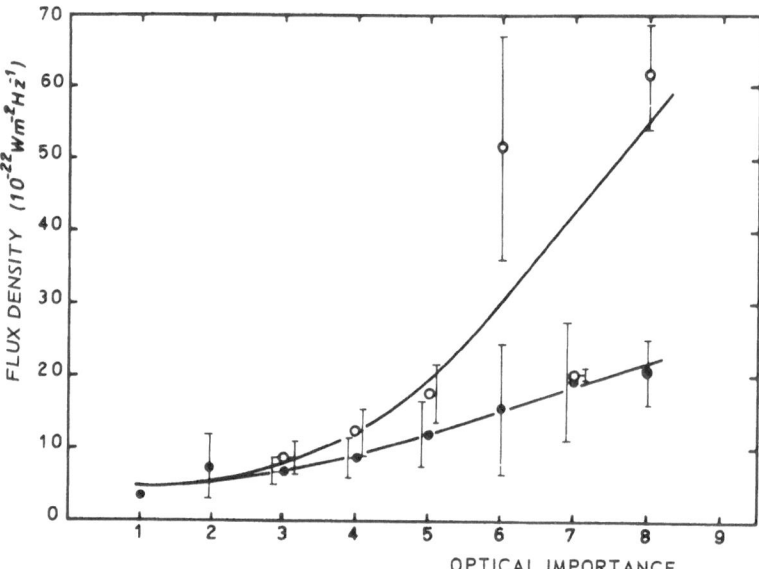

Fig. 5. Mean values of flux density at 3.2 cm as a function of optical
importance for two groups of centers: ● centers which were not associated
with Type IV bursts; ⊙ centers which were associated with Type IV
bursts.

Ed. Note: A more detailed study of these very interesting, ''active'' centers, is reported elsewhere in this volume (Caroubalos).

temperature of the surrounding solar atmosphere is very high, close to the coronal temperature. Also, the centers of activity very often emit nonthermal radiation called radio storms for long periods of time. In the presence of such storms,
it becomes impossible to study the contribution of the thermal radiation of centers of activity. The large interferometer at Nançay has nevertheless revealed
the slowly varying component at a wavelength of 1.77 m, and made it possible to
determine its essential characteristics (Moutot and Boischot [36]).

Even when there are no radio-storm centers on the sun, a recording of the
sun can take a variety of forms, as indicated in Fig. 6, which shows a localized
region of emission, of narrower diameter, superimposed on the emission of the
corona. In some cases, the position of the maximum of the localized emission is
seen to change from day to day as a result of the general rotation of the sun. On
the other hand, sometimes the maximum of the emission remains fixed or turns in
an irregular fashion: this happens when there are many centers of activity present on the sun and their radiations are superposed. This is apparent from the
large values found for the diameters of these sources.

In the case of well-isolated centers, Moutot and Boischot [36] have been
able to determine a set of interesting characteristics, which are briefly reported
below. The altitude found by measuring the apparent speed of rotation of the
sources is of the order of $0.2R_\odot$. It is comparable with the critical altitude at
this frequency. The variation of flux density with the heliographic longitude of
the associated optical center does not fit a cosine law, unlike the observations
at decimetric wavelengths. Indeed, the directivity of the emission centers is
definitely greater. Practically no radiation is observed when the associated
optical plage is more than $10'$ from the central meridian.

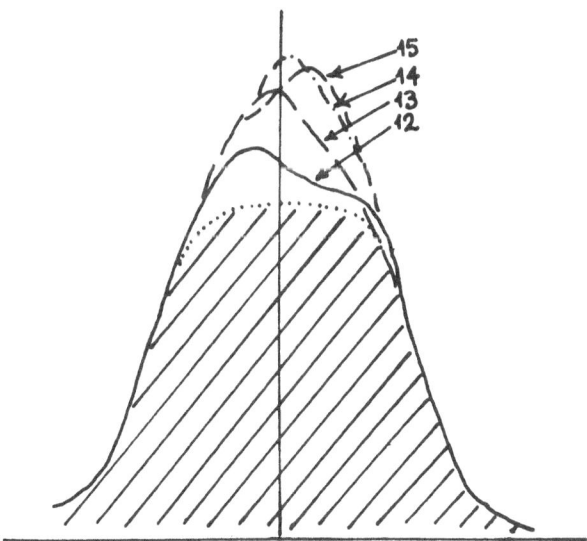

Fig. 6. Radio condensations observed at 1.77 m.

One can explain the directivity of the slowly varying component by assuming that the height of the emissive region is close to the critical altitude. For regions close to the central meridian, the radiation can effectively reach the earth from that altitude and will be intense. On the other hand, for regions near the limb, the rays from the critical level will be refracted away from the earth, and only the weaker emission from greater heights will reach the earth. The apparent diameters of these centers are large, between 10 and 25'. This corresponds to emissive regions 500,000 to 1,000,000 km across.

An interesting relation seems to exist between the flux density Φ and the apparent diameter D. These two quantities are proportional, as indicated by Fig. 7. This could be due to the fact that the brightness temperature is approximately constant — at least for those centers whose study is made possible by the absence of radio storms. The mean brightness temperature is of the order of 1.2×10^6 °K, which is about 400,000 °K higher than the brightness temperature of the quiet sun observed at this same frequency.

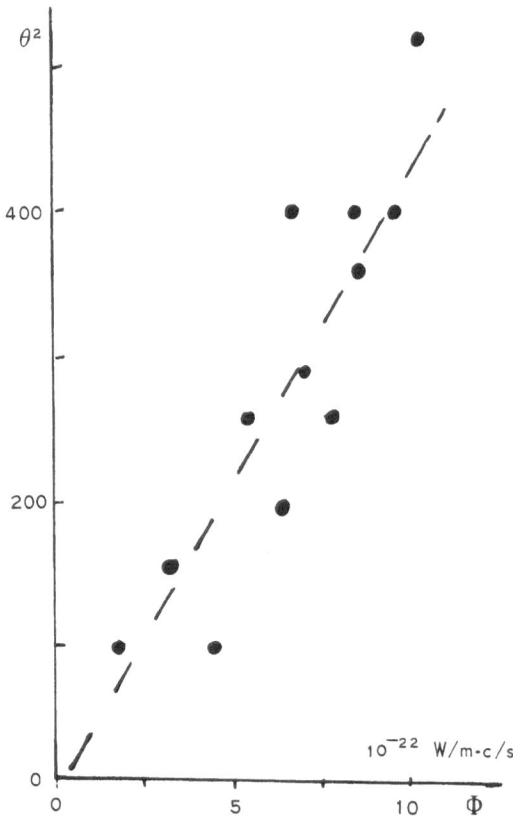

Fig. 7. Relation between the flux density and the
area of radio condensations at 1.77 m.

INTERPRETATION

Thermal Emission Mechanisms

The mechanism most often invoked to explain the slowly varying component, whose radiation is remarkably steady, is the thermal emission from hot dense regions situated above the centers of activity (Waldmeier and Muller [27] and Piddington and Minnett [28]). In addition, one can explain an increase of the optical depth by the existence, above the spots, of the local magnetic field which modifies the conditions of propagation (Denisse [29]).

The brightness temperature along a ray is of the form

$$T = \int_0^\tau T_e(\tau) e^{-\tau} d\tau$$

where T_e is the electron temperature and τ the optical depth. If we admit that the electron temperature is constant in the corona and calculate separately the contribution from the chromosphere, we find that

$$T = T_e(1 - e^{-\tau}) + T_{ch} e^{-\tau}$$

where T_{ch} is the brightness temperature of the chromosphere. The brightness temperature T cannot exceed the electron temperature in the region considered. But, we have seen that for the most intense radio condensations, the brightness temperatures measured at a wavelength of 10 cm can be 3.8×10^6 °K.

The coronal temperatures, deduced from observations of the yellow line (5694 Å), whose presence indicates the existence of particularly active regions, can also reach approximately 3.5 to 4 million degrees (Billings [30]).

On the other hand, Newkirk [32], after determining a model of the electron density of the coronal regions situated above active plages, has calculated the characteristics of the slowly varying component by making use of a coronal temperature of 1.2×10^6 °K, and by neglecting the effects of the local magnetic field. He finds a satisfactory agreement with the observations then available. Nevertheless, his model explains neither the observed spectra nor, obviously, the polarization, since the effect of the magnetic field is neglected.

Finally, various authors, using Newkirk's model of the electron density, but this time without neglecting the influence of the magnetic field, have concluded that a purely thermal mechanism could not explain both the spectrum and the polarization of the slowly varying component (Zheleznyakov [33] and Kakinuma and Swarup [15]).

Indeed, in the presence of a magnetic field, one considers two regimes of propagation, QL and QT (quasi-longitudinal and quasi-transverse). The effect of the magnetic field is to increase the optical depth of the extraordinary mode in both regimes, whereas the optical depth of the ordinary mode stays the same in the QT case and is decreased in the QL case. Since in the absence of a magnetic field the flux emitted from a plasma cannot have a maximum (at radio wavelengths), it is tempting to try to explain the extra flux actually observed between 5 and 8 cm in terms of the magnetic field. However, one should then expect this increase in flux to be accompanied by an increase in the degree of polarization of the radiation, and this is contrary to observations (Kakinuma and Swarup [15]).

In addition, Zheleznyakov [33] has shown that although the form of the theoretical spectrum exhibits a maximum, it does not correspond to the form of the spectrum deduced from observations.

Before rejecting definitely the hypothesis of a purely thermal mechanism, one should look to see how the results would be modified by taking into account an abnormally high and variable temperature in the upper chromosphere (> 5000 km).

Absorption at Harmonics of the Gyrofrequency

It has been suggested that the radiation emitted by an electron at the gyromagnetic frequency and its harmonics is capable of explaining the spectrum of the slowly varying component (Kakinum and Swarup [15] and Zheleznyakov [33]). At a wavelength of 3 cm the optical depth would be important only for the extraordinary mode; on the other hand, at 10 cm it would be large for both modes. Therefore, according to this model, one must effectively expect a decrease in the flux density between 10 and 3 cm, but an increase in the degree of polarization.

REFERENCES

1. A. E. Covington, *Proc. IRE* 36, 454 (1948).
2. F. J. Lehany and D. E. Yabsley, *Australian J. Sci. Res.* 172, 48 (1949).
3. J. F. Denisse, doctoral thesis, University of Paris (1948).
4. W. N. Christiansen et al., *Ann. Astrophys.* 23, 1 (1960).
5. M. Moutot and A. Boischot, *Ann. Astrophys.* 24, 171 (1961).
6. M. Kundu, *Ann. Astrophys.* 22, 1 (1959).
7. M. Pick, *Ann. Astrophys.* 24, 183 (1961).
8. C. Caroubalos, *Ann. Astrophys.* 27, 333 (1964).
9. A. E. Covington, Gladys A. Harvey, and H. W. Dodson, *Ap. J.* 135, 531 (1962).
10. *Progress in Radio Science in Japan* (1963), p. 93.
11. G. Swarup, *Sci. Report No. 13* Contract AF 18(603)-53 (Stanford Electronics Laboratories, Stanford, California, 1961).
12. W. N. Christiansen, N. Labrum, K. MacAllister, and D. S. Mathewson, *Proc. Inst. Elec. Engrs.* 108, 48 (1961).
13. E. J. Blum, *Ann. Astrophys.* 24, 359 (1961).
14. W. N. Christiansen and D. S. Mathewson, *IAU-URSI Symposium on Solar Radio Astronomy* (Stanford University Press, Stanford, 1959), R. N. Bracewell, ed.
15. T. Kakinuma and G. Swarup, *Astrophys. J.* 136, 3 (1962).
16. G. Swarup, T. Kakinuma, A. E. Covington, R. F. Mullaly et al., *Astrophys. J.* 137, 4 (1963).
17. W. N. Christiansen, J. A. Warburton, and R. F. Mullaly, et al., *Astrophys. J.* 137, 4 491 (1957).
18. G. Swarup, doctoral thesis, Stanford University (1961).
19. A. Y. Salomonovich, *Soviet Astron. AJ* 6, 202 (1962).
20. H. Tanaka and T. Kakinuma, *Rept. Ionosphere Res. Japan* 12, 273 (1958).
21. H. Tanaka and T. Kakinuma, *Proc. Res. Inst. Atm.* 7, 79 (1960).
22. M. Pick and J. L. Steinberg, *Ann. Astrophys.* 23, 45 (1960).
23. A. P. Molchanov, *Solnechnye Dannye* 2, 53 (1962).

24. C. Caroubalos, *Proc. NASA Symp. on Physics of the Solar Flares*, held at Greenbelt, October 28–30, 1963. (to be published).

25. C. Caroubalos and M. J. Martres-Tropé, *Compt. Rend. Acad. Sci.* **258**, 830–832 (1964).

26. Y. Avignon, M. J. Martres, and M. Pick, *Compt. Rend. Acad. Sci.* **256**, 2112–2114 (1963).

27. M. Waldmeier and H. Z. Muller, *Astrophys. J.* **27**, 58 (1950).

28. J. H. Piddington and H. C. Minnett, *Australian J. Sci. Res. A* **4**, 131 (1951).

29. J. F. Denisse, *Ann. Astrophys.* **13**, 181 (1950).

30. D. E. Billings, *Ap. J.* **125**, 817 (1957).

31. H. Zirin, *Ap. J.* **129**, 414 (1959).

32. G. J. Newkirk, *Astrophys. J.* **133**, 983 (1961).

33. V. V. Zheleznyakov, *Soviet Astron. AJ* **7**, 5 (1964).

34. G. Swarup and R. Parthasarathy, *Australian J. Phys.* **11**, 338 (1958).

35. J. W. Firor, *IAU-URSI Symposium on Solar Radio Astronomy* (Stanford University Press, Stanford, 1959), p. 136, R. N. Bracewell, ed.

36. M. Moutot and A. Boischot, *Ann. Astrophys.* **24**, 171 (1961).

37. G. Swarup, private communication.

38. J. P. Castelli and J. Aarons, *A Survey of Radio Observations of Solar Eclipses*, this volume, p. 49.

Chapter 5

Radio Frequency Emissions of the Sun in the Centimeter Wavelength Range: The Slowly Varying Sunspot Component

O. Hachenberg

University of Bonn
Bonn, Germany

In the frequency range $1000 < f < 10,000$ Mc/s, the slowly varying sunspot component is the predominant component of solar radio emission. It is superimposed on the steady radiation of the quiet sun; in its maximum phases this component is over three times greater than the radiation of the quiet sun. It has a broad continuous spectrum, which is recognizable in the whole frequency range considered.

Occasionally one finds superimposed on both components — the sunspot component and the steady radiation of the quiet sun — a third radiation component, the radio bursts, which last from about 10 sec to some hours. The radiation flux rises during a burst to one or several maxima and finally returns to preburst level. The burst spectra too are continuous in this frequency range.

Because of the broad-band emission it is possible, using a simple radiometer operating at a discrete frequency, to determine the variation of the radiation from both sunspot component and burst phenomena in this range. Therefore, the monitoring of solar radio emission at discrete frequencies beyond 1000 Mc/s is currently one of the important measurements of the sun.

For a further understanding of the mechanism involved in sunspot-component and burst phenomena, we must be familiar with the spectra. With respect to the broad-band emission, it is sufficient to have intensity values at a series of discrete frequencies and to interpolate the intensity distribution on the spectra. We must be able to measure as accurately as possible the intensity differences at frequencies which are widely separated. Here, again, it is easier to apply simple radiometers to our problem than the sweep-frequency spectrograph.

On the other hand, the necessity of flux measurements at fixed frequencies in the centimeter range is clearly due to the close relationship that exists between the radiation in this range and emissions in the soft X-ray and short-wave ultraviolet ranges. These two types of radiation originate in the low-lying regions of the corona, chiefly in the coronal condensations. This is true for the slowly varying sunspot component, as will be shown; we suppose that it is also true for bursts in the centimeter range. The emission of these two types of radiation depends on the electron velocity and the density in the lower corona and on the coronal condensations.

The relationship is close enough so that it has been possible to use radio radiation measurements to replace the unknown X-ray variations in investigating the variations in the upper atmosphere of the earth. In this way it could be shown that the 20-cm radiation is closely correlated with the ionization of the E-layer of the ionosphere (Denisse and Kundu [1] and Kundu [2]). Not only does the daily variation of mean electron density in the E-layer show a high degree of correlation, but also the distribution of emissive areas over the solar disk is the same for the two types of radiation. The latter has been substantiated by solar-eclipse (Taubenheim et al. [3]) and satellite measurements (Blake et al. [4]).

For centimeter-bursts, a correlation with the ionization in the D-layer was also found, and it has been proved that the time curves of the centimeter-bursts and of the ionizing radiation in the D-layer have a similar nature (Hachenberg and Volland [5], Volland [6], and Frost [7]).

In addition to its ionizing properties, the total energy transport of short-wave radiation is noticeable in the upper atmosphere. The incoming energy leads to a heating of the upper atmosphere of the earth, causing density fluctuations which are closely related to the radio-wave radiation. Priester [8] was the first to call attention to this relationship. As is known, full knowledge of density changes is needed for exact calculations of satellite orbits in order to make the proper corrections for these density fluctuations. From this aspect, the problem of accurately measuring solar radiation is urgently in need of solution.

It is not my purpose here to present all the arguments for the similarity of radio-wave emission and soft X-radiation. The previous remarks should be sufficient to show the importance of measurements at definite frequencies in the range from 1000 to 10,000 Mc. It is also evident that future improvements in accuracy are urgently needed.

EXPERIMENTAL TECHNIQUE

The average error of daily measurements of radiation flux from the sun lies between ±4% if one compares results from different institutes at neighboring frequencies, provided the time intervals from the series of measurements considered are not too long. A rough estimate of the theoretical accuracy shows that, with the above values, we have by no means reached the possible limit.

If, for example, we use an antenna with a gain g = 10,000 (such an antenna has about a 3-db beamwidth of $2°$) and this antenna receives solar radiation at λ = 3.2 cm, the resulting noise temperature at the antenna output would be ~400°K. At λ = 20 cm the radiation flux is approximately 8 times greater.

On the other hand, when we consider that in this frequency range it is not too difficult, using conventional means, to construct a receiver with a sensitivity of the order of $\Delta T = 2°$K, it follows that a theoretical accuracy of 0.5% is available. The difficulty is not to get a small average error in a limited series of measurements; rather, the problem is to maintain a high level of accuracy for long periods of time — say for "the year of the quiet sun" — or better still, for a complete sunspot cycle.

Because of the harmful influences of weather or drift in the apparatus over long periods of time, it is possible for jumps in the measurement series or gradual systematic variations underlying the actual measurements to arise. The cause of these errors is difficult to find and even harder to remove.

In order to keep the accuracy constant for long time intervals, it is necessary that the measuring apparatus be provided with a series of controls. Primarily, the incoming signal must be continuously compared with a good noise generator. For this comparison, the well-known Dicke modulation receiver is especially well suited (Fig. 1). The noise generator (Fig. 1) must be well calibrated and standardized, and should operate at a constant level for a long period. In practice, at $\lambda = 20$ cm one usually uses a noise diode or a gas discharge noise tube with a preset attenuator as noise generators. However, to avoid possible drift, it is better to use a heated resistor for which temperature and impedance matching are continuously controlled. The inertia of the heated resistor makes it practically impossible to completely compensate the incoming power through the noise level of the resistor. Thus, the radiometer is only partially compensated, and amplification fluctuations retain some influence on the experimental results.

With respect to this, the following receiver improvements are recommended. From the phase-sensitive detector take off not only the difference signal (solar radiation minus noise generator) but also the signal of the phase interval in which the noise generator is connected to the receiver input. This second signal indicates the amplified noise level of the noise generator. At constant amplification this signal must remain constant. A change of this signal indicates an amplification fluctuation. It is then possible to use the change in this channel to correct the difference signal in terms of amplification fluctuations. We can also apply this signal as an automatic volume control in the intermediate frequency amplifier as shown in Fig. 1. At the present time at our institute, we prefer to calculate the necessary corrections.

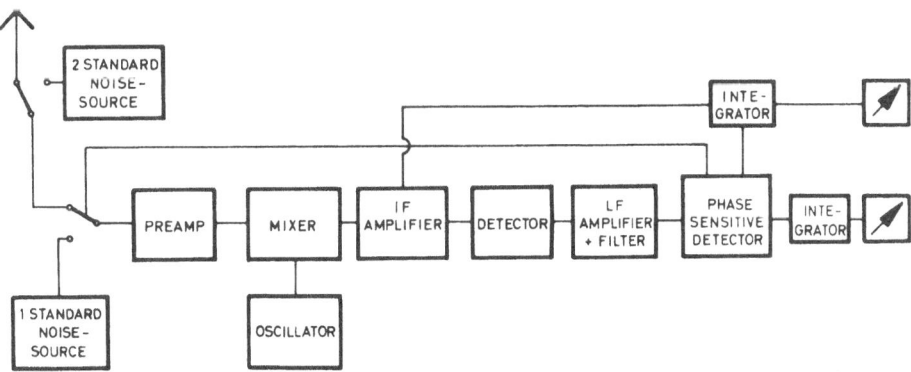

Fig. 1. Modulating-type receiver.

Finally, it is recommended that the apparatus include a second noise generator which can replace the antenna as control for occasional calibration. In the case of radiometers with long feed lines it is extremely important to have a constant noise source as close to the antenna as possible in order to have overall control of the equipment.

Attention should especially be devoted to absolute radiation calibration. If we plot the measured intensity from the individual stations against frequency, in order to see the spectral distribution of the radiation flux of the sun, we readily observe the appearance of an average deviation between 20 and 30% in the absolute values of the radiation. These errors predominantly stem from the inaccurate determination of antenna gain. On the other hand, we can measure with reasonable accuracy the power from the antenna through comparison with the noise level from a heated resistor. The calibration of the antenna can best be conducted at a good standard position which is free from reflection from the earth's surface. Or else, comparison can be made with a standard antenna.

THE MEASUREMENTS

The measurements at λ = 20 cm in Berlin-Adlershof (which are now conducted by L. Mollwo, F. Fürstenberg, and A. Krüger) have been going on since 1954; they cover the sunspot cycle which is now coming to a close. In the frequency range which interests us here — namely, $1000 < f < 10,000$ Mc/s — there are two other measurement series which have likewise run through the whole sunspot cycle. These are the well-known series of the National Research Council, Ottawa, at λ = 10.7 cm, begun in 1948, and the one at λ = 8 cm of the Research Institute of Atmospherics, Nagoya University in Toyakawa, started in 1952. Other series were begun between 1956 and 1958, mostly in connection with the I.G.Y. Here, in addition to the series mentioned above, we shall use the series λ = 3.2 and 15 cm of the Heinrich–Hertz Institute and the series at λ = 3.2, 15, and 30 cm of the Research Institute of Atmospherics, Nagoya, for purposes of comparison.

If we take average values of radiation flux at intervals of half a year, we get our first look at the course taken by solar radiation in the years 1954 through 1964. These average values are plotted in Fig. 2 for λ = 20 cm. For comparison, the average values for the series at λ = 3.2, 8, 10.7, and 30 cm, as well as the average values of the relative sunspot numbers R, are also represented in Fig. 2. Since the long period average of the relative sunspot number and sunspot areas agree (F = 16.7 R), the curve can also be taken to represent sunspot areas. From this graph, we can deduce the following well-known facts:

1. The changes of radiation flux and relative sunspot number with time are extraordinarily similar. This is true for λ = 20 cm and also for the other frequencies in the range considered. In this range we register predominantly the sunspot component of the solar radiation. Outside the boundaries of this range a remainder of the sunspot component is still evident (Firor [9]), but above 10,000 Mc/s its amplitude, in relation to the total radiation of the sun, decreases rapidly to a point where the course of the

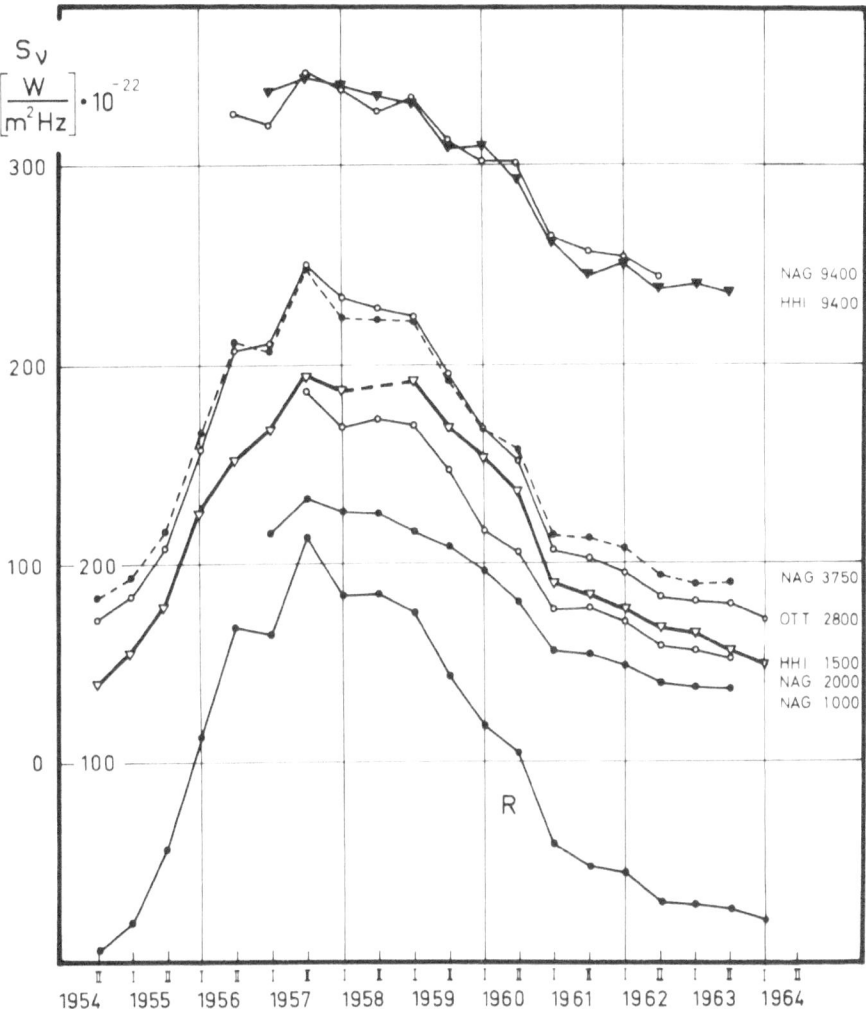

Fig. 2. Half-year means of radiation flux and relative sunspot numbers.

sunspot component is completely masked by experimental errors. Below
1000 Mc the general solar activity progressively covers this component.

2. The absolute amplitude in watts per square meter per cycle per second has a maximum around $\lambda = 10$ cm, but if we consider the amplitude
   relative to the average of the radiation flux at the corresponding frequency, or form the ratio $S_\odot/S_{\odot min}$, this relative amplitude has a maximum near $\lambda = 20$ cm. Table I gives the radiation flux of the quiet sun
   $S_{\odot min}$, the amplitude $\bar{A}$ of the half-year means, and the relative amplitude
   $\bar{A}/S_{\odot min}$.

3. The differences between the curves of the flux density at neighboring
   frequencies also give a rough survey of systematic deviations underlying

TABLE I
Radiation Flux of the Quiet Sun and of the Slowly Varying Component

f, Mc	λ, cm	S_{min}	$\bar{A}$	$\bar{A}/S_{min}$	b_r	$b_r/S_{\odot min}$	$100 a_v$
1000	30	35×10^{-22}	97×10^{-22}	2.8	99	2.8	25×10^{-22}
1500	20	44×10^{-22}	154×10^{-22}	3.5	116	2.6	55×10^{-22}
2000	15	53×10^{-22}	134×10^{-22}	2.5	112	2.1	68×10^{-22}
2800	10.7	69×10^{-22}	179×10^{-22}	2.6	132	1.9	70×10^{-22}
3700	8.0	82×10^{-22}	168×10^{-22}	2.0	129	1.6	69×10^{-22}
9400	3.2	245×10^{-22}	102×10^{-22}	0.42	274	1.1	59×10^{-22}

the different series. These systematic deviations are small: they are, for instance, smaller than 3% for the series from Ottawa at 2800 Mc, and from Toyakawa at 3750 Mc. Instrumental improvements now going on at different observatories for the IQSY will further reduce these systematic errors.

The similarity of the curves of the radiation flux and the relative sunspot number allows us to show that if we plot the average value of the flux density at one frequency against the corresponding average of the relative sunspot number, the points will fall roughly along a straight line. These lines give for the individual frequencies the average increase of the solar radiation with solar activity in the course of a sunspot cycle.

STATISTICAL INVESTIGATIONS OF THE MEASUREMENT SERIES

Much work has already gone on in the field of statistical investigations of daily values of solar radiation flux. (See Pawsey and Yabsley [10], Denisse [11], Christiansen and Hindman [12], Piddington and Davies [13], Covington and Medd [14], Kawabata [15], Waldmeier [16], Tandberg-Hanssen [17], Allen [18], Christiansen et al. [19], Krüger et al. [20], and Tanaka [21].) In the majority of these studies, the correlation of daily values of the radiation with an appropriate parameter of solar activity is investigated. For this parameter of activity relative sunspot numbers, sunspot areas, faculae areas, and magnetic fields are all especially well suited. Although the faculae areas can be taken as the basis of coronal condensation — and therefore have a close relationship with the emission areas of radio radiation — we will give the relative sunspot numbers R preference, as these values can be obtained from day to day with equal quality.

Following a study by A. Krüger, W. Krüger, and G. Wallis [20], we break up the total measured series into half-year intervals and investigate, in each part separately, the correlation between $S_\odot$ and R. For this purpose, we plot the daily values of $S_\odot$ against the corresponding relative sunspot numbers R. Further, we can approximate in this diffuse point diagram a straight line using a method of minimum deviation. Figure 3 shows examples for the half-years 1954 II, 1955 II, and 1957 I.

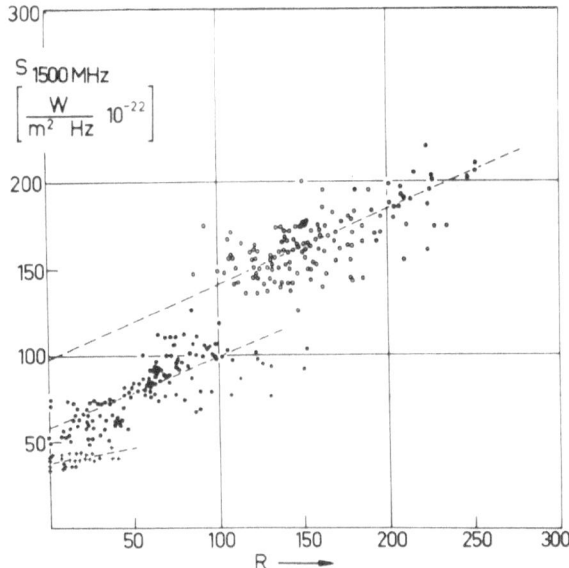

Fig. 3. Correlation diagram between observed solar
radiation flux at a frequency of 1500 Mc/s and sunspot
numbers R. (++) for 1954 II, (··) for 1955 II, (oo) for
1957 I.

We can set

$$S_\odot = a_\nu \cdot R + b_\nu$$

where a_ν gives the slope of this line (regression line) and therefore yields the
average increase of the radiation flux with R, and b_ν is the ordinate intercept of
this line and gives the radiation flux of the sunspot-free sun for $R = 0$.

For different frequencies, we get different results for the value of a_ν, showing it to be frequency-dependent. At $\lambda - 30$ cm, a_ν is small ($a_\nu \approx 0.2$), but
from there on it increases to a broad maximum at $\lambda = 10$ cm, and then at shorter
wavelengths falls off somewhat. In addition, a_ν shows a dependence on the position in the sunspot cycle. The value of a_ν is smallest at the sunspot minimum and becomes greater with increasing solar activity. The progress of the
curve for the sunspot cycle is represented in Fig. 4. The normal progress was
interrupted in 1958 by a peculiar depression at all frequencies for the value of
a_ν. The reason for this phenomenon cannot be clearly explained.

The flux density b_ν of the sunspot-free sun, which is also occasionally
called "the basic component of the slowly varying component" (Tanaka [21])
follows a certain course during the sunspot cycle. At the sunspot minimum b_ν
is identical with the radiation of the quiet sun $S_{\odot \min}$. As the activity increases,
b_ν grows larger. It has its maximum in 1958 II, a year after the sunspot maximum and then decreases as we approach the sunspot minimum (Fig. 5). The in-

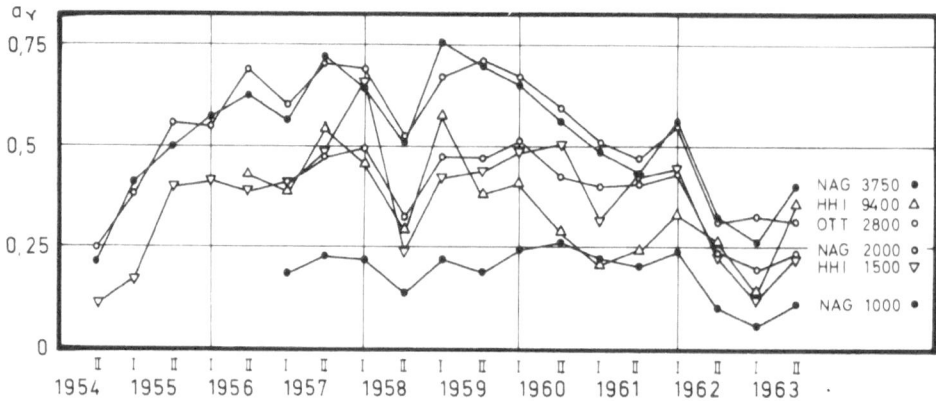

Fig. 4. Variation of a_ν during the sunspot cycle.

crease of b_ν at $\lambda = 20$ cm is from 44×10^{-22} to 120×10^{-22} W/m²-c/s. The maximum values of b_ν, just as the relative amplitude $b_\nu/S_{\odot\,\mathrm{min}}$ derived from the daily values of 1958 II, are also given in Table I.

Hence, during the sunspot cycle the radiation of the sunspot-free sun varies between the values of $S_{\odot\,\mathrm{min}}$ and $b_{\nu\,\mathrm{max}}$. This variability seems uncommon for the sun, whose flux density is extremely constant in the optical range. Attempts have been made to associate this variability with a residue of the radiation from the centers of activity (Piddington and Davies [13]). This has in a way been

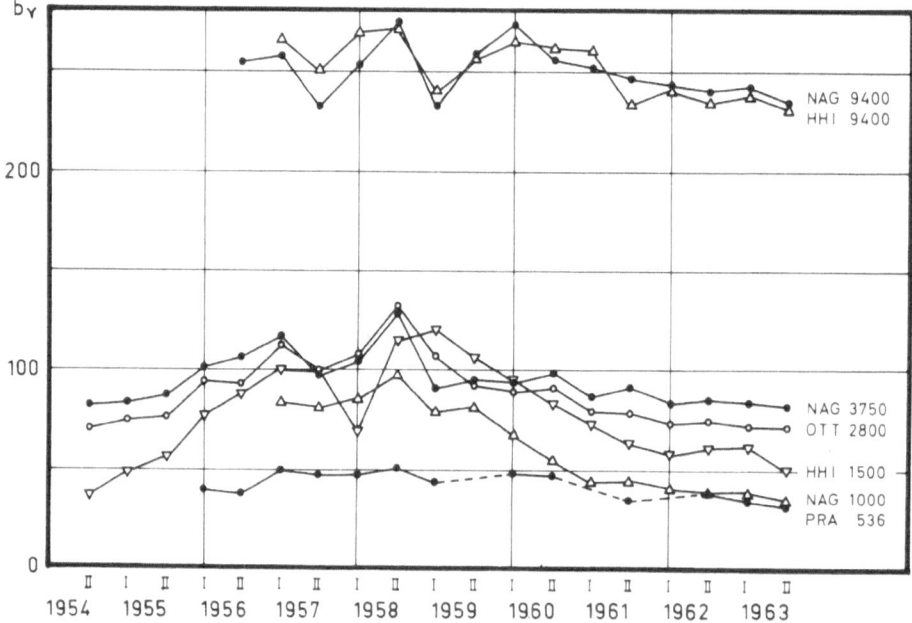

Fig. 5. Variation of b_ν during the sunspot cycle.

successful, but the experiment allows a certain arbitrariness. On the other hand, one should keep in mind that the sunspot-free sun at the time of the sunspot maximum represents a different state from the quiet sun at the minimum, especially if we consider the distribution of the magnetic fields. It may be that b_ν is dependent on the weak magnetic fields distributed at random in the solar atmosphere. Therefore, b_ν does not follow the relative sunspot numbers exactly. In the end, it becomes a question of definition whether we attribute this part of the radiation to the quiet-sun radiation or to the radiation from centers of activity.

THE SIMILARITY BETWEEN DAILY VALUES OF RADIATION FLUX AND RELATIVE SUNSPOT NUMBERS

Statistical investigations also yield the correlation coefficient r, which describes the similarity between measurement series of radiation flux and the relative sunspot number. In Fig. 6, r is represented as a function of the sunspot cycle. At the sunspot minimum r is small, then as the solar activity begins, it climbs rapidly, finally staying at about 0.8 to 0.9 for the greater part of the cycle. In general, the similarity between daily values of centimeter radiation and relative sunspot numbers is very close. From half-year to half-year, r shows a certain degree of fluctuation which occurs in a similar sense at all frequencies. These fluctuations point to the fact that actual differences in the behavior of the radiation flux and the relative sunspot numbers are to be expected. It may be that these variations have their root in the distribution of the centers of activity over the sun's surface. If this distribution is very unsymmetrical — because of the sun's rotation — a marked 27-day period results in $S_\odot$ as well as in R. Then the correlation coefficient has a greater value.

On the other hand, true differences in the behavior of the sunspot component and the relative sunspot numbers can be found. As Covington [22] has shown,

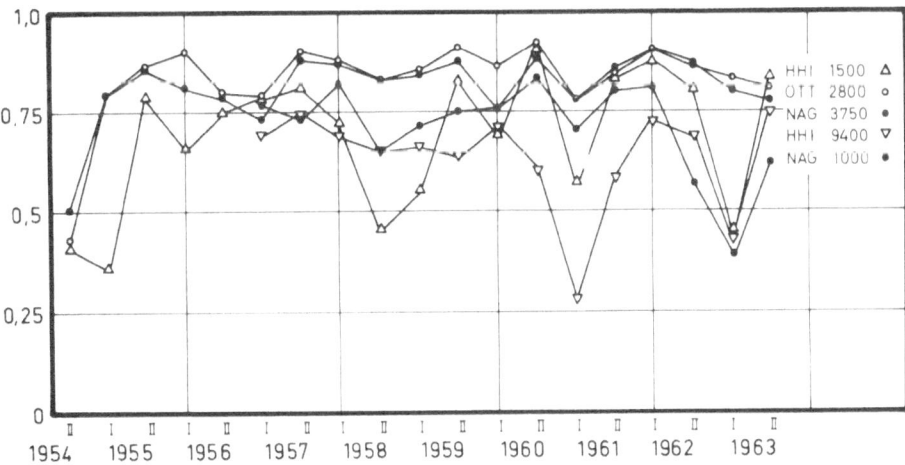

Fig. 6. Correlation coefficients r for 1954 to 1963.

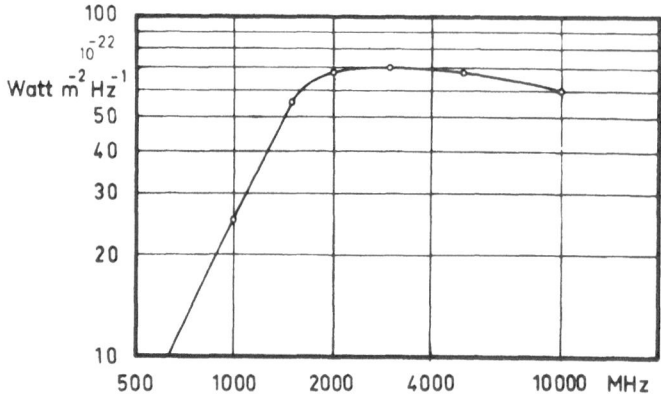

Fig. 7. Mean spectrum of the sunspot component.

radiation in the centimeter wavelength range occasionally comes from polar prominences. In the previous statistical investigations this was not considered; however, these sources of radiation may be the reason for some of the observed peculiarities.

THE SPECTRUM OF THE SUNSPOT COMPONENT

The statistical investigations of the measured series also yield an average spectrum of the sunspot component. The a_ν's have given the slopes for the regression lines and if we form $a_\nu \cdot 100$, we then have the increase in radiation flux from an activity center with $R = 100$. If we do this for the different series, we obtain the radiation flux in terms of frequency. The spectrum shown in Fig. 7 is derived from values of the period July 1, 1957, to June 30, 1958. The intensity increases from the lower frequencies to a maximum between 2000 and 3000 Mc, and then falls off again a bit at high frequencies. Indeed, at high frequencies the curve depends only on the quality of the absolute flux measurement at $\lambda = 3.2$ cm; therefore, the decrease of the intensity at high frequencies is uncertain.

The statistical studies give us an average value of the intensity in a certain time interval, but we see the sunspot component as an emission from individual centers of activity of the sun. The question now arises as to whether or not the spectra of the radiation flux from individual centers of activity would strongly differ from that of the average spectrum.

That differences in the emission from different centers exist, is clearly shown in Fig. 8, which plots the daily values of radiation flux at different frequencies for the period December 28, 1958, to February 10, 1959. On January 8 and January 24, two strong radiating centers appear on the solar disk. At $\lambda = 30$ cm these two centers show approximately the same flux values. However, at 20 cm the center of the 24th is somewhat brighter, and at $\lambda = 3.2$ cm the flux is more than twice as great as on the 8th. In the long series we can find other similar examples.

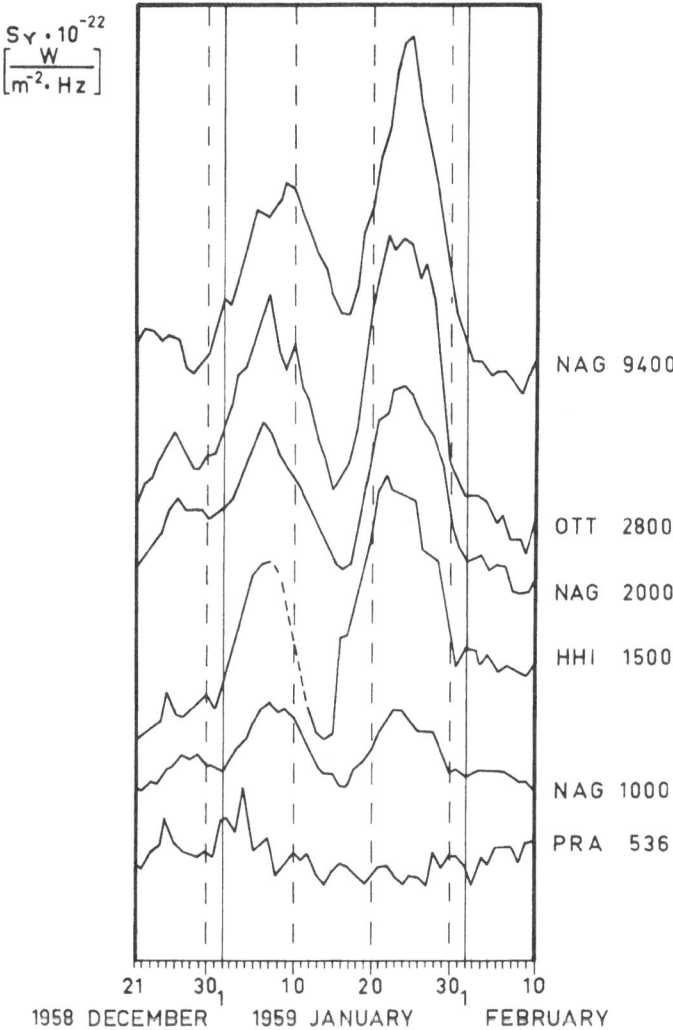

$S_Y \cdot 10^{-22}$

$\cdot \left[\dfrac{W}{m^{-2} \cdot Hz} \right]$

NAG 9400

OTT 2800

NAG 2000

HHI 1500

NAG 1000

PRA 536

21   30   10   20   30   10
1958 DECEMBER   1959 JANUARY   FEBRUARY

Fig. 8. Daily values of radiation flux at different frequencies.

A. Krüger, W. Krüger, and G. Wallis [20] have also investigated the spectra from individual condensations. They used the daily measurements of the flux, deducted the radiation of the quiet sun, and constructed a time – frequency diagram for the sunspot component. The diagram shows that the radiation of the sunspot component generally has a maximum between 2000 and 4000 Mc. However, this representation is also dependent on the absolute calibration of the measurements at 3.2 cm, so the above result can still change.

We generally suppose the sunspot component to be thermal radiation of condensations in the corona. For the case of thermal emission, we can represent the intensity of the emission areas as follows:

$$I_\nu = B(\nu, T_K) \cdot (1 - l^{-\tau\nu}) + l^{-\tau\nu} \cdot B(\nu, T_{Ph})$$

where $B(\nu, T)$ is the Rayleigh – Jeans approximation of the Planck radiation law and $\tau\nu$ is the optical density of the plasma. From the observed spectrum (Fig. 8), it follows that the plasma of the condensation must be optically thin at least for $\nu > 3000$ Mc. We would then have

$$I_\nu = \frac{2\nu^2}{c^2} \cdot \kappa \cdot T \cdot d \cdot \kappa$$

where d is the linear thickness and κ is the absorption coefficient. If we substitute for

$$\kappa_\nu = 1.32 \cdot 10^{-2}(N_e^2/T^{3/2}\nu^2) \ln [427.10^7(T^{3/2}/\nu)]$$

we have that

$$I_\nu = (\text{const } T^{-1/2}) \cdot N_e^2 \cdot d$$

The emission of the optically thin plasma does not depend on the frequency. Above all, there can be no falling off of intensity at high frequencies. In the case of a layered plasma with different temperatures in the individual layers, one can show that with higher frequencies we have an intensity increase but never a decrease. This holds for plasmas without great density fluctuations. We don't want to allow density fluctuations so large that we can no longer count on straight line propagation of the radiation in the plasma.

The observed decrease of density in the mean spectrum of the coronal condensations (when proved by exact measurements above 10,000 Mc) indicates that the emission is not of purely thermal origin. Since the observed higher densities of the material in the condensations are held in magnetic fields, it can be supposed that these magnetic fields also take part in the emission process. It is possible to see in the spectrum the first signs that the emission is produced in part from the synchrotron emission of nonthermal electrons gyrating in a magnetic field.

The explanation of the flat maximum near 10 cm as a contribution of synchrotron radiation leads to certain difficulties however, and we have to devote some energy to establish a convenient model.

The course of the spectrum in the range greater than 10 cm indicates that for these frequencies the plasma of the condensation is predominantly optically thick. At 20 cm the radiation originates essentially in the upper – and most probably hotter – portion of the condensation, while at 3 cm it is the thick nucleus of the lower part of the condensation that contributes most to the radiation. The higher radiation intensity which we occasionally establish for individual condensations in the range of 3 cm indicates that for these cases we have to assume the existence of a thick nucleus in the lower part of the condensation.

THE RADIATION AREAS AND ELECTRON TEMPERATURE
OF THE CORONAL CONDENSATIONS

So far we used statistical investigations of the daily values of the flux density of the slowly varying component to determine the properties and the emission mechanism of the coronal condensations. We obtain a detailed study of the condensations from the interferometer measurements, which give two-dimensional images of the sun and emitting areas. (The results have been discussed thoroughly by M. Pick, so it is sufficient here to give only a general outline.)

If the radiating areas of coronal condensations are known, one can determine the electron temperature of the plasma using the radiation flux of the frequency range $1000 < f < 2000$ Mc/s where the plasma is optically thick. The diameters of the areas, which vary with the frequency of observation, are comparable to those of the faculae; they lie between 1 and $10'$ of arc or $\approx 300,000$ km. If we assume an area to be $8'$ and the radiation flux to be 40×10^{-22} W/m^2-c/s (corresponding to a relative sunspot number $R = 100$), we find the electron temperature of the plasma $T_e = 1.2 \times 10^6$ °K.

From solar eclipse measurements, one finds at 20 cm that the condensation areas are 6 to 11 times brighter than the surroundings (Hachenberg et al. [23]). If we assume that the radiation temperature of the sun at 20 cm is 1.1×10^5 °K, we get a value for the temperature in the condensation, T_e of about 0.7 to 1.2×10^6 °K.

With that, we have again established the result, that the electron temperature does not differ greatly from the temperature of the surrounding corona, but the electron density is about three times higher than in the normal corona.

The electron temperature and density which we derived above are related to the outer layers of the condensations. The inner part — particularly a possibly existing denser core of the condensation — is observable only at higher frequencies. With a high-resolving interferometer at $\lambda = 3$ cm, the diameters of the denser cores in individual condensations could be determined (Kundu [24]), but further steps to determine the density distribution and the electron temperature lead to difficulties. In the inner part of the condensation the influence of the magnetic fields may be important. We must assume that a part of the radiation is produced by synchrotron emission since this is indicated by polarization measurements at 3 cm (Tanaka and Kakinuma [25]). In this case, the simple model of a thermally radiating plasma is no longer applicable.

Detailed observations with high-resolving interferometers at different frequencies in the range considered may help solve these problems. Common observations at different well-equipped observatories have already been done in some cases (Christiansen et al. [26]); in future studies they will be necessary to greatly extend radio condensation measurements.

REFERENCES

1. J. F. Denisse and M. R. Kundu, Compt. Rend. Acad. Sci. 244, 45 (1957).
2. M. R. Kundu, J. Geophys. Res. 65, 3903 (1960).
3. J. Taubenheim and G. Nestorow, J. Atmospheric Terres. Phys. 24, 633 (1962).

4. R. L. Blake, T. A. Chubb, H. Friedman, and A. E. Unzicker, *Astrophys. J.* **137**, 3 (1963).
5. O. Hachenberg and H. Volland, *Z. Astrophys.* **47**, 369 (1959).
6. H. Volland, *J. Atmospheric Terres. Phys.* **26**, 695 (1964).
7. K. J. Frost, Goddard Space Flight Center X – 610 - 64 - 60.
8. W. Priester, *Naturwissenschaften* **46**, 197 (1959).
9. J. W. Firor, *IAU-URSI Symposium on Radio Astronomy* (Stanford University Press, 1959), p. 136, R. N. Bracewell, ed.
10. J. L. Pawsey and D. E. Yabsley, *Australian J. Sci. Res. A* **2**, 198 (1949).
11. J. F. Denisse, *Ann. Astrophys.* **13**, 181 (1950).
12. W. N. Christiansen and J. W. Hindman, *Nature (London)* **167**, 635 (1951).
13. J. H. Piddington and R. D. Davies, *Monthly Notices Roy. Astron. Soc.* **113**, 582 (1953).
14. A. E. Covington and W. J. Medd, *J. Roy. Astron. Soc. Can.* **48**, 136 (1954).
15. K. Kawabata, *Rep. Ionosphere Res. Japan* **8**, 143 (1954).
16. M. Waldmeier, *Z. Astrophys.* **36**, 181 (1955).
17. E. Tandberg-Hanssen, *Astrophys. J.* **121**, 367 (1955).
18. C. W. Allen, *Monthly Notices Roy. Astron. Soc.* **117**, 174 (1957).
19. W. N. Christiansen, J. A. Warburton, and R. D. Davies, *Australian J. Phys.* **10**, 491 (1957).
20. A. Krüger, W. Krüger, and G. Wallis, *Astrophys. J.* **59**, 37 (1964).
21. H. Tanaka, *Proc. Res. Inst. Atmospherics* **11**, 41 (1964).
22. A. Covington, *J. Roy. Astron. Soc. Can.* **57**, 253 (1963).
23. O. Hachenberg, M. Popowa, and H. Prinzler, *Z. Astrophys.* **58**, 36 (1963).
24. M. Kundu, *Ann. Astrophys.* **22**, 1 (1959).
25. H. Tanaka and T. Kakinuma, *Rep. Ionosphere Res. Japan* **12**, 273 (1958).
26. W. N. Christiansen, D. S. Mathewson, J. W. Pawsey, S. F. Smerd, A. Boischot, J. F. Denisse, P. Simon, T. Kakinuma, H. Dodson-Prince, and J. Firor, *Ann. Astrophys.* **23**, 75 (1960).

Chapter 6

Study of the Slowly Varying Component at Three Centimeters as a Function of Solar Activity

Constantin A. Caroubalos*

Ionospheric Institute
National Observatory of Athens
Athens, Greece

INTRODUCTION

Solar centers of activity are associated with localized, relatively stable, sources of radio emission, called "radio condensations," which overlie the plage regions (see the papers by Denisse [1], Castelli and Aarons [2], and Pick [3], in this volume). These condensations give rise to the slowly varying component of solar radio emission. Their characteristics vary with frequency; in particular they are most readily observed at high frequencies.

On the other hand, it is well known that of all the solar centers of activity, only a few are seats of chromospheric flares which accelerate enough particles to sufficient energies to be observable. These particular flares give rise to diverse types of intense nonthermal radio emission and, more rarely, to geophysical phenomena of corpuscular origin, such as sudden commencement geomagnetic storms (SSC), polar cap ionospheric absorption events (PCA), the cosmic ray enhancements observed at ground level (CRE), and Forbush decreases.

We may ask, then, if it is possible to find in these rare centers of exceptional activity, as they are seen at radio wavelengths, some characteristics which distinguish them from the rest of the centers of activity. The interest in such a possibility is evident, since it could provide the basis of a long-term forecast of the radio and geophysical activity coming from the sun.

We shall report here a few points regarding those centers which give rise to Type IV radio bursts and to sudden commencement geomagnetic storms; the strong correlation between these two phenomena is well known from numerous studies (see for example, Sinno and Hakura [4], McLean [5], Simon [6], Bachelet *et al.* [7], de Feiter *et al.* [8], and Caroubalos [9] and [10]).

---

*The present work was carried out in the Observatory of Paris-Meudon during the author's stay in France.

REPETITION CENTERS AND THEIR CONTRIBUTION
TO RADIO AND GEOMAGNETIC ACTIVITY

It has already been established that certain centers associated with intense radio condensation give rise, during their passage across the solar disk, to several radio bursts, including Type IV bursts (Kundu [11] and Pick-Gutmann [12]). We shall call these "repetition centers," or, more precisely, we shall call "repetition centers" those centers of activity which are responsible for at least two Type IV bursts, whether on the same passage across the solar disk or during successive rotations.

It is evident that this definition of repetition centers is necessarily incomplete; indeed, one would certainly find other centers of multiple activity if one could take into account the activity occurring on the invisible hemisphere of the sun. However, it is very probable that the above definition of repetition centers includes almost all of the most important regions of emission.

We have verified that the majority of the centers in question are also at the origin of sudden commencement geomagnetic storms several times during their passage across the solar disk and sometimes during their second or even third passage (Caroubalos [10] and [13]).

Despite the rather small number of repetition centers, their contribution to the total solar activity (Type IV and SSC) is important. For example, considering the sudden commencement geomagnetic storms which occurred during the period from 1958 to 1960, the geomagnetic repetition centers ($n_{SSC} \geq 2$), which comprise only 31% of all the geomagnetically active centers (16 out of 52), are responsible for more than half of the SSC events — about 58% (or 49 out of 85). This corresponds to an average of three events per geomagnetic repetition center.

These observations underline the interest of the study of repetition centers, both for the prediction of certain forms of geophysical activity and for the examination of the physical conditions which give rise to the corresponding solar events.

VARIATION OF THE PROBABILITY OF ACTIVITY AS A
FUNCTION OF THE THREE-CENTIMETER FLUX DENSITY

The radio condensations associated with repetition centers are generally intense at 3-cm wavelengths. In order to make this more precise, we have characterized the centimetric condensations by their mean flux density at 3 cm, as measured with the interferometer at the radio observatory at Nançay (Pick-Gutmann and Steinberg [14]). The mean value is calculated from observations on three days centered on the central meridian passage of the source, in order to avoid perspective effects. If we write Σn for the number of SSC storms coming from solar centers for which the associated radio condensation's emission is contained in a certain interval of flux density, and N for the total number of centers (active or inactive) whose emission lies in this same interval, then the ratio $\Sigma n/N$ is the average number of events per center for this interval of flux density, and by extension, a measure of the probability that such a center will be geomagnetically active. The diagram of Fig. 1 shows the variation of this ratio as a function of the flux density of the associated center, and indicates that the probability of geomagnetic activity increases rapidly when the 3-cm flux density increases above a certain threshold; for the period from 1959 to 1960 this threshold appears to be about 50×10^{-22} W/m^2-c/s. It should be noted that the emission at this wavelength (3 cm) appears to provide a more significant indication of the identity of repetition centers than does the emission at lower frequencies (Caroubalos [10, 13, 15]).

DEPENDENCE OF THE FLUX DENSITY ON THE AREA OF THE OPTICAL CENTER

Numerous studies have shown that the flux density of radio condensations is closely related to the total area of the sunspots in the corresponding center of activity (Christiansen *et al.* [16], Christiansen *et al.* [17], Covington and Harvey [18], Swarup [19], and Covington *et al.* [20]). However, this dependence appears more striking at 21, 9.1, and 7.5 cm than at 3.2 cm (Kakinuma and Swarup [21] and Swarup *et al.* [22]). In fact, at 3.2 cm the dispersion of the values of flux density increases rapidly with the area of the centers. It appears that the area of the group of sunspots is not the only factor involved in the determination of the flux density emitted at 3.2-cm wavelengths by the radio condensation.

We have extended this investigation at 3.2 cm by distinguishing, among all the centers accompanied by sufficiently well-isolated centimetric condensations, those which give rise to a Type IV burst during their passage across the solar disk. Figure 2a shows the dispersion diagram (3.2-cm flux density, $S_{3.2 \text{ cm}}$ vs. "optical importance" of the corresponding center). The optical importance is as indicated in the *Quarterly Bulletin on Solar Activity* and is directly related to the area of the sunspot group.* The arrow attached to several of the points indi-

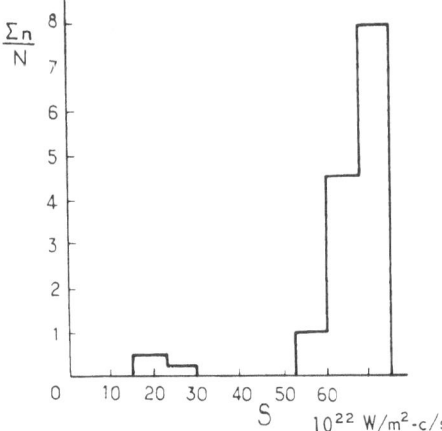

Fig. 1. Variation of the average number of sudden commencement geomagnetic storms (SSC) per center (ratio $\Sigma n/N$) as a function of the 3.2-cm flux density of the corresponding solar radio condensation.

---

*The circles correspond to centers producing Type IV bursts, and the dots to all the other centers.

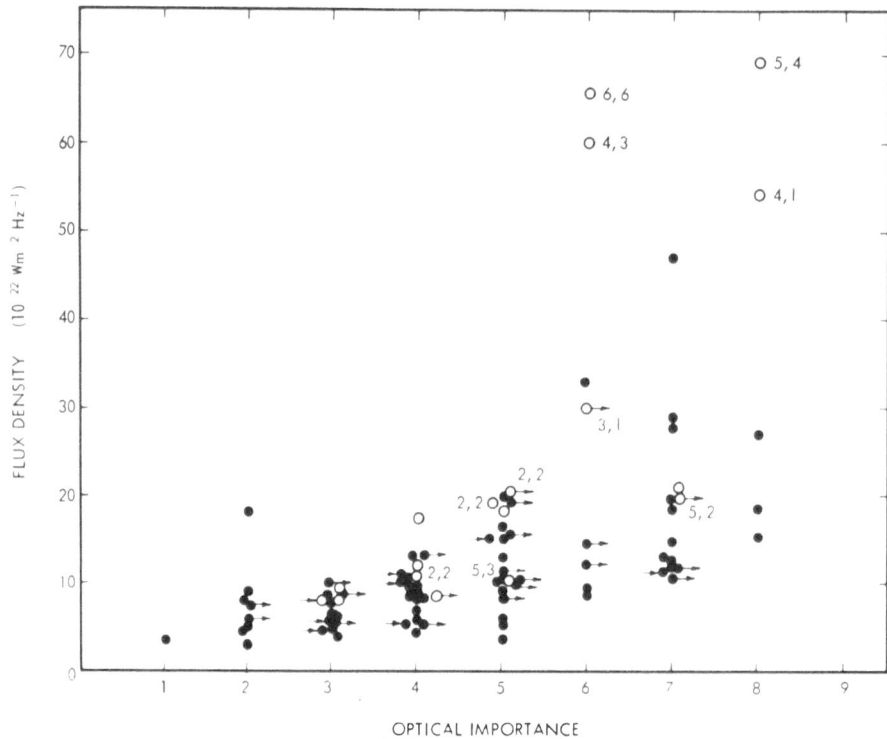

Fig. 2a. Dispersion diagram of 3.2-cm flux density *vs.* the optical importance of the corresponding solar center, where ● is a center of simple optical activity, and ○ a center of radio electric and geomagnetic activity. (The two numbers near the repetition center indicate, respectively, the number of Type IV bursts and the number of SSC storms given by the center.)

cates that the optical importance of these centers may have been underestimated. We see that those centers which are active (Type IV emission and origin of SSC storms) correspond to relatively high flux densities compared with the other centers of the same optical importance.

This increase of the 3.2-cm flux density cannot be attributed to a contribution from the radio bursts themselves since we have taken the precaution of rejecting those dates which were influenced by the Type IV bursts from our calculations of mean flux.

Hence, the large dispersion observed in the diagram of $S_{3.2\ cm}$ *vs.* area may be explained by the existence of two branches: one corresponding to the centers which are likely to give rise to radio and corpuscular activity, and the other including the rest of the centers. This is indicated in Fig. 2b by the two curves which correspond to the circles and the dots, respectively.

For purposes of prediction, the situation can be simplified as follows (Caroubalos [10] and [15]): For a given optical importance, only those centers associated with the most intense centimetric (3.2 cm) emission from radio condensations will be the seat of flares which give rise to Type IV bursts and to geophysical events of corpuscular origin.

ACTIVITY OF THE CENTERS DURING THE PRESENT ELEVEN-YEAR CYCLE

The efficiency of the repetition centers, i.e., the number of events per repetition center, does not appear to be constant during the present solar cycle (at least as observed for the visible hemisphere). It has increased since the maximum of the sunspot cycle. Analogous results have been reported elsewhere for studies of centimetric bursts (Fokker [23]).

We shall first define two characteristic ratios, for which n_1 symbolizes the yearly total number of events (Type IV or SSC) due to the N_1 repetition centers observed during the year considered, and n_2 the number of events associated with the N_2 simple centers of activity occurring during the same period ($n_2 = N_2$ necessarily). The ratio $r = n_1/N_1$ represents the efficiency of the repetition centers in this year, while the ratio $q = N_1/(N_1 + N_2)$ gives the proportion of all active centers (for the year in question) which are repetitive.

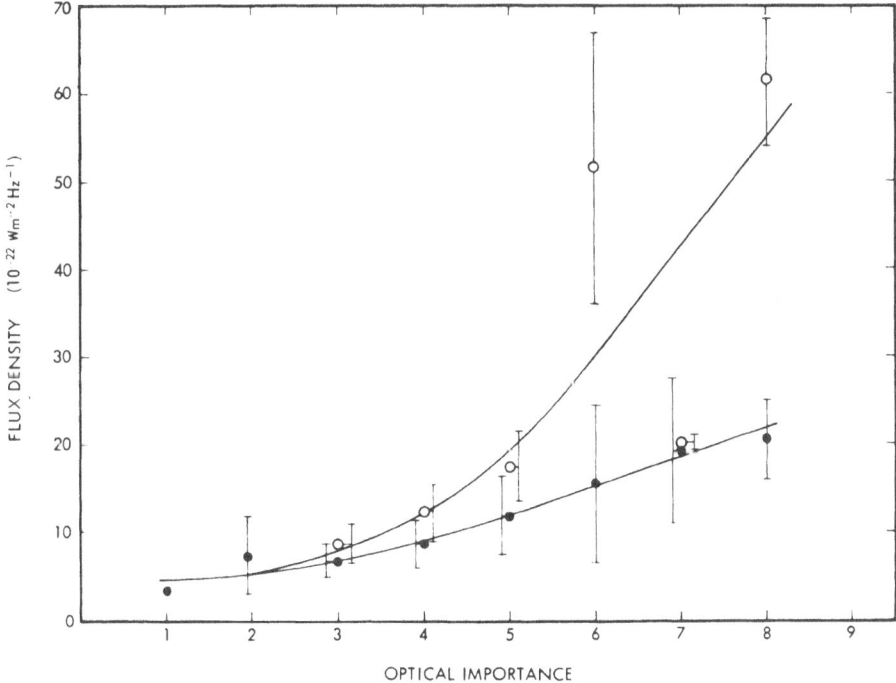

Fig. 2b. Mean curves of the dependence of the flux density on the optical importance for the two groups of centers, where ● is a center of simple optical activity, and ○ a center of corpuscular activity.

The graphs of Fig. 3 show the respective variations of r and q for the two forms of activity (Type IV and SSC) during the course of a great part of the present cycle of solar activity. The diagrams show that the repetitive activity of the sun increased about two years after the sunspot maximum and remained fairly high during the descending part of the cycle. However, it should be noted that it is the ratio r which shows a strong increase during the period in question, while the variations of the ratio q are very weak. Consequently, the effect of the increase of the repetitive activity must be attributed, almost exclusively, to the appearance of very active repetition centers, and not to an increase of their number.

These results closely resemble the singular distribution observed during the solar cycle, both for flares giving rise to high-energy cosmic ray events (CRE) and for the most intense centimetric bursts. It is known that the cosmic ray events observed at ground level (CRE) tend to avoid the maximum phase of the sunspot cycle (Carmichel and Steljes [24]) (c.f. Fig. 3). This effect is purely solar and not a consequence of a modulation due to the conditions in interplanetary space. This is supported by Takakura and Ono [25], who have shown that despite the abundance of major optical flares (3 and 3+) during the maximum of the cycle, the most intense centimetric bursts, which constitute the main indication of particle acceleration, are displaced from the cycle maximum.

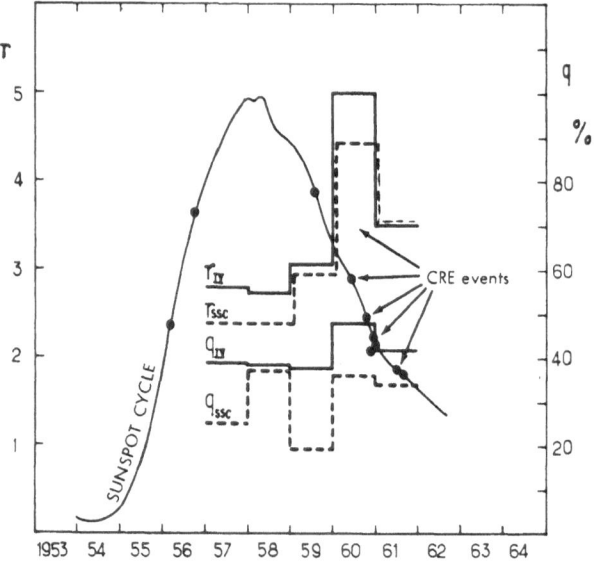

Fig. 3. Variation of the repetition activity of the sun (expressed by the two parameters r and q) during the present solar 11-year cycle. The points placed on the curve representative of the sunspot cycle indicate the appearance of cosmic ray enhancements events observed at ground level (CRE).

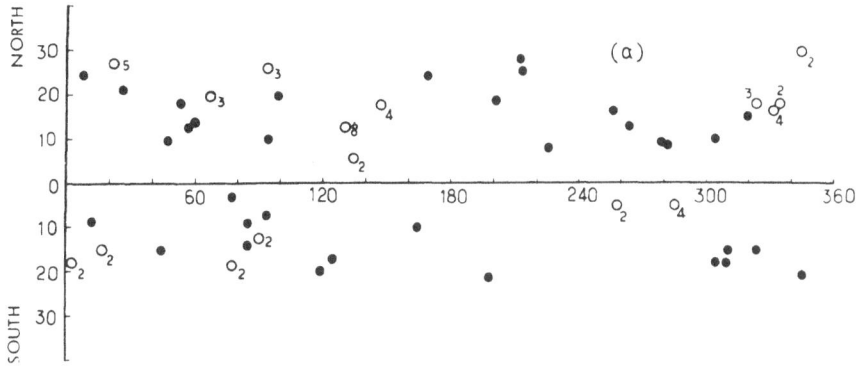

Fig. 4a. Distribution on the sun (in Carrington's coordinate system) of the geo-magnetically active centers. The centers of simple geomagnetic activity are marked by ●, and the repetition centers by O with a number giving the corresponding number of SSC storms.

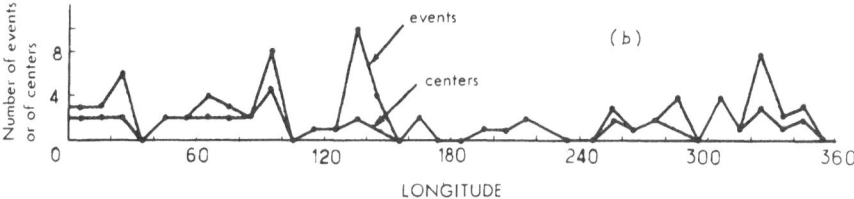

Fig. 4b. Diagram giving, as a function of Carrington's longitude, the number of centers corresponding to the various intervals of solar longitude and also the corresponding number of SSC storms.

We see then that the period of appearance of the most active repetition centers coincides with the period of great acceleration of particles. This effect is easy to understand if we remember that all the cosmic ray events originate in repetition centers (Caroubalos [10] and Fokker [23]). From this we conclude that the conditions for the most efficient acceleration of particles on the solar surface are found in the repetition centers and these conditions appear to persist despite the occurrence of exceptionally violent events.

DISTRIBUTION OF THE REPETITION CENTERS ON THE SUN AND THE SIGNIFICANCE OF THE PRIVILEGED REGIONS

When we study the distribution on the sun of the centers which give rise to the sudden commencement geomagnetic storms, we find the privileged regions which have already been noted elsewhere for other forms of solar activity — coronal green line (Trellis [26]), Type IV radio bursts (Pick-Gutmann [12] and Avignon and Pick-Gutmann [27]), and polar cap ionospheric absorption (Noyes [28], etc.

In Fig. 4a, we have marked the geomagnetically active centers in Carrington's coordinate system. The privileged regions appear when we study the number of events rather than the number of centers. This is indicated in Fig. 4b, where both the number of centers corresponding to the various intervals of solar longitude and also the corresponding number of SSC storms are traced as a function of Carrington's longitude. We see that the repetition centers lie mainly in the privileged regions. The latter are therefore a consequence of the activity of the repetition centers and not the result of the accumulation of isolated centers.

ACKNOWLEDGMENTS

The author is indebted to various members of the Radio Astronomy group of the Meudon Observatory and in particular to Dr. M. Pick, for her very valuable help.

REFERENCES

1. J. F. Denisse, this volume, p. 39.
2. J. Castelli and J. Aarons, this volume, p. 49.
3. M. Pick, this volume, p. 81.
4. K. Sinno and Y. Hakura, *Rep. Ionosphere Res. Japan* 12, 285 (1958).
5. D. McLean, *Australian J. Phys.* 12, 404 (1959).
6. P. Simon, *Ann. Astrophys.* 23, 102 (1960).
7. F. Bachelet, P. Balata, A. Conforto, and G. Marini, *Nuovo Cimento* 21, 248 (1961).
8. L. de Feiter, A. Fokker, Van Lohutzen, and Roosen, *Plan. Space Sci. Rev.* 2, 223 (1960).
9. C. Caroubalos, *Compt. Rend. Acad. Sci.* 255, 2620 (1962).
10. C. Caroubalos, *Ann. Astrophys.* 27, 333 (1964).
11. M. Kundu, *Ann. Astrophys.* 22, 1 (1959).
12. M. Pick-Gutmann, *Ann. Astrophys.* 24, 183 (1961).
13. C. Caroubalos, *Compt. Rend. Acad. Sci.* 256, 5063 (1963).
14. M. Pick-Gutmann and J.-L. Steinberg, *Ann. Astrophys.* 24, 45 (1961).
15. C. Caroubalos, *AAS-NASA Symp. on Physics of the Solar Flares*, NASA, Washington, D.C., p. 204 (1964).
16. W. Christiansen, J. Warburton, and S. Davies, *Australian J. Phys.* 10, 491 (1957).
17. W. Christiansen, D. Mathewson, J. Pawsey, S. Smerd, A. Boischot, J.-F. Denisse, P. Simon, T. Kakinuma, H. Dodson-Prince, and J. Firor, *Ann. Astrophys.* 23, 75 (1960).
18. A. Covington and G. Harvey, *Ap. J.* 132, 435 (1960).
19. G. Swarup, *Sci. Report N:13*, Contract AF 18(603)-53 (Stanford Electronics Laboratories, Stanford, California (1961).
20. A. Covington, G. Harvey, and H. Dodson, *Ap. J.* 135, 531 (1962).
21. T. Kakinuma and G. Swarup, *Ap. J.* 136, 975 (1962).
22. G. Swarup, T. Kakinuma, A. Covington, G. Harvey, R. Mullaly, and J. Rome, *Ap. J.* 137, 1251 (1963).
23. A. Fokker, *Bull. Ap. Inst. Neth.* 117, 84 (1963).
24. H. Carmichel and J. Steljes, *J. Phys. Soc. Japan*, (Supp. A-II), 17, 337 (1962).
25. T. Takakura and M. Ono, *J. Phys. Soc. Japan*, (Supp. A-II), 17, 207 (1962).
26. M. Trellis, *Compt. Rend. Acad. Sci.* 247, 1964 (1958).
27. Y. Avignon and M. Pick-Gutmann, *Compt. Rend. Acad. Sci.* 248, 368 (1959).
28. J. Noyes, *J. Phys. Soc. Japan*, (Supp. A-II), 17, 275 (1962).

Chapter 7

The Solar Flare Phenomenon
as Seen at Radio Frequencies

Donald James McLean*

Observatoire de Paris - Meudon
Paris, France

INTRODUCTION

Observations of the sun in "white" light detect almost exclusively the light emitted by the *photosphere*, which is a thin layer of gas between 200 and 400 km thick. Above the photosphere lies the *chromosphere*, which can be observed during a total eclipse as a faint region of reddish emission around the solar disk when the much brighter photospheric light is masked out by the moon.

Certain of the absorption features — Fraunhofer lines — of the solar spectrum are formed high in the chromosphere, and so observations made through a monochromatic filter which isolates one of these lines can be used to study the chromosphere above the photosphere. Of particular importance in this respect are the lines H_α of neutral hydrogen and the core of the K line emitted by ionized calcium.

Beyond the chromosphere, and extending right out into interplanetary space, is the *corona*, which is an extremely tenuous, extremely hot, gaseous outer atmosphere of the sun. The corona is known to be at a temperature of between one and two million degrees Kelvin, an immediate consequence of which is that hydrogen (the principle component) and helium (the second) are completely ionized. The resulting electron density, deduced from eclipse and coronograph observations, is shown in Fig. 1. Since a plasma such as the solar corona, of electron density N_e (cm^{-3}), does not permit the propagation of radio waves of frequency less than the local plasma frequency f_0 (c/s), where

$$f_0{}^2 = 8.1 \times 10^7 \times N_e \qquad (1)$$

we can be sure that all solar radiation at a frequency f comes from above the level in the solar corona where $f = f_0$. In particular, all emission at frequencies below about 200 or perhaps 300 Mc/s comes from the corona.

The most spectacular feature of observations of the photosphere is the presence of *sunspots* — dark markings with diameters which may exceed $0.05R_\odot \approx$

*On leave from Division of Radiophysics, C.S.I.R.O., Australia.

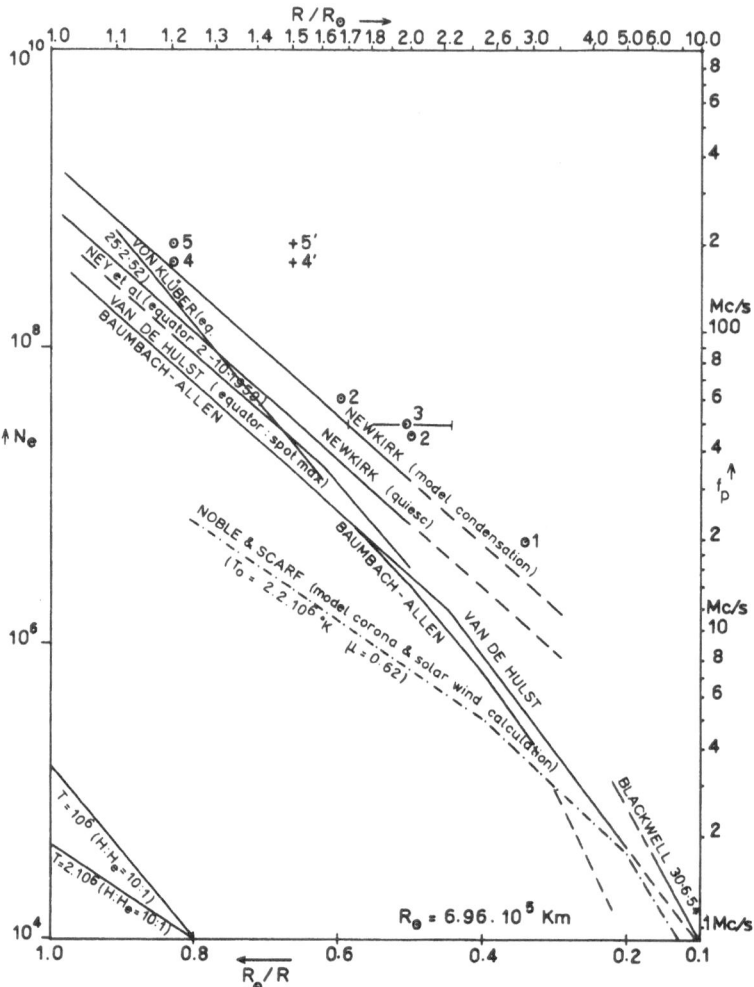

Fig. 1. The results of a number of observations of the electron density of the corona in cm^{-3}. The full curves show the results of various
optical observations of the corona. All except the observations of
Newkirk were made during eclipses. The two curves labeled Newkirk
were deduced from observations made with a coronograph. One is for the
normal corona near sunspot maximum and the other refers to the axis of
an "average" coronal streamer. At least part of the differences observed is due to temporal variations. The dot—dash curve represents a
calculation of the coronal density, aimed at tying this in with solar-
wind observations. The points refer to radio observations of the
heights of occurrence of various types of radio bursts (see list of references at the end of bibliography). Note the very great range of densities involved. The straight lines in the left-hand bottom corner indicate the slopes expected for a *static* corona at the temperatures shown.

4×10^4 km which are carried across the disk, from one limb to the other, as the sun rotates with a period of approximately 27 days, and which have lifetimes of one or two solar rotations. Around the sunspots are observed abnormally bright regions of the photosphere called *faculae*. Probably the most important property of sunspots for our present purpose is that they are the seats of strong magnetic fields, sometimes as much as 3500 G, which are directed into or out of the photosphere. In general, spots of both magnetic polarities are present in one *sunspot group*, the whole being surrounded by a single facular region in which magnetic fields are also observed, but generally less than about 100 G.

Observations of the chromosphere in the light of H_α also show bright regions called *plages*, which lie above the faculae and are of lesser extent.

A *center of activity* is the set of all these disturbances of the normal solar atmosphere: sunspots, faculae, plages, and also filaments and other associated phenomena. The faculae and also quiescent filaments may outlive the sunspots.

In the corona above a center of activity, regions of increased density are observed, probably due to the effects of the magnetic fields of the center of activity. The density in an average *coronal condensation* (or *coronal streamer*) is also shown in Fig. 1.

Cinematographic observations of the sun in H_α (chromosphere) show the existence of *flares*. These appear as bright regions which generally originate within the plage region of a center of activity (flares are brighter than plages). They may extend beyond the plage region and part of a flare may overlie the sunspots. Flares occur in a great variety of sizes, and their morphology and relation to the associated center of activity is complex.

The area covered by a flare is indicated by its *importance* (1–, 1, 2, 3, or 3+). A flare of importance 3+ is defined as having an area > 1200 millionths of the visible hemisphere: such an event is quite rare and has an average duration of about 2 hr. A flare of importance 1 is a much more common event. By definition its area, corrected for perspective shortening, is between 100 and 250×10^{-6} of the visible hemisphere, and its mean duration is found to be about ½ hr.

As we shall see below, practically all solar radio bursts are closely associated with chromospheric flares. Flares are also responsible for a number of geophysical phenomena such as sudden ionospheric disturbances (SID), sudden commencement geomagnetic storms (SSC), and polar cap absorptions (PCA). It is interesting, therefore, to study in greater detail the flare phenomenon if we wish to know more about all these phenomena.

Associated with a flare is a whole complex of phenomena which indicate heating, particle acceleration, and mechanical ejection of gas. Below is a brief list of some of these phenomena and an outline of the interpretations attempted.

OPTICAL FLARE

The flare is not only visible in H_α but in the other Balmer lines and also in a large number (~ 400 – Severny *et al.* [24]) of metal lines which are seen in absorption elsewhere on the disk, but may appear weakly in emission in a flare.

From the absence of Balmer continuum emission, it appears that the temperature must be $\sim 10^4$ °K while Svestka [26] deduces from a study of the profiles of the Balmer lines H_{10} to H_{14} an average electron density of the flare region of about 6×10^{13} cm^{-3}. For another flare, de Feiter and Svestka [7] find that N_e decreases from $\sim 4 \times 10^{13}$ to $\sim 2 \times 10^{13}$ cm^{-3} during the flare.

The interpretation of flare spectra in terms of electron temperatures and densities and other relevant physical quantities is very difficult, but we can hope that these latest results of Svestka's are reliable. Svestka remarks that either the flare is absurdly thin (~ 5 km) or the flare emission comes from filaments which do not cover the whole apparent area of the flare as seen with the inevitably limited resolution of earth-bound instruments. The extra energy of line emission in a very big (3 +) flare is estimated as $\sim 10^{32}$ ergs (Parker [19]) or $\sim 5 \times 10^{31}$ (Ellison [10]). These are extreme cases and more common, smaller flares emit many times less.

VISIBLE MOTIONS

Rapid time sequence $H\alpha$ filtergrams (Athay and Moreton [1]) projected at high speed show clear evidence of rapid propagation of mechanical disturbances (hydrodynamic shock waves?) across the surface of the sun with velocities of 1000 to 2000 km/sec. These are visible close to the flare by their effects on the chromospheric structure, and out to 4×10^5 km from the flare by their sudden perturbation of prominences.

Surges are much slower ejections of matter which are seen to leave the flare at several hundred km/sec along fairly sharply defined, often curved, paths. In $H\alpha$, surges first appear bright (emission), then cool rapidly and appear dark (absorption).

ULTRAVIOLET AND X-RAY EMISSION

Rocket and satellite observations have shown that the $L\alpha$ (Lyman alpha) line of hydrogen in the ultraviolet spectrum does not change appreciably during flares. On the other hand, the X-ray emission is very violently affected. In particular, the flux in the band from 1 to 8 Å is observed to increase ten- or one hundred-fold, which is sufficient to explain the sudden increases of the ionization in the atmosphere which are detected as sudden ionospheric disturbances (SID). The correlation between the detection of X-ray emission between 1 and 8 Å, and SID is very good (Friedman [12]).

Chubb, Friedman, and Kreplin [6] have also observed, using rockets at even shorter wavelengths (<0.5 Å), intense X-ray emission associated with a class 2+ flare. Such hard X-rays can be Bremsstrahlung, emitted by electrons with energies of about 30 kev. The slope of the spectrum fits the hypothesis that these electrons are thermal with a temperature of 10^8 °K (Elwert [11]). The observed flux requires $\sim 10^{37}$ electrons with mean energies of 33 kev; this corresponds to a stored energy of 5×10^{29} ergs (about 1% of the optical emission).

Shklovsky [25] has proposed as a possible alternative that the emission is due to the inverse Compton effect, in which a relativistic electron ($\sim 10^7$ to 10^8 MeV) collides with a pre-existing photon, and transfers a small fraction of its energy to the photon. Shklovsky claims that this mechanism explains the overall spectrum (above and below 1 Å) more adequately, and estimates the energy stored in relativistic electrons as about 10^{30} ergs.

GEOPHYSICAL EFFECTS

Apart from the SID already mentioned as due to X-radiation, the sun ejects high-energy protons and clouds of gas during a solar flare. The former can be observed at the poles (where they can reach the earth along the geomagnetic lines of force), where they cause polar cap absorption events or, in very rare cases when they are particularly energetic (as when they produce an increase in the counting rate of neutron monitors) at intermediate latitudes.

The clouds of gas, or their effects, take about two days to reach the earth, where they give rise to geomagnetic storms. Parker [20] estimates that the cloud which gives rise to a geomagnetic storm carries about 2×10^{32} ergs, which is at least as great as any of the other parts of the flare energy quoted above.

RADIO OBSERVATIONS

We shall now study, in a little more detail than for the above list of phenomena, the radio emissions which occur in association with solar flares. In fact, these emissions apparently include all types of burst emission, and so it is with solar radio bursts in general that we are concerned.

We shall divide these bursts into meter wave bursts, decimetric, and centimetric bursts. The distinction reflects both the author's personal bias and the physical difference between the corona and the chromosphere. This physical distinction comes from the fact, pointed out above, that radiation at frequencies below about 300 Mc/s cannot propagate in plasmas having densities in excess of those at the base of the corona and hence meter wave bursts can only be emitted from the corona.

On the other hand, microwave (i.e., centimetric) and decimetric bursts can originate in the chromosphere or at the bottom of the corona and generally do, since it is there that the flare energy is released and there also that are found the relatively high densities or strong magnetic fields necessary.

It appears (see Wild *et al.* [31] and references therein) that almost all microwave bursts can be considered as the composition of three types: the impulsive burst, the gradual burst (which includes the gradual rise and fall and the post-burst increase), and the microwave component of a Type IV burst (the Type IV μ). All these bursts can be recognized in terms of their intensity, duration, and the rate of rise and fall, and thus can be recognized on single-frequency records. A number of examples are shown in Fig. 2.

On the other hand, a systematic classification of meter wavelength bursts was only made possible with the introduction of dynamic spectra. It is found

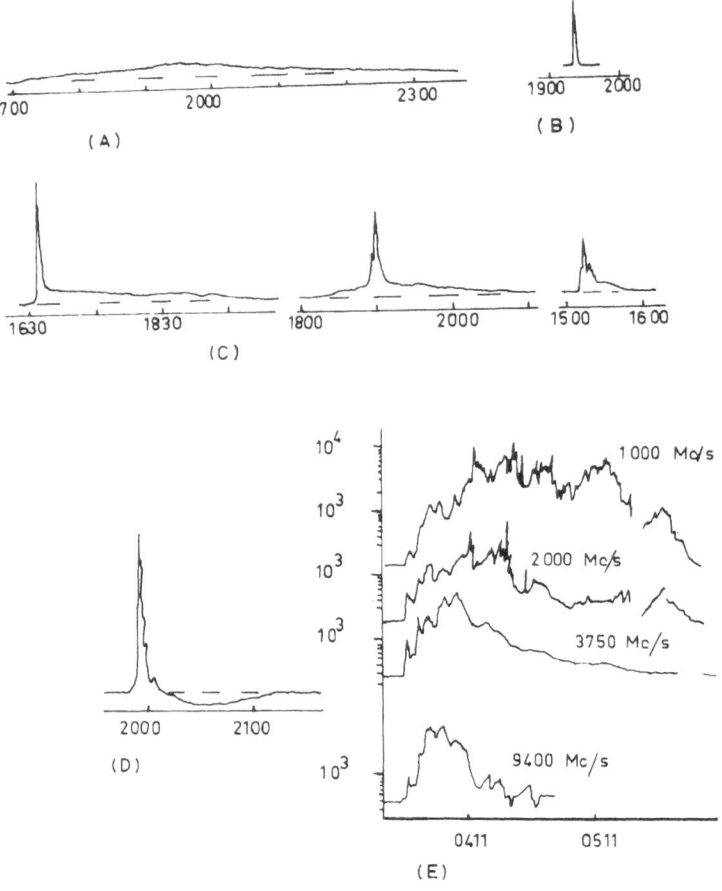

Fig. 2. Examples of the different types of bursts observable at centimeter wavelengths. The curves are all copied from single-frequency observations in which flux varies vertically and time, horizontally. The time markers are indicated for every hour. Curves *A*, *B*, *C*, and *D* were made at a wavelength of 10.7 cm and *E* combines simultaneous observations at the four frequencies indicated. Curves *A* and *B* show, respectively, a gradual burst and an impulsive burst; *C* shows a number of combinations of these two basic types; *D* shows an impulsive burst followed by an absorption phenomenon; and *E* represents a microwave Type IV burst (Type IV μ). See Wild *et al.* [31] for details of original references.

convenient to record such spectra on film with time and frequency varying continuously along and across the film, while the intensity of emission is reproduced as a darkening of the film. With this type of record, essentially all meter bursts can be unambiguously classified in one of about six types as illustrated in Fig. 3. The inset shows the start of an idealized complete meter-wavelength radio flare. We shall discuss this in some detail.

The two most distinctive features of meter wavelength dynamic spectra are the Type II and Type III bursts. Both are relatively sharply defined emissions which are observed to drift systematically from high to low frequencies, with characteristic drift rates, the Type III at about 20 Mc/s per second and the Type II at about 1 Mc/s per second. Both may be accompanied by emission at approximately the second harmonic of the fundamental, although the rapid frequency drift of Type III bursts makes it difficult to distinguish their fundamental from their "harmonic" emission.

The only characteristic frequencies of the corona capable of continuous variation from about 200 to about 10 Mc/s are the gyrofrequency and the plasma frequency. Propagation theory shows that emission at the local gyrofrequency can never escape from the corona and so there remain only plasma oscillations to explain the relatively fine spectra of Type II and Type III bursts. In terms of this idea, the systematic drifts of these bursts are readily explained as being due to outward moving disturbances which excite plasma oscillations at suc-

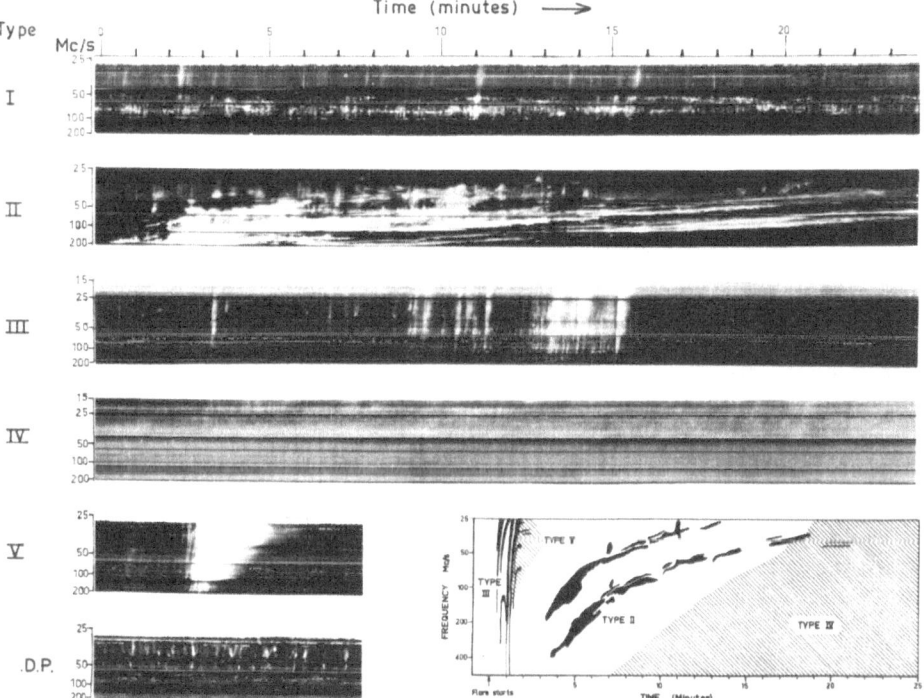

Fig. 3. The types of bursts observable at meter wavelengths, with a spectrum analyzer. For each record, frequency varies vertically and time, horizontally. The intensity observed at each frequency and time is represented by the whiteness or blackness of the record at the corresponding point (white is intense). The inset shows the typical sequence of bursts associated with a flare (in the inset, the black regions correspond to the times and frequencies of intense emission). The region of the inset labeled "Type IV" should be read "Type IV mA" or "Type IV m-moving." (After Wild et al. [31].)

cessively lower frequencies corresponding to the successively lower densities of the regions through which they pass. This interpretation is supported by observations (Wild *et al.* [29] and Weiss [27]) of position as a function of frequency — the systematic shift away from the center of the sun as the frequency decreases is qualitatively in agreement with the plasma hypothesis and in quantitative agreement if the densities are those observed by Newkirk in coronal streamers (or perhaps a little higher). Hence, the plasma hypothesis is widely accepted despite the very serious difficulties involved in an attempted theoretical description of the excitation of and radiation by plasma oscillations.

The velocities of ejection of the disturbances causing Type II and Type III bursts, as deduced from their frequency drift rates, are about 1000 km/sec for the slow-drift, Type II bursts and about 0.5c for Type III bursts.

Although Type II bursts are a very much rarer phenomena than Type III bursts, the two types are closely associated in the sense that most (~60%) Type II bursts are preceded by about 5 min by a group of Type III bursts. If the heights of the two disturbances, deduced from the corresponding frequencies, are extrapolated back in time, they meet somewhere near the chromosphere. Since the association with flares is a very close one, there is every reason to suppose that the two disturbances are ejected from the flaring region at the same moment in the flare's development. The high velocity of the Type III bursts suggests that they are due to individual particle motions while the supersonic velocity of the Type II bursts might be associated with the velocity of a shock front created by the movement of a cloud of gas through the corona.

Groups of Type III bursts are frequently followed almost immediately by a diffuse continuum emission below about 150 Mc/s which lasts for about a minute. This type of continuum emission is called a Type V burst. In compound Type III – Type II bursts, the Type V emission, if it occurs, is frequently finished before the Type II burst commences. It is natural to suggest that the Type V is a different form of emission from the same original disturbance. Wild *et al.* [29] suggested that the Type III emission was excited by relativistic electrons spiraling along the magnetic lines of force in the corona. Irregularities of the magnetic field could scatter these electrons, reducing the pitch of the spiral trajectories of some of them. These scattered electrons would stay for some time in the corona, and emit by the synchrotron process, with the broad spectrum observed for Type V bursts. However, variations of Type V source position with frequency (Wild *et al.* [31]) similar to the variations observed for Type III emission make this interpretation doubtful.

The Type II can also be followed by a "burst" of continuum emission which is called a Type IV. Since the Type II is interpreted as the emission from the shock wave set up by a moving cloud of gas with about the same velocity as the disturbance responsible for a geomagnetic storm, it was natural to associate these two phenomena. However, this appears to be only partially correct, as a careful study (McLean [18]) has shown that a Type II burst which is not followed by a Type IV (continuum) event does not give rise to a geomagnetic storm with sudden commencement (SSC), whereas a geomagnetic SSC is commonly preceded by a Type IV burst (Caroubalos [5] and Bachelet *et al.* [3] and others). Also, those flares which emit the low-energy protons observed at the poles as polar cap absorption events are found to be associated with Type IV bursts, as are the much rarer cosmic ray events in which a flare emits protons which are sufficiently energetic that they can be observed elsewhere than at the poles, despite the reflecting effects of the earth's magnetic field.

These associations have led to a great deal of attention being paid to the Type IV burst. Despite this, the situation is still somewhat confused – the conclusions of different observers appear to be related to the method of observation. The following appears to be fairly well established.

Pick [21] finds a close correlation between the occurrence of intense centimetric bursts lasting more than 10 min and Type IV bursts as originally defined by Boischot [4] at meter wavelengths. The centimetric burst is now called a microwave Type IV burst or Type IV μ. Boischot suggested, and spectral observations confirmed (Haddock [14] and McLean [18]), that Type IV bursts almost always follow Type II bursts, at least if the Type IV burst is sufficiently intense to be observed with normal spectrum analyzers. Pick has also shown that the metric Type IV (Type IV m) should be considered as made up of two parts which are confused when the source is near the center of the sun. However, at the limb, only the first part, which lasts several tens of minutes (Type IV mA) is observable; it is this part which correlates with the Type IV μ (and which has a similar duration). Interferometric observations at meter wavelengths (including Boischot's original observations) show an initial rapid motion of Type IV sources, followed by a stable phase. Wild, Smerd, and Weiss [31] have chosen to call the Type IV mA burst the *moving* Type IV, and the Type IV mB, *stationary*. The emission from the first (moving) part is believed to be synchrotron radiation by MeV electrons, in accordance with Boischot's original suggestion (Boischot [4]), while Denisse [8] has proposed Čerenkov mechanism for the Type IV mB.

Both Type II and Type IV bursts are supposed to be closely associated with ejections of gas clouds through the corona – the Type II by the theory of its emission, the Type IV in particular by its association with geomagnetic storms. The velocities deduced for the two clouds are about 1000 to 2000 km/sec and so it is only natural to try to construct a model in which Type II and Type IV bursts are due to emission by different mechanisms from the same moving gas cloud. One of the first attempts in this direction made by the author is shown schematically in Fig. 4. However, in this model the Type II source lies directly in front of the Type IV source, in direct contradiction to recent observations (Weiss [27, 28]) that the paths of the two sources are very different.

A model which might explain this difference is presented in Fig. 5. In this model, use is made of the theoretical difficulty of explaining how a sharp shock front can form in a gas as tenuous as the corona – collisions being totally ineffective on the scale indicated by the narrow band of frequencies emitted at any one time in some Type II bursts. This difficulty vanishes when there is a magnetic field parallel to the surface of the shock wave (i.e., perpendicular to the gas motion). In this model, then, the shock front is not formed, or is not sufficiently "sharp" to emit strongly, in front of the gas cloud, where the motion is mostly

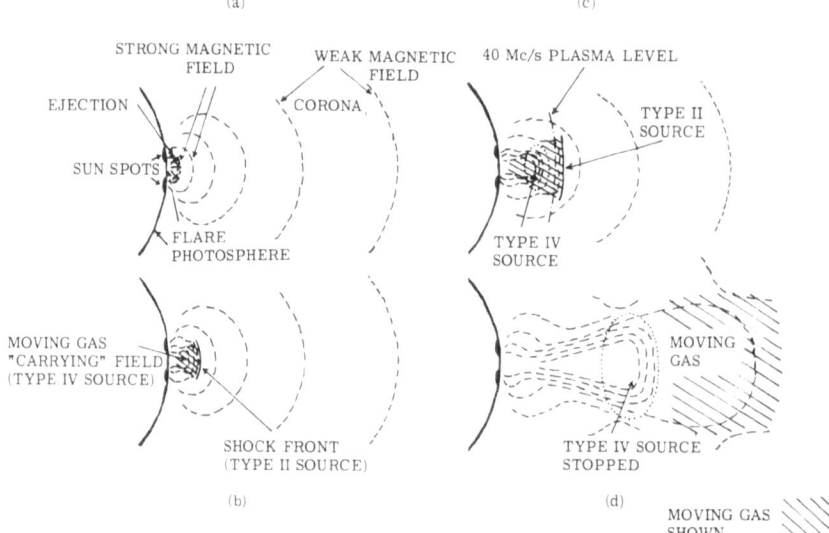

Fig. 4. An early conception of the mechanism of production of Type II and Type IV bursts. (a) A flare occurs in the chromosphere above an active region, and by an unspecified mechanism gives rise to a violent ejection of gas. (b) This cloud of gas carries this gas it encounters ahead of it; since the magnetic field is "frozen" into the gas, it is also compressed and drawn out by the moving cloud. (c) When the shock front, which is supposed to form in front of the supersonic cloud, has passed through the 40-Mc/s level in the quiet corona, the enhanced emission from the region of this front due to the production of plasma waves becomes visible from the earth — these are the Type II bursts. (d) When the part of the cloud carrying the strongest magnetic fields rises high in the corona, it also can be observed from the earth by the synchrotron emission of trapped relativistic particles. This model does not explain the acceleration of the particles. Also, the Type IV source is shown to have been stopped by the magnetic field, while the region of the corona which has been pushed ahead of it continues to give rise to a geomagnetic storm.

along the lines of force. However, at the "sides" of the cloud, where the gas tends to move across the lines of force, a sharp shock front may be expected. This is because the magnetic field, frozen into the gas, prevents any appreciable interpenetration of the moving gas behind the shock front and the stationary gas ahead. Instead, the field and the gas just ahead of the shock front are compressed and accelerated and pass through the shock to become assimilated into the moving cloud behind.

It is from this region that we suppose the Type II emission to come. In Fig. 5, two such regions are shown moving apart, but if we were hidden behind the coronal condensation in which we suppose all this to occur, the other would then be seen to move nonradially away from the flare. On the other hand, the Type IV source, embedded in the moving gas, would move more or less radially, thus explaining the different paths observed for these two related events.

This model also explains the acceleration of the electrons necessary for the Type IV mA synchrotron emission, in the following way. At the point marked A on Fig. 5, an electron will be accelerated by the betatron effect in the increasing magnetic field in the region of the shock. This electron will be accelerated out of the region A and travel along a line of force to the region B, where it will again be accelerated and reflected back to A. This is just a particular case of the well-known Fermi mechanism in which an electron gains a little energy from each of a large number of collisions with a moving cloud of gas, carrying a magnetic field. For an electron to be reflected in the accelerating region, it must have a large pitch angle, and so the travel time from A to B might well be several seconds. Since A and B are moving along the line of force on which our electron is trapped, the whole accelerating process can only continue for a matter of 1 or 2 min. This is probably time enough for a 200-eV superthermal electron to be accelerated to more than 1 MeV.

When the acceleration is over the relativistic electron will remain trapped on its line of force, radiating by the synchrotron mechanism. As the gas continues to rise and expand, carrying the line of force and its electron, the local plasma frequency will decrease, making the electron's emission visible at lower and lower frequencies. Clearly, then, the above description is very hypothetical, but it seems of sufficient interest to justify its presentation here.

According to our description of a complete solar radio flare event, there remains the Type I storm. Wild *et al.* [31] consider that a Type I storm is composed of Type I bursts superimposed on Type I continuum. However, this terminology is not uniformly adopted: Type I is frequently taken to refer only to the burst component. It has really only recently been established (le Squeren [16]) that the start of a Type I storm is also typically preceded by a solar flare. The recognition of this phenomenon was made difficult by the long delay, of the order

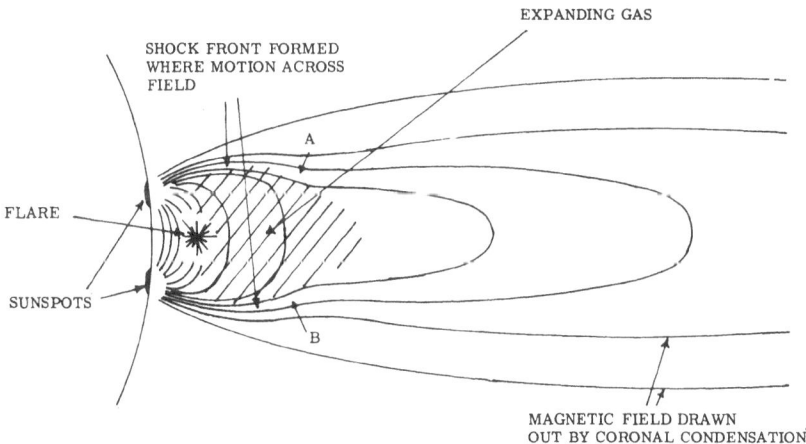

Fig. 5. An alternative to Fig. 4. Instead of the Type II producing shock wave being formed "in front" of the Type IV source, it forms at the "sides," where the expanding gas encounters a transverse magnetic field. Regions A and B are regions of rapidly changing magnetic field; by repeated reflection between A and B, an electron may gain sufficient energy to explain the Type IV mA emission by the synchrotron mechanism.

of an hour, between the flare and the start of the storm. In the light of this fact, there would seem to be little reason to distinguish between the latter part of a Type IV mB burst and a type I storm. For a more detailed discussion of some aspects of Type I emission, see the papers by Fokker, De Groot, and Elgarøy in this volume.

The lack of really characteristic details in microwave bursts makes it difficult to ascribe mechanisms to their emission. The available mechanisms appear to be synchrotron emission and Bremsstrahlung or free – free emission, or a combination of the two. In the case of impulsive bursts, the existence of a closely related X-ray burst (e.g., Kundu [15]) seems to argue in favor of a Bremsstrahlung mechanism.

As we have seen, a solar flare is a violent and complex event. Many aspects remain to be explained; for example, how can the relatively low temperature of 10^4 °K deduced from the optical spectrum be reconciled with the temperature of 10^8 °K deduced from the X-ray spectrum? If, instead, we explain the latter by Shklovsky's theory of X-ray emission by the inverse Compton effect, we must then explain the presence of 10-MeV electrons in the flare region. (These same electrons may explain the Type IV emission, and in any case we must explain the particle acceleration process necessary for PCA and cosmic ray events.)

Studies of flare morphology, in particular by Martres-Tropé and Pick-Gutmann [17], have shown that metric wavelength Type IV bursts can only occur if the associated flare and the center of activity in which it occurs have certain, rather special configurations. Similar results were already known for those flares responsible for cosmic ray events. Such a result obviously contains a very important clue to the nature and origin of those flares which are responsible for metric wave radio emission, but as yet no one has been able to make use of it.

Most people who try to construct a "flare theory" leave the many such detailed problems in abeyance and tackle the fundamental problem of how all the energy is released in the chromosphere. The problem is not that the energy is great by solar standards — the energy released in a great solar flare is only a few percent of the total radiation from the sun every second. On the other hand, it has been calculated that the whole of the thermal energy of chromosphere and corona would not suffice for a large flare.

Since there is no visible perturbation of the photosphere at the time of the flare, it is felt that the energy could not come directly from beneath the photosphere. Hence, it is argued that the energy comes gradually into the chromosphere, where it is stored in some unstable form. Since flares occur in centers of activity, and their position and form appear to be related to the magnetic fields of these centers of activity (Severny [23] and Martres-Tropé and Pick-Gutmann [17]), it is very probable that magnetic fields play an essential role in the flare mechanism.

We shall just mention very briefly two theories which are based on this kind of argument. In the first, formulated by Gold and Hoyle [13], the energy is stored in the torsion of tubes of magnetic field and dissipated by the violent interaction between adjacent tubes if the latter are pushed together. The gas is then heated at the expense of the magnetic field which is partly annihilated.

A recent alternative formulated by Eliot [9] proposes that protons are continually accelerated and stored above centers of activity. A flare occurs when these trapped particles are precipitated into the denser regions of the chromosphere.

REFERENCES

It has not been possible in so little space to do justice to the many people who have done excellent work on various aspects of solar flares. For a more complete study of the flare phenomenon we refer the reader to *Solar Flares*, J. J. and E. V. P. Smith (Macmillan, New York, 1963); while for a review of the radio aspects of solar flares, see J. P. Wild, S. F. Smerd, and A. A. Weiss, *Ann. Rev. Astron. Astrophys.* 1, 291 (1963). Both of these contain extensive, if not exhaustive, lists of references. Below is the list of papers quoted in this paper.

1. R. G. Athay and G. E. Moreton, *Ap. J.* 133, 935 (1961).
2. Y. Avignon, M. J. Martres, and M. Pick, *Ann. Astrophys.* 27, 23 (1964).
3. F. Bachelet, P. Balata, A. M. Conforto, and G. Marini, *Nuovo Cimento* 15, (1960).
4. A. Boischot, *Ann. Astrophys.* 21, 273 (1958).
5. C. Caroubalos, *Ann. Astrophys.* 27, 333 (1964).
6. T. A. Chubb, H. Friedman, and R. S. Kreplin, *Space Research* 4, 759 (1964).
7. L. D. de Feiter and Z. Svestka, *B.A.C.* 15, 117 (1964).
8. J. F. Denisse, *Inform. Bull. S.R.O.E.* 4, 3 (1960).
9. H. Eliot, *Plan. Space Sci.* 12, 657 (1964).
10. M. A. Ellison, *Quart. J. Roy. Astron. Soc.* 4, 62 (1963).
11. G. Elwert, *J. Geophys. Res.* 66, 391 (1961).
12. H. Friedman, *Ann. Rev. Astron. and Astrophys.* 1, 59 (1963).
13. T. Gold and F. Hoyle, *Monthly Notices Roy. Astron. Soc.* 120, 89 (1960).
14. F. T. Haddock, *Proc. IRE* 46, 3 (1958).
15. M. Kundu, *J. Geophys. Res.* 66, 4308 (1961).
16. A. M. le Squeren, *Ann. Astrophys.* 26, 97 (1963).
17. M. Martres-Tropé and M. Pick-Gutmann, *Ann. Astrophys.* 26, 30 (1963).
18. D. J. McLean, *Australian J. Phys.* 12, 404 (1959).
19. E. N. Parker, *P.R.* 107, 830 (1957).
20. E. N. Parker, *Ap. J., Suppl.* 77, 8 (1963).
21. M. Pick, *Ann. Astrophys.* 24, 153 (1961).
22. J. A. Roberts, *Australian J. Phys.* 12, 327 (1959).
23. A. B. Severny, *Soviet Astron. (English Transl.)* 6, 747 (1963).
24. A. B. Severny, N. V. Stechenko, and V. L. Khokhlova, *Astron. Zh.* 37, 23 (1960).
25. J. Shklovsky, *Nature (London)* 202, 275 (1964).
26. Z. Svestka, *B.A.C.* 14, 234 (1963).
27. A. A. Weiss, *Australian J. Phys.* 16, 240 (1963a).
28. A. A. Weiss, *Australian J. Phys.* 16, 526 (1963b).
29. J. P. Wild, K. V. Sheridan, and G. H. Trent, *IAU-URSI Symposium on Solar Radio Astronomy*, Paris (1958), (Stanford Univ. Press, 1959), p. 176.
30. J. P. Wild, K. V. Sheridan, and A. A. Neylan, *Australian J. Phys.* 12, 369 (1959).
31. J. P. Wild, S. F. Smerd, and A. A. Weiss, *Ann. Rev. Astron. and Astrophys.* 1, 291 (1963).

The references consulted for the construction of Fig. 1 were as follows:

OPTICAL

1. Allen, *Monthly Notices Roy. Astron. Soc.* 107, 426 (1947) (Baumbach–Allen Model).
2. Blackwell, *Monthly Notices Roy. Astron. Soc.* 116, 50 (1956).

3. Newkirk, *Ap. J.* 133, 988 (1961).
4. Ney *et al.*, *Ap. J.* 133, 616 (1961).
5. Van de Hulst, *B.A.N.* 11, 135 (1950).
6. Von Klüber, *Monthly Notices Roy. Astron. Soc.* 118, 201 (1958).

RADIO

1. C. A. Shain and C. S. Higgins, *Australian J. Phys.* 12, 357 (1959).
2. J. P. Wild, K. V. Sheridan, and A. A. Neylan, *Australian J. Phys.* 12, 369 (1959).
3. A. A. Weiss, *Australian J. Phys.* 16, 240 (1963).
4. A. M. le Squeren, *Ann. Astrophys.* 26, 97 (1963).
5. M. Morimoto, *Publ. Astron. Soc. Japan* 13, 285 (1961).
6. M. Morimoto and K. Kai, *Publ. Astron. Soc. Japan* 13, 294 (1961).

CALCULATIONS

1. Noble and Scarf, *Ap. J.* 138, 1169 (1963).

Sweep-Frequency Measurements of Solar Bursts

James W. Warwick

High Altitude Observatory and
Department of Astro-Geophysics
Boulder, Colorado, United States of America

For more than a decade, sweep-frequency studies of solar nonthermal radio emission have been carried out in the metric and decametric wavelength ranges. These include information on the polarization and position of bursts and other events. The original phenomenology proposed by Wild and associates at Sydney appears still to be valid; although complications have appeared, they do not vitiate the concepts of bursts generated at and near the plasma frequency by outward traveling disturbances in the corona. A new phenomenon, continuum outbursts associated with violent solar activity, was identified more recently in France. Its description at various observing wavelengths from decametric to centimetric ranges has proceeded apace.

Observations made by the Boulder low-radio-frequency spectrograph and interferometer permit new discussions of these phenomena. For many events, optical data taken simultaneously with high-speed cinematography (one frame per ten seconds) add still further information. Detailed descriptions of the phenomena during a number of events are offered.

We conclude that magnetohydrodynamic shocks provide the means by which the sun appears to expel prominence material in the early phases of flares. The evidence strongly suggests that the observed optical phenomena also represent the early phases of shocks, before they arrive at the coronal levels emitting slow-drift bursts.

INTRODUCTION

The phenomena of solar activity are extremely complex, especially at low radio frequencies. Within a single presentation as limited as this must be, I will therefore forego an exhaustive discussion. Instead, I have chosen to summarize only some features of the most active solar events occurring in conjunction with solar flares.

The strength of the Boulder radio and optical observations of the sun consists in the large number of events for which broad coverage was obtained. The period involved extends back to just after the great flares of July 1959, but contains following that time a high percentage of the events accessible to us. These data are relatively unknown, having appeared only in the "I.G.Y. Solar Activity Report Series" (Number 23, July 1, 1963; World Data Center A, High Altitude Observatory, Boulder, Colorado) (observations from July 24, 1959, to February 28, 1961) and the "Central Radio Propagation Laboratory's F-Series, Part B" (Solar and Geophysical Data) (observations from March 1, 1961, to the present), published beginning with the F-Series of November 1961. The radio spectrograph equipment has been described in several publications, most recently by Lee and Warwick [1]. Construction of this equipment was supported by the United States Air Force Cambridge Research Laboratories. In addition, we operate fixed-frequency interferometers at 8, 18, and 36 Mc/s in Boulder, and total-power 18 Mc/s radiometers in conjunction with Manila Observatory, Rome Observatory, McMath Hulburt Observatory, and the University of Hawaii. Solar data from the Boulder link in this chain were discussed by Warwick and Warwick [2].

Most of the optical data presented here were obtained with a fast-time interval (10 sec between exposures), relatively high-resolution H_α (Lyot) filter plus coronograph at the Climax Station of the High Altitude Observatory. For a few of the events, we used data from other stations (in particular, July 20, 1961, from the Sacramento Peak Observatory) for which we are deeply indebted. The Climax flare patrol is a unique, full-limb, plus disk, equipment in which the limb is recorded with wide passband (3 Å) and the disk with narrow (0.7 Å). Automatic alternation of the pictures is carried out by a programmer, which also deletes the limb exposure on the ninth frame out of each ten and records the image with reduced density on the tenth frame out of each ten. This equipment has been in use since early January 1960. Its construction was supported by the United States National Aeronautics and Space Administration (NASA), which also provided support for the data reductions that are reported herein.

The observational emphasis of this paper will lie on major solar "events," for which we have obtained radio position data interferometrically, as well as combined optical and dynamic spectral data. The positional information, itself obtained as a function of frequency, is especially useful, we find, in that by a special presentation technique in which negative interferometer fringes are "inverted" (converted to positive voltage in the output of the receiver), we can discuss shifts delicately as a function of the spectral variations of the radio burst. We have presented this information in three forms: as simple lines of position at a few times during the event (only for September 6, 1961), as lines of position as a fine-grained function of time during the event, and finally with the theoretical fringes (representing the center of the sun) photographed directly against the dynamic spectral record.

DESCRIPTION OF SEVERAL MAJOR DECAMETRIC EVENTS

Flares can produce a vast array of phenomena, from γ-rays to at least 1.5-Mc/s radiation — far more effects than we can even mention here. On a decametric radio spectrograph and interferometer, features appear which permit easy identification: the emission is usually of high intensity, broad bandwidth, and moderately long duration, 1 hr or more. The general pattern first described by Wild, Murray, and Rowe [3] and complemented by Boischot [4], is typical of deca-metric, as well as metric, bursts.

Figure 1a shows a case in point. At the outset, *slow-drift* (Maxwell's termi-nology) Type II bursts occur – each component, the fundamental and second har-monic, of considerable structural detail. Atypically, this burst is not preceded by fast-drift bursts (Type III) although several do occur later in the event. How-ever, it is followed by a persistent, smooth broad-band continuum that endures over 1 hr. This component was called Type IV by Boischot. The harmonic structure of the slow-drift burst is unusual in our records, confirmed in this case by Fort Davis (Harvard) spectra of the same event. These were kindly lent by A. R. Thompson and A. Maxwell. In Wild's identification of the harmonic phe-nomenon in slow-drift bursts, the harmonic nature of the burst was established by detailed time agreement in the fine structure of the two components. Neither in this case, nor in the Fort Davis record, is that character at all obvious, although detailed structure is manifestly present.

The position of the emission is plotted for this record in Fig. 1b. Consider particularly 30 Mc/s (lower left graph of Fig. 1b), at times earlier than 1605 U.T. The fringes there represent a source about $0°1$ west of the center of the sun. This emission lies in the fundamental of the slow-drift burst. For 35 and 40 Mc/s, from 1600 to 1610 U.T., the emission lies about $0°3$ west of the center of the sun, e.g., the second harmonic lies farther from the center of the sun than the fundamental. There is excellent agreement between our observations of the ap-parent position of the radio emission and the optical position of the flare. That the second harmonic appears farther from the sun than the fundamental, contra-dicts the pattern observed by Smerd, Wild, and Sheridan [5] for a handful of bursts.

The Type IV in this case appears first to drift toward the west, starting at 1610 U.T., from virtually the same position as the harmonic of the II, and subse-quently moving toward the west by another $0°3$. That position arrived at by 1700 U.T., the burst then appears to drift steadily back to the east, arriving at about $0°1$ west of the center in the terminal phase of the event at 1850 U.T.

Bursts of a blobby character appear at various times throughout the Type IV – especially at 1625–1628 U.T., they are strongly shifted in position toward the center of the sun. This kind of burst often appears during Type IV events.

There is a strong kink in the fringes at 1637, which corresponds again to an eastward shift of the emission, in this case unaccompanied by dynamic spectral features. This is not similar to ionospheric refraction or scintillation that we occasionally do observe. For one thing, the interferometer fringes were in this case nearly vertical to the horizon, and the sun was quite high in the sky (46° elevation). Figure 2 shows a case of strong ionospheric scintillations; the fringe displacements tend to recur in a wavelike pattern unlike the shifts on the Type IV.

In summary, an initial outward drift is followed by a final return to the original source position. Superimposed on the general pattern of shifts away

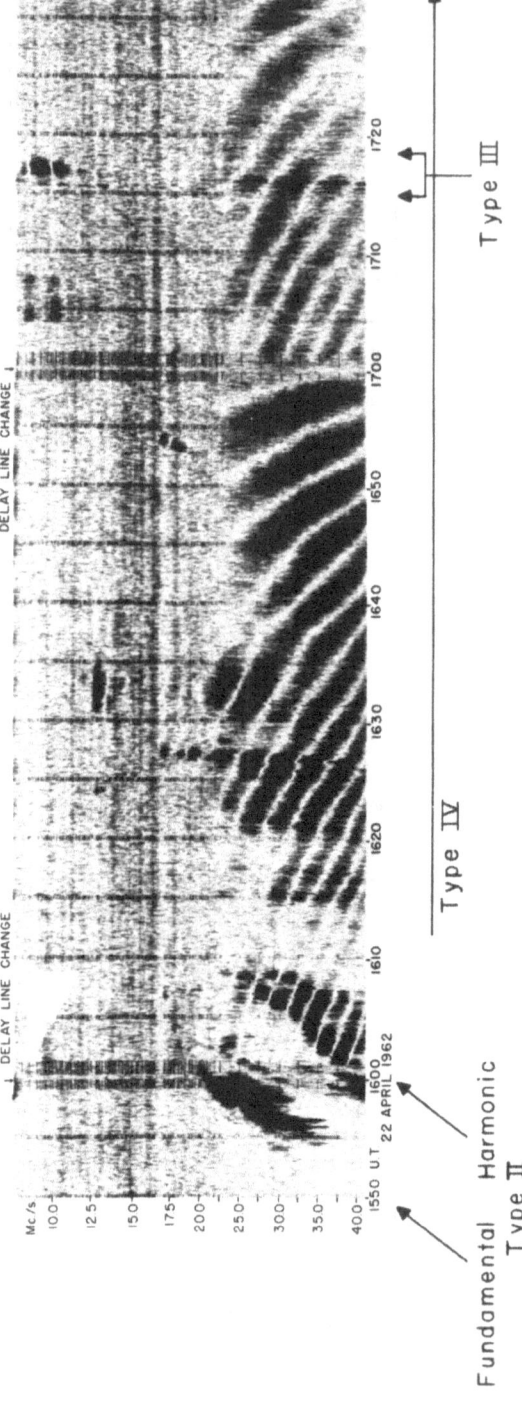

Fig. 1a. The dynamic spectrum of radio emission in connection with a class 1 or 1− flare at 1546 U.T. on April 22, 1962. The flare lay at N12 W48 on the solar disk. The diagonal white streaks are interference fringes, and permit establishment of the position of the emission as a function of frequency and time. Note the positional displacement of bursts on the Type IV emission at 1626 U.T., and other times.

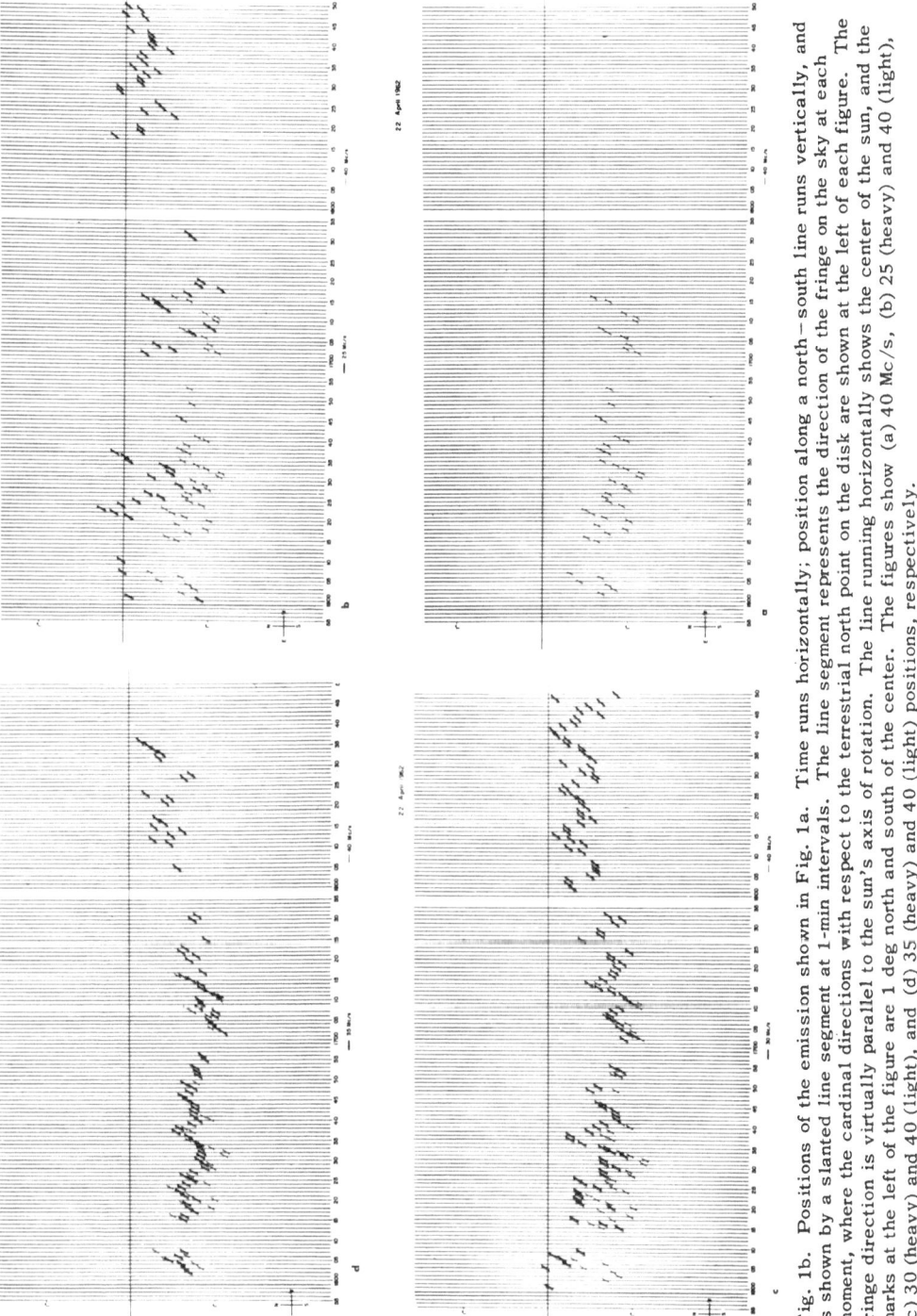

Fig. 1b. Positions of the emission shown in Fig. 1a. Time runs horizontally; position along a north–south line runs vertically, and is shown by a slanted line segment at 1-min intervals. The line segment represents the direction of the fringe on the sky at each moment, where the cardinal directions with respect to the terrestrial north point on the disk are shown at the left of each figure. The fringe direction is virtually parallel to the sun's axis of rotation. The line running horizontally shows the center of the sun, and the marks at the left of the figure are 1 deg north and south of the center. The figures show (a) 40 Mc/s, (b) 25 (heavy) and 40 (light), (c) 30 (heavy) and 40 (light), and (d) 35 (heavy) and 40 (light) positions, respectively.

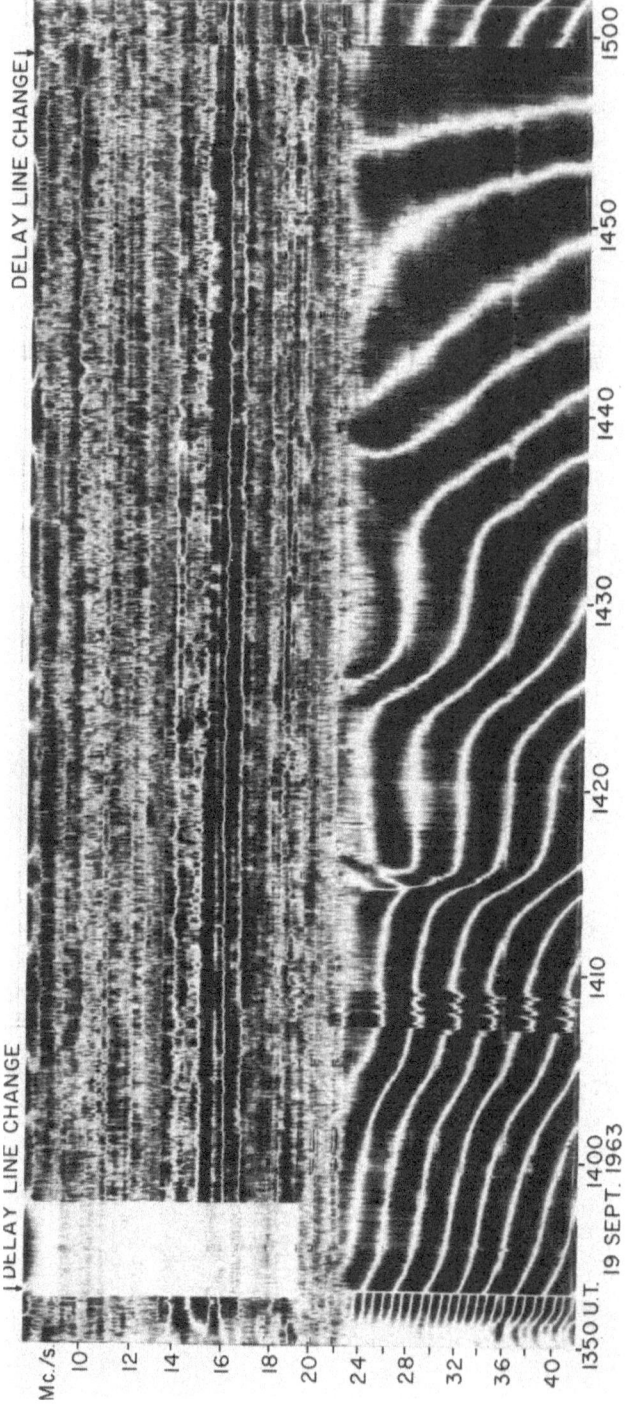

Fig. 2. Solar emission on September 19, 1963. Note the strong wavelike fringe displacements, attributed by us to ionospheric irregularities. Type III bursts at 1407 U.T. appear shifted with respect to the underlying continuum. This shift is presumably of solar origin.

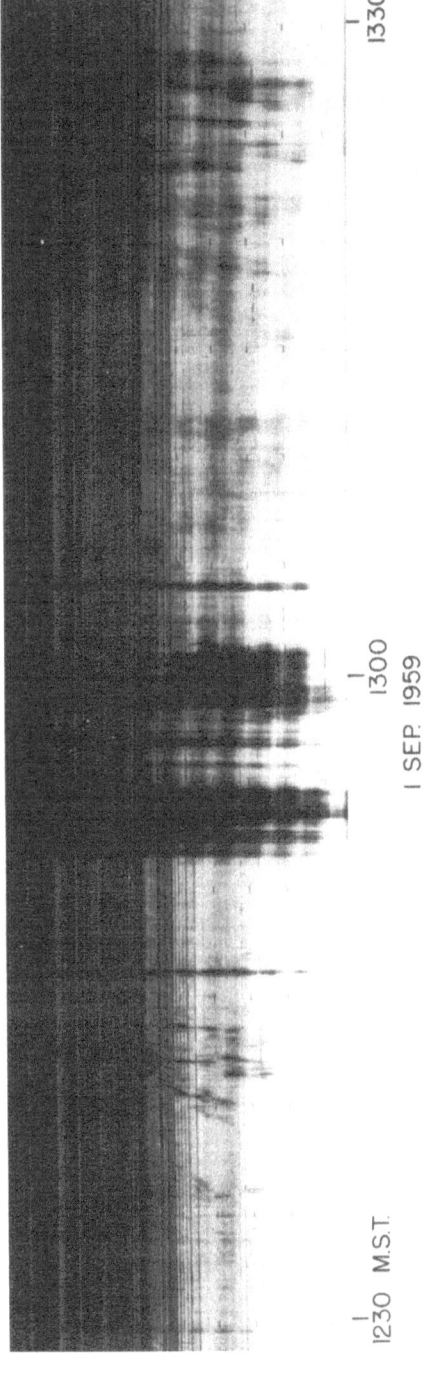

Fig. 3. The faint Type II burst of September 1, 1959, lies between 1235 M.S.T. (1935 U.T.) and 1245 M.S.T. (1945 U.T.). The frequency range is 15 (top) to 33 (bottom) Mc/s. Extremely intense III's occur in the 10 min preceding 1300 M.S.T. (2000 U.T.) and may correspond to another flare than produced the type II (see text). Faint continuum (Type IV?) is present throughout the figure.

from and then toward the sun, there often appear even more complicated shifts.

The morphology of the initial phase of major events is important toward clarifying the physical nature of the process in the atmosphere of the sun. Wild, Smerd, and Weiss [6] emphasize the typical event as the sequence Type III – II – IV. The sequence often occurs with this structure, but major events do also occur in a different pattern – for example, the event of April 22, 1962, which I have just described has no initial Type III. Figure 3 exhibits still another pattern; in this case, a Type II burst, faintly visible from 1935 to 1945 U.T., which is followed by massive flare-produced Type III's over 15 min after the beginning of the Type II. Furthermore, in this case, continuum, which we should probably call Type IV, is developed not only before the III's begin but before the II as well, by 6 min. This event is the probable source of a very small polar cap absorption event (Warwick and Haurwitz [7]). Unfortunately, the case is not unambiguous, since not only was there in progress a class 2 flare which began at 1923 U.T. (and this presumably the source of the II), but a second class 2 flare occurred at 1948 U.T. in another region. Identification then becomes a serious problem.

Figure 4 shows a number of events that illustrate some typical features. In each of these, Type III bursts more or less discretely occur at the onset and are followed by Type II's. We distinguish the Type II and Type III bursts essentially in terms of their frequency drifts. This, of course, is the most basic way to do it; it needs to be pointed out that both types occur in a wide range of drift rates, durations, bandwidths, and relative intensities. The II's are frequently evident in terms of accompanying jagged fringe shifts, as in all of these cases save May 1, 1962. The latter shows a radically complicated fringe structure at 1925 U.T., 30 Mc/s, undoubtedly demonstrating that the sun was then at least a double source, whose elements were distributed with roughly comparable intensity over at least a degree on the sky. The phenomenon is restricted to a small range of frequencies; this severely complicates an analysis. The subsequent Type IV's vary widely in character, some (April 12, 1962) showing many III's superimposed, but coming from a different part of the sun. March 1, 1962, is very smooth for 1½ hr, when there begin to appear a few III's. October 22, 1963, shows consistently many faint III's. However, in each case, the underlying continuum is Type IV: it originates following a major solar burst, comes to a prompt maximum, and decays within 1 or 2 hr.

The Type II phase can often occur without subsequent Type IV (at least at our sensitivity level, about 5×10^{-22} W/m^2-c/s). Figure 5 illustrates a number of Type II bursts, some without either Type III or Type IV bursts. These cases often show typical fine structure. They are minor events, but there is no doubt of their basic II-ness, represented by the distinct slow drifts from high to low frequency. In the case of April 20 and 21,1962, we have isolated minor II's, with III's, on each of the two days preceding the major event of April 22, 1962 (see Fig. 1). The Type II on April 21, 1962, is remarkable for its simplicity and narrowbandedness. On April 20, 1962, a very faint continuum underlies the III's, and may follow the II. On May 23, 1963, the event was observed very early in the morning (0555 local time) and occurred simultaneously with a case of Jupiter's decametric emission. The minor events do not show harmonic structure, but could scarcely be expected to on our records, which do not extend high or low enough in frequency. Both June 5, 1960 and July 8, 1960, and May 23, 1963, show doubling of the structure. In the second case, it is by about 5 Mc/s; in the first case, it is difficult to establish, but is perhaps as much as 7 Mc/s. On March 18, 1961, there is a possible earlier II at 1752; the relation of this structure to the lower frequency II from 1756 to 1802 is of interest.

Type II bursts often consist of a sequence of drifting events that are remarkably similar in intensity, bandwidth, drift rate, and in some cases, fine structure. Figure 6 illustrates the phenomena. October 13, 1960 (Fig. 15) is another case. On August 18, 1961 (Fig. 6), the question arises whether the leading edges of the events at 2051, 2101, and 2114 U.T. would occur at the same time if the frequency coverage extended effectively down in frequency one more octave. Already the leading edges are in roughly 2:1 frequency spacing at 2101 and 2114 U.T. The burst of October 13, (Fig. 15), an even faster drifting one than August 18, 1961, has two components which do not overlap. The question does not arise in this case. On May 25, 1963 (Fig. 6), the bursts at 1632 and 1637 U.T. overlap when the former reaches 19 Mc/s at 1637 U.T., but their shapes are quite different; even the slopes do not look the same. On the other hand, the emission on March 16, 1964, shows the way in which detailed structure in a Type II burst can repeat after (in this case) 25 min. A striking resemblance is apparent in the II from 1558 to 1603 U.T., as compared with the II from 1622 to 1628 U.T.

It is also possible for exceedingly intense, broad-band continuum to arise with neither decisive Type II or Type III bursts in close conjunction. Figure 7 illustrates the event of July 15, 1961, one of a remarkable series of solar activities during that month. Here, Type IV arises ostensibly spontaneously, drifting rapidly (although somewhat more slowly than a Type II drift rate) toward lower frequencies. Strong low-frequency III's, beginning below 25 Mc/s and extending to our low spectral limit at 7.6 Mc/s, occur some 15 min after the IV began. The IV finally drifts down to about 10 Mc/s at about 1615 U.T., and remains visible there for about 1 hr. This event was classified at both Fort Davis and Boulder as Type IV; the accompanying flare was obviously explosive and large, inasmuch as it was classified identically as to time of start and maximum (1508 and 1512 U.T., respectively) at three different stations. The final phase of the event, after about 1745 U.T., is a continuum.

Figure 8 illustrates another of the July 1961 events — here a complete series, III − II − IV, the latter a relatively narrow band which contains considerable complex burst activity. These bursts do not come from the position of the IV, but lie rather to the east or north of it, almost at the apparent center of the sun in many cases. On the other hand, they seem to follow the Type IV bursts down to lower frequencies, even though their positions remain close to the center of the sun. The drift rate of the IV is similar to the rate of the IV on July 15, 1961, a decametric octave in about 22 min. This is notably slower than the slow drift bursts. The occurrence of fine structure moving down toward lower frequencies with the Type IV implies a common source of the two phenomena, but the fact that they occur apparently in distant regions of the sun suggests a

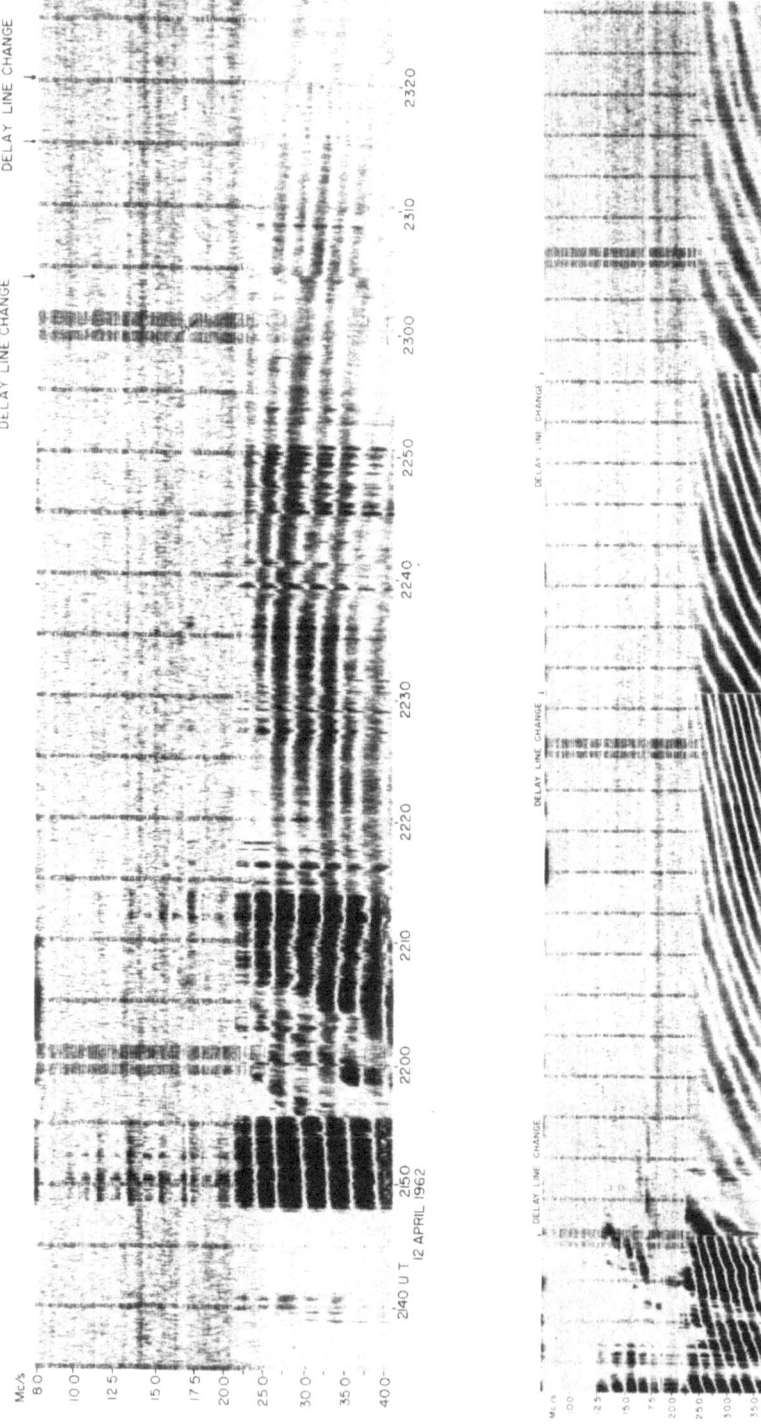

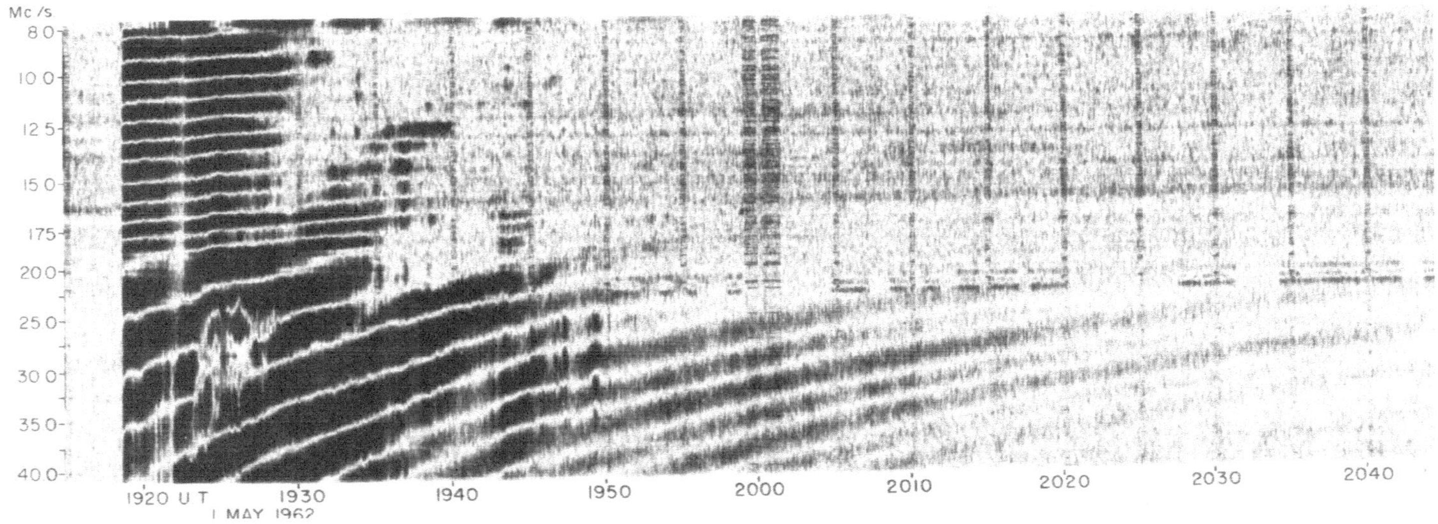

Fig. 4. Four photos — March 1, 1962, April 12, 1962, May 1, 1962, and October 22, 1963. A variety of III~II~IV associations (see text for details).

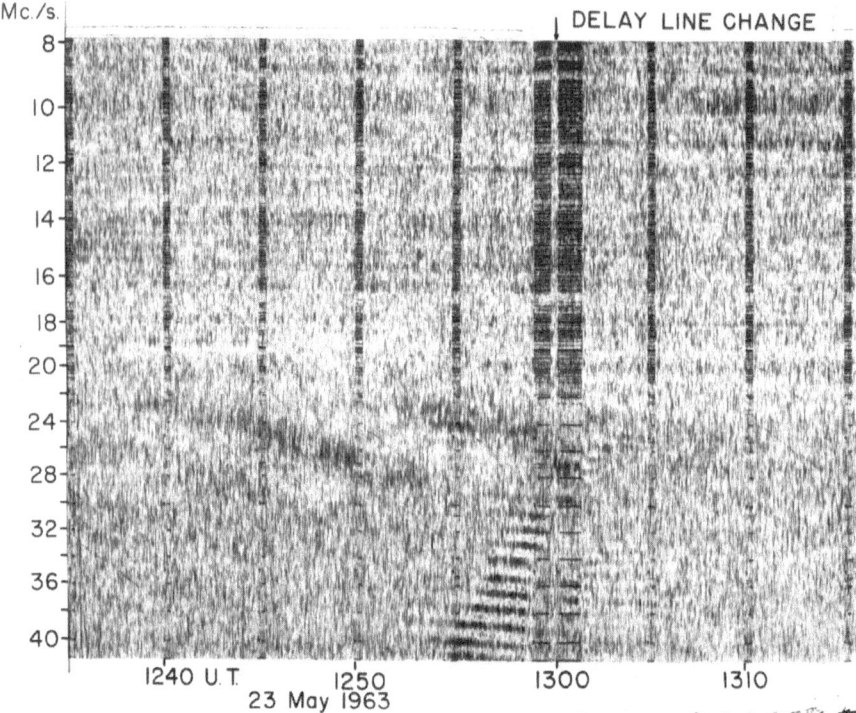

Fig. 5. Six photos — June 5, 1960, July 8, 1960, March 18, 1961, April 20, 1962, April 21, 1962, and May 23, 1963. Type II bursts unaccompanied by Type IV are more intense than the Boulder sensitivity limit. During the first two events the records extend from 15 (top) to 33 (bottom) Mc/s. Add 7 hr to M.S.T. to find U.T.

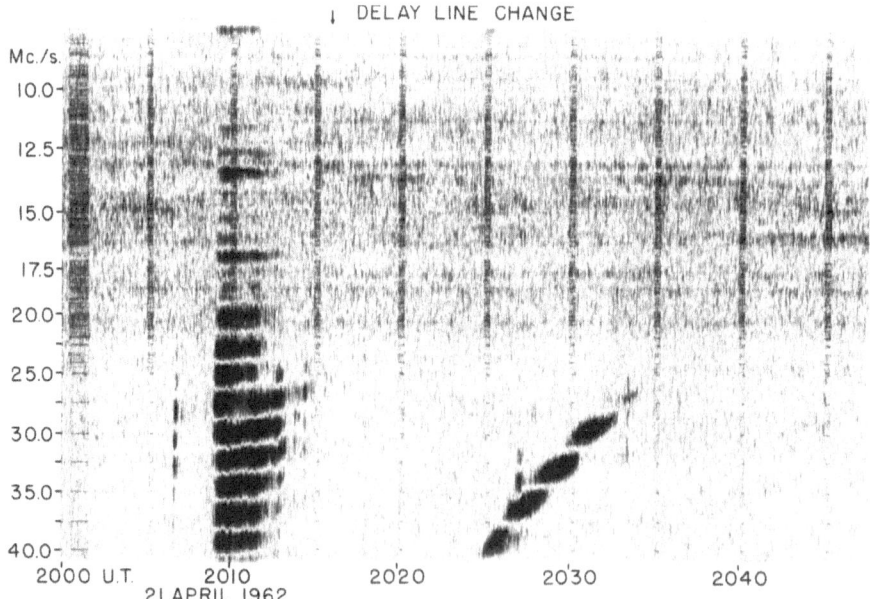

Fig. 5 continued.

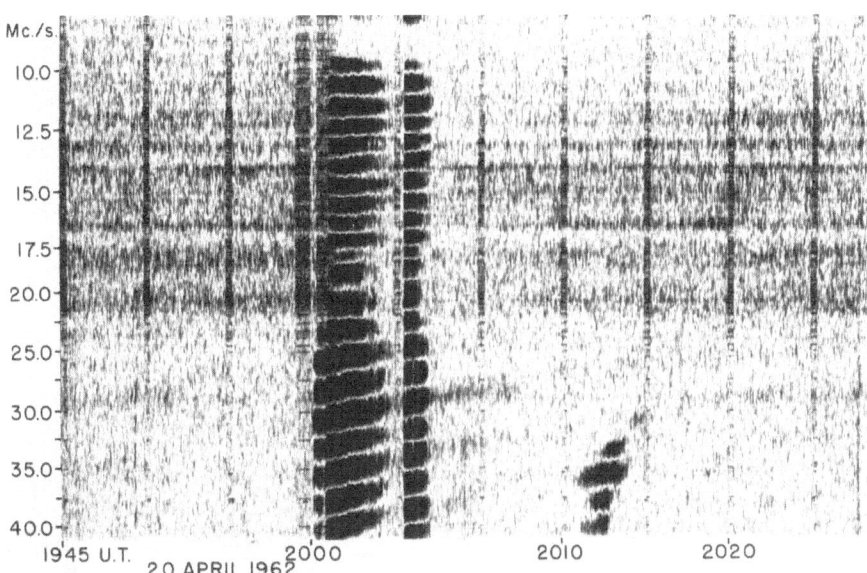

Fig. 5 continued.

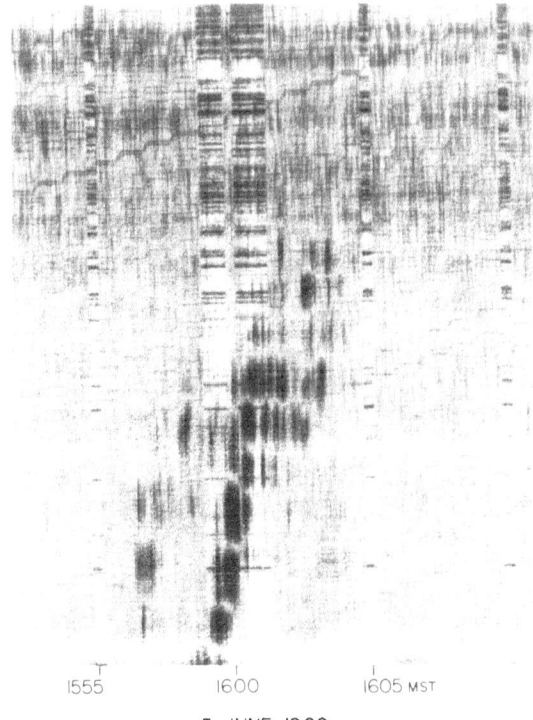

5 JUNE 1960

Fig. 5 continued.

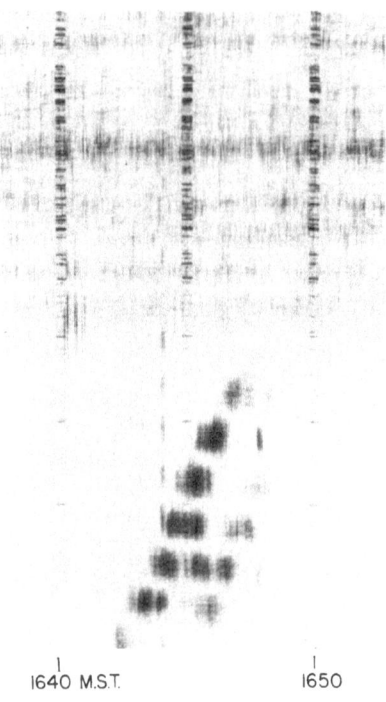

1640 M.S.T.	1650

Fig. 5 continued.

8 JULY 1960

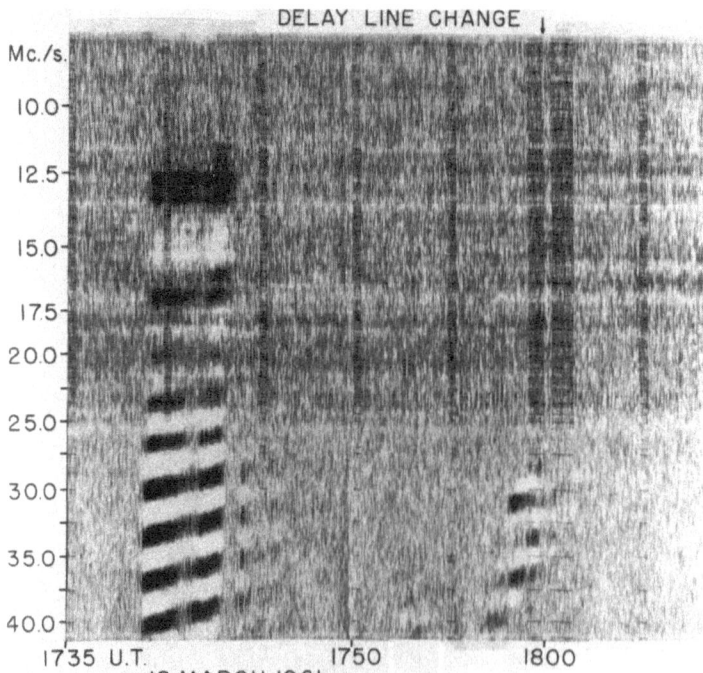

DELAY LINE CHANGE

Mc./s.

10.0—
12.5—
15.0—
17.5—
20.0—
25.0—
30.0—
35.0—
40.0—

1735 U.T.   1750   1800

18 MARCH 1961

Fig. 5 continued.

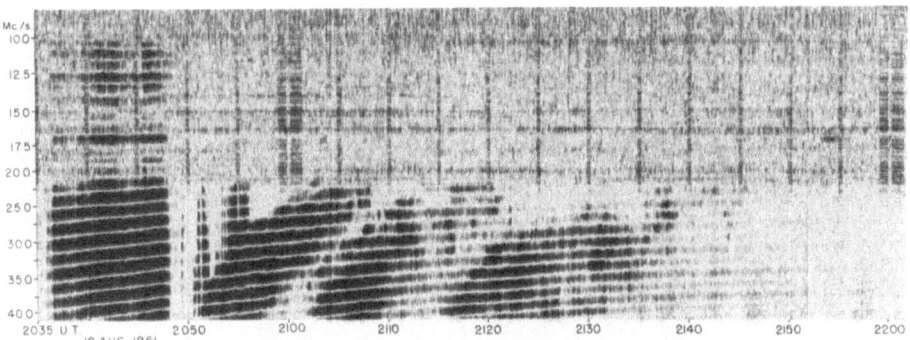

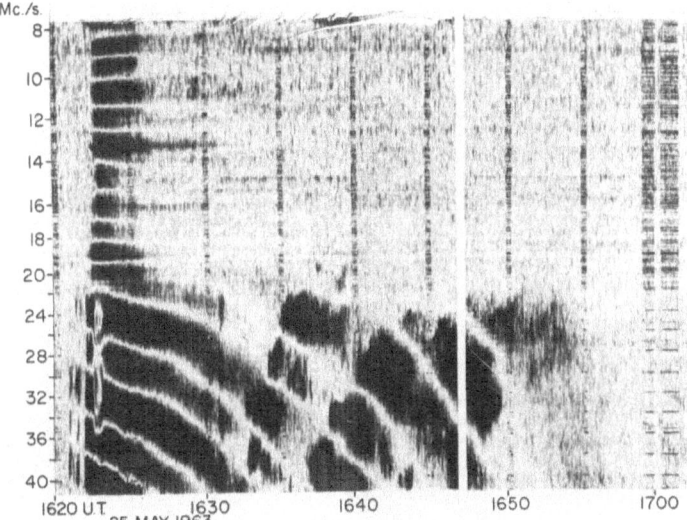

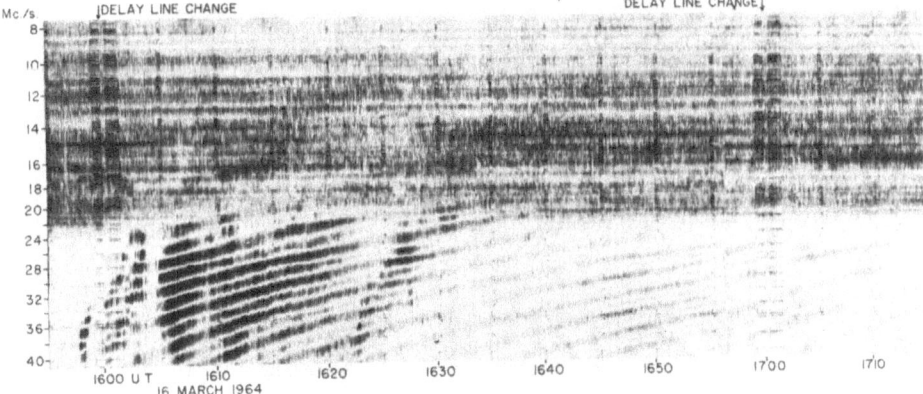

Fig. 6. Three photos — August 18, 1961, May 25, 1963, and March 16, 1964. Events consisting of multiple Type II's. Intensities, bandwidths, drift rates, and sometimes, fine structure duplicate remarkably faithfully (see also Fig. 15).

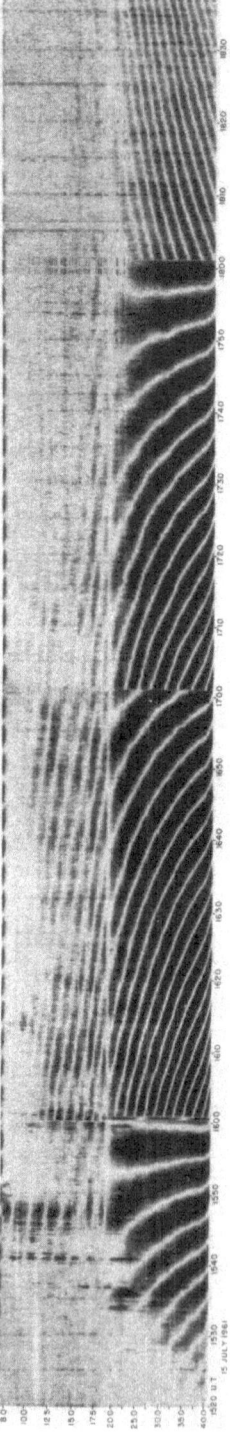

Fig. 7. Type IV emission arising without either preceding III's or II's. However, strong III's do occur at 1537 and 1540 U.T. Note the drift of the onset of the IV toward lower frequencies; also, the appearance after 1745 U.T. of Type III's superimposed on the continuum is characteristic of the formation of decametric continuum after decametric Type IV.

storage and release of the Type IV particles and fine-structure exciters together in a region of space remote from either the Type IV region or the fine structure source. Nevertheless, the frequencies emitted by the two remain synchronous; this represents a paradox, inasmuch as frequency of emission and position of the source must be intimately related. A way out of the difficulty is to assume that the fine-structure emission is observed by reflection from a lower level of the corona. This implies, since a great many (but not all) of the structures are shifted in apparent position toward the center of the sun, that the burst radiation is generated primarily toward the sun's surface. We recall that the bursts superimposed on the Type IV of April 22, 1962 (Fig. 1) were shifted in exactly the same sense.

In the July 20, 1961 case (Fig. 8), the sequence of burst types and positions is as follows: III appears from 1555 to 1602 U.T. to the west of the sun by distances ranging from $1R_\odot$ (40 Mc/s) to $4R_\odot$ (12.5 Mc/s). The III's at a given frequency vary in position by at least $1R_\odot$. II begins at about 1559 U.T. From 35 Mc/s (1600 U.T.) to 22 Mc/s (1603 U.T.), the fringe lies at ever increasing distances to the east of the center of the sun. The variation is from $0.1R_\odot$ to $0.5R_\odot$. From 1603 to 1613 U.T., the fringes at frequencies greater than 30 Mc/s indicate that the source lies about $0.3R_\odot$ to the west of the sun, the distance increasing later in the interval. At the same time, the fringes below 30 Mc/s (but above 20 Mc/s) consistently show the source to the east of the sun, in agreement with the positions indicated in the earlier phase of the Type II. After 1613 U.T., the fringes at all frequencies show the source is to the west of the sun, or near its center. A strong III or III group appears at 1608 U.T., below 20 Mc/s. It is displaced by $3R_\odot$ to the west of the sun throughout the range and shows a rapid outward (westward) displacement. This apparent motion is through a distance of about $1R_\odot$ in 2 min, corresponding to a velocity of 0.05c. This is markedly less than the speed of the III as deduced from its drift rate, about 800 kc/s per second from 20 to 10 Mc/s. The underlying Type IV starts sometime soon after 1610. At each frequency (and roughly by the same amount at each frequency), it moves westward from an initial position $0.5R_\odot$ west of the sun's center. After 1700 U.T., the IV is confined to frequencies largely below 20 Mc/s, at distances of the order of $6R_\odot$ from the sun's center.

The Type III bursts at the outset of the event are almost classic in character. They lie at ever increasing distances from the sun, the lower the frequency, although their positions vary rather markedly one from another. The II consists of a fundamental and second harmonic which are not discernible as separate traces on this record. However, from the consistent positional data, we conclude that the fundamental lies closer to the sun than the harmonic, and actually is slightly to the east of the sun's center at low enough frequencies; this displacement increases the lower the frequency. The sense of the displacement agrees with what we found for the April 22, 1962, event. The harmonic is stable in position, and does not cover a broad enough frequency range to show the character of displacement as a function of frequency. The III or III group at 1608 to 1610 U.T. moves rapidly outward in frequency at 0.05c as time goes on; the implication may be that the burst is excited by a quasi-relativistic stream

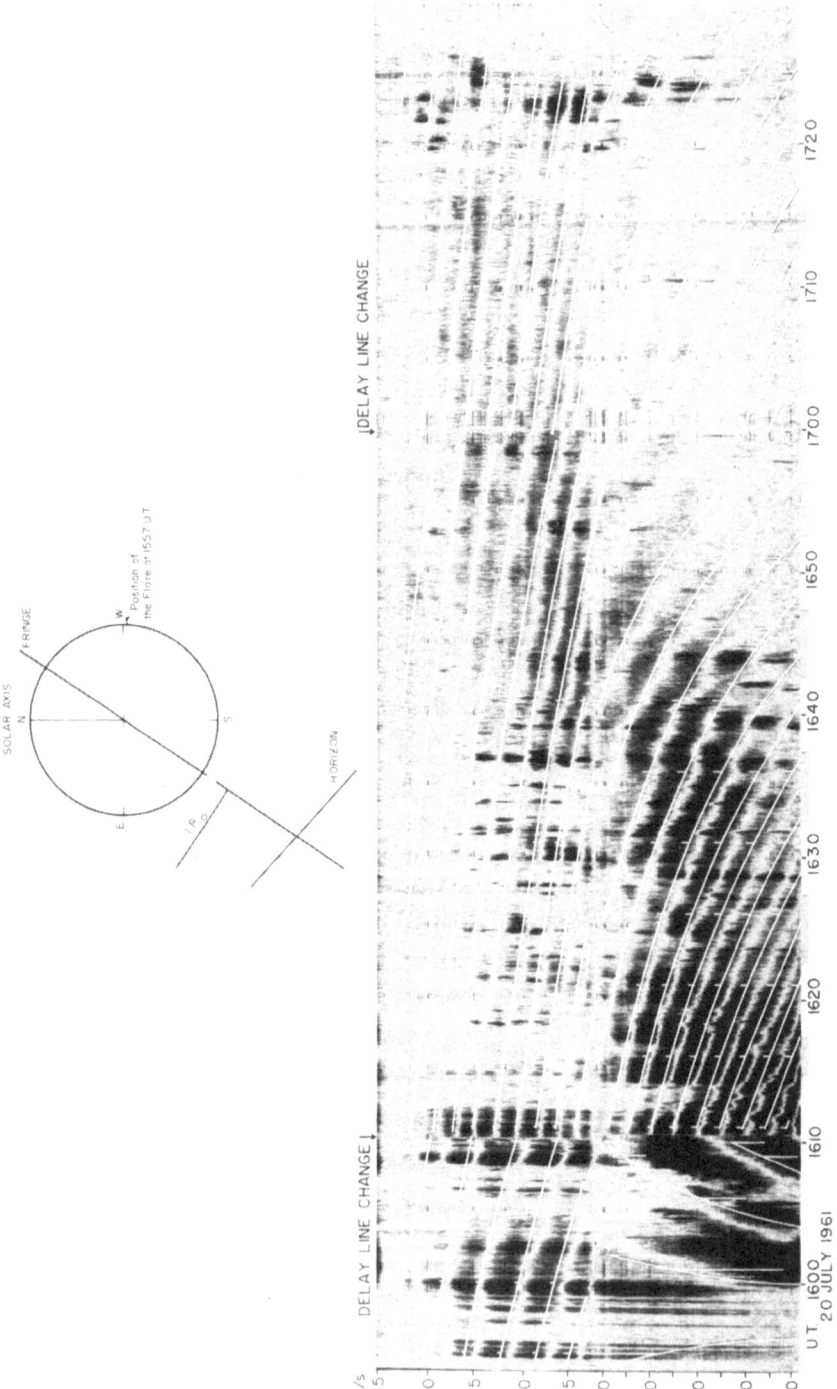

Fig. 8a. The decametric radio event of July 20, 1961. Superposed on the dynamic spectrum are computed fringes establishing the position of the center of the sun. The tabs on each fringe represent one solar radius, on the northern side of the fringe. The exact geometry at 1557 U.T. is indicated at the top of the spectrum; this did not change significantly during the period depicted.

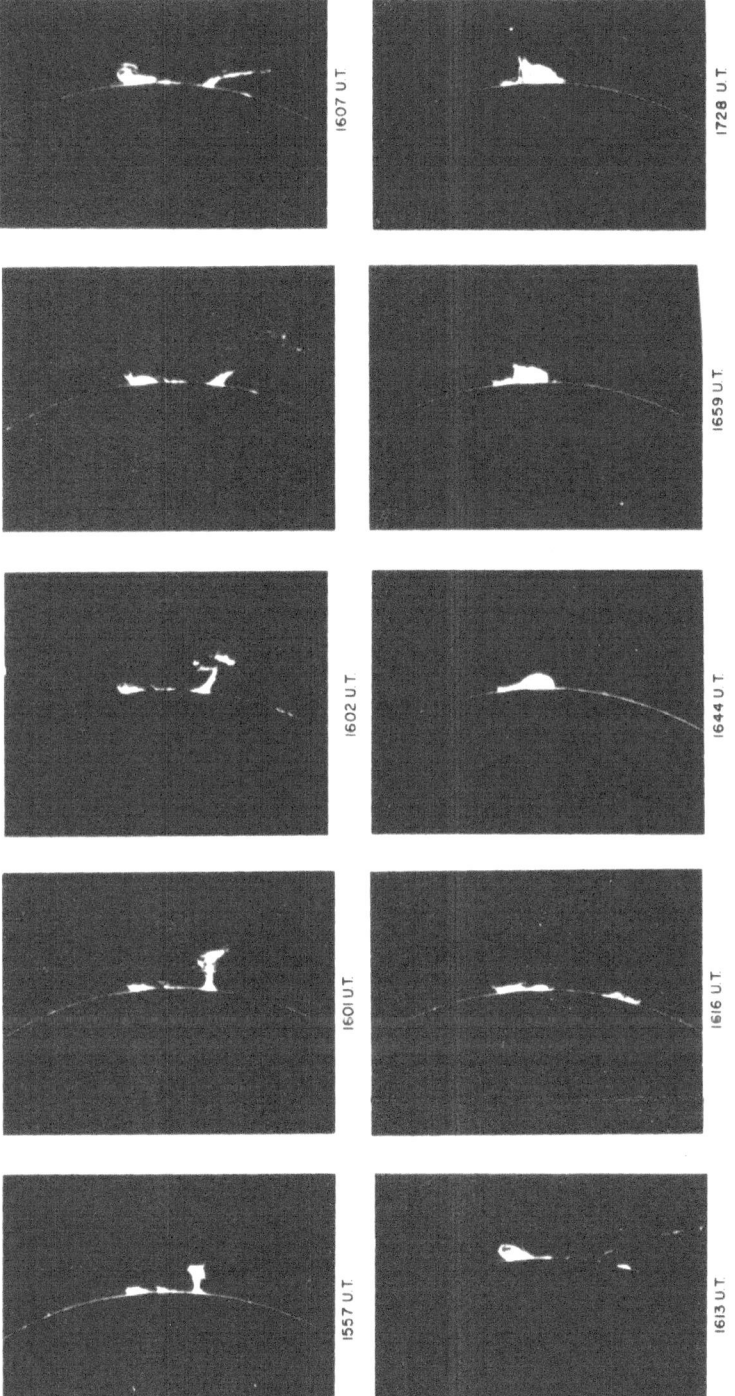

20 JULY 1961

Fig. 8b. The *Hα* optical event associated with the radio emission shown in Fig. 8a. These pictures are reproduced through the courtesy of Howard DeMastus of the Harvard College Observatory and Sacramento Peak Observatory, Sunspot, New Mexico.

passing through a plasma stream which is itself in rapid outward motion. The stream lies at about $3R_\odot$ from the sun, and is moving at $0.05c$. The Type IV is classic, consisting of a broad-band source, covering about one decametric oc-tave. All frequencies are emitted from the same region in space. The region drifts outward at an apparent velocity of about 1500 km/sec.

During September 1961, a series of major events occurred which were geo-physically important. For all of them we have detailed interferometer data, and for two, excellent optical data as well. The first major radio event occurred on September 6, 1961, after about 1800 U.T. It has been discussed by Skerjanec, Wightman, and Warwick [8] and is reproduced again here together with H_α photographs (Fig. 9). In our original report to the CRPL F-Series, the event was

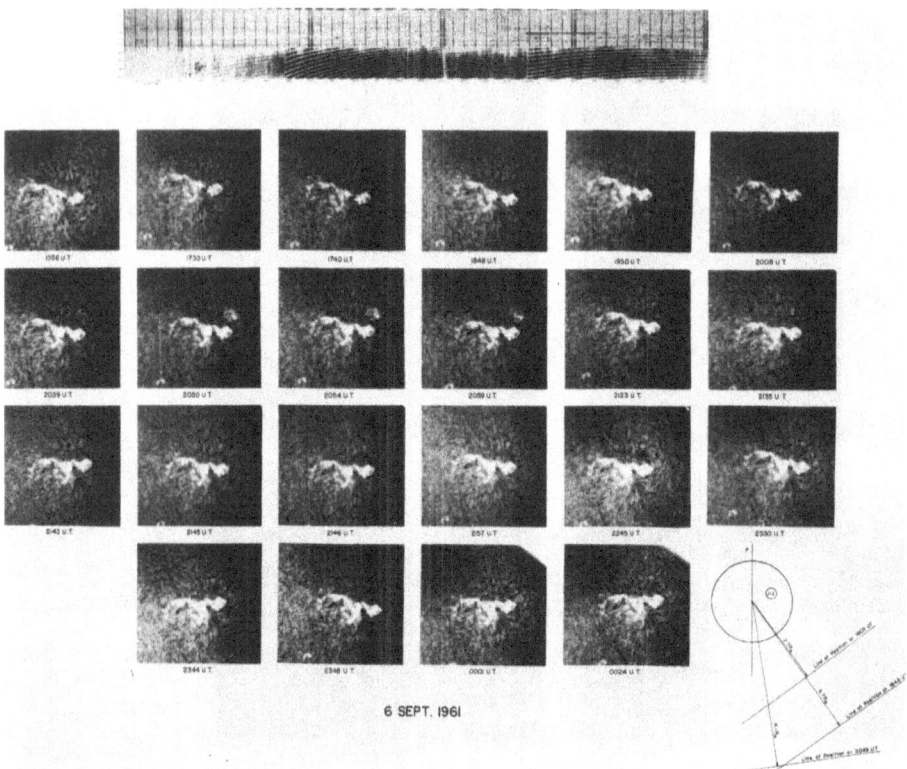

6 SEPT. 1961

Fig. 9. The decametric radio and H_α optical activity on September 6, 1961. The lines of position representing the solar decametric emission are shown in the lower right corner. Times on the decametric spectrum are marked by heavy double or lighter single lines in the top part of the spectrum; the first double line is 1800 U.T., and subsequent ones occur at 1-hr intervals. The emission lies roughly in the range 20 to 41 Mc/s. The H_α photographs exhibit almost constant activity throughout the day; in particular, a small flare is in prog-ress at 1556 U.T., a larger flare starts between 1740 U.T. and 1848 U.T. and is over by 2008 U.T. A still larger flare has started at 2039 U.T. and is accompanied by notable out-lying H_α brightening between 2050 and 2059 U.T. The end of this flare is difficult to establish.

described as "continuum," meaning broad-band, smooth emission, coupled with minor bursts, but not in association with a major Type II or III. This should not be called "noise storm," with which, however, it may often be associated. Noise storm implies the presence of Type I bursts, and they are conspicuously infrequent in decametric spectra. In the present instance, we note that Fort Davis reported a few Type I bursts at metric wavelengths. However, a re-examination of the period between 1804 and 1812 U.T. in this event discloses Type II drift tendencies, similar in appearance to some of the minor II's described in Fig. 5. The following continuum becomes extremely intense, and in Skerjanec *et al.* [8] was reclassified as Type IV on the basis that the II at 1804 U.T. had served as an initiating event. The outward shift is toward the west, and resembles in character other events reported above. On this basis, we believed the radio emission to have had its ultimate source in activity from region "J12." Unfortunately, no major (or minor) optical activity was reported in J12 or anywhere else on the sun at any time on the 6th or even early on the 7th of September 1961.

This fact assumes importance in view of the NASA observation (Bryant *et al.* [9]) of a small proton event beginning earlier than 1200 U.T. on September 7, 1961. One would wish to identify its source, but Skerjanec *et al.* [8], and Bryant *et al.* [9], are in disagreement. They would set the source in a region behind the east limb by a few days at the crucial time early on September 7. Their reasoning was that persistent proton events seem to occur at meridian transit of a region which has previously exhibited proton activity; there was a small and presumably recurrent proton event of this type on September 18, 1961. No initiating activity appears at hand other than the September 7, 1961, proton source which on September 7 must have been $18 - 7$, or 11, days before central meridian transit. Then, 4 days behind the east limb, the region would not have been visible.

We have excellent fast-time rate H_α movies of the sun on September 5, 6, and 8. Large flares occurred on September 5 and 8; particularly on September 5 there was an explosive flare (Athay and Moreton [10]) in region J12. On the 6th, the region produced small flares, dark loop-shaped filaments, and dark surges, but none of this was major activity. On the 8th, it produced a class 2 flare peaking near 1500 U.T. Figure 10a shows a major radio event, including a II–IV sequence, that occurred within 1 hr of the maximum of the J12 flare on the 8th. In preparing our data for this paper, I assumed that this II–IV belonged to the class 2 J12 flare. In fact, the time agreement was rather poor, so we carefully re-examined the optical evidence once more. Strictly on the basis of the time association, we believe that the proper assignment is to the east limb region, probably the one which Bryant *et al.* [9] identify as the proton source on September 7. Drawings of the activity there are shown in Fig. 10b. The position information on this burst acquires special significance for that reason. The theoretical central sun fringes which are also drawn in Fig. 10a indicate that the event lies about $1.0R_\odot$ to $1.5R_\odot$ to the west of the sun. The burst structure during the Type IV phase lies, as in the other events, close to the center of the sun. The notable feature of the fringes, however, is that they indicate no syste-

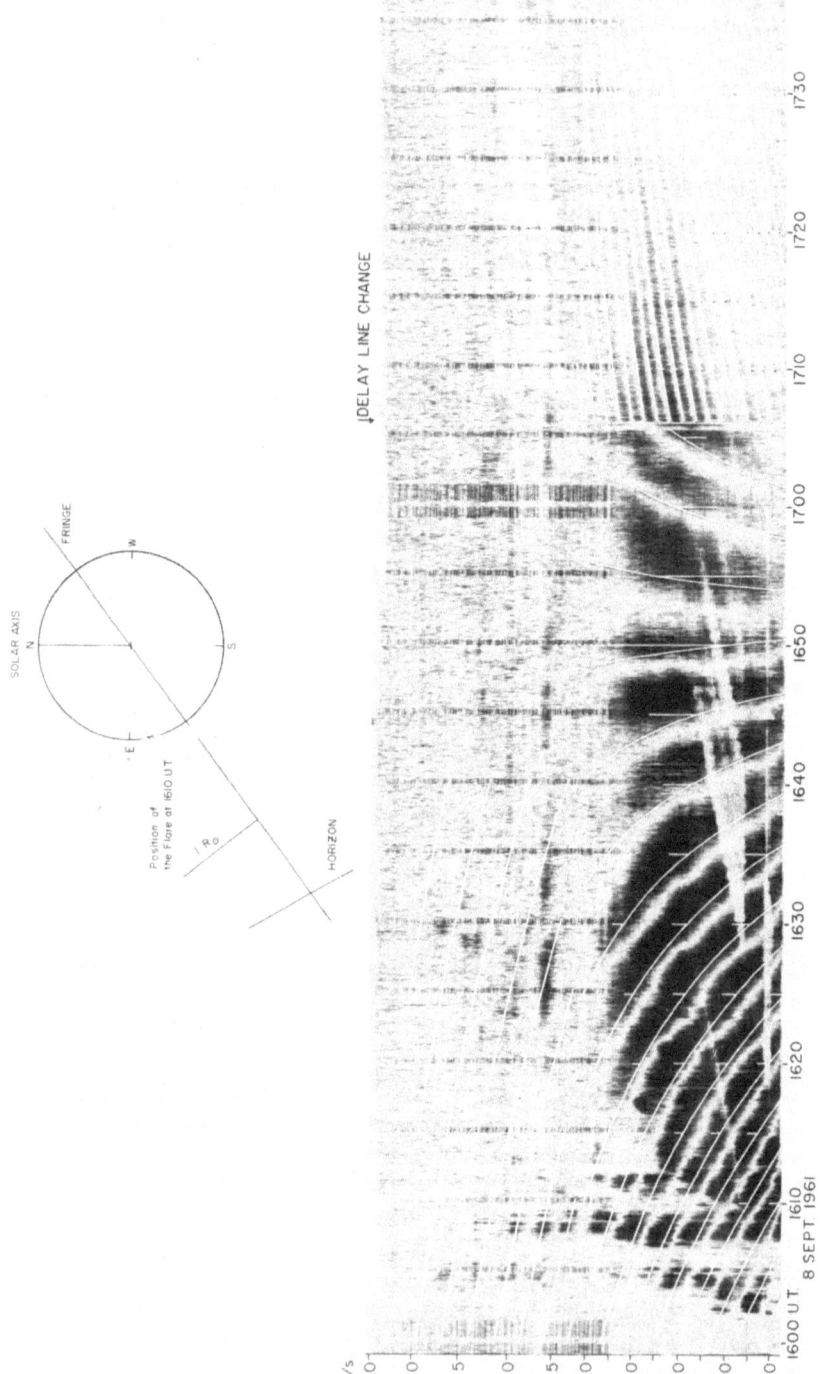

Fig. 10a. The decametric radio event of September 8, 1961. The comments of Fig. 8a apply to this figure as well.

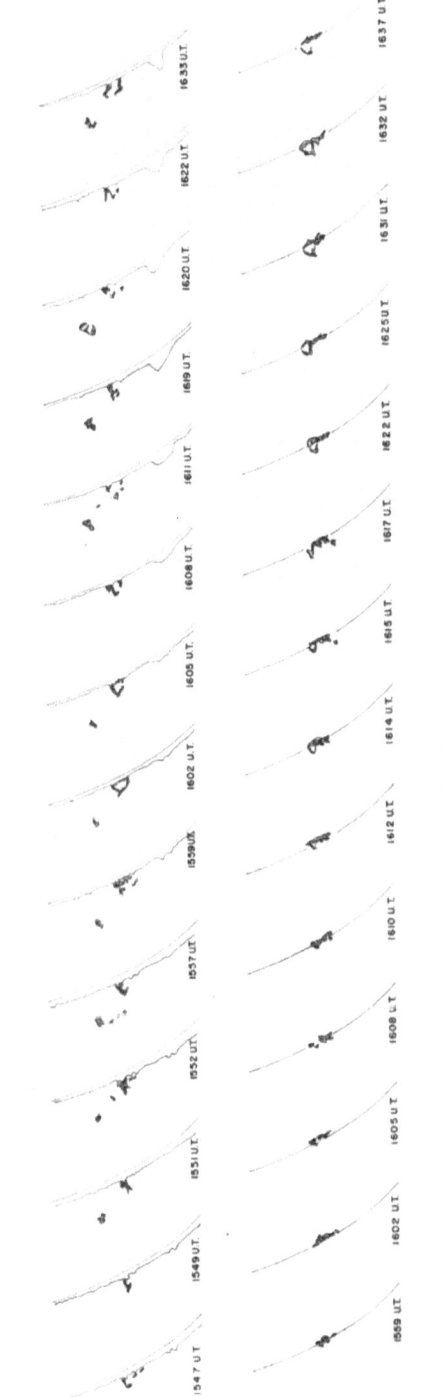

8 September 1961

Fig. 10b. The $H\alpha$ optical event associated with Fig. 10a.

matic displacements (except for the bursts on the IV) of the II or IV throughout
the entire time. Even as a function of frequency, the burst lies at a constant
position. In a qualified sense (see below), it appears that the events of September 10 and 28, 1961, were moving Type IV's, as was the event of September 6
(Weiss [11]). Regardless of the actual west limb position of the September 8,
1961, Type IV, it appears that its motionless character distinguished it from the
west limb region. In every respect save its motion, this Type IV is very similar
on our records to the moving IV's we have recorded (cf. April 22, 1962, Fig. 1).

Region J12 produced major events on September 10, 1961, when it was at the
west limb, and again on September 28, 1961, when it was just to the east of the
central meridian. Our observations at these times were essentially of the north –
south component of the source position.

Figure 11a shows the dynamic spectrum of the very intense burst of September 28, 1961, and Fig. 11b, the position of this emission as a function of frequency. The burst occurred during a time when decametric continuum was in
progress. The dynamic spectrum indicates very intense low-frequency III bursts
from 2214 through about 2233 U.T. An exceedingly intense II with a very steep
drift rate appears at 2233 U.T., at 41 Mc/s. This drift is at about 1 octave per 5
min. In addition, there is underlying strong continuum which reaches 15 Mc/s.
The Type II is hard to pick out owing to the intensity of the event, and would be
undecipherable except for the interferometer fringes. Note the irregular character of the fringes between 2233 and 2240 U.T., at 40 Mc/s. These irregularities drift to lower frequencies at the previously mentioned rate, 4 Mc/s per min.

Figure 11b shows the pattern of positions as a function of time and frequency down to 17.5 Mc/s in the event of September 28, 1961. Apparently, the
emission lay to the north or west of the sun. Since the flare in time coincidence
with our radio event was near the central meridian, we believe the emission actually lay north of the optical sun. Also, if we take the position due west of the
flare, the radio emission would then lie at an extreme distance from the sun.
The pattern of position shifts is obviously similar at all frequencies over the
duration of the event; the observations are widely spread in north – south position at the beginning of the event, but nearly converge by the end of the event at
2400 U.T. We conclude that the overall trend of these observations represents
systematic ionospheric refraction, but that the individual fluctuations on the
curves for the different frequencies are of solar origin. In particular, the motion
to the north and return, through a range of $1.2R_\odot$, between 2214 and 2230 U.T.,
appears to represent the motion of the very intense III's at the outset of the
event. The amplitude of the motion rapidly increases below 20 Mc/s to a range
of about $4R_\odot$. From 2230 U.T. until 2233 U.T., the emission is stable in position. At the latter time, a disturbance moves to the north through about $0.8R_\odot$,
at 40 Mc/s, increasing to about $2R_\odot$ at 25 Mc/s; the disturbance arrives at the
lower frequency at about 2240 U.T. The next decisive shift occurs between
2250 and 2300 U.T., when the emission shifts south by about $2R_\odot$. At 2308
U.T., a superimposed III occurs about $4R_\odot$ north of the centroid of the Type IV
emission. Ionospheric scintillation effects set in after 2320 U.T., and make the
duration of the IV difficult to determine. However, another III occurs at 2343

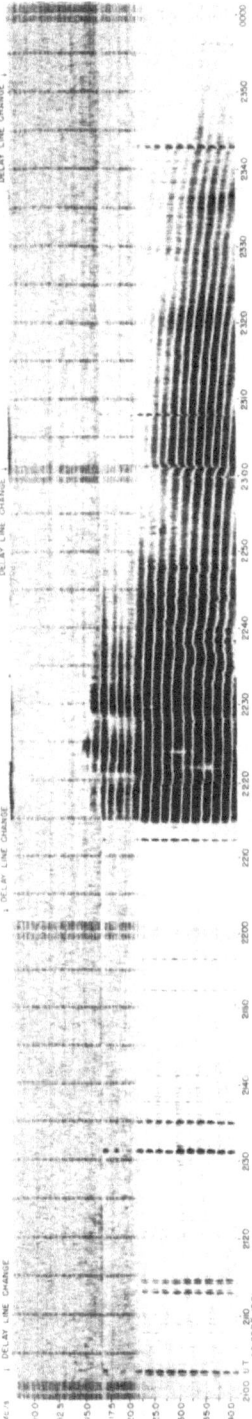

Fig. 11a. The decametric radio event of September 28, 1961 (see text).

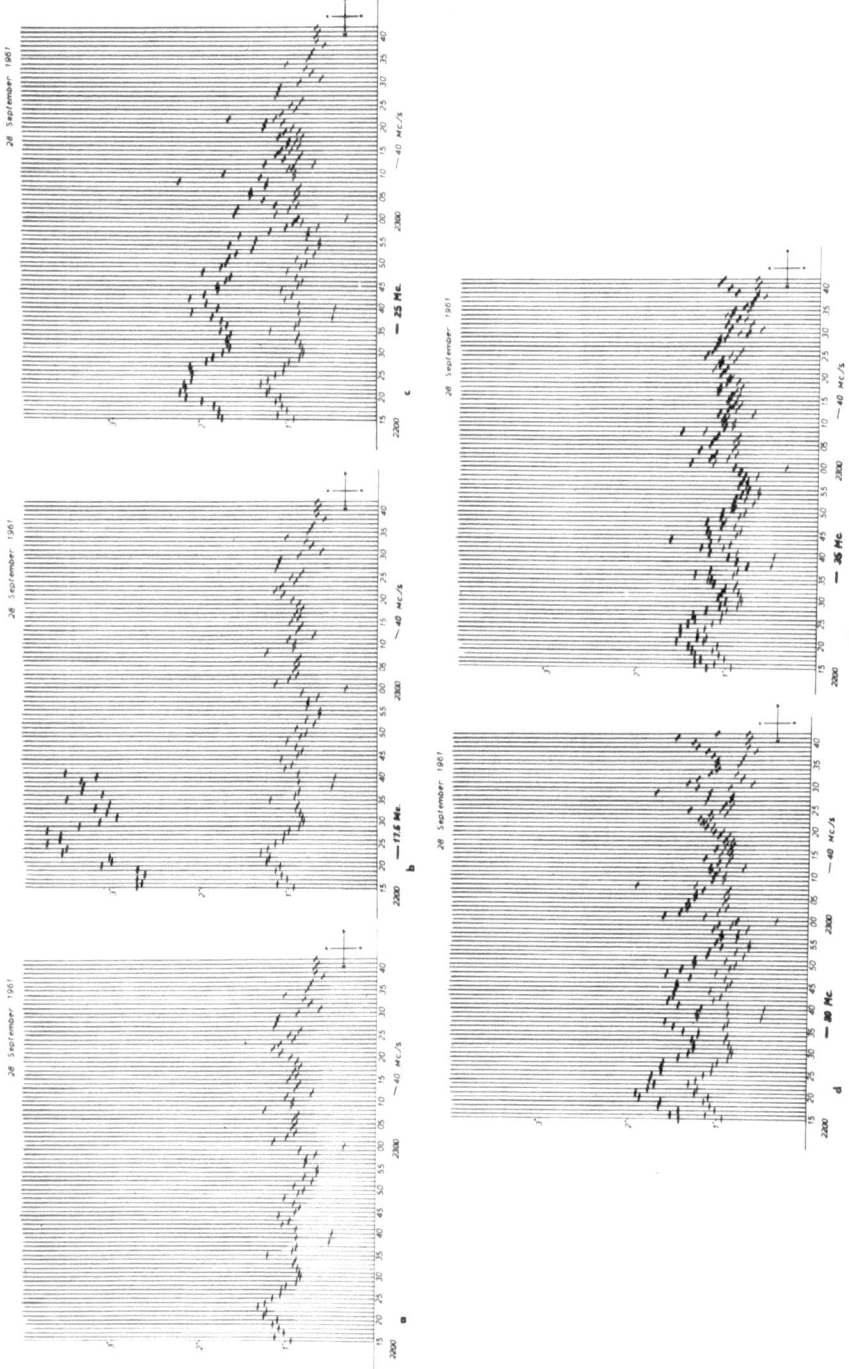

Fig. 11b. Positions of the emission shown in Fig. 11a. The figures show (a) 40 Mc/s, (b) 17.5 (heavy) and 40 (light), (c) 25 (heavy) and 40 (light), (d) 30 (heavy) and 40 (light), and (e) 35 (heavy) and 40 (light). The horizontal axis marks the sun's center; the marks on the ordinate scale are at 1, 2, and 3° north of the sun's center.

U.T., and is shifted to the north by an amount comparable to the III at 2308 U.T.

The shifts in position in the III's and in the intense emission at the outset of the event, as well as in the II, cannot be the result of ionospheric refraction. These shifts are surprisingly large, and reaffirm the importance of two-dimensional interferometry of solar radio emission.

The radio emission of September 10, 1961 (Fig. 12) was similar in motion and dynamic spectrum to the event of September 28, 1961. Here, separation of ionospheric refraction was more difficult, but suggests motions similar in sign and magnitude during corresponding phases of the events. Note in particular the strong shifts in position before 1955 U.T., as compared with later than 1958 U.T. (at 40 Mc/s).

The records of September 10 and 28, 1961, are similar in terms of the intensity and continuity of the events throughout the early phases, when III's and II's virtually disappear in the very intense continuum. In 1961, we recorded one more event with these properties, on November 10. Figure 13 shows representative H_α optical and decametric radio phenomena recorded at the High Altitude Observatory during the event. The west limb is notable for the simultaneous occurrence of continuum and bursts, that is, Type IV and II. From a position about 3 or $4R_\odot$ to the west of the sun, at 1440 U.T., the emission appeared to move to the center of the sun by 1448 or 1452 U.T., depending on the frequency. This early shifting of the emission may be a product of ionospheric refraction effects, since both its amplitude and frequency vary strangely with time. After 1452 U.T., the burst appears to develop slow-drift appearance, and also a strong western motion. The initial displacement of the II is about $4R_\odot$ to the east of the sun at 1452 U.T. and 25 Mc/s. By 1500 U.T., the Type IV emission is moving rapidly westward at all frequencies, at a rate of about 3600 km/sec. This motion is interrupted by narrow-band bursts at 1504 and 1523 U.T., and by a broad-band kink in the fringes at 1517 U.T. The drift rate quoted above is an average for the interval between 1500 and 1525 U.T. The kink may or may not be solar in origin.

PHENOMENOLOGY OF COMBINED OPTICAL AND RADIO OBSERVATIONS

It may be of further use to present some discussion of particular events. For example, Fig. 14 shows combined optical and radio data for the large event of August 11, 1963. In its initial phase, a very rapid, ascending prominence occurs. Its transverse speed is about 400 km/sec. In our decametric spectra, the Type II first appears at 1905 U.T., 40 Mc/s. The tip of the H_α emission at that time lies at $0.4R_\odot$ above the west limb. This is high enough to be the 40 Mc/s plasma level according to the Baumbach–Allen model, and suggests the possibility that we are observing optically the region of Type II fundamental emission on this occasion. A well-known difficulty arises as to which model of the corona to use to determine the height of the Type II from its frequency (Maxwell and Thompson [12]). In the present case, observationally we agree with the Baumbach–Allen densities; the H_α emission is much too low to attain the appropriate heights of the Newkirk streamer model, several-fold higher in den-

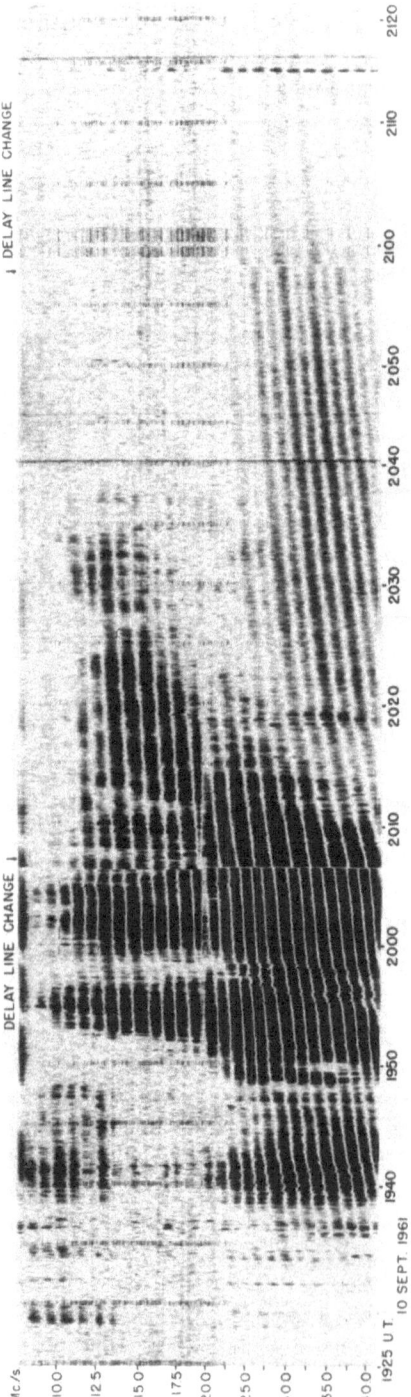

Fig. 12. The decametric radio emission of September 10, 1961.

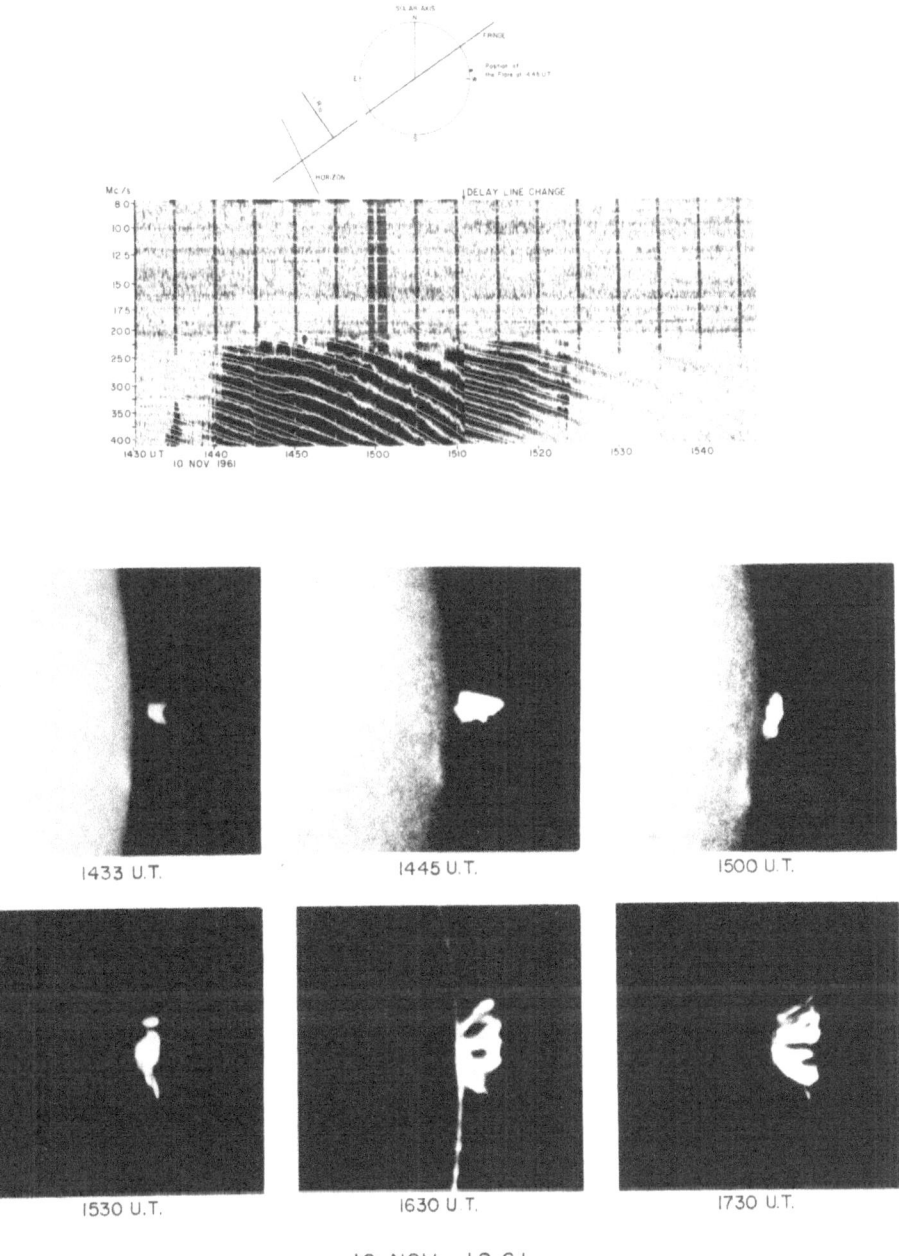

Fig. 13. The decametric radio emission and H_α activity of November 10, 1961. The remarks of Fig. 8a concerning the computed fringes apply here also. Note that the Type IV is over before the maximum phase of the post-flare loops.

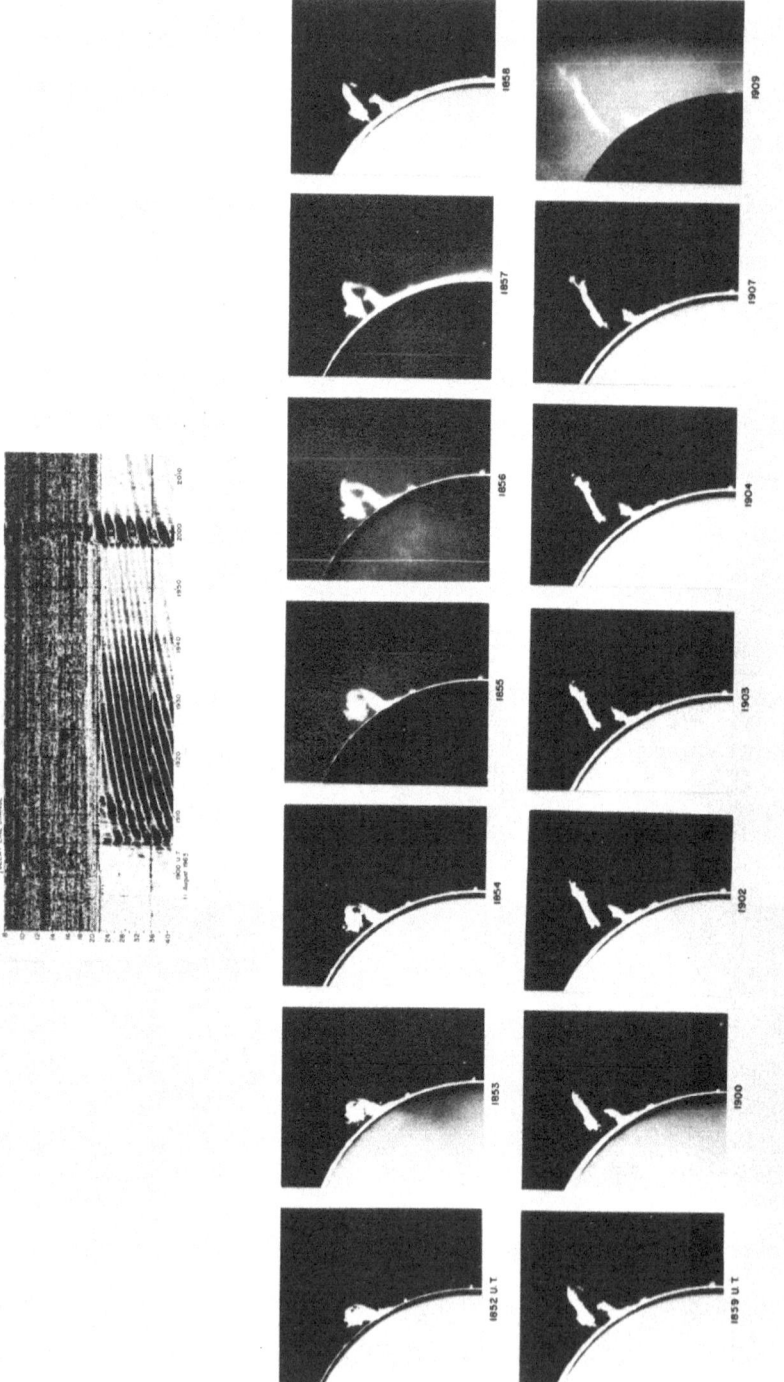

11 August 1963

Fig. 14. The decametric radio emission and $H\alpha$ activity of August 11, 1963.

sity. Type II traverses a height range of 280,000 km, according to the Baumbach – Allen formula, in about 300 sec, a speed of 930 km/sec. The H_α prominence speed is less than half this value. The interferometer fringes in this
event are not as useful as they might be, since they are inclined at about 55° to
the solar axis of rotation, rather than lying parallel to it. The position angle of
the prominence was 19° north of the west point on the limb. The slow-drift motion in the fring system appears to be from $0.3R_\odot$ south to $1R_\odot$ north. If we assume radial motion, the II moves from the east side of the sun, at $0.5R_\odot$, to the
west, at $1.7R_\odot$.

The apparent H_α speed is much less than the Baumbach – Allen speed, which
in turn has been believed to be much less than the true disturbance speed in the
corona. One can consider average heights on different frequencies of various
radio sources observed interferometrically. These tend to be much higher than
the plasma heights. Also, the electron density above active centers may be enhanced by a considerable factor over values for the quiet corona.

The electron corona observed (Heynekamp [13]) by the K-coronameter before
the event of November 10, 1961, was not enhanced over the region. The final
interferometric height, at 20 Mc/s, of the slow-drift burst of August 11, 1963, is
less by a small amount than the height of the 20 Mc/s plasma level given by the
Baumbach – Allen model, when no account is taken of coronal refraction. It may
be just as reasonable, in an event accompanied by a major prominence like this
one, to derive the distance scale from the optical phenomenon occurring simultaneously in very nearly the same part of the sun's atmosphere, as to average
values over many other events of widely different character. If we use the prominence to set the distance in the present case, we must extrapolate the prominence speed, from $0.4R_\odot$ outward at 400 km/sec for 300 sec; the 20 Mc/s plasma
height is then found to be at $1.6R_\odot$, compared to the directly measured height
(assuming radial motion of the II) of $1.7R_\odot$. The hypothesis of association between the ascending prominence and the slow-drift burst appears to be confirmed in this case. It should be noted that the base-line range of frequencies
over which our measures hold is an octave lower than the lowest of the other results reported by Maxwell and Thompson.

Figure 15 shows the correlation between Type II bursts and an ascending
prominence on October 13, 1960. On this occasion, the ascending prominence
rose at about 580 km/sec. From its launch at 1901 U.T. to the first appearance
of Type II at 40 Mc/s, requires 22 min elapsed time, during which the prominence motion would carry it through 765,000 km (assuming we are observing the
fundamental). While almost twice as high as the Type II level for August 11,
1963, the prominence heights and the Type II frequencies of Fig. 15 are consistent with the ten-fold higher densities concluded by Maxwell and Thompson.
It is probably significant that the height measures to which they refer were
largely made at sunspot maximum, while the August 11, 1963, event occurred
close to minimum.

The July 20, 1961, event (Fig. 8) shows still another Type II — ascending
prominence association. In this case, the fastest visible material has reached
about 150,000 km above the limb at 1605 U.T. This is far behind the Type II

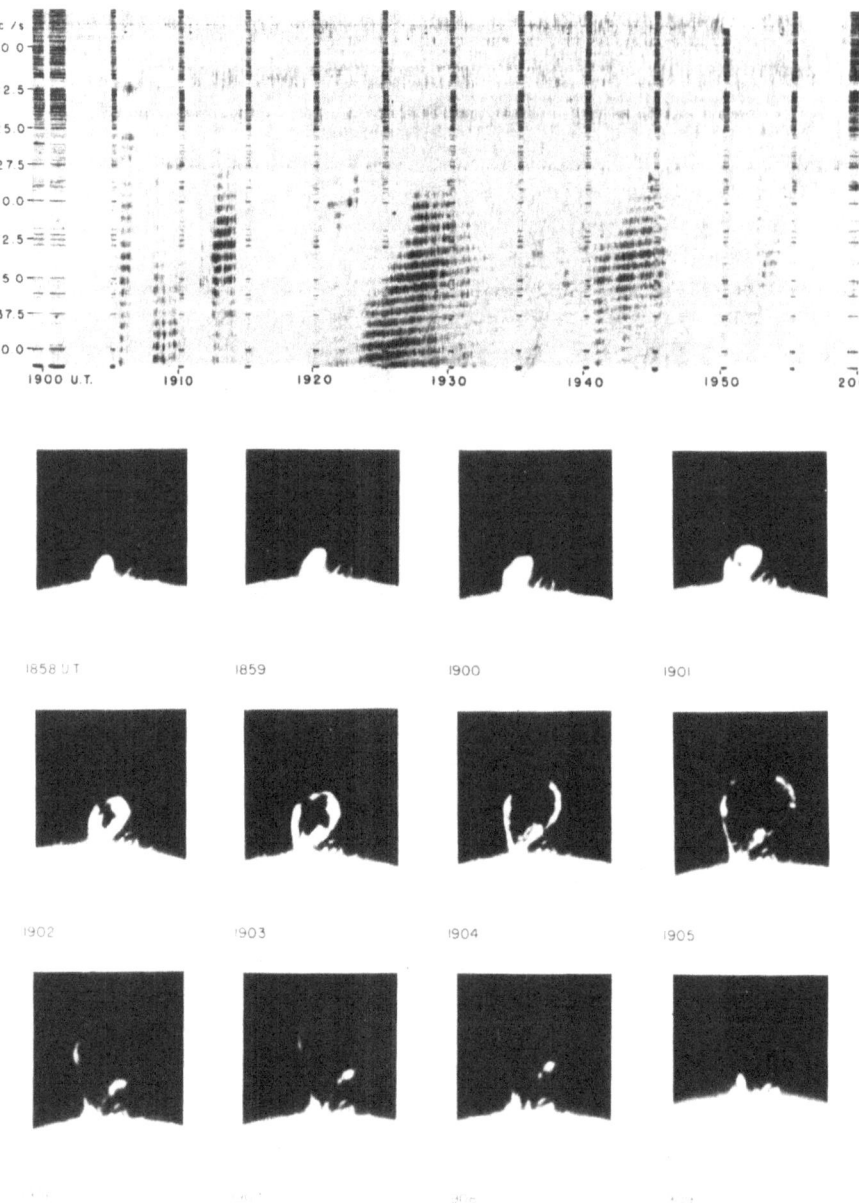

13 October 1960

Fig. 15. The decametric radio emission and H_α activity of October 13, 1960.

burst, already at the 25 Mc/s level by that time. This spray prominence moves at no more than 230 km/sec. It might conceivably be the source of the bursty Type IV we observed drifting to lower frequencies, which appears first soon after 1615 U.T., at 40 Mc/s. The spray prominence at that time should, by extrapolation at constant velocity, have reached about $3R_\odot$ to the west of the sun. It could have been high enough to pass through the region of subsequent Type IV emission.

The October 23, 1962 event (Fig. 16) is important as the source of a proton event observed by Mariner II in interplanetary space. The event is a good one from the point of view of position finding; for example, our fringe geometry is favorable at that time of day for east – west shifts, and also, the equipment was briefly changed in its mode of operation during the event so that negative (white) and positive (black) fringes would be separately identifiable. Unfortunately, our reduction hasn't yet been completed. At the outset, the ascending prominence takes on a looplike configuration which ultimately disappears at its highest point and leaves behind two legs. The highest point rose between 1648:50 and 1649:10 U.T., by 17,900 km, corresponding to 900 km/sec speed. Its position at 1648:50 U.T. is 179,000 km above the limb. By 1655 U.T. (the beginning time of the Type II at 40 Mc/s), the emission would, at this rate, have risen to 492,000 km. Clearly, the fastest parts of this prominence may well have been the physical source of the Type II.

PHYSICAL INTERPRETATION

Origin of Type II's and Ascending Prominence

The time correspondences between fast ascending prominences and Type II bursts suggest basic physical similarity between the two phenomena. We find a broad range of heights at which a given frequency is emitted, whether it be identified as first or second harmonic, but the range is generally consistent with one or another coronal model. It may be possible in this way to construct an electron density model of lower regions of the corona, where the prominence remains visible. In case an extrapolation of the prominence motion is required, this procedure would seem a bit arbitrary.

However, rather than this, it is of greater interest to examine the optical morphology of the events to discover possible physical mechanisms involved in the Type II – prominence association. There are a few sprays (Warwick [14]), such as the July 20, 1961, event but a more typical II-associated prominence shows considerable coherent structure at first glance reminiscent of the arches and arcs of stable loop prominences that often follow major flares. These arcs often move out through the corona with persistent shapes (Fig. 15). Interpreted literally, an event such as this might seem to represent an arc-shaped line of magnetic force expanding into the corona. The kinetics of this event especially resemble a balloon being inflated.

It is important to understand that our optical data cannot present a complete description of the three-dimensional distribution of emitting material. Instead,

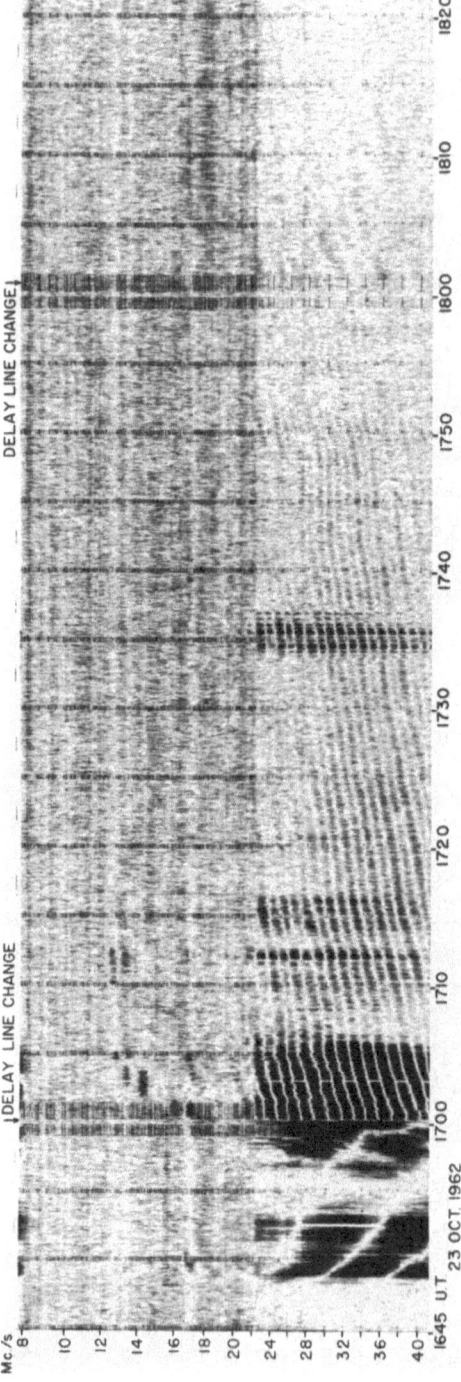

Fig. 16a. The decametric radio emission on October 23, 1962. At two points, 1653 and 1703 U.T., the negative fringes appear white. This is done to assist in the reduction of positions from the data.

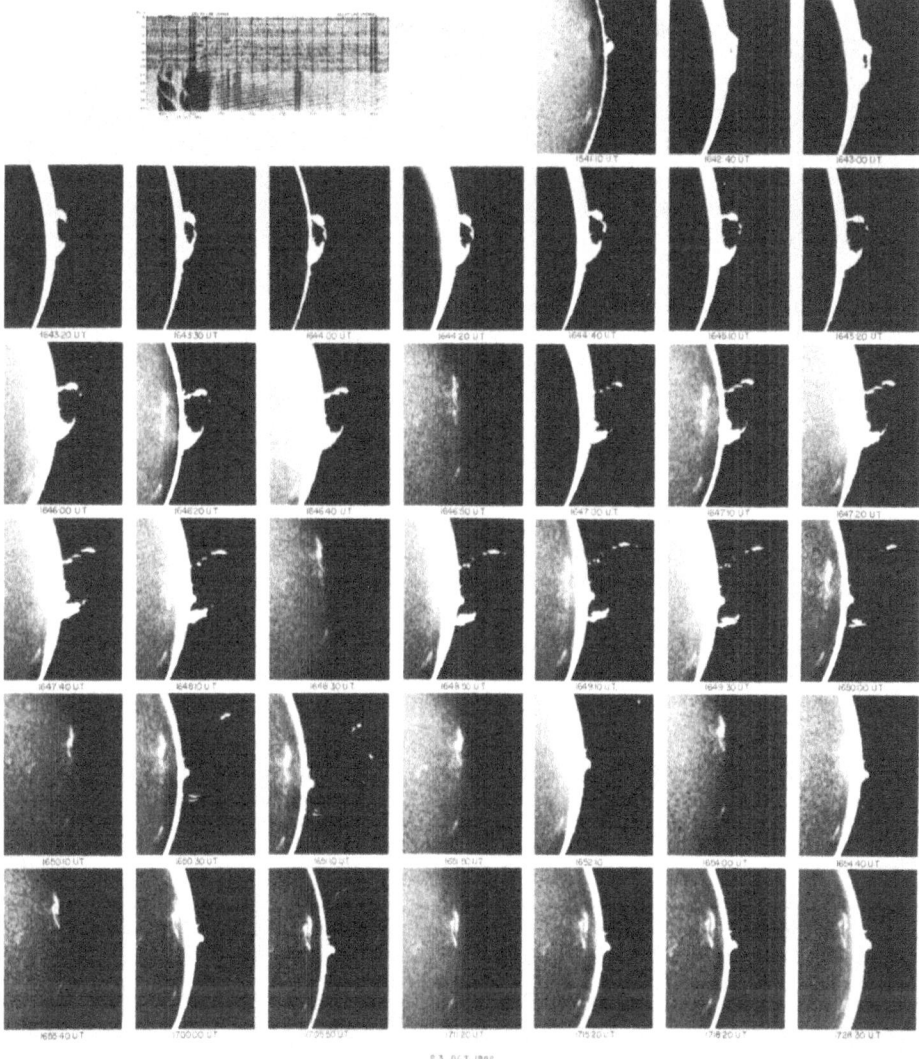

Fig. 16b. The H_α emission associated with the radio emission in Fig. 16a.

because of the limited H_α band-pass of the filter used to make our pictures, we observe only material lying near a cut through the prominence, along a plane at right angles to the line of sight.

Suppose then, that a portion of the sun's atmosphere, in the chromosphere or very low corona, is set in motion by some violent force. The gas is abruptly heated, and begins expanding into the atmosphere. At the heart of the bubble there would be a field-free region. The motion of the prominence gas would sweep away the overlying magnetic field and plasma out to the height at which the gas pressure decreased to match the environmental magnetic-field-plus-gas pressure.

Regardless of the work the bubble does against environmental pressure, it must entrain the fields initially threading it. There will be a complicated dynamical interaction in which heat is transformed to tensional energy along the lines of force. An upper limit of the efficiency follows from the assumption that the initial energy is totally available as translational energy.

After being set in motion at velocity v_0, the blob reacts inertially with the field threading it. Its dimensions parallel to v_0 will shrink and the lines of force, connected to the surface of the sun in the direction opposite to v_0, will lengthen. The magnetic flux through the sides of the blob, of area A, is AB, and is constant. The field strength B increases as A decreases. The tension along the lines of force is $AB^2/8\pi$; the blob would have to stop moving at some point where B increases to a large enough value. This does not happen, because the blob itself resists compression as a result of gas pressure, NkT. We set $NkT = B^2/8\pi$.

The blob mass is M, and the equation of motion in the incompressible case is

$$M(d^2x/dt^2) = -2AB^2/8\pi$$

where x is the displacement away from the sun. The pressure of the blob is

$$(M/\mu)(kT/lA) = B^2/8\pi$$

where l is the side length of the blob and μ is the mass of the hydrogen atom. Therefore

$$(d^2x/dt^2) = -2kT/\mu l$$

For a hydrogen number density 10^{12} cm^{-3} (chromosphere), and $T = 3 \times 10^6$ °K (corona), the acceleration is -8 km/sec^2, 30 times larger than solar gravity and far greater than observed decelerations. The distance at which the blob is stopped by tension depends on its initial speed, say 400 km/sec; then

$$d = \frac{1}{2}(v_0^2/a) = 40,000 \text{ km}$$

Since $NkT = B^2/8\pi$, we find that $B = 100$ G, a reasonable figure for this part of the sun's atmosphere near flares.

The structure of the field, after it has been dragged along by the blob, must be a flat ribbon, at its extremities anchored to the sun. Unless they initially share the impulse which started the blob moving (i.e., were in effect part of the blob), regions of the chromosphere over and under the blob, as well as to its sides, will not participate in the motion. In these regions, the lines of force must slip off the blob, twist slightly around its base, and return virtually to their initial configurations. The elasticity of the lines of force is along the arc, not across them from the blob to the adjacent motionless material. The analogy is then not to a balloon under inflation, which has two-way stretch, but rather to a thin ribbon, like a garter.

Primarily on account of the very strong tension, this model does not seem a useful one to represent the ascending prominences connected with flares. Its weakness lies in the initial abrupt energy input required to drag out the lines of force. The field strengths are so large in active centers, that at solar densities, temperatures, and velocities, the atmospheric dynamics are nearly completely under the control of the magnetic field.

A more attractive possibility regards the magnetic field as the basic parameter, which varies as a result of changes in the current systems anchored in the photosphere. If we may judge from the generally stable topology of prominences, modifications in the photospheric currents are by and large adiabatic, occurring slowly compared with characteristic times for chromosphere and corona. A rapid change in photospheric currents would entail, if there were no solar atmosphere, a readjustment of the external fields within the time taken for light to pass across the region of space involved. When plasma is present, any abrupt change in the field source propagates as a discontinuity in B at roughly the Alfvén velocity $v_A = B/\sqrt{4\pi\rho}$, where ρ is the plasma density. This is a hydromagnetic shock. I propose it as the explanation for the rapid, ascending prominences associated with Type II bursts.

From this point of view, the Type II burst itself becomes another manifestation of the shock wave. Of course, this idea is widely accepted anyway, because the inferred speeds of Type II's seem to agree generally with the predicted coronal Alfvén speeds.

The shocks can travel in any direction in the coronal plasma. To infer the probable magnetic topology in the II-associated prominences, we reason from the stability of atmospheric magnetic fields before and after flares (Warwick [15]). This fact is demonstrated by the persistence of H_α configurations in active centers as prominences come and go. The stability implies that whatever changes do occur, occur in the current strengths. We may then properly represent an atmospheric field change by a change in the dipole moment of a hypothetical current source. This change occurs in the photosphere where magnetic and gas pressures interact strongly.

The consequence of the change is that kinks in the lines of force propagate out into the corona. The locus of the kinks at an instant of time then represents the locus of H_α emission. The topology of the change is such that within a certain sphere, the lines of force have their new concentration, and outside of it, their old concentration. The lines of force in the two configurations are connected by roughly tangential magnetic field segments that intercept the old and new field lines at sharp angles. These segments rise, parallel to the sun's surface.

In this geometry, the greatest jump in direction of the lines of force occurs for the shock propagating along the direction of the dipole axis. We therefore conclude that the II-associated prominences would tend to occur in magnetic field regions where the lines of force lie perpendicular to the sun's surface.

Such a configuration is also favorable to the escape of the particle streams that excite Type III emission, and to the still higher-elevation, relativistic electrons that produce Type IV emission.

Origin of the Type IV's

The IV's remain a mystery. No direct optical evidence for them exists, although *K*-coronameter tracings have been made in some cases (e.g., November 10, 1961).

It is perhaps relevant to point out that the decametric IV's on a few occasions degenerate into the events I have called decametric continuum. An example of this sort of continuum is shown in Fig. 17. The notable features are two: there is a continuum of radio emission present at all times (down to a time resolution of ½ sec), and a massive number of Type III bursts occurs on this continuum. Continuum may persist for several days, and tends to associate with the central meridian passage of an active center. It may move through wide regions of the corona.* The low-frequency continuum is not identical to noise storms on higher frequencies, say above 50 Mc/s, but very often associates with noise storms.† We have observed the degeneration of IV's into continuum events, and on other cases, such as September 6, 1961, have observed what appear to be transition cases between Type IV and continuum. There appears to be a strong physical connection between the two.

If the decay toward lower frequencies on July 20, 1961 (Fig. 9a), represented a lowering of the magnetic field strength in the region of radio emission, we might derive a rough model of magnetic field as a function of time-height for this event. At $6R_\odot$, the emission lies at 10 Mc/s. It arrived at this height, after 30 min of drifting, from 20 Mc/s, where it lay at about $1R_\odot$. The frequency has changed by two-fold, corresponding to a halving of the magnetic field strength. This halving radically disagrees with the fall off in dipole field strength between $1R_\odot$ and $6R_\odot$. We must conclude that the magnetic field is actually transported out to these heights, even though we regard the dragging óut of chromospheric fields by mass motions as an impossibility. By transport of the field, we do not indicate motion occurring at the time of a flare. If the pre-existent magnetic field structure at high altitudes over active centers ducts degenerating Type II shocks arriving at those heights, then we expect to see the decrease in radio frequency follow the magnetic field strength in a coronal streamer or ray. The ray presumably begins at a height where $NkT \simeq B^2/8\pi$, and is from that point outward structured by the steady, outward flowing coronal gas.

---

* See Boischot, Lee, and Warwick [16]. The event of June 3, 1959, reported there as moving, did not in fact move. An error was made in the data reduction. The events of May 11, 12, and 13, 1959, were moving continuum; the motions on this occasion are confirmed by Erickson [17].

† See Wild, Smerd, and Weiss [6], Fig. 3. Their example of Type I represents decametric continuum in the 25 to 50 Mc/s range. This particular example is strikingly illustrated by many Type III's, but complete freedom from storm (Type I) bursts. The latter are, however, abundant from about 60 Mc/s upward in frequency.

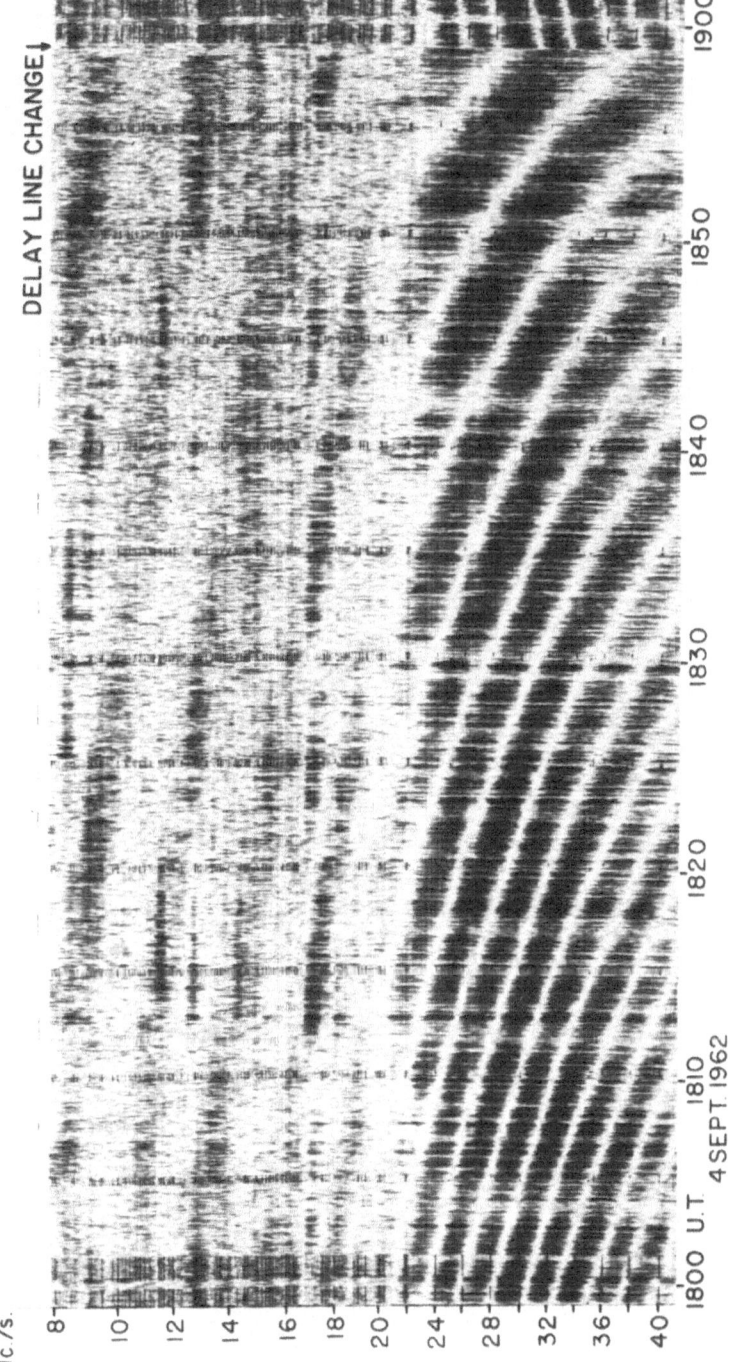

Fig. 17. Decametric continuum observed on September 4, 1962. Note the extremely high rate of occurrence of Type III bursts.

That description is consistent with a number of observed solar phenomena. These are: the recurrence tendency of subrelativistic proton streams upon passage of active centers across the disk (Bryant *et al.* [9] and Gregory and Newdick [18]) and especially decametric continuum events. The latter indicate the several days long production and escape of quasi-relativistic electrons and protons from active centers. These phenomena together imply the continuing presence of an escape route into the high solar corona and from the corona. The large, impulsive increases in escaping particles represent occasions when shocks succeed in transporting particles to the level of the corona where $NkT \simeq B^2/8\pi$.

ACKNOWLEDGMENT

This research was supported by the United States Air Force Cambridge Research Laboratories and the National Aeronautics and Space Administration.

REFERENCES

1. R. H. Lee and J. W. Warwick, *Radio Sci. J. Res. NBS/USNC - URSI* **68D**, 807 (1964).
2. C. S. Warwick and J. W. Warwick, "Paris Symposium on Radio Astronomy," ed. R. N. Bracewell (Stanford Univ. Press, Stanford, California, 1959), p. 203.
3. J. P. Wild, J. D. Murray, and W. C. Rowe, *Australian J. Phys.* **1**, 439 (1954).
4. A. Boischot, *Ann. Astrophys.* **21**, 273 (1958).
5. S. F. Smerd, J. P. Wild, and K. V. Sheridan, *Australian J. Phys.* **15**, 180 (1962).
6. J. P. Wild, S. F. Smerd, and A. A. Weiss, *Ann. Rev. Astron. Astrophys.* Volume 1 (Annual Reviews, Inc., Palo Alto, California, 1963).
7. C. S. Warwick and M. W. Haurwitz, *J. Geophys. Res.* **67**, 1317 (1962).
8. R. Skerjanec, D. A. Wightman, and J. W. Warwick, *Inform. Bull. Solar Radio Observatories* **13**, 5 (1963).
9. D. A. Bryant, T. L. Cline, U. D. Desai, and F. B. McDonald, *Phys. Rev. Letters* **11**, 144 (1963).
10. R. G. Athay and G. E. Moreton, *Ap. J.* **133**, 935 (1961).
11. A. A. Weiss, *Australian J. Phys.* **16**, 526 (1963).
12. A. Maxwell and A. R. Thompson, *Ap. J.* **135**, 138 (1962).
13. C. Heynekamp (High Altitude Observatory, Boulder, Colorado), private communication.
14. J. W. Warwick, *Ap. J.* **125**, 811 (1957).
15. J. W. Warwick, Proc. of a symposium held at the Goddard Space Flight Center, Greenbelt, Maryland, October 28–30, 1963 (ed. W. N. Hess, NASA Publication SP-50, Washington, D. C., 1964), p. 441.
16. A. Boischot, R. H. Lee, and J. W. Warwick, *Ap. J.* **131**, 61 (1960).
17. W. C. Erickson, *Phys. Rev. Letters* **3**, 365 (1959).
18. J. B. Gregory and R. E. Newdick, *J. Geophys. Res.* **69**, 2383 (1964).

Chapter 9

Noise Storms

A. D. Fokker

University of Utrecht
Utrecht, The Netherlands

A noise storm is the most common solar radio phenomenon at metric frequencies. Noise storms occur associated with most of the major sunspot groups. Generally, a noise storm manifests itself on a broad band of metric frequencies. The lowest frequencies at which noise storms occur are near 50 Mc/s, the highest frequencies up to which storm activity commonly extends lie between 300 and 400 Mc/s. A noise storm typically consists of a background continuum component of enhanced radiation and a component consisting of many short-lived narrow-band bursts. In this paper, we give a survey of the various characteristics of noise storms.

GENERAL DESCRIPTION OF THE STORM PHENOMENON

The relative importance of the enhanced background continuum and the burst activity varies widely from one noise storm to the other. A variability index, ranging from 0 to 3, is used to characterize the amount of burst activity. Storms of variability 1 commonly present spread clusters of storm bursts or a more or less continuous succession of small bursts; storms of variability 3 present a continuous succession of strong bursts that overlap each other to a large extent (see Fig. 9). During such violent storms, the scale of the bursts is continually changing.

For moderate storms, the background continuum intensity is a few times the intensity of the quiet sun ($\sim 8 \times 10^{-22}$ W/m^2-c/s at 200 Mc/s). During severe storms, the continuum background amounts to several times the quiet sun level and it may be more than ten times as high. In storms of great variability, many bursts have peak intensities that greatly surpass the background continuum intensity.

A very typical characteristic of all noise storms is the tendency of storm bursts to occur in clusters. The clustering tendency of spread bursts was investigated by de Jager and van't Veer [1], who succeeded in finding a statistical law that describes the way in which these bursts are distributed in time.

The Occurrence of Noise Storms

Noise storms occur only when one or more sunspots are present on the sun's disk. In general these are the major sunspots of Brunner types E, F, or G. If only major spots within five days from central meridian passage are considered, it is seen that the probability for having strongly enhanced metric radio emission (200 Mc/s) is roughly proportional to their number:

	Percentage of days with strong noise activity
On days with one major sunspot.....................	10
On days with two major sunspots....................	22
On days with three or more sunspots (average 3.4)....	34

The percentage of association of noise activity with sunspots increases with the area of the spot and its magnetic field strength. Strong noise activity is produced by 15% of the E-type spots and by 26% of the F-type spots. There is no correlation between noise activity and flare activity of sunspots (Fokker [2]); centers of activity which produce many flares may exhibit only relatively little noise activity and vice versa.

The background intensity level can be defined as the lower limit to which the intensity occasionally drops, intermittently between portions of burst activity. During a storm, this level varies only slightly (by some 10%, say) in the course of minutes. Over periods of the order of 1 hr, the background intensity may change by a factor of 2, or sometimes more. The enhanced radiation generally covers a large frequency range, of the order of 100 Mc/s or more. No steep variations of intensity with frequency occur.

Bandwidth and Duration of Storm Bursts

Early radio-spectrographic observations of the burst component (Wild [3]) revealed that storm bursts have a bandwidth of the order of a few Mc/s. It is normal for a storm burst not to show a frequency drift. The dynamical spectrum of a typical storm burst is designated as Type I. The spectrum of storm bursts has been studied in detail by Elgaröy [4] and by de Groot [5]. Elgaröy used the radio-spectrographic technique to obtain frequency sweeps over the range 190 to 215 Mc/s. The output was displayed by intensity modulation on an oscillograph screen and photographed on moving film. Intensity contours were derived by photometering the film. Single frequency records at seven nearby frequencies, 2 Mc/s apart, were obtained by de Groot, who was able to record bursts of very small intensity with a time resolution of 0.015 sec. Series of observations were made first at frequencies near 400 Mc/s, later at frequencies near 270 Mc/s. Values of bandwidth and duration of storm bursts have been specified as follows:

	Elgaröy	de Groot	
	200 Mc/s	270 Mc/s	400 Mc/s
Bandwidth	3 to 5 Mc/s	5 to 8 Mc/s	6 to 9 Mc/s
Duration	0.3 to 0.7 sec	0.1 to 0.3 sec	0.1 to 0.3 sec

There is also a class of storm bursts with an exceptionally short duration, which de Groot [6] calls "spike" bursts. Such bursts were also reported by Elgaröy [7]. These bursts have a lifetime of the order of 0.1 sec. They have about the same bandwidth as normal storm bursts and occur mostly during periods of considerable noise activity, sometimes in association with Type IV solar radio events. An example of a multifrequency record of Type I bursts is given in Fig. 1.

The Relation Between Storm Bursts and the Background Continuum

An old suggestion is that the background continuum may be due to a mere superposition of a great many small storm bursts (Wild [3], Vitkevitch [8], Takakura [9], and Kai [29]). The following facts seem to lend support to this suggestion: (a) the background continuum and the storm bursts commonly present the same sense of circular polarization; and (b) the source regions of storm bursts and continuum radiation generally coincide (see section entitled "The Structure of the Sources").

This question was dealt with by Takakura [9], who introduced an amplitude distribution of elementary pips by which he could account for both the "steady" level and for the occurrence of separately visible storm bursts. From an analysis of slow-speed records, Fokker [2] obtained some evidence which was also slightly suggestive of the superposition hypothesis. However, the feasibility of a statistical model does not yet prove the validity of the hypothesis.

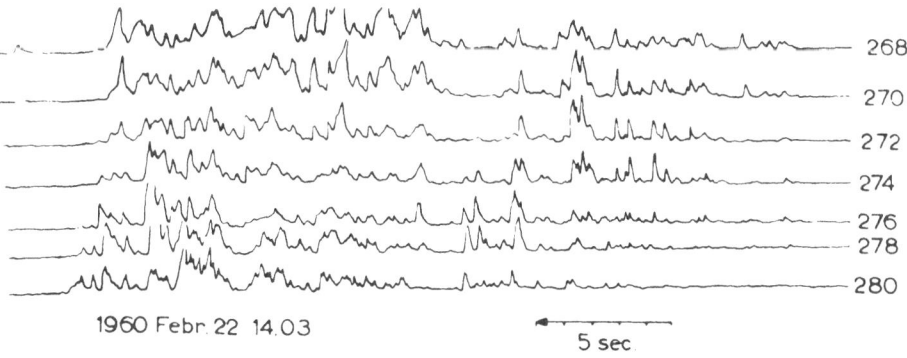

1960 Febr. 22 14.03   5 sec

Fig. 1. A group of Type I bursts as recorded at adjacent frequencies (courtesy T. de Groot).

As a matter of fact, the burst component itself is often a superposition of many pips that overlap each other to a large extent, but this does not imply that the whole of the continuum is due to overlapping elementary bursts. The author is inclined intuitively to think that this is not the case. No doubt there is an intimate connection between the enhanced continuum radiation and the storm bursts, but we cannot yet say what the true nature of this connection is.

Overlapping of Bursts

From quick-run records of violent noise storms, it is evident that there is a very considerable overlap of bursts. Several individually nondistinguishable bursts obviously contribute to a given intensity peak on the record. The distribution of amplitudes of the recorded peaks was studied by Fokker [2]. This distribution fits closely to a distribution $P(A)$ of the form

$$P(A) = \frac{A}{a^2} e^{-A/a}$$

where A represents the burst amplitude (see Fig. 2). This law apparently applies to each curve that results from the superposition of a sufficiently large

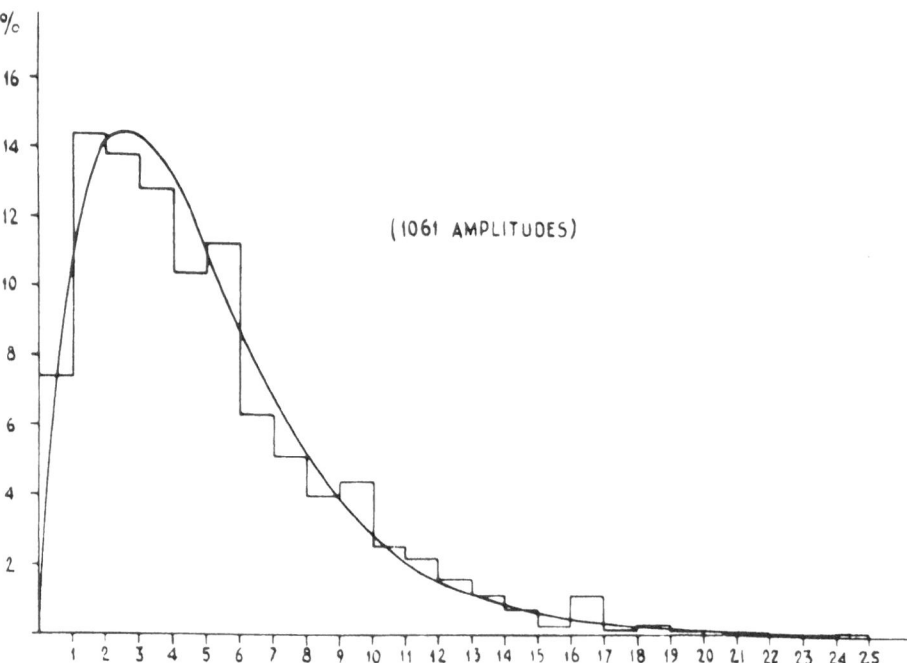

Fig. 2. Amplitude distribution for bursts in lively noise storms. The distribution was obtained from a number of homogeneous fragments of different storms. The mean amplitude for each fragment was set equal to 5.

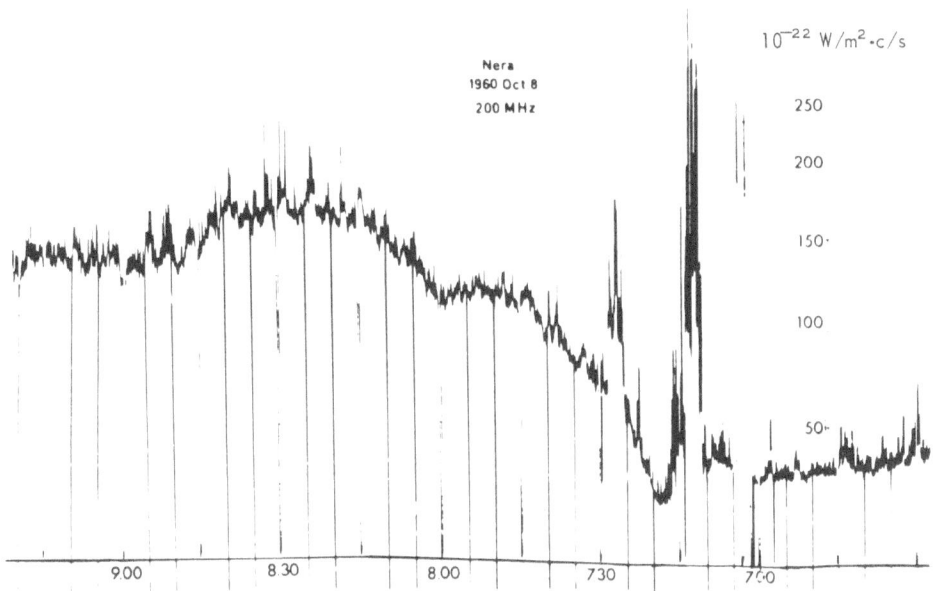

Fig. 3. Reinforcement of background continuum, triggered by a flare of importance 1. A small microwave outburst occurred at 07.11. The remarkable dip immediately after the initial outburst is real.

number of elementary pips. It does not depend on the form of the amplitude distribution of such pips.

Commencements of Noise Storms

A period of noise activity always coincides with the passage of one or more sunspot groups over the solar disk. When near the limb, a sunspot is not likely to contribute to noise activity, but a large spot near the center of the disk almost always does so. Thus, periods of noise activity commonly begin when, due to the sun's rotation, a given sunspot on the eastern hemisphere has proceeded sufficiently far toward the disk's central part.

During a period of noise activity, irregular variations in the intensity of the storm and sometimes relatively steep reinforcements occur. Le Squeren-Malinge [10] found the commencements of reinforced noise activity to be statistically related to the occurrence of flares in the same center of activity to which the noise storm is related. A superposed epoch diagram of flare occurrences before and after the reinforcement shows a considerable excess of such flares in the hour preceding the reinforcement. The flares which were identified by le Squeren-Malinge as probably related to the storm commencement are not necessarily important flares; many of them are class 1 flares. A striking reinforcement of the background continuum is shown by Fig. 3.

Peculiar Storms

On some rare occasions, storms have been observed that present intensity variations that differ markedly from those of the normal Type I storms. A very remarkable event was that of November 4, 1957 (Fokker [2] and Avignon et al. [11]). Lasting about 6 hr, this storm presented rapidly oscillating intensity variations. Moreover, the intermittent occurrence of dips in intensity, which resembled negative bursts, was very striking. Most of the intensity variations were probably of solar origin. At frequencies of 55 and 550 Mc/s, the sun was perfectly calm. The storm was observed at 169, 200, and 250 Mc/s.

Other cases of oscillating intensity variations and negative bursts were reported by Dröge et al. [12] to have occurred on 240 and 460 Mc/s. Some of these occurred during the November 1960 period of considerable Type IV activity.

Another queer storm occurred in the period August 23–27, 1959. At 200 Mc/s, no short-lived bursts were presented, but worldwide irregular intensity fluctuations with a period of 0.1 to 0.2 min occurred on August 23 and 24 (Fokker [13]). At 55 Mc/s, fierce oscillations occurred intermittently. On spectral records in the range 48 to 165 Mc/s, taken at Freiburg on August 25 and 26, slowly drifting bands were observed (Rabben [14]). Lack of correlation of these intensity fluctuations between different observatories seems to indicate an ionospheric origin (Simon [15]). Scintillation effects of solar radio emission on these dates were observed all over the world. It is remarkable that special circumstances apparently existed near both the sun and the earth. That these might be related to each other is an interesting possibility.

INTERFEROMETRIC LOCATION OF SOURCES

The Height of Sources

Position determinations of sources at meter wavelengths were first made by Payne-Scott and Little [16] at 97 Mc/s. The positions were derived from two-element interferometer observations at spacings of 85 and 100 wavelengths. Payne-Scott and Little found that the sources of enhanced radiation are generally associated with a particular sunspot group and that their positions shift progressively from east to west as a result of the sun's rotation. Positions far outside the solar limb were sometimes found, indicating that the sources are situated at heights of several tenths of a solar radius in the corona. Early two-element interferometer observations were also made by Owren [17] at 200 Mc/s.

The multielement interferometer technique was first applied to solar observations at meter wavelengths at the Nançay observatory (Boischot [18]). The Nançay instrument consists of 32 parabolic mirrors. Operating at 169 Mc/s, the interferometer has a beamwidth of 3.'8. During some 2 hr around local noon, the sun, on passing through the interference lobes, is scanned several times. Each scan reproduces the distribution of the sun's strip-integrated brightness. Thus, sources at different distances from the north–south diameter can be told apart. The initial observations were described by Boischot [18]. The results of the daily scans are published in the form of diagrams in the *Quarterly Bulletin on Solar Activity* and in the *CRPL Solar-Geophysical Data* of the National Bureau of Standards (Fig. 4).

In general, interferometer positions determined with a two-element interferometer operating at 255 Mc/s (Nera observatory) agree closely with the posi-

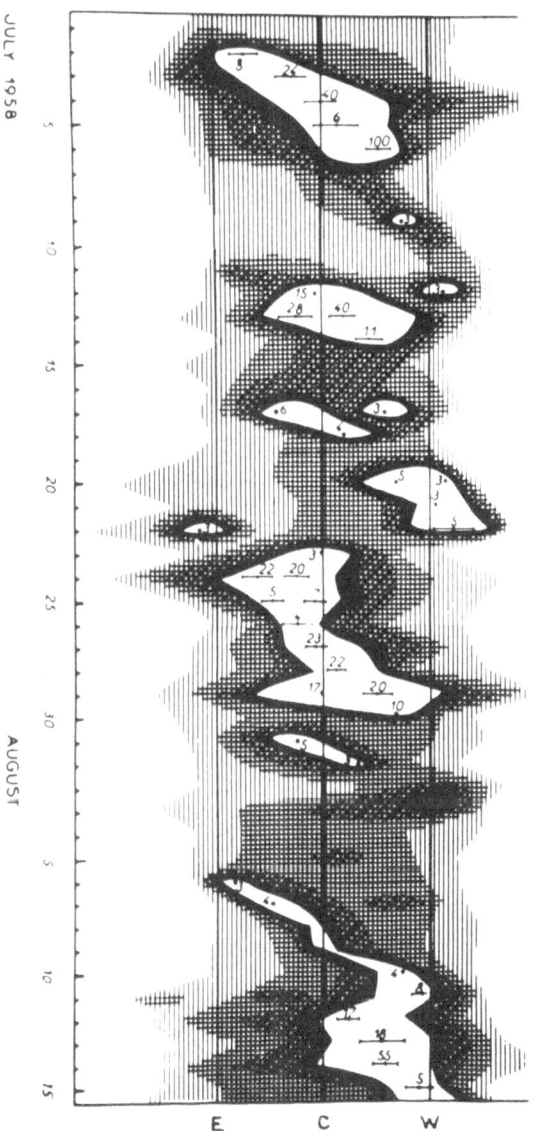

Fig. 4. Example of a diagram representing the scans obtained with the 169-Mc/s interferometer of Nançay observatory.

tion of the strongest source as indicated by the 169 Mc/s scans.

The progression of the radio position of a source, from east to west, is by no means regular, even when the noise activity is associated with only one major sunspot group. Considerable lateral displacements of the radio source with respect to the optical center of activity obviously occur. Only near the limb can one make an estimate of the height, from the deviation of the source position from the spot position. An average height for many sources can be derived from the mean daily rate of progression of sources when on the central part of the disk. Such determinations by Fokker [2] and by Suzuki [19] yield a height of about 0.3 of a solar radius for sources radiating at 200 to 250 Mc/s (see Fig. 5). Le Squeren-Malinge [10] derived from the 169 Mc/s observations a mean height of only $0.15R_\odot$. The difference from the former result might be due to the difference in observational techniques, but it is not clear why the results differ so much.

For sources near the limb, the mean height follows from a comparison of radio and sunspot positions. A somewhat greater height is found nearer the limb than in the central part (see Fig. 5). Near the limb, le Squeren-Malinge found a height of $0.5R_\odot$ for 169 Mc/s. Fokker [2] found limb sources to be situated systematically farther to the outside at 169 Mc/s than at 255 Mc/s. The mean ratio of apparent source heights was found to be 1.5.

Two-Dimensional Location

In March 1960, a north–south branch was added to the 169-Mc/s Nançay interferometer. This branch consists of eight parabolic mirrors and provides a minimum beamwidth (for the sun's northernmost declination) of 8'. By the introduction of suitable phase delays, 15 different lobe systems are obtained which

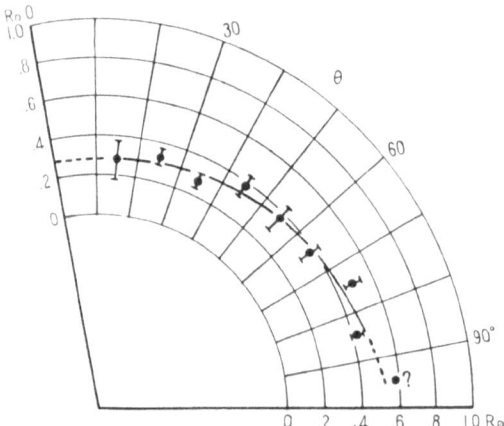

Fig. 5. Apparent height of 200-Mc/s sources at different longitudes (after Suzuki [19]).

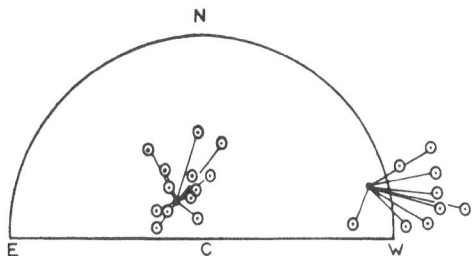

Fig. 6. Positions of radio sources associated with different sunspots at the same
heliographic position (after Le Squeren-
Malinge [10]).

are mutually shifted by $\frac{1}{16}$ of the lobe interval (56'). Fifteen simultaneous records are run, each record corresponding to a particular lobe system. From a
comparison of the amplitudes by which bursts or continuum enhancements appear on these records, the north–south position can be obtained. In combination
with the east–west interferometer, two-dimensional position determinations are
thus obtained.

A diagram prepared by le Squeren-Malinge illustrates nicely how the radio
positions compare with the sunspot positions (Fig. 6). The radio source is at
leisure to be situated anywhere around the sunspot position. The average radio
position is of course situated farther from the center than the corresponding
spot. In a few cases, le Squeren-Malinge observed two radio sources that obviously were associated with one and the same sunspot group.

Directivity, East–West Asymmetry

The directivity of the radiation of sources of storm radiation can be demonstrated by looking for the percentages of sunspots that are observed to be noise-
active on different parts of the solar disk. These percentages decrease strongly
toward the limb, as shown by Fig. 7. One may also plot the number of sources of
storm radiation as observed at different distances from the north–south diameter
(Fig. 10).

From such statistics, in combination with the number–intensity relation for
Type I bursts, Morimoto [21] estimated that the radiation intensity in a direction
inclined at 45° to the normal is reduced by a factor of about $\frac{1}{20}$. Fokker [2] compared number–intensity plots for sources within and further out than two-thirds
of a solar radius from the north–south diameter and found an intensity ratio $\frac{1}{6}$.
The estimate of Morimoto pertains essentially to the Type I bursts, and Fokker's
estimate to the background continuum. The two estimates seem to imply that
storms in the central part of the disk present greater variabilities than limb
sources, but such an effect has not been established; rather, an indication of
the opposite was noted by le Squeren-Malinge.

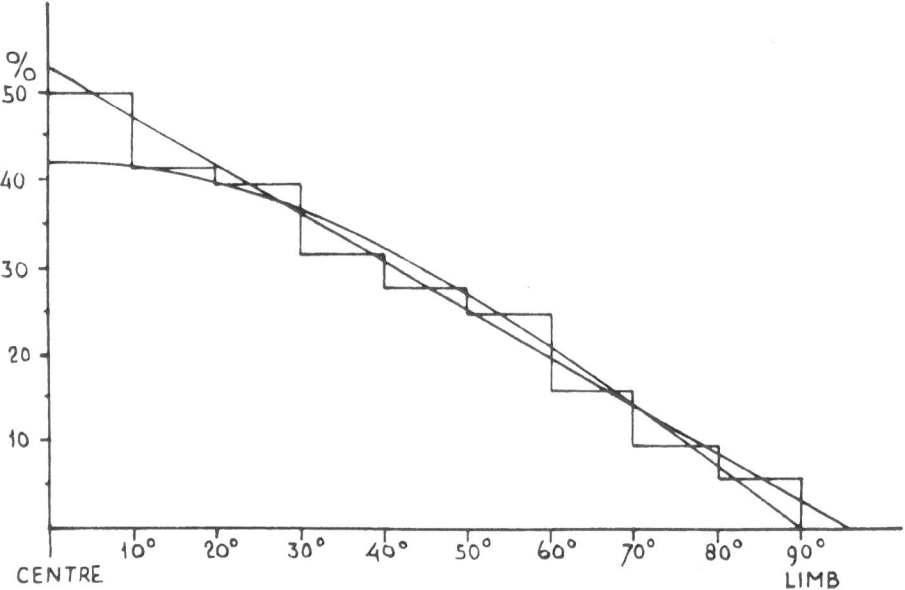

Fig. 7. Percentages of sunspots that are noise active, for different zones of the solar disk.

We note that the term attenuation, or limb darkening, does not seem to be very applicable to the situation. When a source reaches high in the corona, it is more likely to be visible when near the limb. But its visibility may also be governed by specific coronal structures that do or do not prevent the escape of radiation in certain directions. So the visibility of a source seems to be due, to a certain extent, to chance conditions.

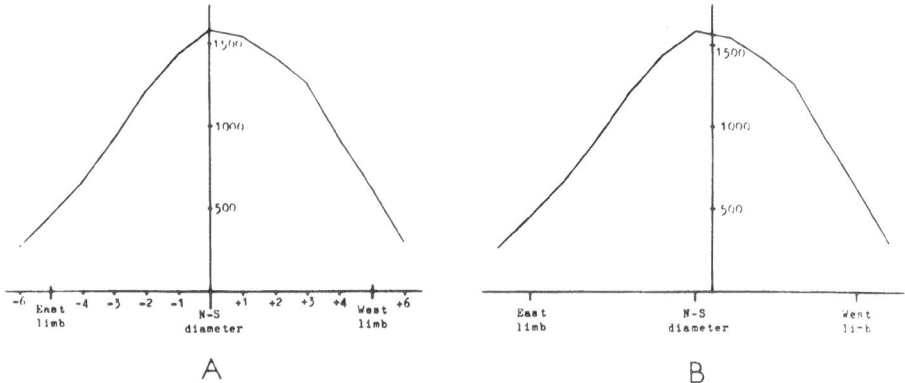

Fig. 8a. Relative frequency of occurrence of bright sources at different distances from the sun's north – south diameter.

Fig. 8b. The same diagram, shifted by 0.10 of a solar radius to the east (after Fokker [20]).

Combining data on the center-limb distribution with the presumed height from which radiation can escape into a given direction, le Squeren-Malinge [10] derived a model for the average distribution of emissivity with height in the source.

The existence of an east–west asymmetry is now firmly established. The asymmetry was first noted by Fokker [2] and was subsequently corroborated by Suzuki [19] and by le Squeren-Malinge [10]. The effect is illustrated by Fig. 8, which shows the relative frequency of occurrence of the "white" regions on the Nançay diagrams, at different distances from the sun's north–south diameter. The excess of sources on the western hemisphere can be accounted for by as- suming the mean emission diagram to have a tilt of about 4°. This effect seems to indicate that the coronal structures are slightly tilted, in the sense that they are lagging behind the rotating sun.

THE POLARIZATION OF NOISE STORMS

Phenomenology

It is normal for noise storms to be strongly circularly polarized. For a given source of storm radiation, the sense of polarization doesn't change during the passage over the solar disk. However, unpolarized storms do occur occasionally. An unpolarized storm has characteristics similar to a polarized storm (see Fig. 9). On the average, unpolarized storm bursts seem to have a somewhat

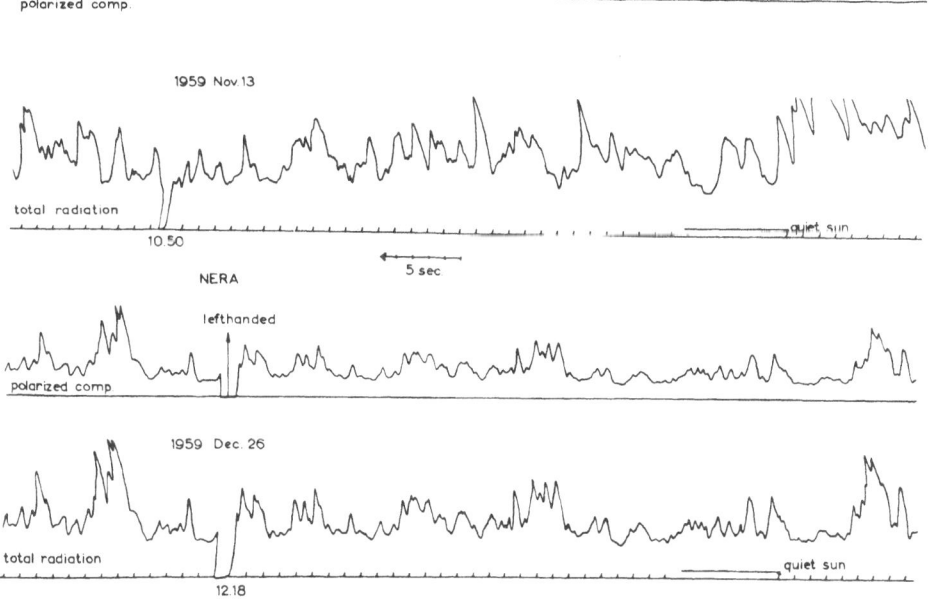

Fig. 9. Example of an unpolarized storm (up) and of a polarized storm (down).

longer duration than polarized bursts. Storm bursts of very short duration are never observed to be unpolarized. The mean bandwidth of polarized and un-polarized storm bursts is the same (de Groot [5]). Contrary to polarized storms, unpolarized storms tend to originate near the sun's limb rather than in the central part of the solar disk (Suzuki [19]). This is clearly illustrated in a diagram prepared by le Squeren-Malinge (Fig. 10).

The sense of polarization of a noise storm is correlated with the hemisphere, north or south, where the source is situated. During the last solar cycle, the sources on the northern hemisphere produced predominantly left-handed polarization, southern sources were mostly right-hand polarized. The magnetic field that predominates in the corona above an activity center probably derives from the strongest magnetic field associated with the activity center; this is generally the field connected with the preceding sunspot of a bipolar group. This implies that the polarization sense observed is the one which corresponds with the ordinary magneto-ionic mode.

In the presence of more than one noise-active region, a mixed type of polarization is sometimes observed, indicating that opposite senses of polarization are being contributed by different sources. The normal result of the superposition of two oppositely polarized storms is a decrease of polarization of the background continuum and the presence of oppositely polarized storm bursts (see Fig. 11). In the great majority of cases, mixed polarization occurs when more than one center of activity is present. In such a case, the 169 Mc/s scans of the Nançay observatory often indicate two source regions, but a few cases have been established in which mixed polarization was produced by only one noise-active region.

In some cases the polarization behavior does not correspond with one of the normal patterns. For instance, it has been observed that the bursts all had an identical sense but widely different degrees of polarization. There is also a type of activity which contains bursts with small variable polarizations of both senses of rotation. On a few occasions a progressive development of polarization has been observed: a storm which is initially unpolarized subsequently develops strong polarization. On one occasion, storm-burst activity, after having been

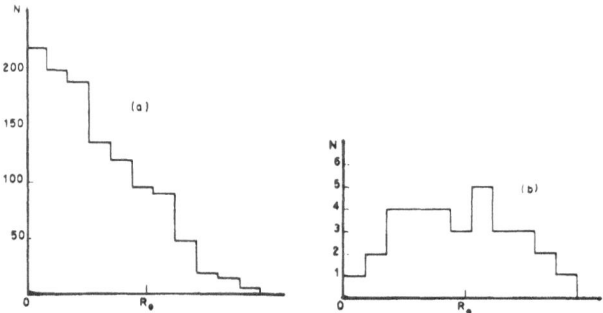

Fig. 10. Distribution of the distances of sources to the north—south diameter for (a) polarized storms and (b) unpolarized storms (after le Squeren-Malinge [10]).

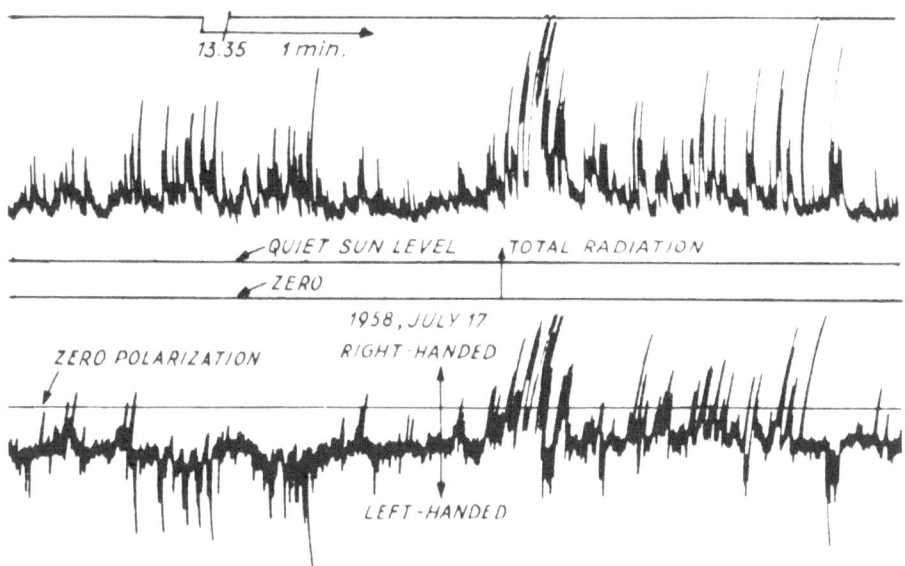

Fig. 11. Example of mixed polarization.

fully polarized, became unpolarized for about 1½ hr and then became polarized again, without changing its character.

Discussion

From the fact that polarized and unpolarized storm bursts are very similar, we may infer that the property of polarity is not inherent in the mechanism of generation of a burst, but that the state of polarization is entirely due to effects of propagation of the radiation through the corona. In the presence of a magnetic field, there are two modes of radio propagation: the ordinary (o) and the extraordinary (x) mode. These modes are in the general case elliptically polarized with opposite senses of rotation (see e.g., Ratcliffe [22]). The refractive indices are given by the expression

$$\mu^2 = 1 - \frac{X}{1 - Y^2 \sin^2 \theta / 2(1 - X) \pm \{[1/2Y^2 \sin^2 \theta / (1 - X)] + Y^2 \cos^2 \theta\}^{1/2}}$$

where the plus sign applies to the o-mode;

$$X = \omega_p^2 / \omega^2 \qquad \omega_p^2 = 4\pi N e^2 / m$$

$$Y = \omega_H / \omega \qquad \omega_H = eH / mc$$

and θ represents the angle between the direction of propagation and the magnetic field.

That storm radiation is commonly strongly polarized seems to indicate that it originates at heights not far above the level where the refractive index for the

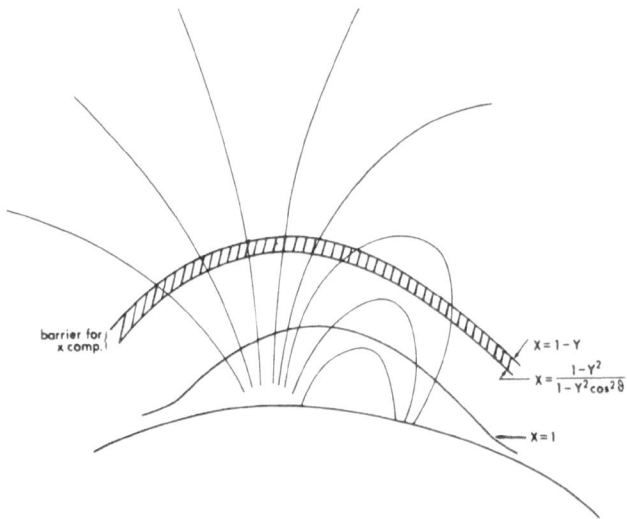

Fig. 12. Model indicating the levels of escape for the ordinary and extraordinary magneto-ionic component.

ordinary magneto-ionic component equals zero. At this level, the plasma frequency equals the radio wave frequency ($X = 1$). In the presence of a magnetic
field only the ordinary mode can propagate from this level on.* Above the level
$X = 1$, there is a stop region for the extraordinary mode. This stop region extends from the level where $X = (1 - Y^2)/(1 - Y^2 \cos^2 \theta)$ to the level where
$X = 1 - Y$. Only from above the level where the refractive index for the x-mode
is zero ($X = 1 - Y$), can both magneto-ionic modes propagate outward (see Figs.
12 and 13). Any radiation from below this level will escape in the o-mode only.
Generally, the pertinent polarization is elliptical, but on passing through regions
of decreasing X and Y, the radiation becomes more nearly circularly polarized
(quasi-longitudinal propagation).

Case I. Mixed Polarization

Since it is likely that magnetic fields of both polarities exist in the different
parts of the source region of a noise storm, one may wonder why mixed polarization
does not arise from one single noise active region more often. This can be
understood by the following argument.

Since the characteristic polarization of a given mode has to fit with the
direction of the magnetic field, the sense of rotation of the polarization ellipse
reverses in that part of the ray's trajectory where the magnetic field is transverse. Rays from different parts of a noise active region, on traveling in the

---

* There is a possibility of having appreciable absorption of the o-wave by gyro resonance
at the level where $Y = 0.5$, if this level is situated above the level where $X = 1$.
However, for propagation at small angles with the magnetic field, gyro resonance is
much reduced (cf. Ginzburg and Zheleznyakov [23]).

direction of the earth, thus will acquire all the same rotation sense of their polarization ellipses; this sense is the one belonging to the polarity of the magnetic field as it is witnessed at great distances from the sun (see Fig. 14a).

However, if we let the electron density in a magneto-ionic medium approach zero, it is evident that a circularly polarized radio wave will not "spontaneously" change the shape of its polarization ellipse or its sense of rotation, if the direction of the magnetic field changes along its path. If the electron density and the magnetic field strength continuously decrease along the ray path, there will be some point beyond which the magneto-ionic parameters have such small values that the state of polarization is no longer affected by changes in the direction of the magnetic field. The two magneto-ionic modes do not, then, propagate independently of each other; strong coupling exists between them. The criterion for strong coupling was derived by Budden [24], who stated that a certain coupling ratio Q, which is a function of X, Y, and of their derivatives to the path length along the ray trajectory, should be greater than 1. We are especially interested in the values Q takes for transverse propagation. According to Cohen [25], in the case of transverse propagation we have the following relation between the wave frequency f_t, for which $Q = 1$, and the magneto-ionic quantities

$$f_t^4 = 10^{17} NSH^3$$

where S is a scale measure defined by $(1/S) = (d\theta/ds)$ (s is the path length along the ray trajectory). For a radio frequency f, we have strong coupling if $f \gg f_t$, and no coupling if $f \ll f_t$.

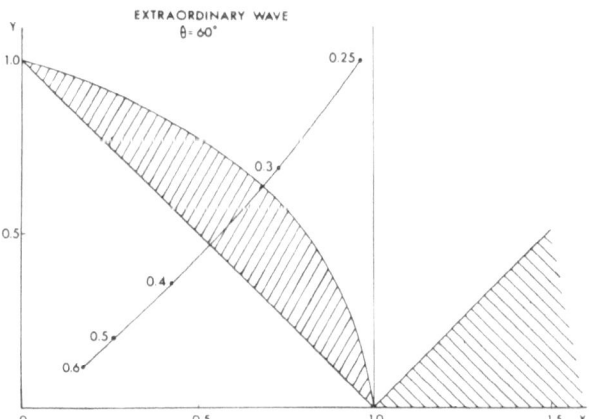

Fig. 13. Location of forbidden regions for the extraordinary magneto-ionic component in the (X, Y) plane. The slightly curved diagonal line represents a possible ray trajectory, with the heights as fractions of a solar radius indicated (after Kai [29]).

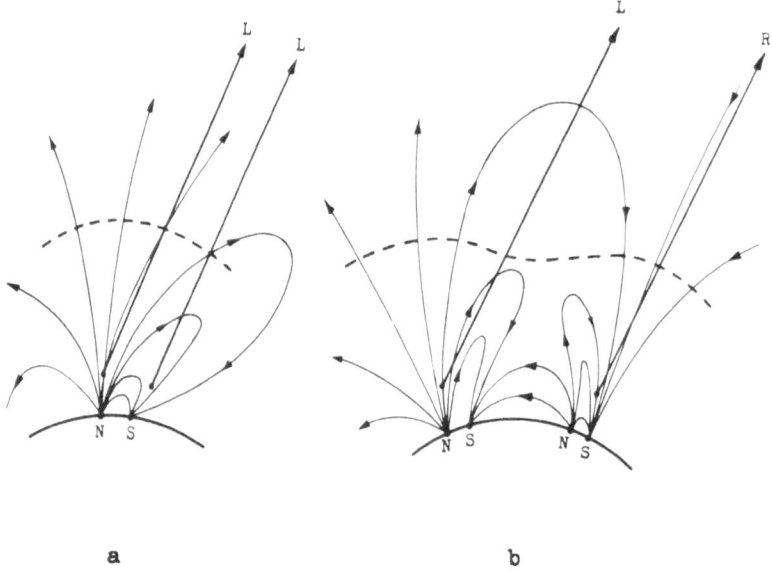

a   b

Fig. 14. Illustration of a possible magnetic field configuration in connection with (a) uniform polarization from one activity center and (b) mixed polarization from two different activity centers. The supposed location of the polarization limiting region is indicated by a dashed line.

Along a given outward-bound ray through the corona, the value of f_t decreases continually. Those points for which $f = f_t$ roughly define the polarization limiting level. Beyond this region the sense of rotation of a polarized wave will no longer change even if the magnetic field reverses somewhere along the ray path. Whether or not mixed polarization will arise, from either one source region or two different centers of activity, depends on the location of the polarization limiting region and the magnetic field configuration. Figure 14a illustrates how one bipolar sunspot region may give rise to uniform polarization; Figure 14b shows how mixed polarization may result from two different centers of activity. Figure 15 applies to the rare case of mixed polarization from one and the same center of activity. In these figures, the supposed location of the polarization limiting region is indicated by a dashed line.

Case II. Unpolarized Storms

If the source is situated well above the stop region for the x-mode, the two magneto-ionic modes can both travel outward, and we shall observe radiation which is virtually unpolarized. If a source is to be visible from near the limb, it must also be situated sufficiently high. So, once we are dealing with a limb source, there is a better chance for its radiation to be unpolarized than when a source is observed nearer to the center of the disk; this is in agreement with the observations.

Another possibility for storm radiation to be unpolarized was pointed out by Suzuki [19]: If in the polarization-limiting region the ray trajectory of the o-wave is transverse to the magnetic field, the characteristic (linear) polarization will be fixed from there on. Subsequent differential Faraday rotation will then depolarize the signal as it is received with a conventional bandwidth. The prevalence of one sense of polarization in radiation from a particular center of activity suggests that the geometry of the magnetic field is quasi-unipolar. Transverse propagation of the observed radiation is thus more likely to take place when the source is near the limb than when it is in the central part of the disk. However, it seems somewhat questionable whether the pertinent condition is fulfilled for the radiation from all the different points of the source. Rather, one would expect to observe, in addition to a certain number of unpolarized bursts, bursts with various, possibly small, degrees of polarization. Actually, noise activity of such a type is sometimes observed; however, in a typical unpolarized noise storm all the bursts are fully unpolarized.

In connection with the first one of the above-mentioned possibilities, Takakura [26] remarks that unpolarized bursts may also occur when a noise storm is near the center of the disk, but that they will then escape attention because of the strong preponderance of polarized bursts that originate in lower levels of the same source region. With this in mind, the author looked specifically for unpolarized bursts which occur during polarized storms, inspecting the polarization records of the Nera observatory. In about 30% of the polarized storms a very few unpolarized storm bursts were actually found (see Fig. 16),* but such unpolarized

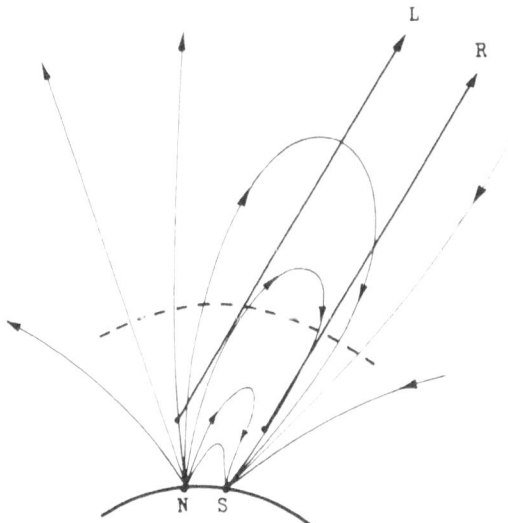

Fig. 15. Illustration of how mixed polarization may
be produced in one center of activity.

---

*Some uncertainty arises from the possibility that Type III bursts (unpolarized) are mistakenly identified as Type I bursts.

bursts occur only very sporadically. Moreover, there is no indication that marginally discernible unpolarized storm bursts are present in any substantial quantities. So there is no evidence that many smaller unpolarized bursts are drowned under the multitude of polarized bursts. The bursts that are identified as unpolarized, when taken alone, would never constitute a true storm like the real unpolarized storms. Moreover, one is never sure whether perhaps such occasional unpolarized storm bursts arise from somewhere else on the solar disk.

Since about 15% of the storms with great storm bursts (variability 2 or 3) are unpolarized, it should not be a great exception to find a nonnegligible admixture of recognizable unpolarized bursts during a polarized storm. The lack of such evidence seems to indicate that all the bursts of certain storms are observed to be unpolarized when viewing the source from some specific, generally oblique, directions; the same bursts, when viewed from other directions, might appear polarized. Due to small changes in the magnetic field configuration, or due to the sun's rotation, a certain kind of noise activity could change from unpolarized to polarized, as is sometimes observed.

THE STRUCTURE OF THE SOURCES

For the understanding of the noise storm phenomenon, it is of course of basic importance to know more about the extent and brightness distribution of the sources of both the background continuum and the storm bursts, and of the positions of the individual storm bursts.

The Background Continuum

The extent and brightness distribution of the source of the relatively steady background continuum can in principle best be determined with a multielement

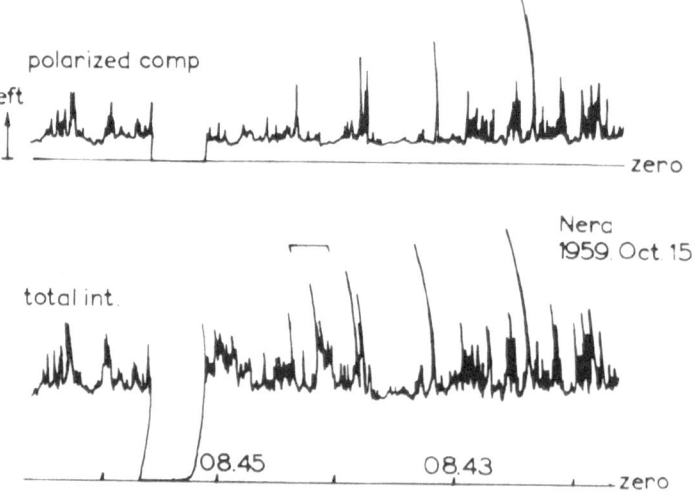

Fig. 16. Sporadic unpolarized detail in a polarized storm.

interferometer with a great resolving power. From the 169-Mc/s scans of the Nançay interferometer (beamwidth 3.8), the diameter of sources larger than 2' can be determined. The brightness distribution can only be determined if the source has an extent well above the beamwidth.

Information on the angular extent of a source can also be obtained with a two-element interferometer, provided no more than one source is present. The measurement consists in determining the "visibility" of the interferometer signal. The power P one obtains from a two-element interferometer after adding the voltages from the two aerials is given by

$$P = T + A \cos wt \tag{1}$$

where w is the phase shift in radians per unit time due to the rotation of the earth. The ratio $V = A/T$ is the visibility. The interferometer signal $A \cos wt$ can be recorded separately from the total radiation T by means of a phase-switching device.

We define $F(\epsilon)$ as the one-dimensional brightness distribution which results from integration of the source's brightness along north–south strips; ϵ is the angular distance of a strip to a reference north–south line. If we assume $F(\epsilon)$ to be symmetrical around the origin, A is given by the expression

$$A = \int F(\epsilon) \cos (2\pi n \epsilon) d\epsilon \tag{2}$$

where n is the spacing of the antennas expressed in wavelengths, while

$$T = \int F(\epsilon) d\epsilon$$

For a circular source of radius ρ with a uniform brightness distribution, we have

$$V = 2J_1(a)/a$$

where $a = 2\pi n \rho$. For a circular source with a Gaussian brightness distribution $f(r) \sim e^{-(r/\rho)^2}$, we have

$$V = e^{-a^2/4}$$

By measuring the visibility V at a given spacing n, we can derive the value of ρ for a supposedly uniform or Gaussian source.

By measuring V at different spacings, one obtains a number of points on the visibility curve $V(n)$. From such measurements one can derive the brightness distribution, still assuming the source to be circular.

Two examples of a brightness distribution and the corresponding visibility curve are given in Fig. 17. Various sets of 3 points represent measurements at spacings 390, 570, and 960 wavelengths ($\lambda = 1.5$ m) which were made at the Nera observatory. An example of a record at these three spacings is given in Fig. 18.

TABLE I

Model	WS, ρ	Gaussian, e^{-1} radius (570λ)
September 4, 1961	1.4	1.3
March 25, 1961	0.6	0.5
January 26, 1961	1.5	1.4
August 30, 1958	1.9	1.6
Model	F	
July 18, 1961	2.3	1.9
July 12, 1961	1.7	1.6
July 11, 1961	2.2	1.9
January 10, 1959	3.8	2.7

In Table I, we list the values of ρ, as indicated in Fig. 17, and the e^{-1} radius which follows for a Gaussian source from the visibility at the spacing 570λ. The model WS is one given by Weiss and Sheridan [27] to represent visibilities observed for Type II bursts; the model F was given by Fokker [2] and is a superposition of 4 Gaussian distributions. Le Squeren-Malinge [10] found the diameters of sources at 169 Mc/s to range from about 1 to 9' of arc, the mean value being 3.8.

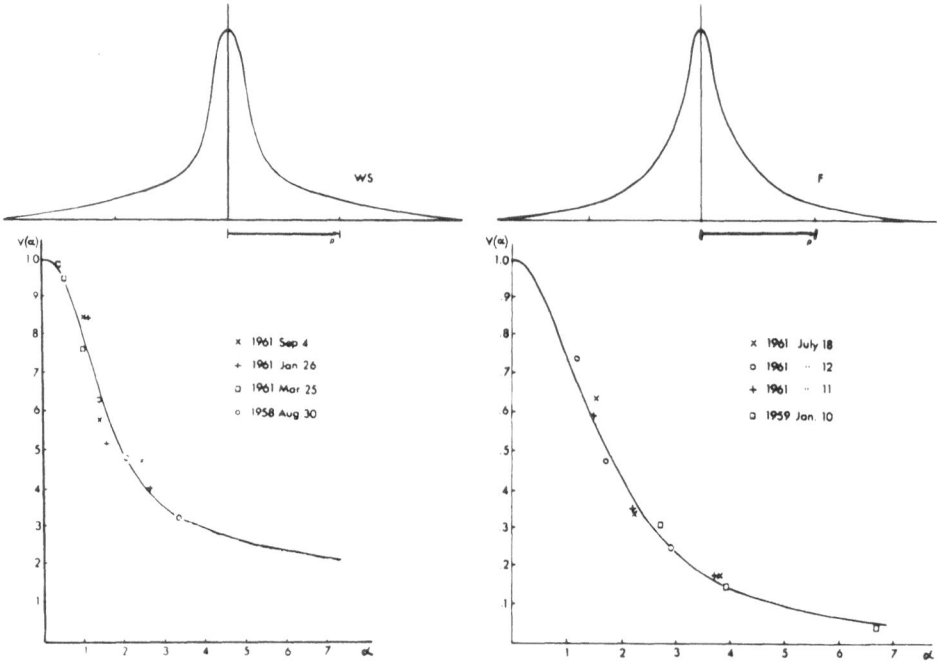

Fig. 17. Model circularly symmetric brightness distributions and corresponding visibility curves.

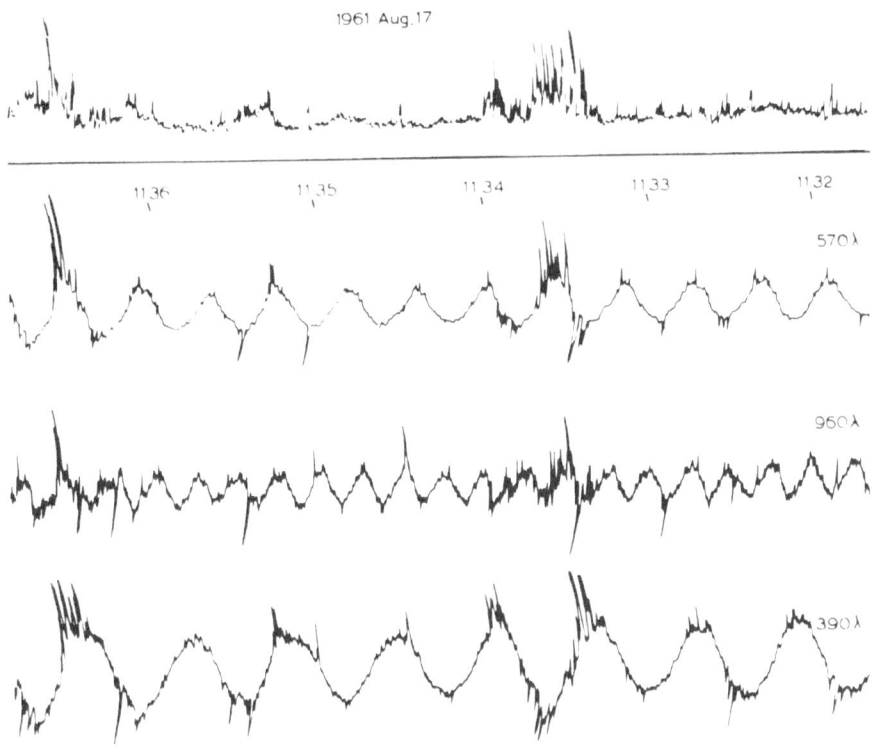

Fig. 18. Example of interferometric records at three different spacings
(frequency 200 Mc/s).

The Storm Bursts

From a single interferometer record we cannot derive the position of such a
transient event as a storm burst, since we do not know in what phase of the interference cycle it occurs. If we knew for certain that the source has zero angular
extent [in that case $A = T$, cf. equation (1)], we could derive the phase by taking the ratio of the amplitudes by which the burst appears on the interferometer
and the total radiation record. However, it is probable that $A/T < 1$ because of
the source's angular extent. The phase can be determined indeed by recording a
second interferometer channel, in which the interferometer signal is delayed by
90°. Such a delay is introduced by having one of the antenna signals pass
through an extra ¼λ piece of cable. The two interferometer records thus yield
the signals $I_1 = A \cos wt$ and $I_2 = A \sin wt$. By measuring the deflections I_1
and I_2, we obtain both the phase wt and the amplitude A. On comparing the
phase of the burst with the phase of the background continuum signal at the time
of the burst, we can readily determine the position of the burst source relative to
the continuum source.

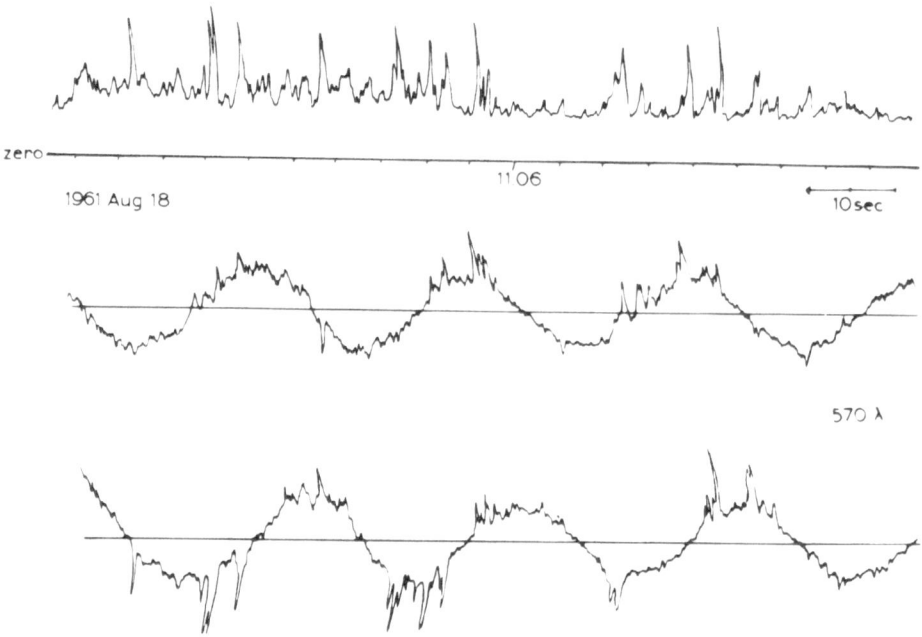

Fig. 19. Example of two interferometric records with a mutual phase difference of 90°
(frequency 200 Mc/s).

Such, or similar, observations have been made at 81 Mc/s with a spacing of
270λ at the Mullard Observatory (Högbom [28]) and at 200 Mc/s with a spacing of
340λ at Tokyo Astronomical Observatory (Suzuki [19] and Kai [29]) and with
spacings 390 and 570λ at the Nera Observatory. An example which shows two
mutually phase-delayed interferometer records and the corresponding total radiation record is given in Fig. 19. An example of observational results of Suzuki is
given in Fig. 20, while Fig. 21 gives some histograms of the positions of bursts
relative to the source of background continuum. The approximate extent of the
background continuum is also indicated in Fig. 21.

Generally, burst positions are found to be scattered over a range of 1 to 3'
of arc, sometimes the scatteredness exceeds 3'. In general, the bursts are situated within the source of background continuum, but they are not always distributed symmetrically around its center (see Fig. 21).

From a comparison with records of the radio source Case A at times of ionospheric scintillation, Suzuki concluded that most of the scatteredness of storm
bursts is real. He noted the interesting fact that the more intense storms tend to
present a smaller scattering range of the bursts.

Attempts to determine the angular size of sources of storm bursts were made
by Högbom [28] and by Fokker [30]. Högbom, working at 81 Mc/s, found the mean
diameter of burst sources to be in the range from 5 to 7'. Fokker obtained at 200
Mc/s burst diameters (e^{-1}, Gaussian) of the order of 3', but this value should be

considered somewhat provisional. The spread in diameters of individual bursts of the same storm seems to be of the order ±15%. Kai [29] did not obtain absolute values for the A/T ratio, but he did find this ratio to be systematically smaller for the continuum source than for the burst source.

So we arrive at a picture of a noise-active region in which we see bursts flashing up in different parts within a "steady" source of background noise. The angular extent of the bursts is probably somewhat smaller than the extent of the continuum source. The latter has a considerable central brightness concentration.

DISCUSSION

The first question which should be answered in connection with noise storms is whether or not a storm burst is to be considered as a true discrete event. We may justify this way of putting the problem by the following argument.

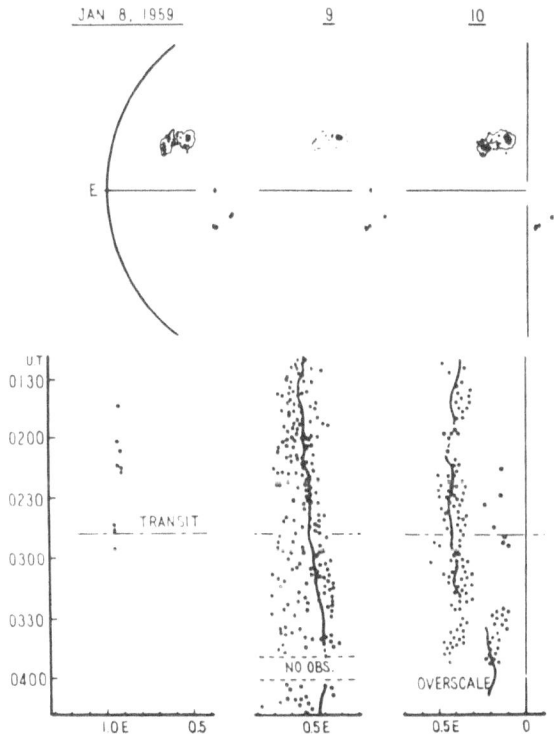

Fig. 20. Positions of storm bursts on three successive days as compared with positions of the corresponding sunspot group. The two sources on January 10 (lower right) produced oppositely polarized bursts (after Suzuki [19]).

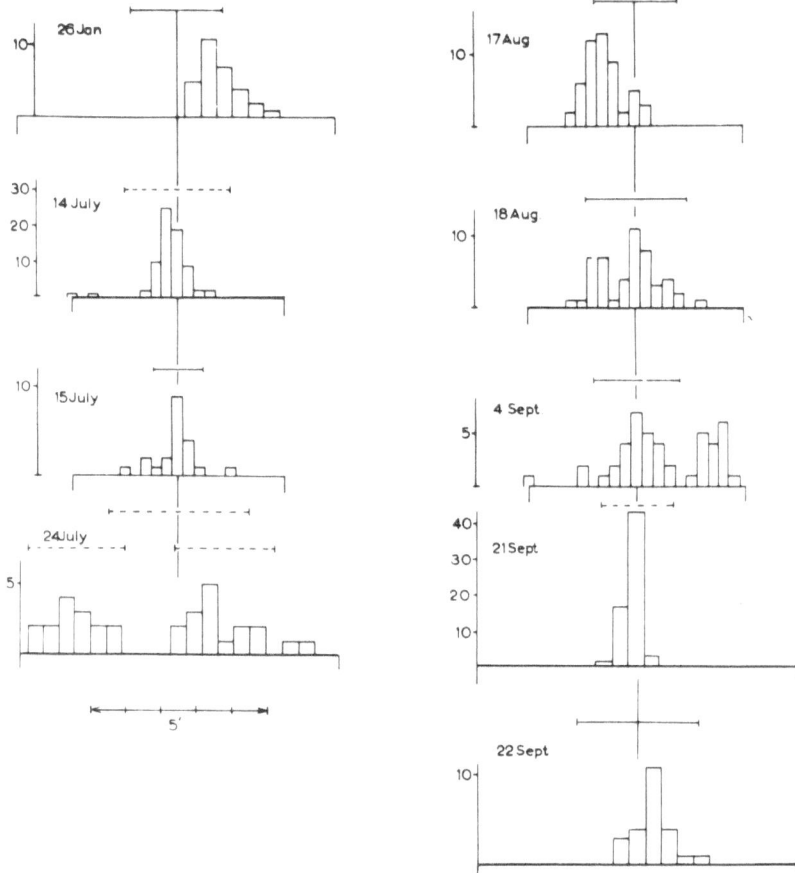

Fig. 21. Histograms of burst positions as compared with position and approximate extent (indicated by horizontal line segments) of the background continuum source (observations made in 1961 at the Nera observatory on 200 Mc/s).

1. One may consider a noise storm as the superposition of many individual
pips. Each pip may be the result of a specific physical process in a well-defined
point in the corona. One may think, for example, of a discrete stream of fast particles which pass the level where $X = 1$ at a place where there is a steep density gradient. Such a disturbance is a discrete event, and what we observe is
just the superposition of the radiation effects of many such events. The author
obtained a fluctuating signal, which strongly resembles the intensity fluctuations
common to noise storms, by constructing a model superposition of many individual
pips (Fokker [2]).

2. One may consider a noise storm as being essentially a fluctuation phenomenon. For instance, ionospheric scintillation is such a phenomenon. One
scintillation is the accidental result of the arrival of waves from slightly differ-

ent directions with continually changing mutual phase differences; it does not at
all correspond to a discrete physical event or to the superposition of such events.
The author constructed a model in covering a square with many randomly moving
circles. The surface of the unobscured part of the square varied in a way analogous to the intensity changes during a noise storm (Fokker [2]).

It is the author's opinion that it is not yet possible to decide between these
two alternatives. But whatever possibility we may favor, it seems likely that
storm bursts have their origin close to the level where $X = 1$. Here, plasma
oscillations occur with a frequency equal to the observed radio frequency. Also,
fluctuation effects are likely to be strong, since the refractive index is close to
zero.

That the level of radio emission is strongly linked with the (mean) coronal
electron density is evidenced by the whole of height determinations at different
frequencies (Morimoto [31]), which yield a rough proportionality between the radio
frequency and the square root of the electron density. Locating the origin of 200
Mc/s emission at a height of $0.3R_\odot$, we find that the corresponding electron density $N_e = 5 \times 10^8$ is about 12 times greater than the density in the coronal
model (maximum) of van de Hulst [32] and about 3 times greater than the density
in Newkirk's [33] steamer model. This seems to indicate that the radio emission
originates in coronal elements that have a high density relative to their surroundings.

Because of the linkage with electron density, storm radiation seems to be related to plasma processes. Of course, one immediately thinks of plasma waves;
the dispersion relation is

$$\omega^2 = \omega_p^2 + \overline{v_t^2} \times k^2 \tag{3}$$

where v_t is the thermal velocity of the electrons, and $k = 2\pi/\lambda$. It is thought that
plasma waves in the corona are excited by the passage of a stream of fast electrons. For the phase velocity v of plasma waves we have

$$v_f = \frac{\omega}{k} = \frac{(\overline{v_t^2})^{1/2}}{(1 - \omega_p^2/\omega^2)^{1/2}}$$

Whenever a superthermal electron moves with a velocity $v > v_f$, it will excite
Čerenkov plasma waves of different frequencies. A wave with a given frequency
propagates in a specific direction, making an angle ψ with the electron velocity
vector, where $\cos \psi = v_f/v$ (see Cohen [34]).

The way in which plasma waves give rise to radio emission is a problem.
There is no magnetic field associated with a plasma wave; there is only a periodically alternating electric field. However, in the presence of inhomogeneities of the
plasma density or magnetic field, there may be some coupling between the plasma
and electromagnetic modes (Field [35]). The conversion efficiency is greatest in
the vicinity of the level where $\omega = \omega_p$, but it has a small value, of the order of
10^{-7}, as estimated by Ginzburg and Zheleznyakov [23].

A second possibility is the mechanism of scattering of the plasma wave on small-scale random fluctuations of either the neutral plasma density or of the space charge. For the efficiency of conversion of the energy of plasma waves into electromagnetic radiation at the same frequency, by scattering on neutral fluctuations, Ginzburg and Zheleznyakov [36] derived the expression

$$\epsilon = \frac{4\pi e^4 N \mu}{3m^2 c^3} \tag{4}$$

where μ is the refractive index. Without entering into the physical concepts behind this mechanism (see Cohen [37]), we give a simple argument which happens to lead to very nearly the same expression as equation (4).

Due to the plasma wave, all electrons will be subject to a periodically fluctuating acceleration a. Consequently, each electron emits electromagnetic radiation at a rate

$$p = \frac{2e^2 a^2}{3c^3}$$

where $a = -eE/m$, and E is the instantaneous electric field strength associated with the plasma wave. The volume emissivity P is

$$P = (2e^4 E^2 N \mu)/(3c^3 m^2) *$$

Since the energy density of the plasma wave is $E^2/4\pi$, the conversion efficiency ϵ' is

$$\epsilon' = \frac{8\pi e^4 N \mu}{3m^2 c^3} \tag{5}$$

For $N = 5 \times 10^8$ (corresponding with 200 Mc/s), we have $\epsilon' = 5 \times 10^{-6}$. Equations (4) and (5) differ by only a factor 2.

Radio emission will be available only for frequencies above ω_p. The emission is mainly due to those electrons which have a thermal motion in a direction opposite to the direction of propagation of the plasma wave. These electrons experience an electric field fluctuating at a rate slightly higher than the frequency of the plasma wave. The value of μ^2 for the emitted radiation is consequently greater than 0.

Radiation at approximately the harmonic of the plasma frequency can be produced by so-called combination scattering of plasma waves on charge fluctuations. This kind of scattering was invoked by Ginzburg and Zheleznyakov to account for harmonic radiation of Type II and Type III bursts.

---

* The refractive index μ enters similarly as it does in the expression for the emissivity of Bremmstrahlung (cf. Westfold and Smerd [38]).

Scattering on neutral fluctuations may be invoked both with regard to the fundamentals of Type II or Type III bursts, and to the noise storm. The characteristic duration of Type I bursts is related by Takakura [26] to the lifetime of the beam of fast electrons, which is determined by the redistribution time t_D set by the collisions with thermal electrons. A duration of the correct magnitude is found for electron velocities of about 1.5×10^4 km/sec.

One may also think of high-energy particles with a much longer lifetime that are trapped in magnetic fields (Denisse [39]). On leaking downward to deeper levels, such particles would excite plasma waves by the Čerenkov effect. Diffuse radio radiation would result from conversion, by coupling or by scattering, of the plasma waves into the electromagnetic mode. The storm bursts might be produced by special processes (specular reflection?) near the level where, for both the plasma and electromagnetic waves, the refractive index becomes zero.

If the radiation of both the background continuum and the storm bursts arises from the conversion of plasma wave energy into radio emission, then the storm-burst phenomenon calls for a very sudden release of a considerable fraction of the plasma wave energy. Whatever the mechanism of such a release, the conditions for a rapid conversion are most likely to be fulfilled close to the level where $X = 1$. Conversion here may take place by mode coupling, as suggested by Denisse [39]. Another circumstance which may be of interest is the following. A downward traveling plasma wave will, on approaching the level where $X = 1$, be subject to a rapidly decreasing refractive index. Normally, the wave will be refracted away from the direction of the density gradient and ultimately it will be reflected. However, in the presence of a magnetic field the normal plasma wave exists only in the longitudinal mode, with the wave normal parallel to the magnetic field. So the magnetic field tends to prevent the refraction of the plasma wave which otherwise would have taken place. The excess of radio emission might be due to the very processes which ensue from this conflicting situation.

We now make an estimate of the energy involved in storm radiation. Let us suppose we are dealing with a background continuum with an intensity of 50×10^{-22} W/m^2-c/s. Let us consider all the radiation that is emitted in the frequency range 170 to 200 Mc/s and let us suppose this radiation to originate in the height range of 0.30 to 0.35$R_\odot$, an interval of 35,000 km. If we assume the emission to take place in a solid angle of π steradians, a total power of 1.06×10^{17} ergs is radiated per second. Allowing for absorption of the radiation by a factor 10 and assuming a conversion efficiency for plasma waves of 10^{-6}, the corresponding total energy of the plasma waves is 1.06×10^{24} ergs. This may be compared with the thermal energy density of the coronal plasma which, for $N_e = 4 \times 10^8$ cm^{-3} and $T = 10^6$ °K, is 16.6×10^{-2} ergs/cm^3. The required total energy of plasma waves is thus equivalent to the thermal energy content of a volume of 6×10^{24} cm$^3 \cong (2000$ km$)^3$. If we take the lateral extent of the source region to be $40,000 \times 20,000$ km, the total volume concerned with the emission in the 30-Mc/s range is $40 \times 20 \times 35 \times 10^9$ km$^3 = 28 \times 10^{27}$ cm^3. The energy concentrated in the plasma waves which pervade this volume would thus be of the order of 0.0003 of the thermal energy content.

A final remark pertains to the extent of the source region of a storm burst. Supposing the radiation to originate in a region where X is close to 1, the group velocity in the source region $v_g = c(1 - X)^{\frac{1}{2}}$ is greatly reduced. For $X = 0.96$, we have $v_g \approx 0.2c$. Allowing for a rise time of 0.1 sec, this would imply that for a burst in the center of the solar disk the source region should not be thicker than 6000 km. For a burst at a certain longitude, the source region is oblique to the line of sight. Because of the difference in propagation time from the two extremities, there is an upper limit to its lateral extent; for a longitude of 30°, this limit can be estimated at about 20,000 km. Apart from the conditions set by the propagation time, it is difficult to conceive of a disturbance that excites such a large region practically instantaneously. Therefore, we cannot but assume the source region to be considerably smaller than 1′ in angular extent.*

The observed size of about 3′ at 200 Mc/s and of 5 or 6′ at 81 Mc/s must therefore be largely due to scattering of the radiation by coronal irrgularities. On reasonable assumptions, one may expect the following proportionality to hold true

$$t^2 \sim \int_{source}^{earth} \frac{\overline{N}^2(s) \times s^2 \times ds}{R(s)}$$

where t is the angular extent of the source, s is the path length from the source, and R is the distance to the sun's center. This relation actually implies that the angular extent increases with decreasing frequency.

CONCLUSION

Observations of noise storms have yielded information on several characteristics of storm radiation: (1) bandwidth and duration of storm bursts, (2) polarization behavior, (3) directional characteristics (including east–west asymmetry), and (4) structure of sources. Yet we still do not know (1) what the connection is between the background continuum and the storm bursts or (2) whether storm bursts represent discrete physical events of fluctuation phenomena; and we have only vague ideas on (1) the plasma processes involved and the conversion mechanism, and (2) the role which the magnetic fields play in the generation of noise.

From the theoretical side, no satisfactory answers have been given yet, although some possibilities have been offered which are worth exploring. We do not expect the development of plasma physics to provide satisfactory explanations soon, but our concepts are likely to improve.

From the observational side, our picture of the storm phenomenon can still be extended and refined. Interferometric observations on the structure of storm sources combined with polarimetry, at a number of different frequencies, are likely to give valuable results. A careful analysis and combination of observational material may yield some clues or hints toward the interpretation of the fascinating and puzzling phenomena we observe.

---

* Takakura [26] chooses the lateral extent of the source to be $(1.4 \times 10^5 \text{ km})^2$, which we think is impossibly large.

REFERENCES

1. C. de Jager and F. van't Veer, *Rech. Astr. Obs. Utrecht* 14, 1 (1958).
2. A. D. Fokker, *Studies of Enhanced Solar Radio Emission at Frequencies Near 200 MHz*, thesis, Leiden (1960).
3. J. P. Wild, *Australian J. Sci. Res.*, A 4, 36 (1951).
4. O. Elgaroy, *Astrophys. Norvegica* 7, 123 (1961).
5. T. de Groot, thesis, Utrecht Univ. (1965).
6. T. de Groot, *Inform. Bull. of Solar Radio Obs.* 9, (1962).
7. O. Elgaroy, *Inform. Bull. of Solar Radio Obs.* 9, (1962).
8. V. V. Vitkevitch, *Intern. Astron. Union, Symp.*, No. 4 (1957), p. 363.
9. T. Takakura, *Pub. Astron. Soc. Japan* 11, 55 (1959).
10. A. M. le Squeren, *Ann. Astrophys.* 26, 97 (1963).
11. Y. Avignon, A. Boischot, and P. Simon, *Ann. Astrophys.* 21, 243 (1958).
12. F. Dröge, F. Holweger, and A. Unsold, *Inform. Bull. of Solar Radio Obs.* 6, (1961); F. Dröge and P. Reimann, *Inform. Bull. of Solar Radio Obs.* 8, (1961).
13. A. D. Fokker, *Inform. Bull. of Solar Radio Obs.* 1 and 2 (1960).
14. H. H. Rabben, *Z. Astrophys.* 55, 73 (1962).
15. P. Simon, *Inform. Bull. of Solar Radio Obs.* 3, (1960).
16. R. Payne-Scott and A. G. Little, *Australian J. Sci. Res.*, A 4, 508 (1951).
17. L. Owren, *Radio Astron. Rep. No. 15* (Cornell Univ., 1954).
18. A. Boischot, *Ann. Astrophys.* 21, 273 (1958).
19. S. Suzuki, *Ann. Tokyo Astron. Obs.* 7, 75 (1961).
20. A. D. Fokker, *Bull. Astron. Inst. Neth.* 17, 214 (1963).
21. M. Morimoto, *Publ. Astron. Soc. Japan* 15, 46 (1963).
22. J. A. Ratcliffe, *The Magneto-Ionic Theory and Its Applications to the Ionosphere* (Cambridge Univ. Press, 1959).
23. V. L. Ginzburg and V. V. Zheleznyakov, *Soviet Astron.* 3, 236 (1959).
24. K. G. Budden, *Proc. Roy. Soc. (London)*, Ser. A 215, 215 (1952).
25. M. H. Cohen, *Astrophys. J.* 131, 664 (1960).
26. T. Takakura, *Publ. Astron. Soc. Japan* 15, 462 (1963).
27. A. A. Weiss and K. V. Sheridan, *J. Phys. Soc. Japan, Suppl. A-II*, 17, 223 (1962).
28. J. A. Högbom, thesis, Cambridge Univ. (1959).
29. K. Kai, *Publ. Astron. Soc. Japan* 14, 1 (1962).
30. A. D. Fokker, unpublished.
31. M. Morimoto, *Ann. Tokyo Astron. Obs.* 8, 125 (1963).
32. H. C. van de Hulst, *Bull. Astron. Inst. Neth.* 11, 135 (1950).
33. G. Newkirk, Jr., *Astrophys. J.* 133, 983 (1961).
34. M. H. Cohen, *Phys. Rev.* 123, 711 (1961).
35. G. B. Field, *Astrophys. J.* 124, 555 (1956).
36. V. L. Ginzburg and V. V. Zheleznyakov, *Soviet Astron.* 2, 653 (1958).
37. M. H. Cohen, *J. Geophys. Res.* 67, 2729 (1962).
38. K. C. Westfold and S. F. Smerd, *Phil. Mag.* 41, 831 (1949).
39. J. F. Denisse, *Inform. Bull. of Solar Radio Obs.* 4, (1960).
40. A. Boischot and M. Pick, *J. Phys. Soc. Japan, Suppl. A-II* 17, 203 (1962).

Chapter 10

Narrow-Band Studies of Solar Bursts

Öystein Elgaröy

The Institute of Theoretical Astrophysics
Oslo, Norway

INTRODUCTION

Solar bursts of radio noise in the meter and decimeter wave region are the products of dynamic phenomena in the corona. Before satisfactory models of these phenomena can be evolved, it is necessary to collect a large amount of relevant observational material. In the radiofrequency range, various types of radiometers, polarimeters, panoramic receivers, and interferometers have been used to secure such material.

In studies of solar bursts, a knowledge of the dynamic spectra of the emissions is essential. Dynamic spectra also formed the basis of the classification of bursts introduced by Wild and McCready in 1950 [20].

Through the study of records from wide-band panoramic receivers, an insight into the large-scale features of the emissions is gained, whereas much of the fine structure is lost. It was felt that the fine structure, in cases where such structure existed, might have a more direct bearing on the way in which the bursts originate, and thus to the physical state in the source, than the large-scale structure. Therefore, narrow-band spectrographs were constructed and applied in solar radio investigations.

OBSERVATIONAL EQUIPMENT

Two different types of receivers are used for narrow-band spectrometry. The first one, the multichannel spectrograph, receives the radiation simultaneously in several channels which are closely spaced in frequency. The bandwidth of each channel is, of course, comparable to or less than the spacing between adjacent channels. A block diagram of a multichannel receiver is shown in Fig. 1. The output is usually fed to pen recorders. Such receivers have a good sensitivity, and it is easy to measure the intensity variations of the received signals. The drawback is that a great amount of work is involved in the construction of dynamic spectra from the records. Also, it is necessary to interpolate the intensity between the observing channels, and this contains a measure of uncertainty because bursts of

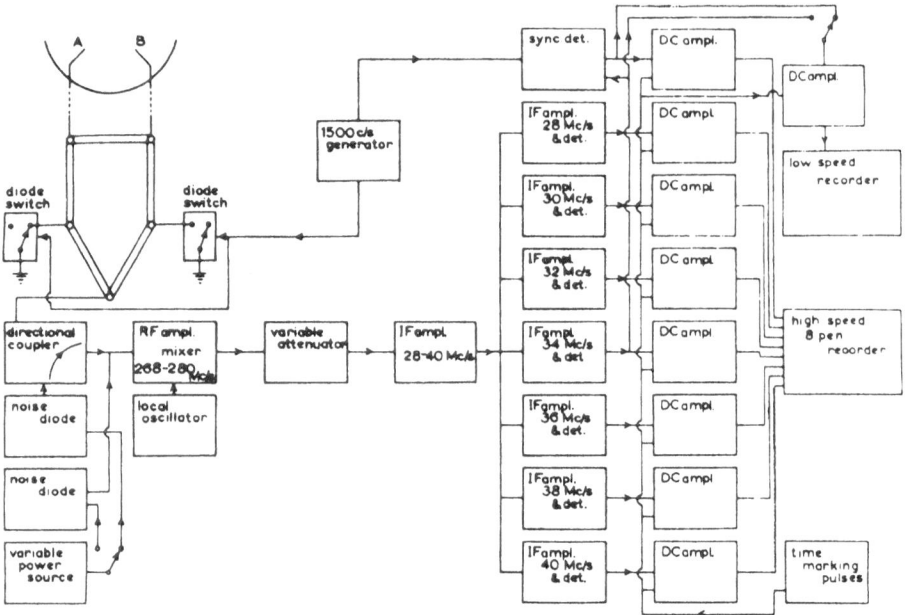

Fig. 1. Block diagram of a multichannel spectrograph and polarimeter.
[de Groot, *B.A.N.* 502, (1960)].

bandwidth comparable to, or smaller than, the frequency difference between neighboring channels might be present. Multichannel receivers have been used especially by observers in the USSR and the Netherlands.

The second type of receiver is the swept-frequency receiver, in which the frequency band under observation is scanned over and over again with a single channel. The demand for high resolution in time and frequency poses some special requirements on this type of receiver, but parts of the construction can also be made
quite simple since the swept-frequency range need not be large. A block diagram
of a narrow-band swept-frequency receiver is shown in Fig. 2.

A two-stage broad-band RF amplifier precedes the first mixer and serves together with a relatively high IF frequency (42 Mc/s) to reject image signals. The
frequency sweep is made in the local oscillator by variable permeability techniques. In the first IF amplifier the bandwidth is 1 Mc/s, but it is cut down to
0.3 Mc/s in the second IF amplifier which has a center frequency of 15 Mc/s. The
frequency sweep is performed 50 times per sec. This has proved to be sufficient
in the majority of actual observations. A linear frequency variation with time is
wanted in the local oscillator, as is a fast fly-back and a constant output voltage.

The optimum bandwidth of a swept-frequency receiver is determined by the
scanning rate. If the bandwidth is extremely narrow, the output pulse essentially
consists of the transient response of the IF amplifier. Its duration is then in-
versely proportional to the bandwidth. The resolution will therefore improve when
the bandwidth is widened. At the other extreme, if the bandwidth is very wide the

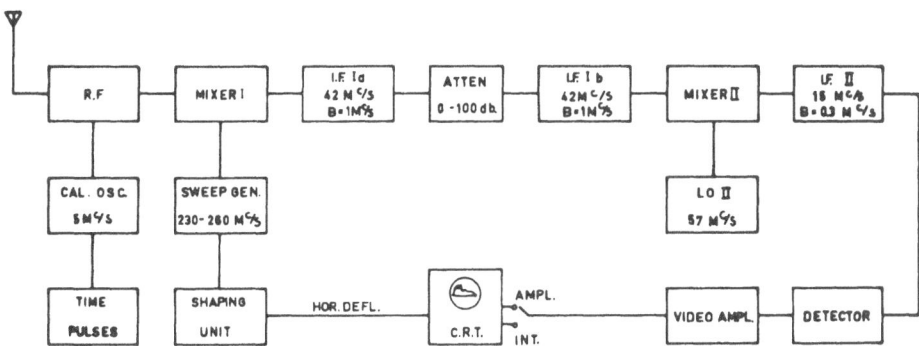

Fig. 2. Block diagram of a swept-frequency receiver.

output pulse is simply a trace of the passband characteristic, and its duration is proportional to the bandwidth. The scanning rate of swept-frequency receivers for solar work is relatively low, and optimum resolution will be very narrow. As we decrease the bandwidth, noise fluctuations in the output increase. The result is that the limit to the resolving power is not set by the scanning speed, but is determined by the necessity of suppressing noise fluctuations in the output. The bandwidth of 0.3 Mc/s has proved to be a good value which roughly corresponds to the reading accuracy on the records. The output of swept-frequency receivers is usually displayed on intensity-modulated cathode ray tubes and recorded on con-

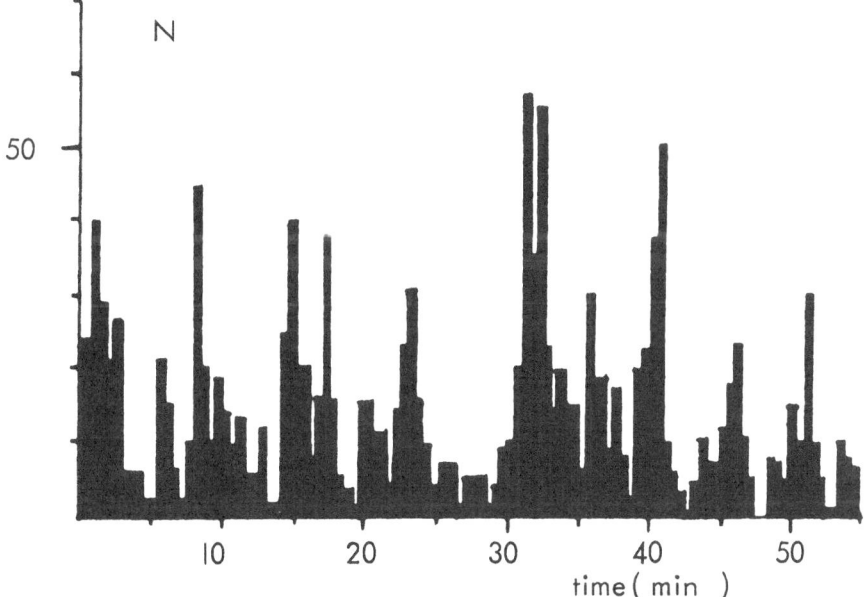

Fig. 3. The time variation of the number of bursts recorded per 1/2-min interval in the frequency range 220 to 195 Mc/s.

tinuously moving film. High-resolution swept-frequency receivers have been built in the U.S.A., the USSR, Japan, Finland, and Norway.

SPECTRAL OBSERVATIONS OF STORM BURSTS

Solar noise storm radiation often looks very complex when observed with a narrow-band spectrograph. The most prominent feature is the large number of short-lived, almost monochromatic bursts. In the frequency interval from 230 to 205 Mc/s more than 100 bursts may occur per min during storms of high variability, but such large numbers of bursts are not generated without intermission. There is a typical tendency for the burst frequency to fluctuate, as can be seen from Fig. 3. Occasionally the burst activity shows a regular periodicity, but more common are irregular variations. The individual bursts in the "showers" which last some 5 min seem to be independent of each other. This is different from the clusters of bursts in which 5 to 10 bursts occur in rapid succession, all of them lined up along the same frequency, or with a systematic change of frequency. It is likely that the bursts in a cluster are associated with one another as suggested by their behavior on the spectral records.

Finally, turning our attention to the individual bursts, it is readily found that their shapes in the frequency – time plane differ widely (Elgaröy [2]).

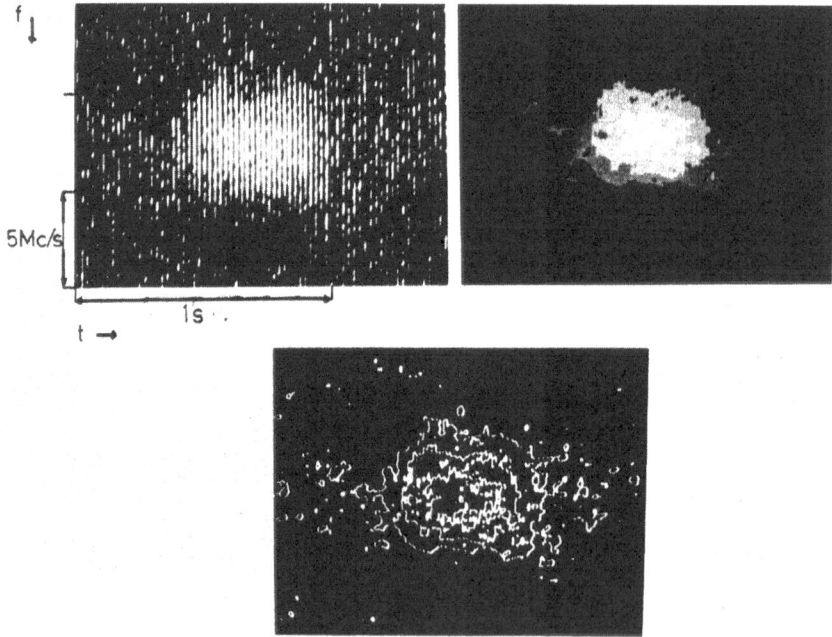

Fig. 4. Type I(s) burst and isophotes.

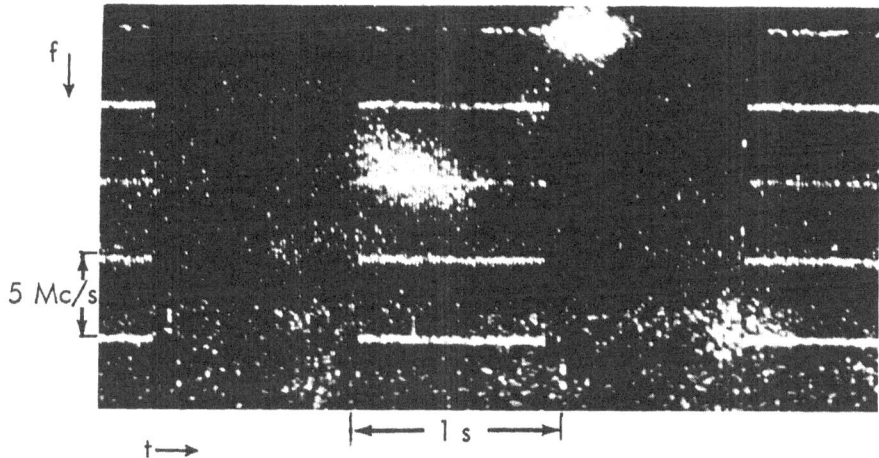

Fig. 5. Burst of Type I(r).

DYNAMIC SPECTRA

Although the dynamic spectra of Type I bursts take many forms, some are quite characteristic and are frequently encountered.

The most prominent dynamic spectrum in noise storms is that of Type I(s) bursts with narrow bandwidth (a few Mc/s), short duration (some tenths of a second), and stable center frequency. This is the type of spectrum originally thought to be characteristic of all storm bursts. The properties of the bursts can be better seen when isophotes are drawn. This usually entails a considerable amount of work, especially if very fine details are wanted. Therefore, an isodensitometer is very useful in the analysis of the records. The isodensitometer displays the isophotes on the screen of a cathode ray tube which can be photographed. Isophotes can be drawn for any film density in the material to be treated. Figure 4 shows a Type I(s) burst together with isophotes from the isodensitometer.

There are also storm bursts with frequency drift from higher to lower frequencies or vice versa, which may be denoted Type I(d) and Type I(r). We can determine the frequency drift velocity from the displacement with time of the center frequency. In the case shown in Fig. 5, the drift velocity amounts to about 5 Mc/s per second. There is no apparent difference in the bandwidth and duration of this burst as compared to bursts of Type I(s). As judged from our measurements, the only difference between bursts of Types I(d) and I(r) is the direction of the frequency drift. Both types may occur shortly after each other on the records.

The frequency drift velocity of Type I(d) and I(r) bursts is in the range of some Mc/s per second. This value falls between the velocities of Type III bursts, which are about one order of magnitude larger, and those of Type II bursts, which are roughly one order of magnitude lower.

Some Type I bursts have a high-frequency drift velocity of about the same magnitude as found for fast-drift bursts of spectral Type III. However, the bandwidth and duration are closer to the values which are characteristic of Type I bursts, as seen from Fig. 6. We introduce the notations I(fd) and I(fr) for Type I bursts with fast frequency drifts and fast reverse frequency drifts, respectively. Such bursts often tend to have shorter duration and larger bandwidth than other storm bursts. Vitkevitch, Gorelova, and Lozinskaya [16] have observed some bursts which are similar to the ones described here. The duration of the bursts was 1 sec or less, the bandwidth was of the order of 12 Mc/s, and the frequency drift velocity about –20 Mc/s per second. The observing frequencies were between 70 and 150 Mc/s.

Quite a few storm bursts have a curved shape in the frequency-time plane. If this effect is pronounced, we get a miniature U burst. There are also bursts on the records which do not fit into any of the classes mentioned. In some cases, the reason for this is that neighboring bursts partly overlap and complicate the picture.

It is not clear yet if the same variety of burst spectra is met with in the frequency region above 250 Mc/s. A swept-frequency receiver operating in the band from 340 to 310 Mc/s has therefore been put into operation at the Oslo Solar Observatory.

The frequency of occurrence of the different types of storm bursts has been determined. Discarding bursts that were weak or obscured by interference, the following results were found from observations around 200 Mc/s on three days of storm emission.

	Percentage		
Date	I(s)	I($d + fd$)	I($r + fr$)
August 13, 1959	49	21	30
September 10, 1959	71	13	16
September 11, 1959	77	12	11

On August 13, about one burst in two showed some kind of frequency drift. On the two other days, the corresponding proportion was roughly one in four. The average proportion of stable bursts to drifting bursts is somewhere around 2:1. This relation probably changes during a noise storm and may also vary for different storms. It has thus been noticed that occasionally there is a significant increase in the percentage of storm bursts with a fast-frequency drift velocity.

The study of high-resolution burst spectra has shown that the phenomenon of frequency drift is quite universal in meter wave bursts. Although most storm bursts belong to the I(s) class, a significant portion of them differ from this type. This must be accounted for in any satisfactory theory of the origin of the bursts.

I should like to remark that the distinction between bursts of Type I(d) and I(fd) is useful in the description of noise storms, but may not be real. It is not unlikely that there is a continuous transition from very low to very high values of the frequency drift velocity. This needs further study.

Duration

Several difficulties are met with in the determination of the duration of storm bursts on single-frequency records on account of the complexity of noise storms

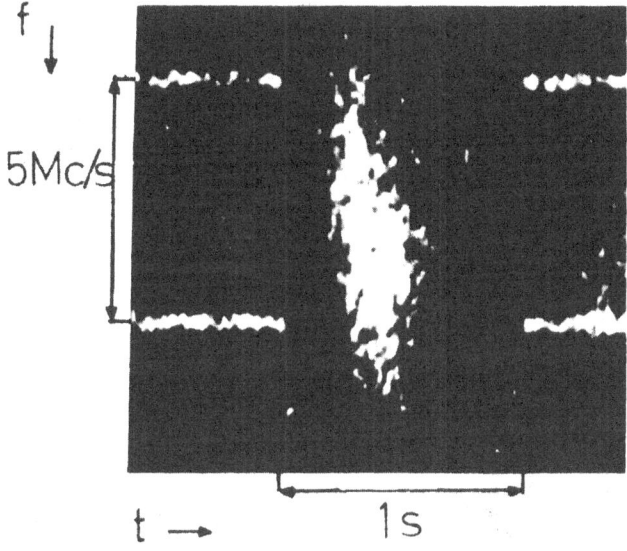

Fig. 6. Burst of Type I (fr).

and the great similarity of Type I and Type III bursts on such records. Observations made with high-speed recording equipment in the 200-Mc/s range have nevertheless given fairly good results. They seem to agree on a burst duration of about 0.3 to 0.4 sec as measured between half-power points.

It is laborious to determine the half-power duration of bursts on spectral records of the intensity-modulated type, but the lifetime, defined as the time during which the burst can be distinctly discerned from the background continuum, is easily measured. The lifetime will, of course, be longer than the half-power duration. Results from an analysis of the burst lifetime on three successive days of storm emission are shown in the histograms of Fig. 7. There is a considerable decrease in the number of bursts with lifetimes of 0.2 sec or less; similarly, lifetimes longer than about 1 sec are rare. The majority of the bursts have lifetimes between 0.3 and 0.7 sec. There is no appreciable difference in the distribution curves of the three successive days. Computing the mean values, we find 0.63, 0.53, and 0.65 sec. These results are consistent with the previous measurements of half-power duration from single-frequency records.

On some occasions very short-lasting bursts occur. Such activity was, for instance, particularly pronounced in the noise storm of August 15 through 19, 1961. The short-lived "flashes" of radiation were found singly, in groups of 2 or 3, or in clusters lasting from 0.5 to 1 min. On August 17, for instance, 23 bursts were recorded on the spectrograph film between 14.16 and 14.17 U.T. Their mean lifetime was only 0.13 sec.

The two histograms in Fig. 8 illustrate the difference between flash burst activity and normal storm burst activity. The respective mean lifetimes calculated

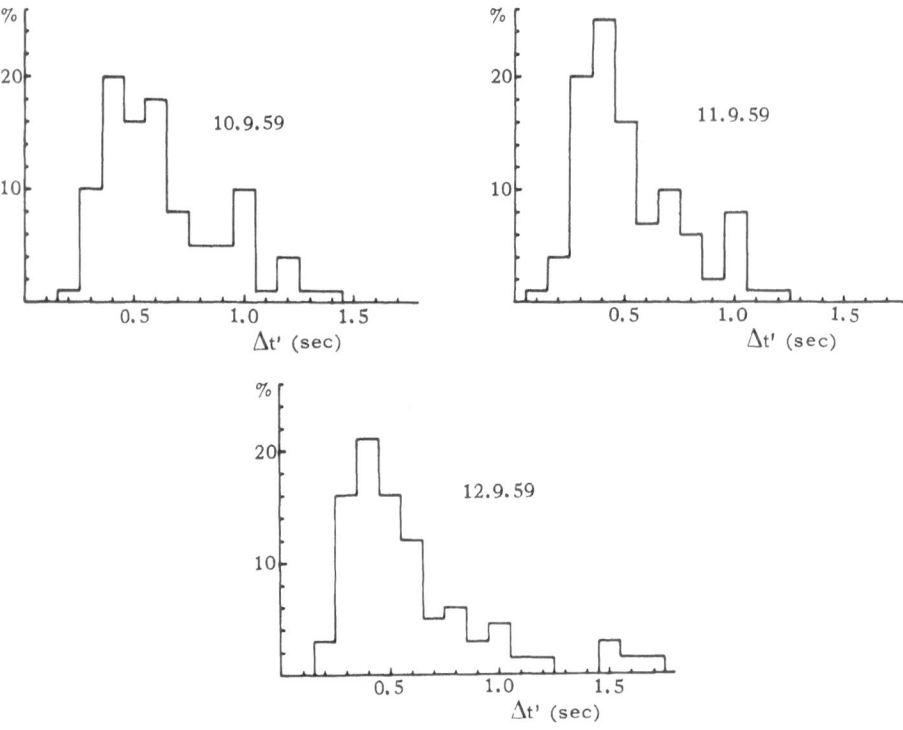

Fig. 7. The distribution of bursts of different lifetimes on three days of storm emission.

from the two distributions are 0.18 and 0.50 sec. The shortest lifetime measured for the flash bursts was only 0.06 sec, which is close to the instrumental limit of resolution.

It is important to know if the burst duration varies with the observing frequency. To investigate this, one should preferably make simultaneous observations in different frequency ranges. This has not yet been done, and one is left with the less satisfactory possibility of comparing data which have been collected from different noise storms and by various methods. When this is done, we find that there is only a slow variation in the burst duration as a function of the frequency. In the 300 to 400 Mc/s range, the half-power duration is about 0.18 sec (de Groot [4]), around 200 Mc/s it amounts to 0.3 sec, and at 100 Mc/s it has increased to roughly 0.5 sec (Vitkevitch, Gorelova, and Lozinskaya [16]). This result is of value for the interpretation of the burst phenomenon.

Line Profile and Bandwidth

The shape of the spectral profile of storm bursts can be determined from photometer tracings along the frequency axis of intensity-modulated records and from direct photographs of amplitude-modulated scans. It may, of course, also be determined from multichannel records. Before a good determination of the profile is made, one has to correct for several instrumental effects which may easily disturb the results of the analysis. Errors may arise from nonlinear frequency scale, variable gain over the observed frequency band, saturation of the receiver, receiver time constant, and receiver bandwidth.

A very accurate analysis of 22 bursts observed at the Oslo Solar Observatory showed that 11 were symmetrical, 2 had a low-frequency cutoff, and 9 had a high-frequency cutoff (Fig. 9). The profiles of the symmetrical bursts showed good agreement with a Gaussian curve. Further investigations showed that perhaps the majority of Type I bursts have only small deviations, if any, from a symmetric pro-

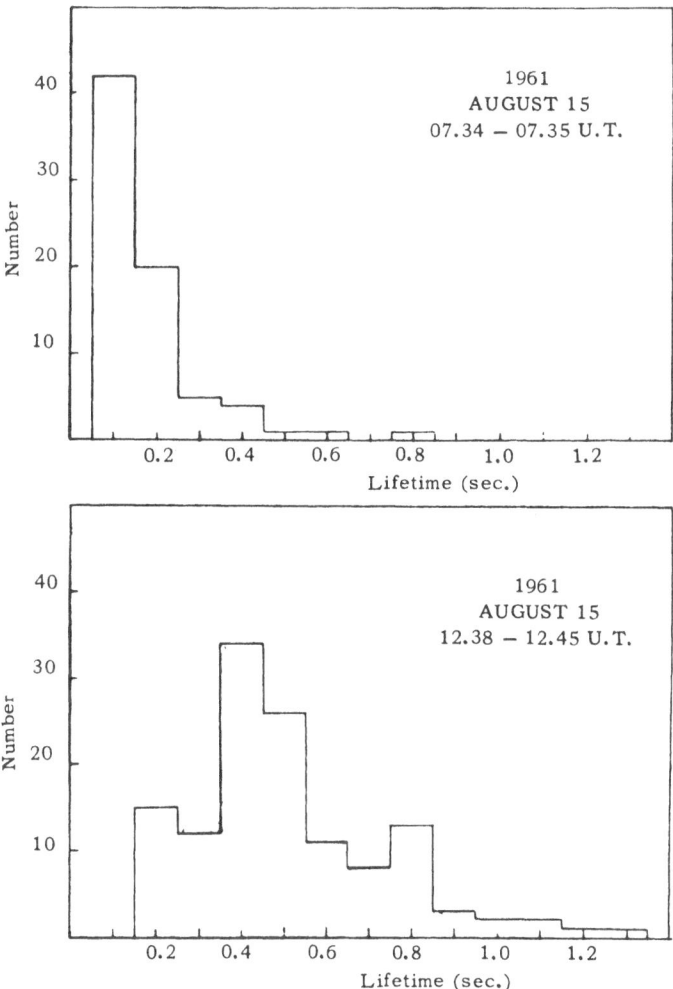

Fig. 8. Histograms showing the difference in duration of ''flash'' bursts and normal storm bursts.

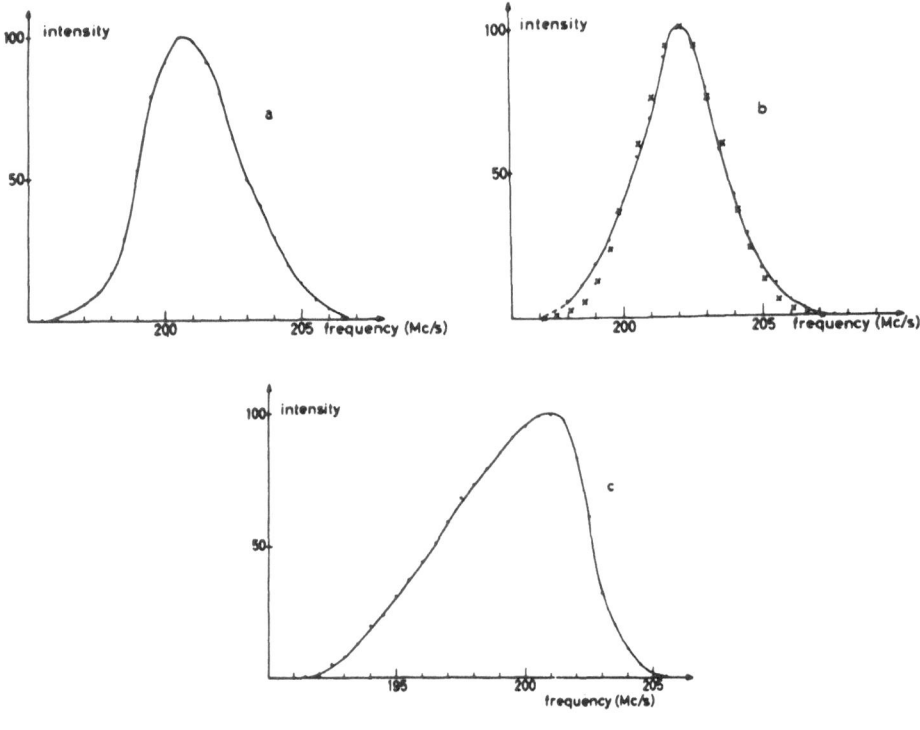

Fig. 9. Line profiles of storm bursts.

file. Attempts were made to find bursts with systematic transitions from asymmetry at the beginning to symmetry at a later stage. Examples of such transitions were indeed present, but there were also instances of transitions in the reverse order. In cases of frequency drift, the drift direction had no influence on which side the cutoff in the frequency spectrum occurred. In the 300-Mc/s region, de Groot [4] found that single storm bursts had no frequency drift and that the intensity could be written as $I(f,t) = I_1(f) \cdot I_2(t)$, where I_1 and I_2 were symmetric profiles.

The frequency profiles of the bursts show that the bandwidth is small. This was long known because simultaneous records separated by only a few Mc/s were often very different. In Fig. 10, a histogram of the number of bursts of different half-power bandwidths is shown. The bandwidths were determined from spectral records in the 200 Mc/s range. Half-power bandwidths between 2 and 5 Mc/s are normal. Small changes in bandwidth occur, thus on 4 different days of storm emission mean bandwidths of 4.2, 3.2, 4.1, and 2.7 Mc/s were found.

The question of a possible connection between burst intensity and bandwidth is not yet settled. Vitkevitch [15] made a statistical determination of the bandwidth of 200-Mc/s storm bursts recorded with a two-channel receiver. He divided the bursts into two groups. One group comprised intense bursts with amplitudes above 50% of the mean burst level. The other group contained bursts that were weak, but could be definitely recognized on the recordings. In his material, Vit-kevich found a mean bandwidth of 6 Mc/s for small bursts and 12.3 Mc/s for the strong ones. This difference is very large, considering that Wild [19] found no definite correlation between amplitude and bandwidth on his spectral records.

By using data from different observers, we can determine the dependence of the bandwidth on the observing frequency. In the 100-Mc/s range, Wild [19] found a bandwidth (measured between points of quarter-maximum intensity) of 4 Mc/s to be the most common. If a Gaussian profile is assumed, the ratio between quarter-power and half-power bandwidth is

$$\frac{\Delta B\,(\tfrac{1}{4})}{\Delta B\,(\tfrac{1}{2})} = \left(\frac{\ln \tfrac{1}{4}}{\ln \tfrac{1}{2}}\right)^{\tfrac{1}{2}} = 1.4$$

The value of 4 Mc/s measured between points of quarter-maximum intensity then correspond to a half-power bandwidth of 2.9 Mc/s. On the higher frequencies be-tween 300 and 400 Mc/s, de Groot [4] found a half-power bandwidth of 6 Mc/s.

Present evidence thus indicates an increase in the bandwidth of the storm bursts with increasing frequency. The results are illustrated in Fig. 11. For the frequency interval from 100 to 400 Mc/s, the results are approximated by the formula

$$\Delta B = \frac{f}{100} + 2 \,[\text{Mc/s}]$$

Echoes in the Solar Corona?

Several investigations have been made to detect echoes of short-duration solar radio emissions. Payne-Scott [11], who observed on 60 and 19 Mc/s, noticed that a number of unpolarized bursts showed two humps, the first one being the more intense. This was tentatively interpreted as an echo phenomenon, and Jaeger and Westfold [24] showed theoretically that echoes could be expected from the regu-lar corona. Spectral observations by Wild, Murray, and Rowe [25] made it clear that the double-humped bursts most likely were of spectral Type III, and that the second part was the second harmonic and not a delayed repetition of the first part. Later, J. A. Roberts [12] described a spectral type of radio burst which he called

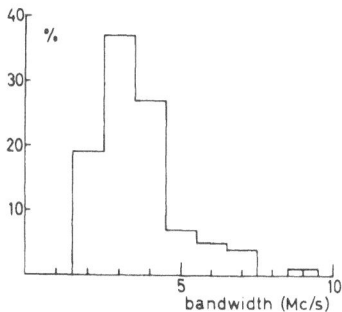

Fig. 10. Histogram of the distribution of bandwidths of 200 Mc/s storm bursts.

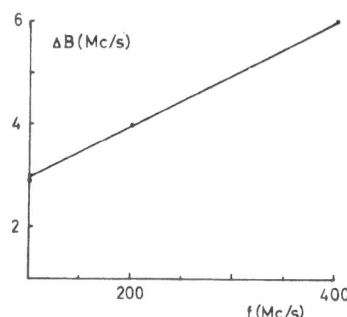

Fig. 11. The bandwidth of storm bursts
as a function of frequency.

reverse-drift pairs. Such bursts are of short duration and are confined to the longer
meter wavelengths. A reverse pair has two elements, the second being a repetition
of the first after a delay of 1.5 to 2 sec. Roberts suggested that the second element was an echo of the first, reflected from lower levels in the solar corona. The
observed delay between the elements could be quantitatively explained if the burst
radiation occurred at the second harmonic of the coronal plasma frequency and the
density gradient were 1.5 times steeper than in the Baumbach – Allen model.

A statistical search for echoes was made by de Jager and van't Veer [7].
They picked out singly occurring, well-pronounced bursts and counted the number
of preceding and following bursts in a 5-sec interval on their 200-Mc/s high-speed
records. If there were echoes in the material, more bursts should be found after
than before the central one. They found no significant difference, and concluded
that there was no evidence of echoes in the quiet parts of the records.

When the dynamical spectra of the bursts are recorded, it should be possible
to recognize individual echo events. The genuine echo phenomenon, caused by
two waves emanating from the same source but propagating along different paths,
should comprise a leader and a delayed weakened "ghost" which shows the same
features as the first element.

The general impression one gets from a visual inspection of a large number
of swept-frequency records is that echolike cases are rare events. But this may
be incorrect if the echo delay time is less than about 0.3 sec; then the eye will not
detect the echo.

It is obvious that the echo hypothesis would be strengthened if it could be
established that details in the first element were repeated in the second one. In
some cases this seems to be true. There are several examples in which double
bursts of quite similar characteristics occur. Figure 12 shows some possible
echoes. Delay times are usually between 0.5 and 0.7 secs. The swept-frequency
observations have not yet proved that echoes of Type I bursts exist, but they offer some support to such a suggestion.

Further progress awaits the simultaneous observation of the dynamic spectrum and the position on the sun. If echoes exist, the reflected ray has a position different from that of the direct ray, being systematically nearer to the center of the solar disk.

Burst "Extensions" and "Bridges"

Storm bursts often reveal peculiar features. One may for instance find "extensions" near the beginning and the end of the burst. In Fig. 13, it appears as if a disturbance of weak intensity around 195 Mc/s drifts a couple of megacycles to lower frequencies where it explodes in an intense burst, and then fades back to 195 Mc/s and disappears. One may speculate whether, by using equipment of still better sensitivity, it would be possible to detect some sort of connection, a "bridge" between different bursts. Some evidence in favor of such an effect is often found on the records. An investigation of this phenomenon is now being undertaken. If the effect is real, it is very important.

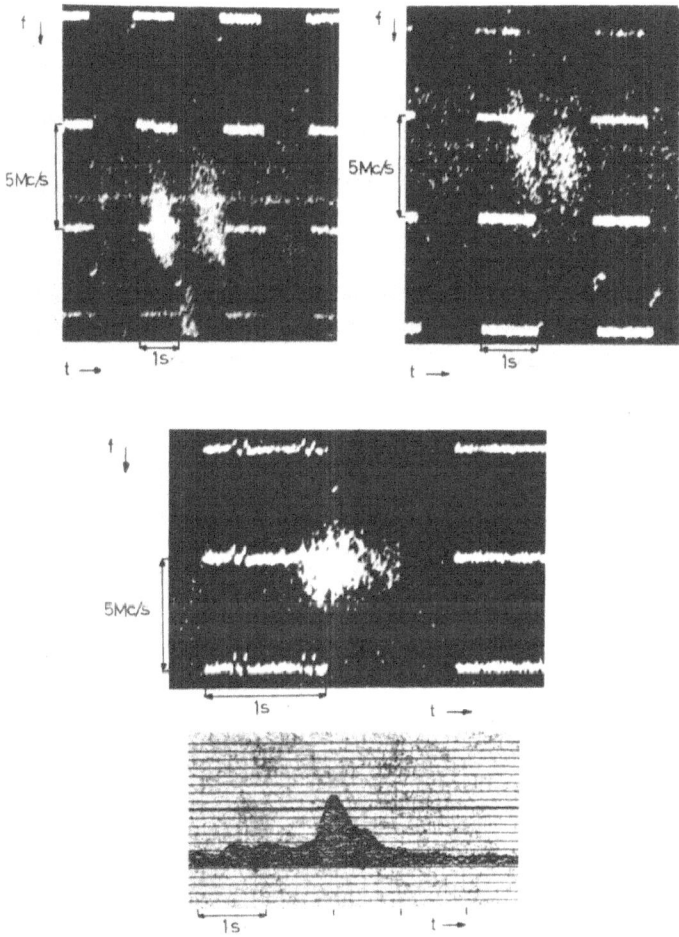

Fig. 12. Examples of possible echo events.

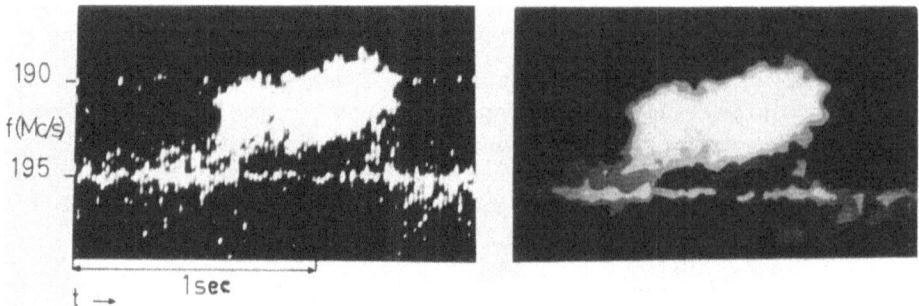

Fig. 13. Burst with "extensions."

Some noise storms have a very diffuse and complex character and it is difficult to describe the bursts in terms of duration, bandwidth, and dynamic spectrum.
Here much exploratory work remains to be done.

REMARKS ON THE ORIGIN OF TYPE I BURSTS

Several theories have been proposed to explain the noise storm phenomenon.
Here we shall restrict the discussion to some of the problems connected with the
burst component of the storms. The properties of the bursts which must be accounted for in any satisfactory theory are: (a) the short duration (0.15 to 0.5 sec),
(b) the narrow bandwidth (3 to 6 Mc/s), (c) the frequency drift of some of the bursts,
(d) the high degree of circular polarization, (e) the confinement of the burst emission to frequencies between about 50 and 400 Mc/s, and (f) the very high brightness temperature of the burst sources (10^{10} °K).

The high brightness temperature of the bursts excludes thermal radiation as a
mechanism for the burst generation. A further restriction on possible mechanisms
is imposed by the monochromatic character of the bursts. It seems as if only two
possibilities are left open, either a plasma mechanism or gyro radiation.

The problems involved in the explanation of the bursts in the plasma model
may be schematized in the following way:

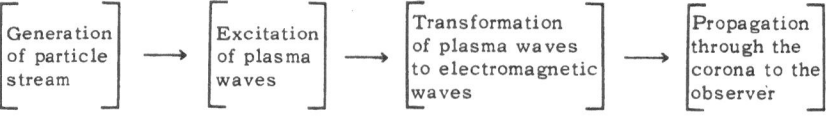

In the plasma theory, the region where the plasma waves are generated must
be the same as that where they are transformed into electromagnetic waves, because plasma waves attenuate rapidly in time and space. It is also clear that the
electron beam must have a short lifetime since the duration and the drift range of
the bursts are small. Thus, it seems reasonable to assume that the electron stream
is generated and the plasma waves excited and transformed into transverse waves
at the same place in the corona, i.e., in the burst source.

What determines the duration of Type I bursts? It is tempting to assume that the burst duration is determined by the damping time of the plasma oscillations. Turning to the observations for guidance in this question, we find, first, that the burst duration is too long compared to the damping time of the plasma waves and, secondly, there is no pronounced variation of the duration of the bursts as a function of the frequency. This is contrary to the behavior of the collision frequency which is proportional to the square of the plasma frequency. The burst duration is therefore mainly determined by the lifetime of the electron stream. The electron stream breaks down due to four processes, which are associated with the following characteristic times (Takakura [14]):

1. Deflection time t_D due to collisions with thermal electrons.
2. Energy exchange time t_E due to collisions with thermal electrons.
3. Energy loss time t_1 due to excitation of coherent plasma waves.
4. Energy loss time t_2 due to excitation of incoherent plasma waves.

The shortest of these is the one of importance for the duration of the bursts.

When the velocity of the electrons in the beam exceeds the most probable speed of the thermal electrons, the deflection time t_D is less than the energy loss time t_E. Takakura [14] has further shown that t_D might be less than t_1 and t_2. The deflection time t_D is therefore decisive for the burst duration. It is given by (Spitzer [13])

$$t_D = \frac{m^2 v_0^3}{8 \pi e^4 N \ln \Lambda [\Phi(v_0/v_t) - G(v_0/v_t)]}$$

where m is the electron mass; e, the charge of proton; N, the electron density; v_0, the velocity of electron stream; and v_t, the velocity of thermal electrons.

For $T \approx 1 - 2 \times 10^6 \, °K$ and $N \approx 10^8 - 10^9 \, cm^{-3}$, $\ln \Lambda \approx 20$. If $v_0 > 2 v_t$, $\Phi(v_0/v_t) - G(v_0/v_t)$ is 0.9 to 1.0. The expression therefore reduces to

$$t_D \approx 2 \times 10^{-3}(m^2 v_0^3/e^4 N) \qquad \text{or} \qquad t_D \approx 2.5 \times 10^{-12}(v_0^3/f^2)$$

Inserting durations of 0.18, 0.3, and 0.5 sec for the respective frequencies of 400, 200, and 100 Mc/s, we find stream velocities of 2.2×10^9, 1.7×10^9, and 1.3×10^9 cm/sec.

The most probable speed of thermal electrons $v_t = \sqrt{2kT/m}$ amounts to 5.5×10^8 cm/sec, when $T = 10^6 \, °K$. Thus, we find that

$$\left(\frac{v_0}{v_t}\right)_{400} \approx 4, \qquad \left(\frac{v_0}{v_t}\right)_{200} \approx 3, \qquad \text{and} \qquad \left(\frac{v_0}{v_t}\right)_{100} \approx 2.4$$

On all the frequencies $v_0 > v_t$, and plasma waves might be excited by the stream.

With these velocities for the electron beam, we may calculate the frequency drift velocity. Using Newkirk's model for the electron density above an active region, we find that the frequency drift velocity in the 200 Mc/s region should amount to roughly 10 to 20 Mc/s per second, which is in fair agreement with the observations. In a time of about a third of a second, the burst should drift about as much as its own bandwidth. We have noted that in several cases a higher frequency drift velocity is found. Then, the velocity of the electron stream might be higher, or, equally likely, the density gradient might be steeper than in the model, which, of course, only represents an average value.

The bandwidth of the bursts may be determined by the distribution of electron densities in the source.

It has also been assumed that the bandwidth is due to thermal Doppler broadening, (de Jager and van't Veer [7]). For a temperature of 10^6 °K, we would have

Frequency (Mc/s)	Bandwidth (Mc/s)
100	2.3
200	4.5
400	9.0

It is seen that the variation in bandwidth with frequency is larger than observed, and Doppler broadening cannot be the only mechanism responsible for the bandwidth if the temperature is roughly constant at the different source heights.

How is the electron stream created? It may be caused by shock waves or, as recently proposed by Takakura [14], by the collision of two hydromagnetic waves propagating in opposite directions along a magnetic field. In Takakura's model, one can calculate the speed of the Alfvén waves which is necessary to accelerate thermal electrons to the velocities which we have found. It can be shown that the necessary speed is about $v_0/3$, and thus we have

$$v_m = \frac{H}{\sqrt{4\pi\rho}} \approx \frac{v_0}{3}$$

where $\rho = M_p N$, and M_p is the mass of the proton.

Inserting numerical values, one obtains a field strength of 15 G in the 80 to 100 Mc/s source, about 50 G in the 200 Mc/s source, and near 150 G at 400 Mc/s. The existence of such field strengths in the corona is probably possible at the observed heights of noise storms.

The field strengths which were found seem to be sufficient to explain the circular polarization of the Type I bursts as a propagation effect. If radiation is to escape from the corona, it must be generated above the level of zero refractive index for the ordinary mode. The height of this level is determined by the electron density distribution in the corona. In the presence of a magnetic field, there is another level above the first, at which the refractive index for the extraordinary wave is zero. Radiation which is generated in the region between the two levels is polarized, because only the ordinary mode escapes. In order to polarize the escaping waves completely in the ordinary sense, the magnetic field in the 200 Mc/s radio source must be greater than 29 G. This condition may be easily satisfied. Unpolarized bursts might originate in dense clouds at a higher altitude.

An interesting hypothesis was put forward by Twiss and Roberts [23] concerning the possibility of explaining the bursts as caused by gyro radiation. They proposed that the bursts were due to gyromagnetic radiation which was amplified by a negative absorption coefficient in the medium it traversed. The amplification is necessary in order to explain the high brightness temperature associated with the observed flux. (This mechanism of stimulated emission is the same as is applied in masers.) A necessary condition for the amplification is that there is an over-population of the states of higher energy. Thus, both in the plasma model and the gyro model it is necessary to have streams of electrons with energy in excess of the mean thermal energy. The problem of acceleration of electrons in the corona is therefore of great importance.

One major difficulty in the interpretation of the noise storm phenomenon is that the observed properties of the radiation are the products of both the generation and propagation of the radiation through the corona, and it is very difficult to determine by observation what is due to propagation and what is due to the generation. Denisse [1] has pointed out the importance of the propagation, and he has proposed a theory in which storm bursts are explained as a propagation effect. Fast particles are ejected from chromospheric flares. Some of the particles are trapped in suitable magnetic field configurations in coronal streamers. The trapped particles diffuse slowly along the streamers and excite plasma waves. These waves are transformed into radio waves when they propagate along magnetic lines of force in the direction of increasing density, i.e., toward the solar surface. They are then coupled to the ordinary mode which propagates in the same direction, but which may be reflected in dense regions. On this assumption, noise storm bursts may be explained as sporadic specular reflections (focusing) in small frequency bands. The reflected waves may be observed to have highly variable intensities because the specular reflections may be very directive. Some support for this theory has been provided by observations by le Squeren [22].

When considering the different theories for the generation of Type I bursts, one should face the fact that several of them may be right. For instance, it seems reasonable to explain some of the very complex noise storms in terms of the model of Denisse, whereas others may be more effectively described in the Takakura or the Twiss and Robert models.

SPECTRAL OBSERVATIONS OF TYPE III BURSTS

The wide-band swept-frequency receivers permit unambiguous classifications and extensive studies of the large-scale features of Type III bursts, but it is useful to supplement this material with information from high-resolution radio spectrographs.

In many cases, the recorded bursts are well defined without significant fine structure. Such bursts are easily recognized as spectral Type III. In other cases, the records give a more complex impression, and it might be necessary to obtain additional information from wide-band swept-frequency observations.

Frequency Drift

It is relatively simple to determine the frequency drift velocity of Type III bursts from the records. This might be done in several ways. One method is to scan the bursts along the time axis close to the lowest and highest recording frequencies and determine the drift velocity df/dt from the time displacement of the maximum intensity on the photometer scans. Another method is to display the records on millimeter graph paper and determine the frequency drift velocity from the inclination of the leading edge of the burst with the time axis.

A histogram of the number of bursts of different drift velocities is shown in Fig. 14. The mean drift velocity as calculated from the observations (Elgaröy and Rödberg [3]) is

$$\frac{df}{dt} = -(60 \pm 10) \text{ Mc/s per second} \quad (\text{rms error})$$

The histogram shows a large spread in the values of the observed frequency drift velocity. The distribution curve has a maximum around −40 Mc/s per second and a "tail" extending to −150 Mc/s per second. It is also noteworthy that there is a marked cutoff near −20 Mc/s per second.

Wild [18] determined the frequency drift velocity of Type III bursts in the frequency range from 70 to 130 Mc/s. He found that the drift velocity was approximately proportional to the observing frequency, and was given by

$$\frac{df}{dt} = -\frac{f}{k}$$

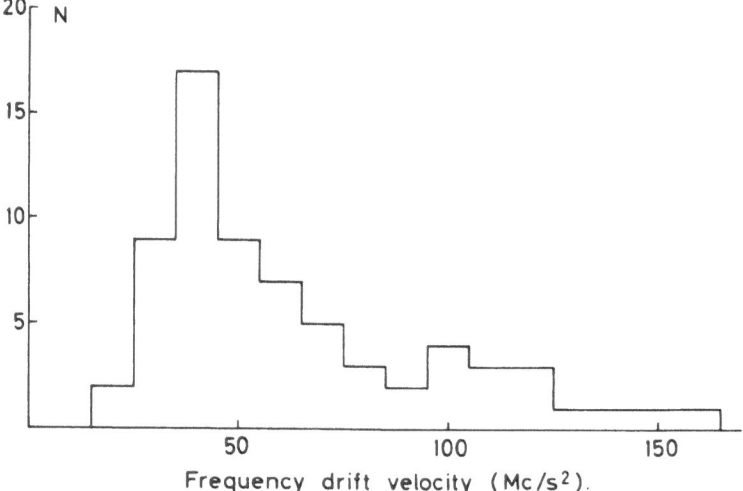

Fig. 14. The distribution of Type III bursts of different frequency drift velocities (negative) in the 200-Mc/s range.

where k is a constant which varies from one burst to another. The mean value of k determined from a sample of 10 bursts was 4.5 sec. On inserting the mean value of df/dt as determined from the observations with the narrow-band swept-frequency receiver, we find $k = (3.6 \pm 0.6)$ sec, which is relatively near the value found by Wild. The value of k appears to decrease with increasing frequency. This trend of a decreasing k with increasing frequency was also found by Maxwell, Howard, and Garmire [9]. This is, moreover, the trend to be expected if the bursts are caused by oscillations which are excited by the outward traveling of a disturbance in the solar corona. If the exciting agency propagates at constant speed in a corona where the electron density follows Newkirk's model [10], one expects that

$$\frac{(\Delta f/\Delta t)_{100}}{(\Delta f/\Delta t)_{215}} \approx 0.32$$

which is near the observed ratio of 0.37.

If we know the variation of electron density with height in the solar corona and the position of the burst on the sun, we can find the connection between the velocity of the exciting disturbance and the observed frequency drift velocity. When we further know the distribution of the number of Type III bursts with the position on the sun, and the true velocity distribution of the exciting disturbances, we can calculate the expected distribution of frequency drift velocities as observed on a particular frequency. When this is done, we find good agreement between the observed distribution in Fig. 14 and the calculated curve when exciting velocities between 0.2 and 0.5c are supposed to be equally likely.

The Time Profile of Type III Bursts

The time profile is interesting to study because it can give some information about the mechanism of generation and the physical conditions at the place of origin of the oscillations.

Time profiles of Type III bursts were studied by Payne-Scott [11] and de Jager and van't Veer [7], but the investigations suffer from the fact that even if it is reasonable to believe that most of the analyzed bursts belong to special class III, one may have included Type I bursts. Recently, a study of records from the narrow-band swept-frequency receiver at Oslo was made by Elgaröy and Rödberg [3].

The mean time profile of 13 Type III bursts was determined. The result is shown in Fig. 15. The statistical uncertainty, given by the rms error, is indicated by the length of the bars. Defining t_p as the rise time from a chosen intensity level to maximum intensity and similarly t_f as the time spent in decay from maximum intensity to the chosen level we find that at a level of $I/I_{max} = 10\%$, for instance, $t_p = 0.48$ sec and $t_f = 0.54$ sec.

The observed average time profile of the bursts can be calculated on the basis of simple assumptions. The method was first used by de Jager and van't Veer [7]. We assume that the burst is excited by a disturbance which moves rapidly out through the corona, exciting at each level oscillations at the appropriate plasma frequency. The oscillations have an abrupt start and are damped out according to

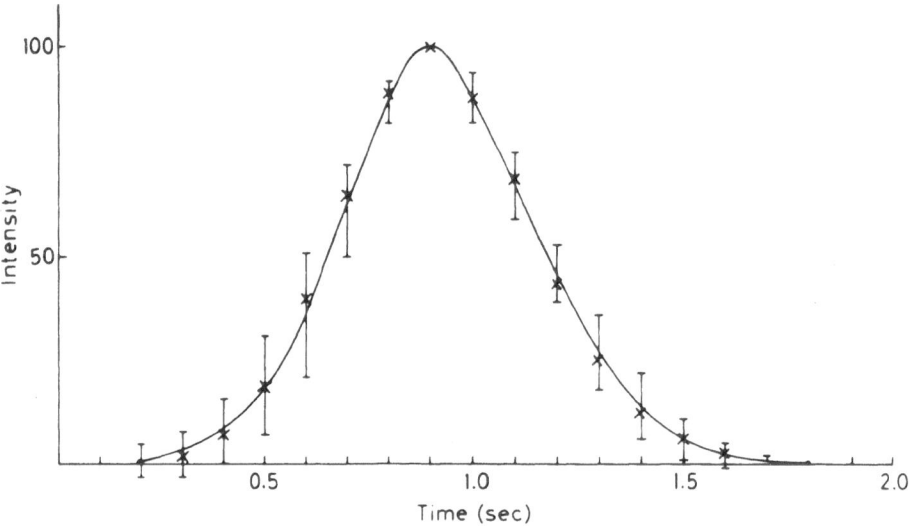

Fig. 15. The mean time profile of Type III bursts in the 200-Mc/s band.

an exponential law, exp $(-\nu t)$, where t is the time and ν is the damping constant. At any time, the excited oscillations have a frequency spectrum of Gaussian shape, following a function of the Type I proportional to exp $-[(f - f_m)/\Delta]^2$, where f_m is the frequency of the maximum intensity and Δ is half the e-width. The frequency drift velocity is given by $df/dt = -a$, and a is assumed to be constant in a relatively small frequency range around the observing frequency f_0. The time at which f_0 is excited is zero, and the time axis is so defined that $f = f_0 - at$. The intensity of the burst at the frequency f_0 and at the time T is then given by

$$I(T) = C \int_{-\infty}^{T} e^{-(at/\Delta)^2} \cdot e^{-\nu(T - t)} dt$$

where C is a constant. Introducing $z = at/\Delta - \nu\Delta/2a$, we can show that the time profile is given by

$$I(T) = C' e^{-\nu T} \left[1 \pm \frac{2}{\sqrt{\pi}} \int_{0}^{|z|} e^{-z^2} dz \right]$$

where the plus sign should be used when $z > 0$ and the minus sign when $z < 0$.

When the frequency drift velocity a, the bandwidth Δ, and the damping constant ν are known, one can easily calculate the profile. The mean frequency drift velocity was found to be $-(60 \pm 10)$ Mc/s per second, but the bandwidth and the damping constant are unknown. It is easily seen from the records that the bandwidth is large, and Δ is certainly greater than 10 Mc/s. The expected value of the ratio a/Δ is thus likely to lie in the region $a/\Delta < 6$.

Profiles with different values of a/Δ and ν can be constructed and compared with the observed mean time profile. It is then found that for values of $a/\Delta = 3.5$, and $\nu = 8$, the agreement with the observations is extremely good, as can be seen from Fig. 15, where the crosses refer to the calculated profile. If one takes into account the uncertainty in giving the observed profile, the following range of possible values of a/Δ and ν is found, giving profiles in reasonable agreement with the observations.

$$5 > \frac{a}{\Delta} > 3 \qquad\qquad 9 > \nu > 5$$

If we take a mean frequency drift velocity in the 200-Mc/s range of -60 Mc/s per second, the value $a/\Delta = 3.5$ gives a half-power bandwidth for Type III bursts of

$$h = 2\sqrt{\ln 2}\,\Delta = 28$$

Therefore

$$h \approx 30 \text{ Mc/s}$$

This value seems quite reasonable.

Weiss and Sheridan [17] published a curve showing the mean time profile of 12 simple Type III bursts observed on 40 Mc/s. If we try to fit a theoretical curve to the observed points, we find good agreement with values of $a/\Delta = 0.35$ and $\nu = 0.25$. A reasonable value of a in this frequency region seems to be 8 Mc/s per second. We then obtain a half-power bandwidth of $h = 38$ Mc/s.

The bandwidth of the natural oscillations in the corona is probably very small, as shown by Malville [8]. Thus, in order to produce the large bandwidths necessary to explain the observed time profile of the Type III bursts, one may assume that the bandwidth is determined by the dimensions of the exciting agency. In Newkirk's model for an active region, the electron density along the axis is given by

$$N_e(\rho) = 8.26 \times 10^4 \times 10^{4.32/\rho} \text{ [electrons/cm}^3]$$

where ρ is in units of the solar radius $R_\odot = 7 \times 10^5$ km. The plasma frequency is given by

$$f_p = 0.9 \times 10^4 \sqrt{N_e} \text{ [c/s]}$$

Differentiating with respect to ρ, we obtain

$$\frac{df_p}{d\rho} = \frac{-1.29 \times 10^7}{\rho^2} \times 10^{2.16/\rho}$$

From this expression we can calculate $d\rho$ when df_p is given, and thus find the depth of the exciting disturbance. Using $\rho \approx 1.15R_\odot$ at 215 Mc/s and $\rho \approx 1.80R_\odot$ at 40 Mc/s, we find

$$d\rho_{215} \approx 0.04R_\odot \qquad \text{and} \qquad d\rho_{40} \approx 0.6R_\odot$$

The large source depth favors the suggestion that the exciting agency of Type III bursts is a stream of particles ejected at very high speed from the sun. Such a mechanism was first proposed by Wild, Roberts, and Murray [21]. The increase in the source dimensions on the way out through the corona can be explained if the exciting particles are ejected with a velocity dispersion. Hughes and Harkness [5] have arrived at similar results in a study of wide-band records from Fort Davis.

In the analysis of the time profiles, we found damping constants of 8 and 0.25 sec^{-1} at 215 and 40 Mc/s, respectively. If the damping constant is identified with the collision frequency at the place of origin of the radiation, we find a coronal temperature of $\approx 2 \times 10^6$ °K. This is in agreement with Hepburn's [6] streamer temperature of 2.3×10^6 °K found from optical eclipse observations.

CONCLUSION

During noise storms, the frequency of occurrence of Type I bursts (storm bursts) fluctuates. The individual bursts seem to be independent of each other, but some which occur in clusters of 5 to 10 members are probably in some way related.

Type I bursts are nearly monochromatic with $(\Delta B/f) = 2\%$. The mean duration of the bursts increases from about 0.15 sec at 400 Mc/s to 0.5 sec in the 100-Mc/s range.

Observations with a high-resolution swept-frequency receiver in the 200-Mc/s range show that storm bursts with different dynamic spectra occur. Multichannel observations at 300 and 400 Mc/s seem to give somewhat different results. Echoes in the solar corona have been looked for, but have as yet not been firmly detected. Theories for the burst generation are still only vague, and the problems involved need further consideration.

High-resolution radio spectrographs may also be used for investigations of Type III bursts. Some observations at the Oslo Solar Observatory show that at 200 Mc/s $(df/dt) = -(60 \pm 10)$ Mc/s and a bandwidth of 25 to 30 Mc/s is deduced indirectly. High resolution spectrographs may also be used for observations of other spectral types of solar emission. Much work remains to be done in this area.

There is a need for more high-resolution radio spectrographs operating simultaneously in different frequency ranges. For the correct interpretation of the records, wide-band spectra should also be available. Better insight into the storm-burst phenomenon may also be gained through a combination of high-resolution observations with simultaneous interferometric two-dimensional determinations of the burst positions.

REFERENCES

1. J. F. Denisse, *Inform. Bull. European Solar Obs.* No. 4 (1960).
2. Ö. Elgaröy, *Astrophys. Norvegica* 7, 5 (1961).
3. Ö. Elgaröy and H. Rödberg, *Astrophys. Norvegica* 8, 9 (1964).
4. T. de Groot, *B.A.N.* 15, 502 (1960).
5. M. P. Hughes and R. L. Harkness, *Ap. J.* 138, 239 (1963).
6. N. Hepburn, *Ap. J.* 122, 445 (1955).
7. C. de Jager and F. van't Veer, *Rech. Astr. Obs. Utrecht* 14, 1 (1958).
8. J. M. Malville, *Ap. J.* 136, 266 (1962).
9. A. Maxwell, W. E. Howard, and G. Garmire, *Sci. Report N:14*, Contract AF 19(604)1394 (1959).
10. G. Newkirk, *Ap. J.* 133, 983 (1961).
11. R. Payne-Scott, *Australian J. Sci. Res.* 2A, 214 (1949).
12. J. A. Roberts, *Australian J. Phys.* 11, 215 (1958).
13. L. Spitzer, *Physics of Fully Ionized Gases* (Interscience Publishers, New York, 1956).
14. T. Takakura, *P.A.S.J.* 15, 4(1963).
15. V. V. Vitkevitch, *Dokl. Akad. Nauk. S.S.S.R.* 101, 229 (1955).
16. V. V. Vitkevitch, M. V. Gorelova, and Lozinskaya, *Soviet Astron. — AJ* 4, 595 (1960).
17. A. A. Weiss and K. V. Sheridan, *J. Phys. Soc. Japan, Suppl. A-II,* 17, 223 (1962).
18. J. P. Wild, *Australian J. Sci. Res.* 3A, 541 (1950).
19. J. P. Wild, *Australian J. Sci. Res.* 4A, 36 (1951).
20. J. P. Wild and L. L. McCready, *Australian J. Sci. Res.* 3A, 387 (1950).
21. J. P. Wild, J. A. Roberts, and J. D. Murray, *Nature* 173, 532 (1954).
22. A. M. le Squeren, *Ann. Astrophys.* 26, 97 (1963).
23. R. Q. Twiss and J. A. Roberts, *Australian J. Phys.* 11, 424 (1958).
24. J. C. Jaeger and K. C. Westfold, *Australian J. Sci. Res.* 3A, 376 (1950).
25. J. P. Wild, J. D. Murray, and W. C. Rowe, *Australian J. Phys.* 7, 439 (1954).

Chapter 11

Solar Noise Measurements
By The Riometer Technique

Michael Anastassiades

Ionospheric Institute
National Observatory of Athens
Athens, Greece

Several years ago many observatories throughout the world organized programs of systematic measurements of ionospheric absorption by observing variations of the cosmic noise background. This Riometer technique, developed during the International Geophysical Year, assumed that the radiated power emitted by the cosmic noise background was constant. Any variation in the amplitude of the received signal would then be produced by the change in ionospheric opacity, and for this reason the device was named the Riometer (Relative Ionospheric Opacity Meter).

The Ionospheric Institute of the National Observatory of Athens has had a program of systematic measurements of ionospheric absorption of this type for several years. Since 1959 several papers on ionospheric absorption in middle latitudes have been published by the Institute, but the most important contribution of Riometers for our latitudes was the study of extraordinary solar events.

Absorption during these solar events in latitudes such as Athens rarely exceeds 1 to 1.5 db, while at high latitudes values of several decibels are frequently observed. The high-latitude absorption is associated with aurora. In the scatter of Riometers over the globe, high-latitude installations are prominent. On occasion they measure absorption of 20 db or greater on the 30-Mc/s standard Riometer. In this range of frequencies the ionosphere is not completely transparent for an extraterrestrial signal.

When solar events are to be studied by the Riometer technique, more comprehensive information can be obtained by using more than one frequency. This multifrequency Riometer technique can extend our knowledge of the effects of solar events on the ionosphere, and certain useful results can be derived from its application. The quiet sun makes a negligible contribution to the total cosmic noise received on the earth (at least when observed with a low-gain antenna). During disturbed periods solar noise can become several times greater than the normal cosmic noise level. Flares which occur during disturbed periods of the sun produce several types of radio emissions, which are classified according to their dynamic spectra into several types of bursts.

In radio astronomical research, radio emission from flares is investigated by observations at many frequencies. It is common practice to use special receivers operating between 100 and 10,000 Mc/s. This means that metric, decimetric, and centimetric bands are explored. Receivers operating on lower frequencies, i.e., 20 or 30 Mc/s, are used to investigate ionospheric absorption primarily, rather than solar activity. Even lower frequencies are employed in propagation experiments which are influenced by solar activity. For example, VLF techniques contribute greatly to research on solar–terrestrial effects.

Ionospheric phenomena are systematically avoided by radio astronomers when investigating solar events. In the first place, the presence of the ionosphere complicates even further the solar phenomena, which are by their very nature complicated enough. However, the response of ionospheric layers to a solar flare can yield valuable information on the nature of solar activity. A large bibliography exists on solar events recorded on frequencies greater than 100 Mc/s. However, few papers deal with sudden cosmic noise absorption (SCNA) caused by solar activity.

The use of the multifrequency Riometer technique on metric and decametric bands makes it possible to correlate SCNAs with the same solar event obtained employing receivers operating on higher frequencies. This extension to the lower frequencies is important because it may be possible to provide information on the evolution of a radio burst in the outer solar corona. For this purpose, it is necessary to have complete data on the behavior of the filter formed by the ionospheric layers.

We would like to present some results derived from the close collaboration between the ionospheric, the radio astronomical, and the optical solar service group of the National Observatory of Athens (Anastassiades *et al.* [1]). The Ionospheric Institute used two 50-kW vertical sounding panoramic recorders for routine work as well as recorders for selected parameters of the ionosphere. Radio astronomical research is pursued through a number of Riometers operating on 20, 27.6, 58, and 108 Mc/s, and a solar radio telescope operating on 2980 Mc/s. Finally, the optical solar service is able to observe the sun for more than 300 days a year and get excellent images due to the clear properties of the sky of Attica. Auxiliary services consist of the magnetometer stations and satellite-tracking facilities in Athens and Crete for the study of scintillations, as well as for large irregularities and total electron content. A cross-modulation device in Crete operates to give electron profiles of the *D*-region.

Figure 1 is an example of correlation of measurements made by this instrumentation, except for the neutron monitor data. All observations were made in Athens in September, 1963. From all these observations we concluded that the responsible solar center 6964 was a repetition center which produced at least three flares accompanied by corpuscular emission. This corpuscular emission was responsible not only for the magnetic storms but also produced typical ionospheric disturbances. The solar radio telescope recorded strong bursts on 2980 Mc/s for all three of the above-mentioned flares. The optical observations of the covering and uncovering of the spots contained in the center by the bright filamentary branches of the flares correlated well in time with the variations in the recordings on 2980 Mc/s. Finally, the successive ionospheric storms appeared with their negative phases of differing

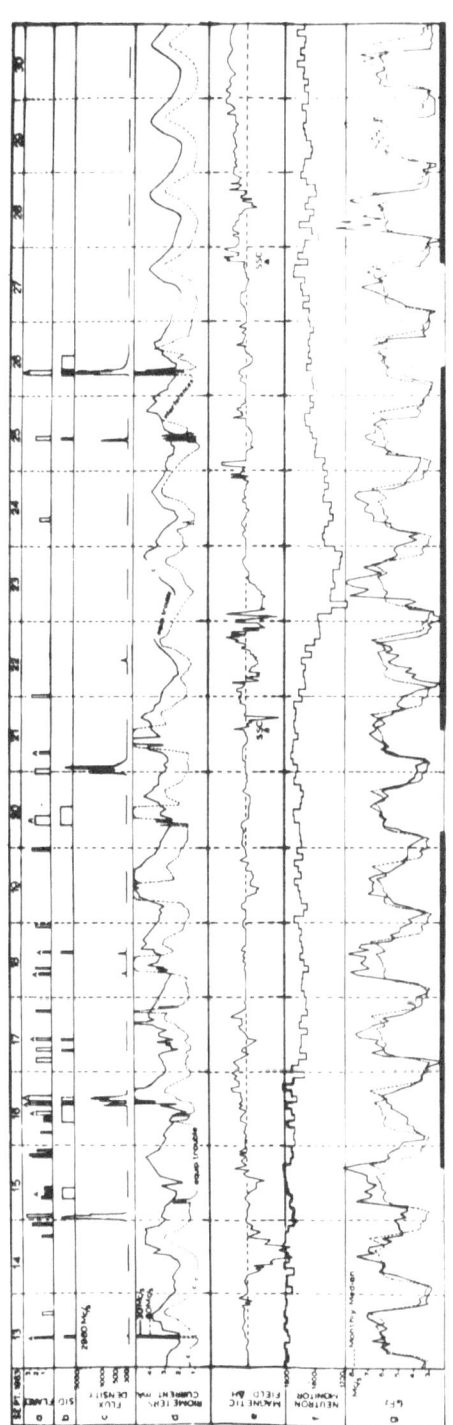

Fig. 1. Optical and geophysical data obtained during the period September 13 to 30, 1963.

duration. During the first of the storms a decrease in total electron content was observed.

STUDY OF SOLAR- TERRESTRIAL EVENTS BY
THE RIOMETER TECHNIQUE

The electron density increase in the D-layer produced by X-radiation emitted by some flares is recorded as a SCNA by Riometers working on 30 Mc/s and on lower frequencies. SCNAs can be observed even at 60 Mc/s, but their magnitude is not as pronounced as for lower frequencies. Sudden cosmic noise absorption events are produced by specific ionospheric layers and it is rather simple to determine which layers are responsible for the absorption (Anastassiades and Giouleas [2]). This determination can be ascertained by the employment of the vertical incidence sounding technique. By comparing results from the above-mentioned two techniques, one can observe the product aN (the recombination coefficient times the number of electrons per unit volume) in the absorbing layer, as shown by Riometers, and compare it with the product obtained from the vertical sounding technique (Fig. 2). A first-order explanation of this discrepancy is that the Riometer technique is sensitive to all ionospheric layers involved, whereas in the vertical sounding technique only the lower layers are involved. Appleton and Piggot [3] have indicated the importance of the recombination coefficient in calculating the dependence of absorption on the solar zenith angle, as has Sarada and Mitra [4]; Anastassiades [5] has given our own group examples of the variation of the recombination coefficient with ionospheric height. The well-known relationship $\log \rho \propto (\cos x)^n$, relating zenith angle with absorption can be and was checked by Riometers and was found valid. The exponent determined by the cosmic noise technique was found equal to the values for summer, equinoxes, and winter, obtained by other methods.

All of these measurements can be made when we have an SCNA. We can, in addition, obtain some indications of the refractive index of the interplanetary medium, by employing the multifrequency Riometer technique, with at least three Riometers operating on three-spaced frequencies. When an SCNA is recorded at two different frequencies, we can determine the velocity of ascension in the solar corona of a shock wave front or of a cloud of ejected particles from a flare. The Air Force Cambridge Research Laboratories installed a number of Riometers, located in different latitudes but along the same longitude. Such a network is formed by Riometers installed in Accra (Ghana), Addis Ababa (Ethiopia), Athens (Greece), Hermanus (South Africa), New Delhi (India), Sacramento Peak, New Mexico, (U.S.A.), Bedford, Mass. (U.S.A.), and Great Whale River (Canada). All Riometers are of the same type, operating on the common frequency of 30 Mc/s. Goldman and Horowitz [6], using results from this network, investigated the effect of two solar flares which took place on April 15, 1963. The dependence of absorption on the solar zenith angle was investigated for all stations. The zenith angle variation of the stations was from $10°$ (Accra) to $76°$ (Bedford) for the 1123 U.T. flare and $34°$ (Bedford) to $80°$ (Athens) for the 1614 U.T. flare. The well-known relationship between zenith angle and absorption gave an exponential value for n of the order of $\frac{3}{2}$

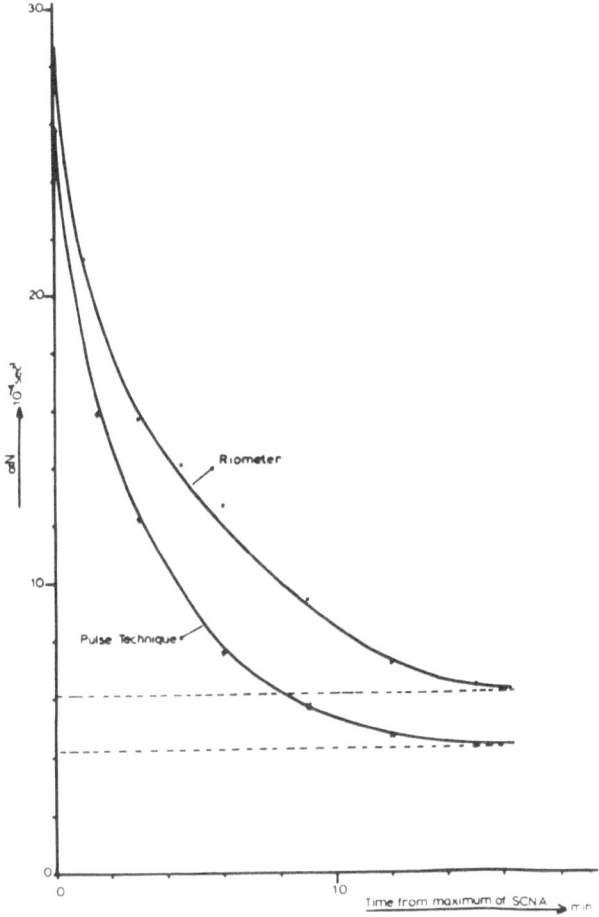

Fig. 2. Variation of aN with time during the decay period
for the flare on May 23, 1961.

(Fig. 3). Such a network can provide additional information on the nature of flares and can distinguish solar flares emitting purely electromagnetic radiation from flares emitting particle emission. Riometers detect mainly flares of electromagnetic nature, while flares with corpuscular emission produce special effects at high latitudes.

FORMS OF SCNA'S AND TYPE OF BURSTS

Extra solar activity as recorded on the ground depends on two diverse effects: Radio frequency radiation increases the signal level, and X-rays and at times corpuscular radiation increase the absorption. In the 20 and 30 Mc/s fre-

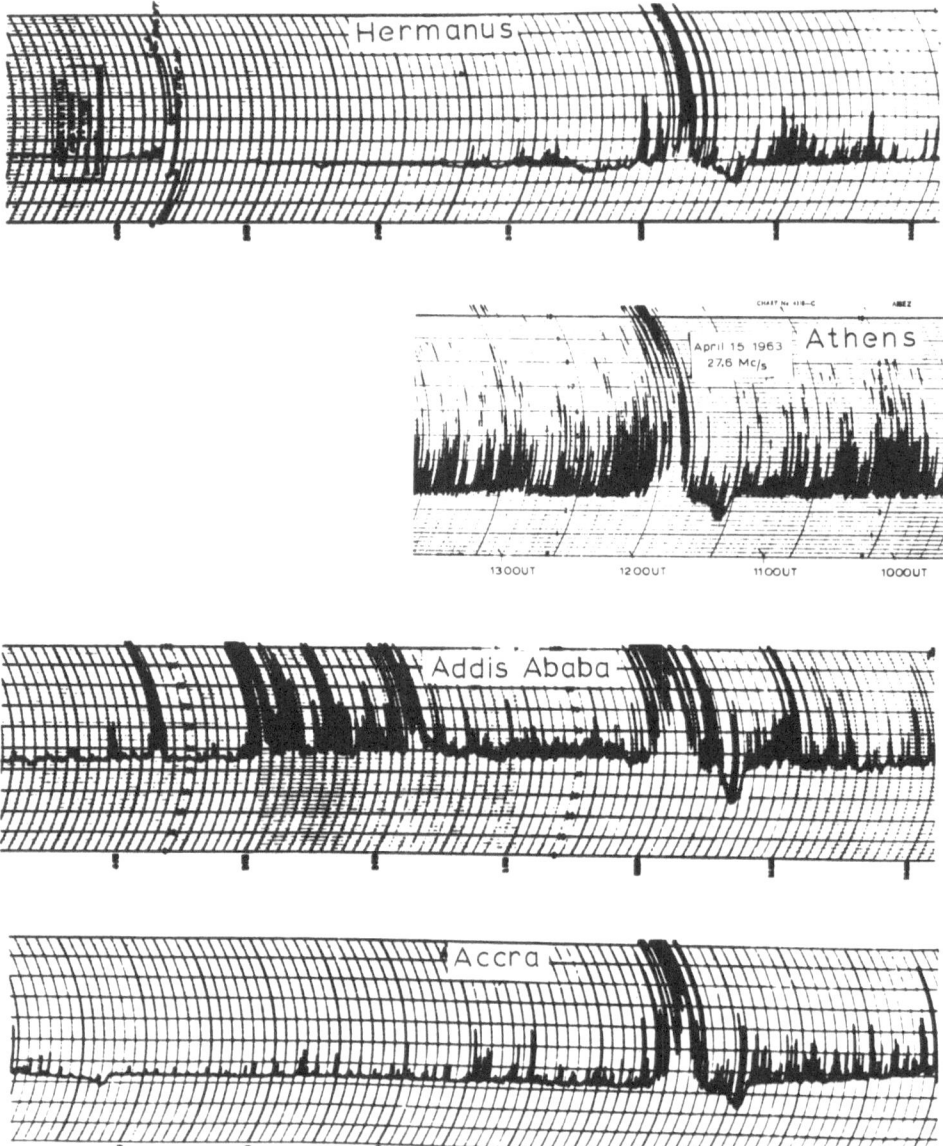

Fig. 3. Riometer effects on 30 Mc/s at different sites due to solar flares
of April 15, 1963, at 1123 U.T.

quency range, the latter effect, the SCNA, predominates. Riometers operating
on higher frequencies show predominantly bursts of Type II, III, and IV, due to
the low ionospheric opacity for those frequencies.

We can now divide SCNAs into three specific types — A, B, and C (Fig. 4).
Several variants of the three basic forms are observed, but we can always refer

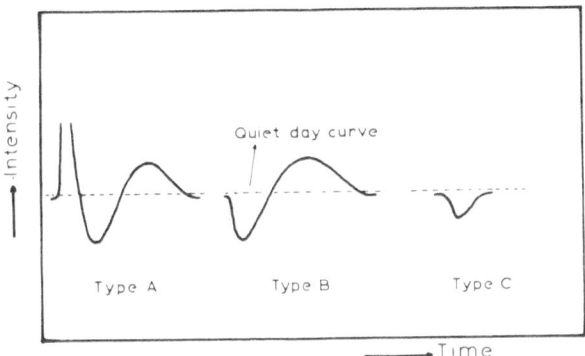

Fig. 4. Forms of typical SCNAs.

them to the basic types. Type A SCNA begins with a burst of Type II, followed by absorption and a continuum increasing the level of cosmic noise for several minutes or hours. Type B SCNA begins with absorption and is followed by a continuum of increased noise. Finally, Type C takes the form of a decrease followed by an increase in the noise level similar to a letter V. Figure 4a presents typical recordings of the above types as obtained by Riometers. Usually SCNAs of Type A are associated with bursts of Type II–IV. Only the presence of the ionosphere deforms the typical shape of such a recording with the appearance of absorption. The absorption effect, by decreasing the noise level, shortens the Type IV continuum recordings by an amount relative to the importance and nature of the flare. SCNAs of Type B are related to strong X-ray emission associated with the first phase of the evolution of a flare. The radio emission associated with this type of SCNA seems to be of short duration. Type C SCNAs are produced by flares with strong X-radiation but not accompanied by important radio emission on metric and decametric bands.

In studying SCNAs produced by flares, we can follow the evolution of solar radiation to regions of the outer corona, where radiation in this low-frequency band originates. An SCNA of Type A shows that the shock wave front associated with Type II burst was propagated to the outer solar corona. If we have at our disposal higher frequencies produced by a flare, we can form a picture of its evolution. If, for example, a Type II–IV flare burst is observed at higher frequencies without a 30-Mc/s SCNA, the shock wave front did not arrive at regions where decametric wavelength radiation could escape from the corona. On the other hand, when an SCNA is recorded without any centimetric or metric radiation, we can assume that the effect was localized in the chromosphere. SCNAs of Type B are rather rare. Of the total number of SCNAs recorded during the last years, a great number were of Type A and fewer of Types C and B.

STUDY OF TYPICAL FLARES PRODUCING SCNA'S

In order to make the behavior of flares producing SCNAs clear, let us consider a few cases of special interest.

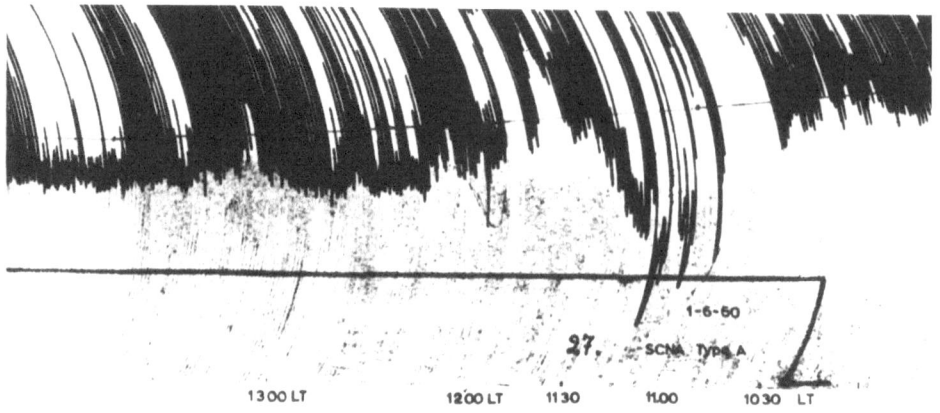

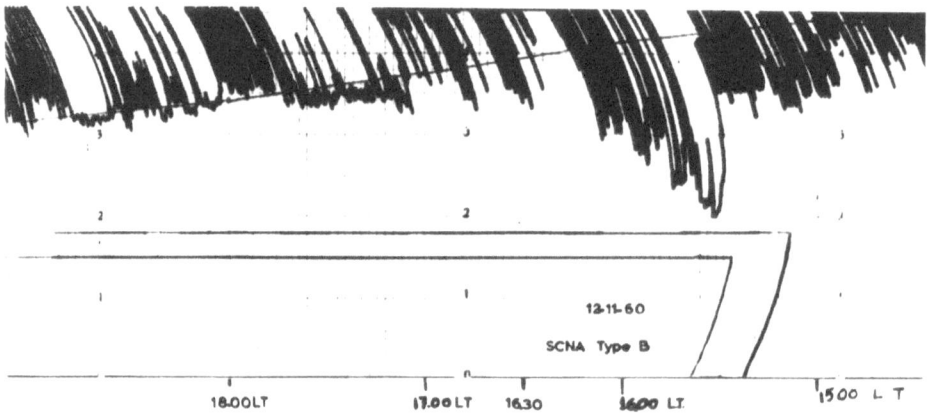

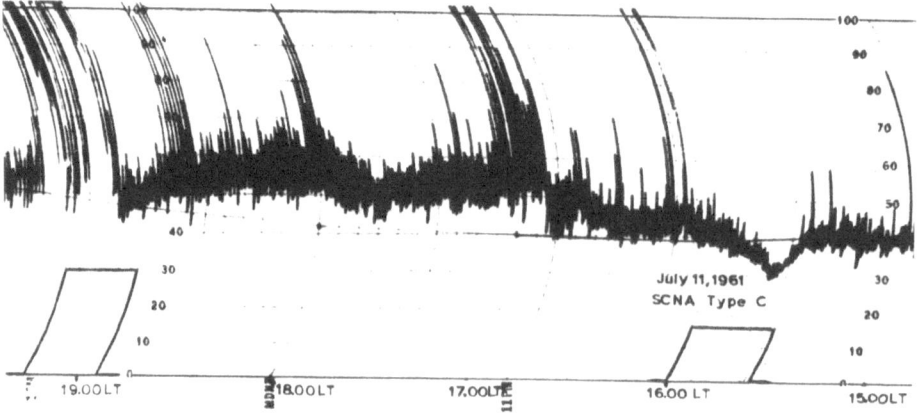

Fig. 4a. Different types of SCNAs.

If we examine the day of September 16, 1963, when the solar center of activity 6964 produced at least three characteristic flares of optical importance greater than 1, we shall find each one of different character and evolution. The solar center's prehistory started before September 16 and we can follow this in Fig. 1. At 1018 U.T. of September 16, an SCNA Type C was recorded by Riometers operating on 30 and 60 Mc/s. The solar radio telescope on 2980 Mc/s and the Riometers on 108 Mc/s did not show any increases. It is obvious that the flare of 1018 U.T. was not accompanied by detectable radio frequency emission and produced only X-radiation. The SID on the continuous recording of vertical sounding supported this hypothesis (Fig. 5).

A little later at 1303 U.T. another flare produced completely different effects. A rather short burst was recorded on 2980 Mc/s and an increased noise level on 108 Mc/s, while an SCNA of Type C was recorded on 30 and 60 Mc/s by Riometers. A sharply defined SID on the vertical sounders indicated the influence of X-rays. (Fig. 6).

One hour and thirty minutes after the start of the above-mentioned flare, at 1436 U.T., another flare produced a large burst on 2980 Mc/s of complex form; the duration was greater than 1 hr. On 108 Mc/s, the noise level, which was already high, increased even more, forming a two-step continuum, while on the 30 Mc/s Riometer instead of an SCNA, a Type II burst was observed. Finally, the continuous recording of the vertical sounding showed a much less marked SID than during the two earlier flares (Fig. 7).

We can see by the above example that in a rather short time a solar active center produced three flares of completely different character. We can follow the evolution of each one by using not only recordings of solar noise at higher frequencies, but in addition using data influenced by the reaction of the ionosphere. By following this technique we can see that the 1018 U.T. flare's radiation did not have any detectable radio frequency emission, but was accompanied by X-rays. The situation changes when we examine the flare of 1303 U.T. We can proceed to a more complete study of this flare since in addition to the radio frequency recordings on several frequencies we have the optical evolution on H_α. Figure 8 shows the curves of H_α, the centimetric component, and the "ionizing radiation" issued from this flare. The ionizing radiation cannot be UV or Lyman-α because as Friedman [7] showed, only X-ray radiation is able to produce an SCNA. The very well-marked sharp SID confirms this assumption. We can check this result by comparing the two Type C SCNAs recorded on decametric and metric wavelengths. The maximum of absorption indicated by these units coincides with the maximum of the relative electron density. The coincidence of the first maximum of the H_α curve with the ionizing radiation shows that X-rays are produced only during the first phase of the flare evolution. Savitch [8] arrived at the same conclusion from a number of observations of flares during the period of high sunspot activity. All other maxima of H_α are not correlated with maxima of either the ionizing radiation or the centimetric component. This latter correlates more closely with X-ray radiation than with variation of H_α. We can see, for example, at 1312 U.T. a second maximum on centimetric wavelengths which corresponded to a Type C SCNA and an SID, but it is not correlated with the second maximum of the H_α curve. Essentially

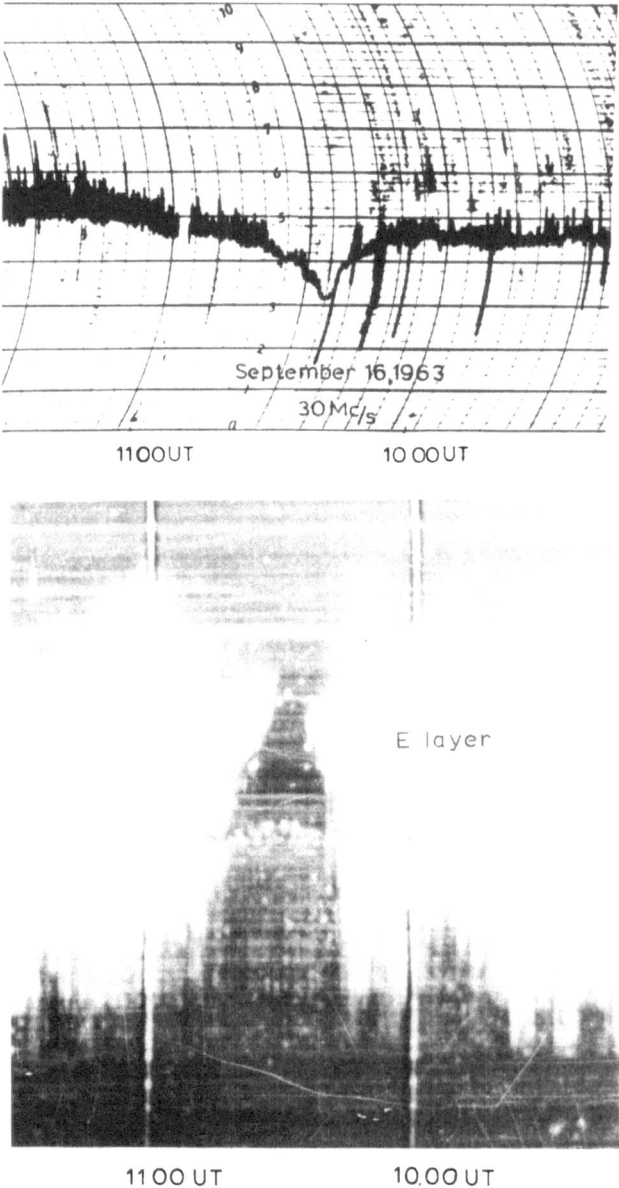

Fig. 5. Evolution of the flare of September 16, 1963, at 1018 U.T.

this result is in good agreement with the conclusions of Kundu [9] and de Jager and Kundu [10], who have shown the closer correlation of the centimetric component and X-ray radiation rather than the centimetric component and H_α.

Examining recordings of the flare 1303 U.T. on different frequencies, we can note distinctive events in the flare by comparing it with the earlier flare of 1018

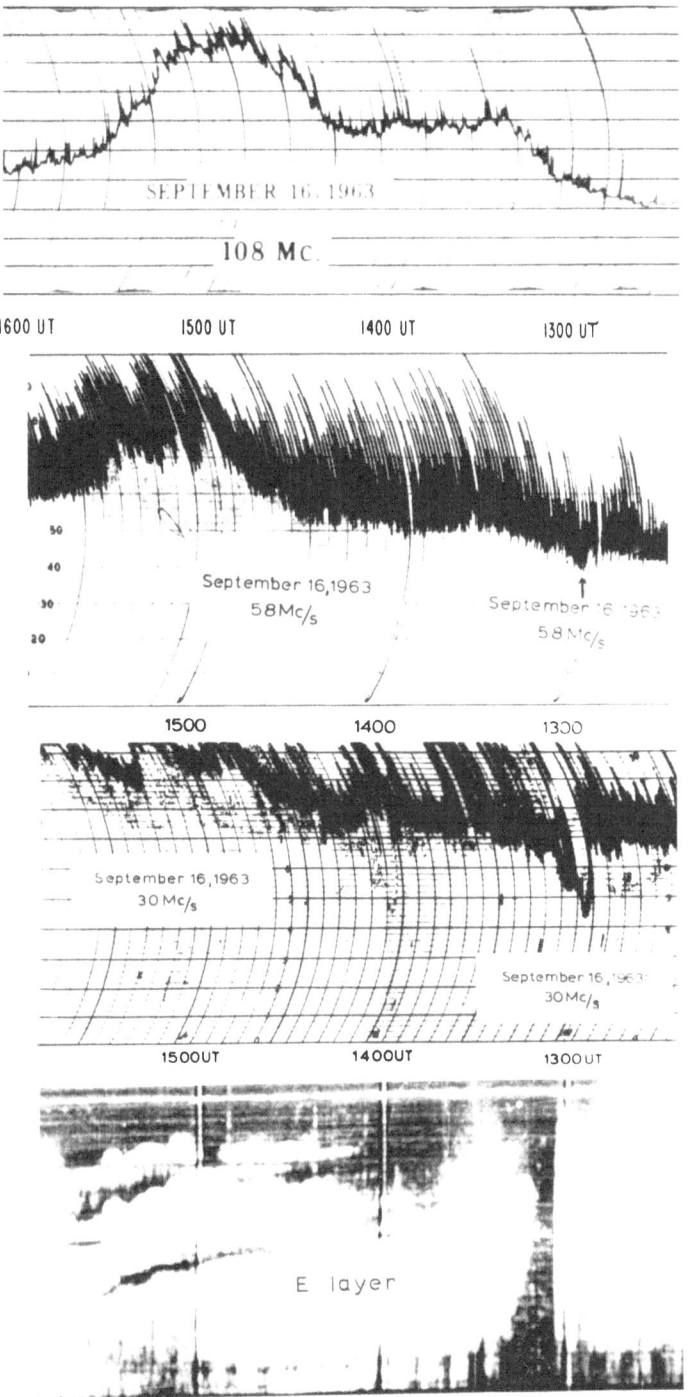

Fig. 6. Evolution of the flare of September 16, 1963, at 1303 U.T.

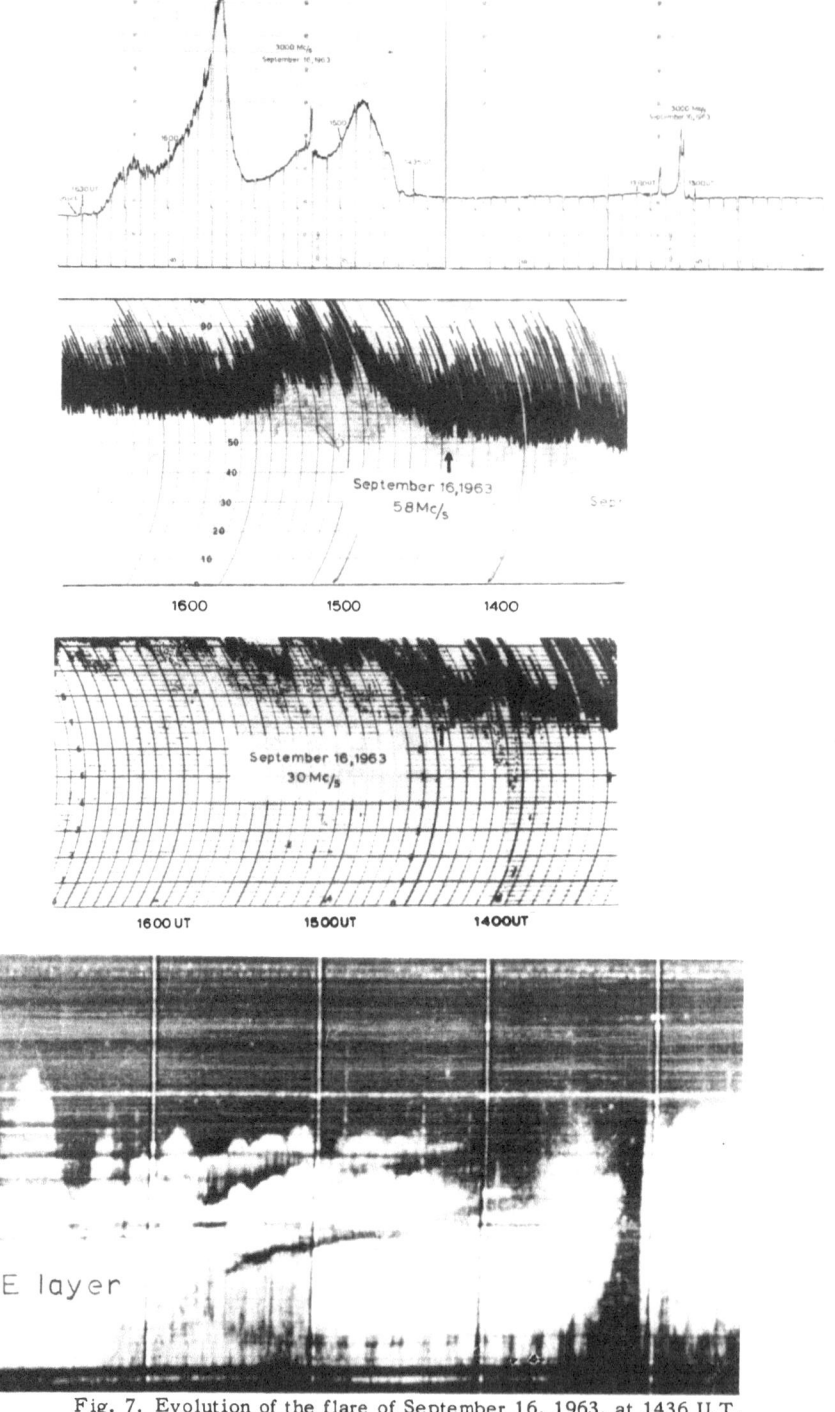

Fig. 7. Evolution of the flare of September 16, 1963, at 1436 U.T.

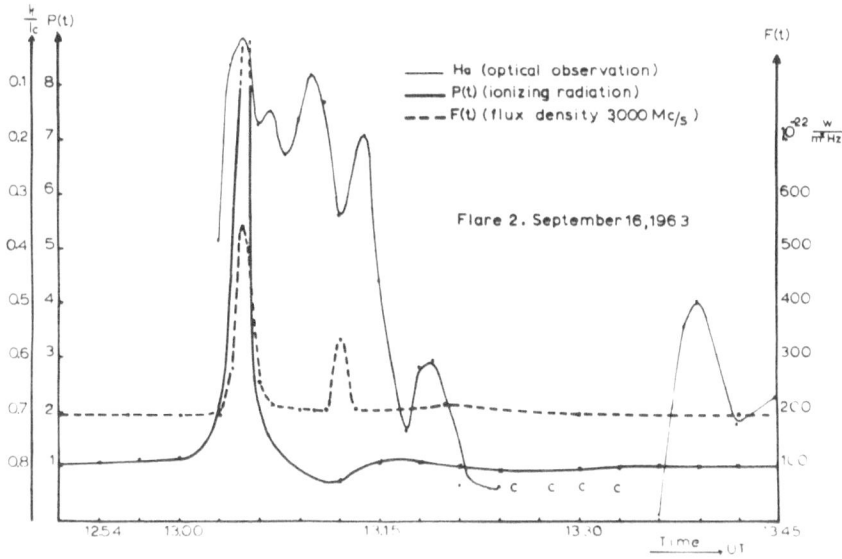

Fig. 8. Flux density (3000 Mc/s), ionizing radiation, and H_α vs. time due to solar
flare of September 16, 1963, at 1303 U.T.

U.T. X-ray radiation was strong and the latter flare ejected material which produced a shock wave front followed by a cloud of particles trapped in the magnetic
field and emitting synchrotron radiation. On 108 Mc/s synchrotron radiation produced
an enhancement of noise level without any effect from plasma oscillation radiation. On lower frequencies of 60 and 30 Mc/s, only SCNAs of Type C were recorded. We can then assume that during this flare of 1303 U.T. the material
ejected did not produce any effect in regions of the solar corona where 60 and
30 Mc/s radiation escapes.

Finally, let us examine the third flare produced during this same day, at 1436
U.T. H_α measurements are not available, but we have broad-band recordings divided into two parts, centimetric bursts and continuum in the metric. A burst was
also recorded in the decametric band. The ionosounder showed an SID which is not
too clear and is of a special form.

A good example of a typical eruption producing corpuscular radiation is that
of September 26, 1963. Its corpuscular nature was shown by a geomagnetic SSC
storm and a Forbush decrease. The flare produced characteristic effects on all
frequency bands. An SID recorded by the continuous recording of the ionosounder
is the only indication that in its early phase the flare produced X-ray radiation
(Fig. 9). The form of the metric and decametric bursts is the classical Type II–IV.
One can easily observe the doubling in the Type II burst; the Type IV continuum
lasted at least 1 hr. This flare showed a classical evolution of events. Material,
ejected from the active center during the flare explosion, rose in the corona, gen-
erating the shock wave. Because it is highly ionized and conductive, the jet had
trapped within it a part of the local magnetic field of the active center. On the

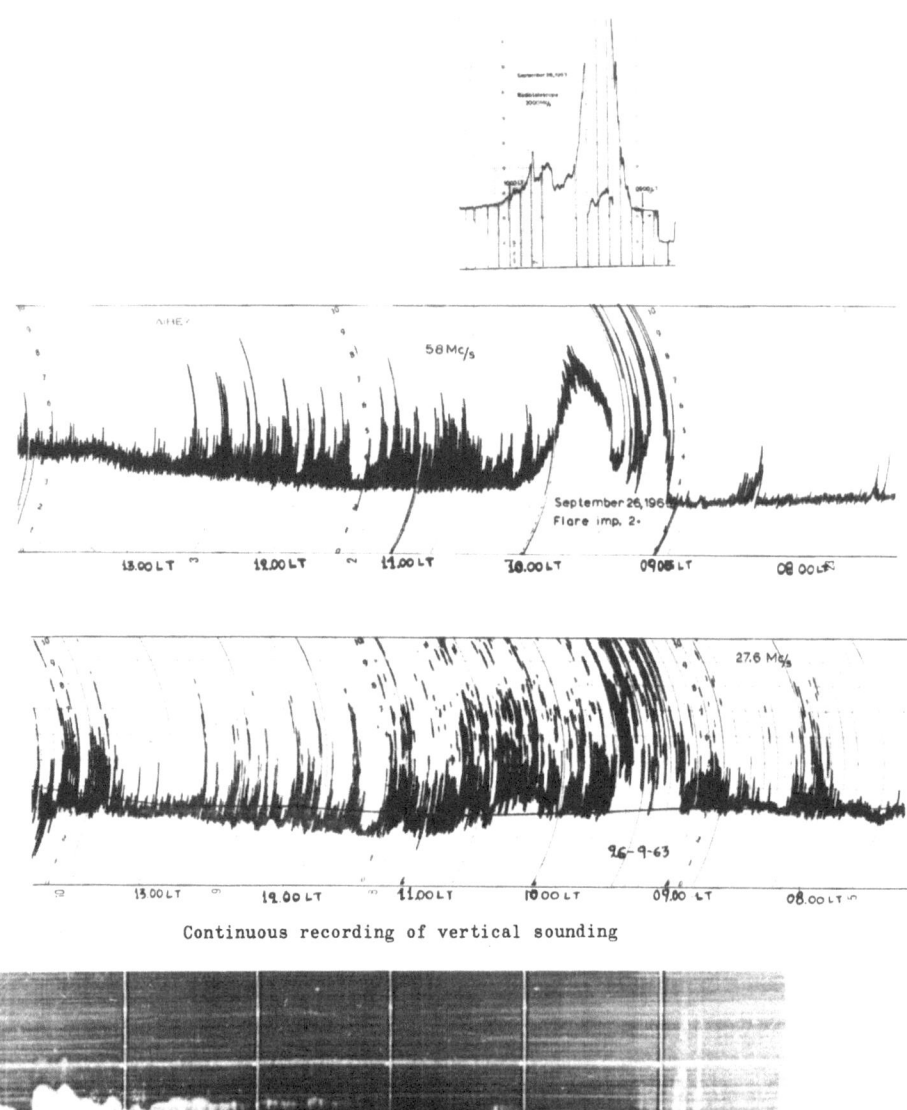

Continuous recording of vertical sounding

Fig. 9. Evolution of the flare of September 26, 1963, at 0701 U.T. (0901 L.T.).

other hand, electrons and other particles of high energy, created during the eruptive phase, are captured in the magnetic field and cannot escape. When this cloud
rises into the corona, the shock wave which preceded it and which had a much
higher velocity will arrive at the layers where metric waves are radiated in its
passage; plasma oscillations which formed the Type II emissions were recorded.
Type IV continuum which followed was due to synchrotron emission from trapped
electrons, which progressively lost their energy by radiation.

When examining the evolution of the September 26, 1963, flare, we can see that
a strong X-radiation produced a complete blackout of the ionosphere (Fig. 9) as is
shown by the continuous recordings of the ionosounder. However, in spite of the
high ionospheric opacity, recordings on 60 and 30 Mc/s did not present any absorption effect and a typical Type II–IV burst was recorded on these frequencies. This
means that the solar radio event was exceedingly strong. We can state that such
behavior characterizes solar events followed by corpuscular emission of relatively
high-energy particles. Recordings of Type II–IV bursts without SCNAs on 60 and
30 Mc/s in the presence of a blackout detected by ionospheric measurements can
be considered as special degenerated Type A SCNAs, indicating that the shock
wave was active even on the layers of the outer corona. In such cases, when recordings of Type II bursts are obtained on three different Riometer frequencies,
we can estimate to a first approximation the order of magnitude of the velocity of
ascension of the shock wave. This is done by comparing the times of the first
rise of Type II bursts recorded on those frequencies. The value of the mean velocity of ascension for the September 26 event was of the order of 1000 km/sec.

In some rare and highly exceptional cases, the above-mentioned evolution of
flares followed by corpuscular emission is modified. When relativistic particles
are emitted, combined absorption effects of X-radiation and particles are present.
Let us examine, for example, the flare of November 12, 1960. This flare, and its
effects produced by plage 5925, was one of the most completely studied solar
events recorded to date. In our measurements, Riometer recordings at 60 Mc/s
showed a burst of Type II, while the 30 Mc/s records showed an SCNA of Type B.
The presence of an SID showed strong X-radiation from the flare recorded on the
continuous recording of the ionosounder. However, the unusual feature of this
flare was that it produced a remarkable increase of cosmic ray radiation which
was measured at ground level. We can then assume that both the SID and the
SCNA are due to the combined effects of X-ray radiation and particles. In fact,
particles arriving at the earth's atmosphere within a few minutes after the optical
flare started to produce absorption exceptionally high for our latitude. It would
be quite normal, as in the above-mentioned case of the flare of September 26, to
have Type II–IV recordings on 60 and 30 Mc/s. The modification and the appearance of a SCNA of Type B is due to particles of very high energy. It is very
probable that shock waves also reached the outer corona, but it is not possible to
prove it by using this same method because of the extreme opacity of the ionosphere.

As a general rule we can observe that on metric bands we have SCNAs of
Type A, and on decametric wavelengths SCNAs of Type B. These in turn must
be correlated with the electron density throughout the ionosphere. Basically the

ionospheric response is more definite in the decametric band, compared to metric waves. It is very difficult to give general rules on the correlation between SCNAs and their causes. It is not possible to state that certain forms of SCNA are related to flares of specific properties. There are a number of varieties of forms and types Some examples given here are only typical cases, but a large number of varieties exists.

Only long experience can eliminate the doubts about the types of SCNAs and lead to some information on solar activity. The contribution of Riometers to this study will be valuable. For the study of specific solar events, the close collaboration within an ionospheric group of radio astronomers and optical astronomers might form a new approach for solar research.

ACKNOWLEDGMENT

The research reported in this document has been sponsored in whole by the *Air Force Cambridge Research Laboratories*, under Contract AF 61 (052)-261, through the European Office of Aerospace Research (OAR), United States Air Force.

REFERENCES

1. M. Anastassiades, D. Ilias, C. Caroubalos, C. Macris, and D. Elias, *Nature* 201, 357 (1964).
2. M. Anastassiades and P. Giouleas, *Scientific Report No. IIA 005* (Ionospheric Institute, National Observatory of Athens, June 1963).
3. E. Appleton and W. R. Piggot, *J. Atmospheric Terres. Phys.* 5, 141 (1954).
4. K. A. Sarada and A. P. Mitra, *Results of Cosmic Noise Observations at Delhi on 22.4 Mc/s* (private communication).
5. M. Anastassiades, *Radio Astronomical and Satellite Studies of the Atmosphere* (North Holland Publishing Co., Amsterdam, 1963), p. 238, J. Aarons, ed.
6. S. C. Goldman and S. Horowitz, *Nature* 199, 1147 (1963).
7. H. Friedman, *The Sun's Ionizing Radiation: Physics of the Upper Atmosphere* (Academic Press, New York, 1960), J. A. Ratcliffe, ed.
8. N. A. Savitch, *Some Ionospheric Results Obtained During the I.G.Y.* (Elsevier, Brussels, 1959), p. 130, W. J. G. Beynon, ed.
9. M. R. Kundu, *J. Geophys. Res.* 66, 4308 (1961).
10. C. de Jager and M. R. Kundu, *Space Research, Vol. III* (North Holland Publishing Co., Amsterdam, 1963), p. 836.

Chapter 12

Radio Frequency Emission of the Sun in the Centimeter-Wavelength Range: Microwave Bursts

O. Hachenberg

University of Bonn
Bonn, Germany

The bursts of the centimeter-wavelength range are widely different from the bursts of the meter-wavelength region. They are relatively simple and the temporal course of the radiation flux is relatively smooth. They don't show the strongly marked sharp peaks (halfwidth $\leq$ 1 sec) which are typical for the metric region. Whereas the meter-wave bursts can be classified into the well-known Types I–V, the centimeter-wave bursts don't allow such a physically significant classification. We must treat them as a special class of bursts, which we call the microwave bursts (Wild *et al.* [1]).

Microwave bursts have a broad continuous spectrum. This has been verified by the comparison of measurements at different discrete frequencies (Hachenberg [2] and Takakura [3]). This fact has also been supported by sweep-frequency spectrograph observations (Kundu and Haddock [4] and Thompson and Maxwell [5]).

THE RELATIONSHIPS BETWEEN MICROWAVE BURSTS, FLARES, AND SID

Microwave bursts are closely associated with the H_α flares of the optical spectrum (Hachenberg and Krüger [6] and Harvey [7]). In particular, those flares which have a broadening of H_α greater than 3.5 Å give rise to a microwave burst (Hachenberg [8]). Usually the maximum phase of the bursts coincide with the flash phase of the flares. On the other hand, a high percentage of flares which give rise to a microwave burst is also accompanied by an SID (sudden ionospheric disturbance) (Hachenberg and Krüger [6] and Harvey [7]). It can also be shown that the temporal variations of the radiation flux of the bursts and of the ionizing radiation in the ionosphere have a similar course (Hachenberg and Volland [9], Volland [10], and Anderson and Winckler [11]). Thus, we must assume that the centimeter-wave emission of the bursts is closely connected with the X-ray generation of the flares.

THE BURST RECORDS

By reason of the broad-band spectrum it is possible, using only four or five simple radiometers operating at four or five discrete frequencies within the frequency range considered, to get fairly complete observations both of the temporal course of the radiation flux and of the spectrum of the bursts. Such single frequency observations of the solar radiation flux at 1500, 2000, 3000, and 9400 Mc/s have been systematically carried out at the Heinrich-Hertz Institut, Berlin-Adlershof. During the I.G.Y., 970 bursts have been identified on the solar patrol records. The results have been compiled in the "Burst Catalog of the Heinrich-Hertz Institut." This catalog presents the material for a part of the following statistical investigation.

The duration of the bursts lies between 10 sec and some hours, with a strongly marked maximum at 2.5 min (Krüger [12]). The maximum radiation flux of the bursts lies normally between 5×10^{-22} and 200×10^{-22} W/m^2-c/s. Occasionally a burst is observed that reaches an intensity of some thousands of flux units. The distribution of the maximum flux is given in Fig. 1. The falling off of the curves below 10×10^{-22} is due to the limit of sensitivity of the radiometer (Krüger [12]). The total energy of a burst is obtained by integrating the flux over the duration of the burst. The distribution of the total energy/cycle bandwidth and square meter is represented in Fig. 2 (Krüger [12]).

If we accept the fact that the intensity in the spectrum of the bursts is for a major part of the frequency range independent of frequency, then we can use Fig. 2 as a starting point for estimating the energy that is emitted by a burst in the centimeter-wavelength range. If we take for instance a burst of the energy of 10^{-18} W/m^2-c/s, this burst gives a total energy in the frequency range considered of 10^{-8} W/m^2.

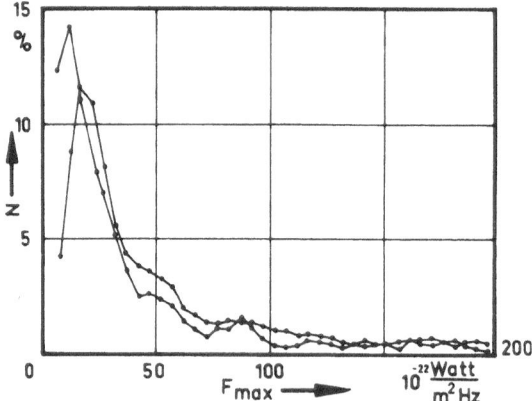

Fig. 1. Relative frequency of burst maxima during I.G.Y.
(Upper curve at λ = 20 cm; lower at λ = λ = 3.2 cm.)

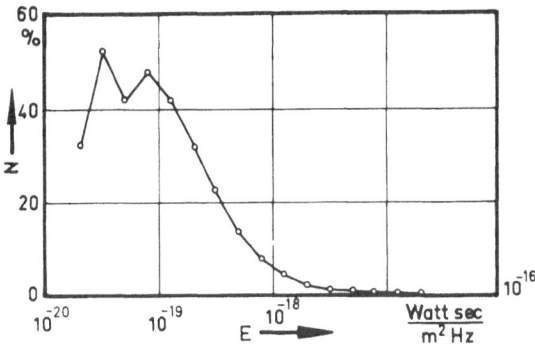

Fig. 2. Frequency distribution of the total amount of
energy E per $[m^2 \, c/s]$ received from centimeter bursts.

TYPE OF BURST FORMS

In the frequency range considered, the bursts have a broad-band spectrum,
and therefore over a large portion of the range they follow a relatively similar
course. For this reason, it is possible to set criteria for the classification of
burst forms which are applicable to the whole range. Since the I.G.Y., the well-
known classification of Covington (Dodson *et al.* [13]) has been adopted for the
whole frequency range. This classification divides bursts into ten different
groups according to the details of their structures.

For this present discussion, we may combine some of these burst groups for
simplicity of handling the material. For our purpose, we can list the following
three general groups.

1. The impulsive bursts. This type consists of simple peaks with a duration
   of 10 sec to about 5 min. Figure 3a gives records of a burst of this type
   at three different frequencies. The halfwidth of these bursts usually decreases slightly with increasing frequency. Simple combinations of such
   peaks may also be included in this group (Fig. 3b). The impulsive bursts
   often coincide with the flash phase of H_α flares and with the appearance
   of the "first part events" in the meter-wavelength range, that is Type III
   bursts.

2. The gradual rise and fall type. This type has similarly a simple form; the
   radiation flux rises slowly to a maximum and then returns slowly to preburst level in a uniform fashion. The process lasts about 10 to 30 min. In
   Fig. 4a, an example of a normal burst of the gradual rise and fall type is
   represented. The frequency of occurrence of this type increases somewhat
   toward decreasing wavelengths. One is inclined to look at this as a special type, but if we consider the course of these bursts in the whole spectral range, we can occasionally establish that a burst which belongs to the
   gradual rise and fall type at λ = 3 cm, has a very complex structure at
   longer wavelengths (λ = 20 cm). Figure 4b gives records of a burst of

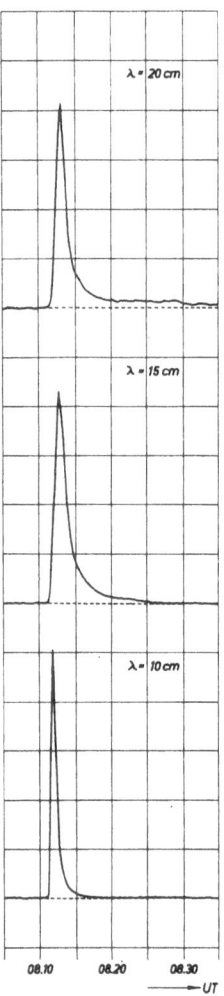

Fig. 3a. Records of a simple impulsive burst
(September 7, 1957) at different wavelengths.

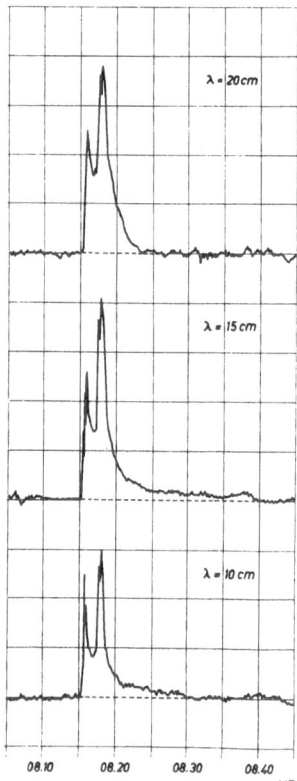

Fig. 3b. Impulsive burst (July 12, 1958). (M-burst
after Takakura [3]).

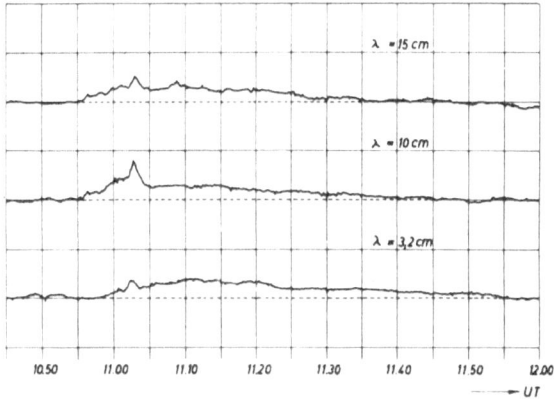

Fig. 4a. Burst of gradual rise and fall type
(December 2, 1957).

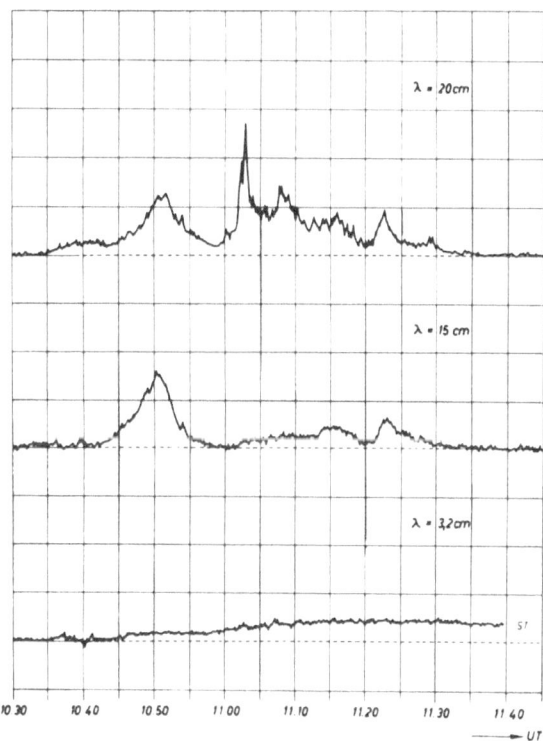

Fig. 4b. Burst (December 20, 1958) is at $\lambda = 3.2$ cm of
gradual rise and fall type. At $\lambda = 15$ cm and $\lambda = 20$ cm,
it has a very complex structure.

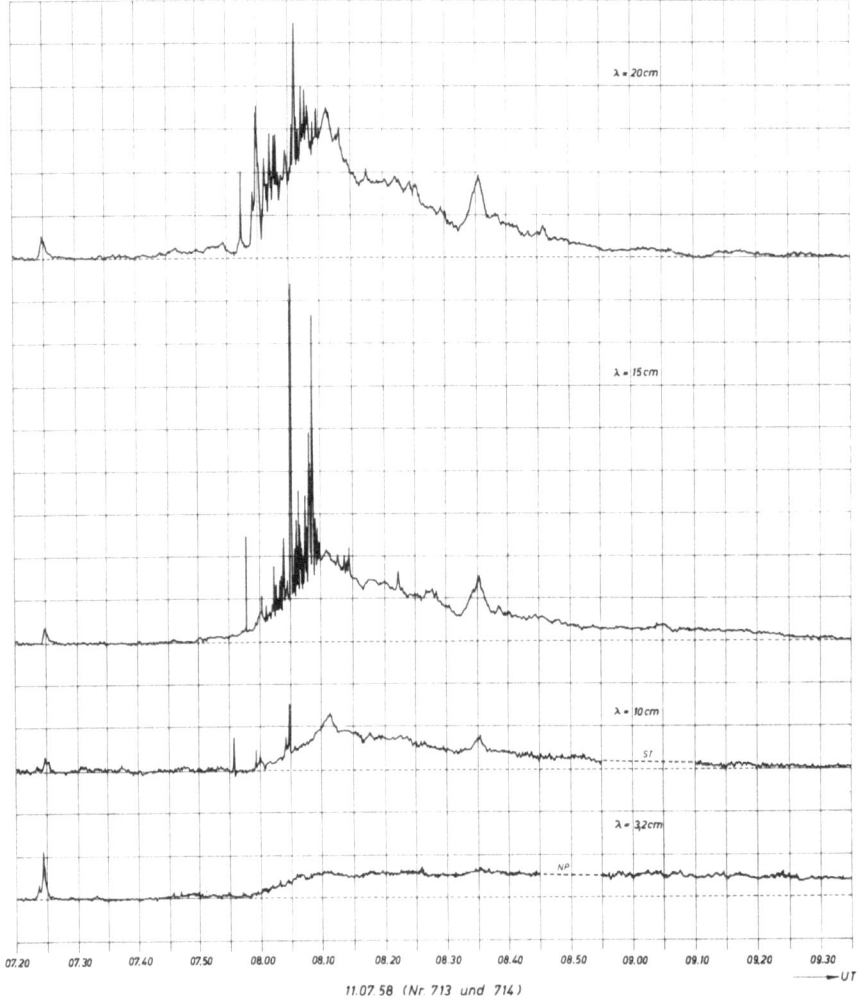

Fig. 4c. Burst of gradual rise and fall type. At λ = 15 cm, many sharp peaks
are superimposed.

this type at different frequencies. Thus, we cannot always consider this
type as a special one.

3. Complex burst forms and large outbursts. These bursts have a complicated
    structure, in some cases consisting of a combination of groups (1) and (2)
    (see Fig. 5a), or in others follow a course which can no longer be sepa-
    rated into the fundamental forms of groups (1) and (2) (see Fig. 5b).

The groups are not sharply separated against one another. For instance, there
are bursts with a duration between 5 and 10 min which could be classified either
into group (1) or into group (2). On the other hand, the decision whether a burst

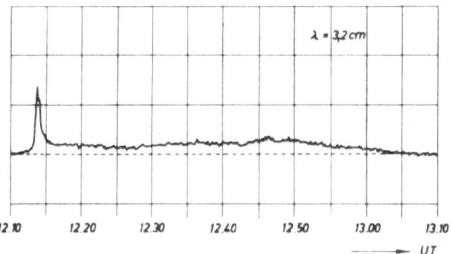

Fig. 5a. Complex burst – impulsive burst
superimposed on gradual rise and fall type.

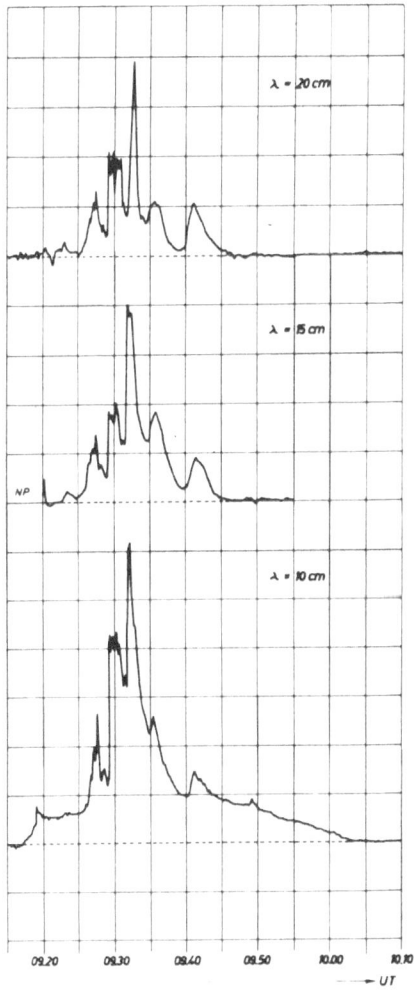

Fig. 5b. Complex burst (December 19, 1957).

belongs to the groups (1) or (3) is only a question of definition. There are no physically significant differences between these groups.

Attempts to find any correlation between centimeter burst forms and the well-known burst types of the meter-wavelength range produced few results. The impulsive bursts are usually accompanied by Type III bursts of the meter-wavelength range. The Type II and Type IV events are almost always connected with the third group – however, the reverse, that all the long enduring complex bursts can be approached as centimeter-Type IV, is by no means true.

THE TIME – FREQUENCY DIAGRAMS OF BURSTS

We get a clearer picture of the overall process of a burst if we construct a time – frequency diagram with the aid of individual records at the different frequencies. This sort of representation corresponds to the photographs we get from a sweep-frequency spectrograph, except that here we represent the radiation flux by lines of equal intensity (Takakura [3], Krüger [12], Kakinuma and Tanaka [14], and Fokker [15]). In Figs. 6 – 8, examples of such time – frequency diagrams are represented. These have been extended to include the meter-wavelength range in order to show the connection with Type IV events (Hachenburg and Krüger [16]).

Figure 6 gives the time – frequency diagram of a burst of the impulsive type, which emits exclusively in the centimeter-wavelength range. The increase of intensity at high frequencies is clearly recognizable. The fact that the time of beginning and maximum are the same at all frequencies can also be seen. Both the

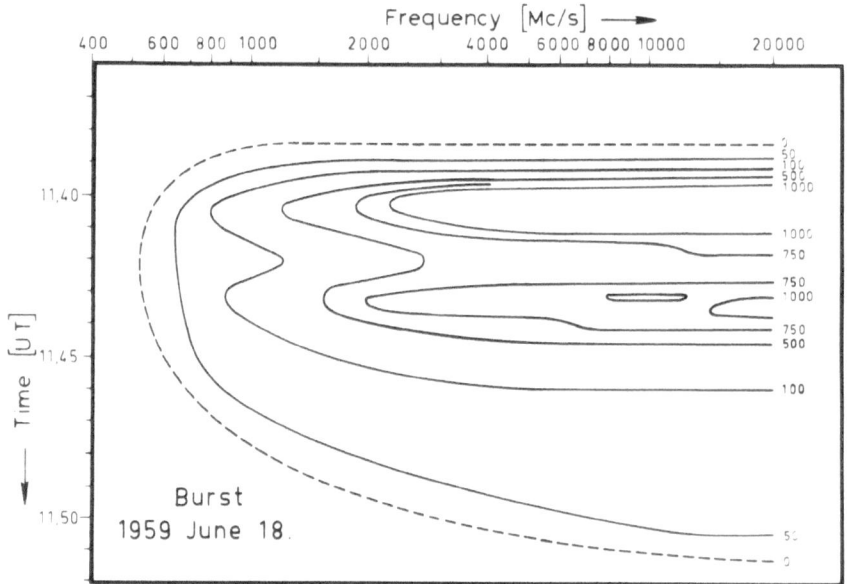

Fig. 6. Time—frequency diagram of a burst of impulsive type, which emits exclusively in the centimeter-wavelength range.

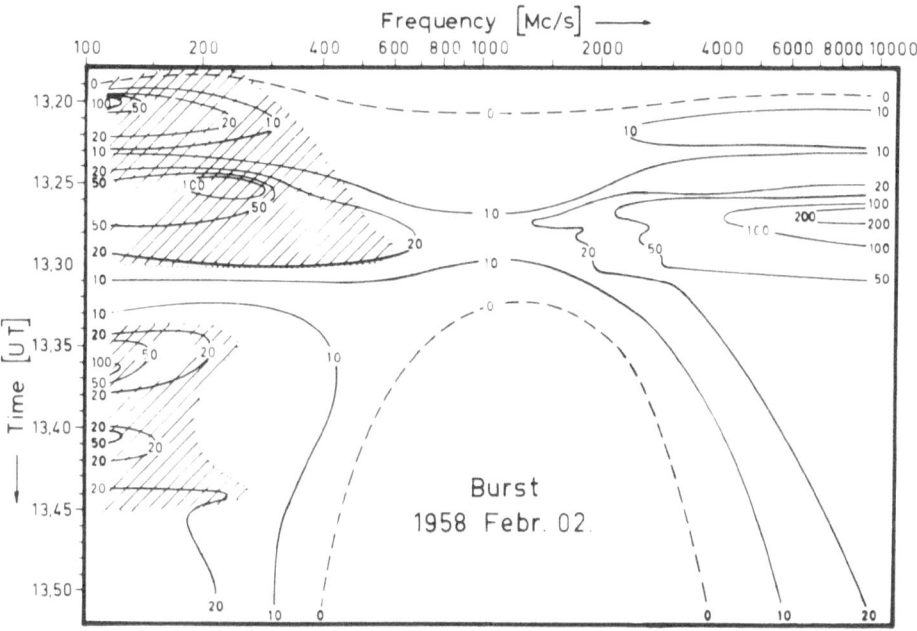

Fig. 7. Time—frequency diagram of a burst of February 2, 1958.

steep radiation increase before the maximum and the subsequent gradual decrease
can be seen from the extent of the bunching together or spreading out of the contour lines of equal radiation flux.

Figure 7 shows an outburst of February 2, 1958, in which an emission in the
centimeter-wavelength range is coupled with a continuous emission in the meter
range. The continuum of the meter-wavelength range has the appearance of a wide
emission band. What is striking in this burst is a low activity in the range from
400 to 2000 Mc/s. A part of the meter range continuum in which short period and
narrow-band peaks are superimposed on the continuum is designated by shaded
areas. We find a weak connection between the appearance of the continuum and
these Type I peaks.

In Fig. 8, the time—frequency diagram of a very complicated burst is represented. It is evident that in this case the emission in the meter-wavelength range
and in the considered centimeter band consists of a number of broad emission bands.

From the investigation of a number of such time—frequency diagrams, we
came to the following general conclusions:

1. The continuum of burst radiation in the centimeter range shows in many
   cases a gradual increase of intensity toward high frequencies. At 10,000
   Mc, this increase occasionally continues, or else settles down to a constant intensity in which it can remain all the way into the millimeter-
   wavelength range. This spectral type overlaps in many cases the group
   of impulsive bursts (and also that of the M-bursts introduced by Takakura

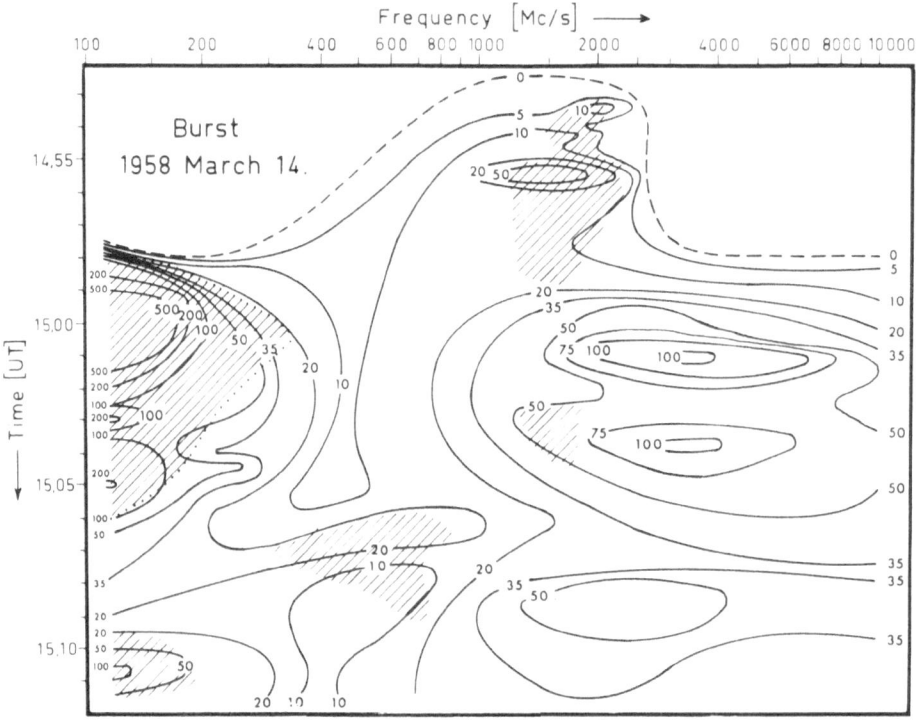

Fig. 8. Time – frequency diagram of a complex burst.

[2]. However, we can see that long enduring burst forms can also have a similar spectrum.

2. Next, in the centimeter range, there often is an emission in the form of broad bands, which have a certain similarity to the bands in the meter-wavelength range. When bursts in this type have a certain strength and also belong to the long enduring group, we consider them as centimeter Type IV bursts.

From a series of well observed bursts, the midpoints of the emission bands as well as the width of the bands at $I = 0.75\,I_{max}$ were determined. Table I gives the results of a number of bursts (Hachenberg and Krüger [16]). The midpoints of the bands show a wide spread over the whole frequency range. It was also found, that the band width greatly increases as the frequency becomes greater and can approach values up to 6000 Mc.

STATISTICAL INVESTIGATIONS OF BURST SPECTRA

After we have recognized both types of burst spectra, it is perhaps still of importance to gain an overall view of the frequency of appearance of the different

TABLE I

Microwave Burst Emission Bands

Date	Time (U.T.)		$L*$ (Mc/s)	$B*$ (Mc/s)	$J*$ ($\cdot 10^{-22}$ W/m^2-c/s)
1957, September 9	8^h	15^m	5000	6000	60
11	13	27.5	3000	—	13
18	11	20.5	2000	750	175
20	12	40	4000	4000	65
20	13	19	900	900	100
23	14	54	5000	2500	25
26	13	50.5	4000	3500	45
26	14	38	2000	—	18
November 11	14	19	1900	1500	150
December 19	09	33	4000	3500	350
1958, January 20	14	58	4000	4000	350
23	09	11.5	4500	6500	30
February 8	08	41	2800	2000	10
March 11	08	27	2000	700	22
14	15	01	4000	3500	165
April 2	15	31.5	2300	1300	13
15	12	27.5	5000	4000	25
29	11	54.5	2500	1600	75
June 9	09	51	4500	4500	18
25	16	28	3500	4500	92
July 2	06	59.5	4000	2500	16
8	13	38	2500	2500	48
26	08	58	5000	6000	115
August 4	07	43.5	4500	5000	32
25	10	01	4500	4000	75
September 12	09	12.5	4500	5000	85
22	07	45	2000	1000	85
December 5	09	25.5	3000	5500	22
7	10	52.5	2000	1200	164

* L = midpoints of the bands, B = bandwidth, and J = radiation flux.

spectra. It would, however, be a little wasteful to construct a time–frequency diagram for each burst. For the statistical inspection of the data, it is sufficient to consider the bursts through the spectra of their maximum phase. The observed spectra can be subdivided into three groups (Hachenberg and Krüger [16]):

A. Spectra which show an increase toward higher frequencies, or which go over into a section in which the intensity is independent of frequency.
B. Spectra which increase at meter wavelengths; these have a strong continuum predominantly in the meter range.
C. Spectra which have in the considered frequency range a clearly marked maximum.

Of course, combinations of the above groups appear to a certain extent, but we shall not go into that aspect.

Table II gives the results of the statistical investigations. Of the 970 bursts observed at Adlershof during the IGY, the spectra of 734 were clearly determined.

TABLE II

Spectrum of Bursts in the Centimeter Range

Spectrum	Group	Description	Emission mechanism	Frequency in I.G.Y., %	Remarks
$f \longrightarrow$	A	Intensity increases with frequency	Thermal, part of a synchrotron radiation	56	
$f \longrightarrow$	B	Intensity decreases with frequency	Synchrotron radiation	10	Belonging (among others) to centimeter Type IV burst
$f \longrightarrow$	C	Intensity has a maximum in centimeter range	Synchrotron radiation	26	Long-enduring events: centimeter-Type IV?
$f \longrightarrow$	A B	Superposition of spectra A and B	See A and B	8	

Of these, more than half show a spectrum with increasing intensity as we go to higher frequencies. In 26% of the cases, a maximum exists in the centimeter frequency range. Bursts with intensity increasing toward longer wavelengths were less frequenty (10%). A small number show a combination of A and B, and have a minimum in the considered range. Comparable results were found by T. Kakinuma and H. Tanaka [14].

An attempt to bring together the different spectral types with the forms of the bursts has brought out the result that in the groups of bursts of spectral Type A as well as B, all burst forms are present. No burst form has shown a preference for one or the other spectrum. It is only in the statistical average that some weight is given one way or the other — for example, the simple burst forms (impulsive bursts) which give a certain preference to the spectral type A with a gradual increase as we go to higher frequencies.

CONCLUSION

The investigation of burst spectra shows us that we must differentiate between two basic spectral types: burst spectra with a gradual increase toward high frequencies; and burst spectra with more or less sharply defined emission bands. For these, we will have to assume two correspondingly different mechanisms of emission.

It is possible to associate the burst spectra of monotone increase predominantly with thermal emission from flare areas. See (Hachenberg and Wallis [17],

Kawabata [18], Gelfreich [19], and Hirth [20]).   We are then led to an equivalent temperature of 10^5 to 10^8 °K. One support for this assumption is the fact that centimeter radiation appears together with excessive X-radiation, and both follow a similar course with time.

For burst spectra which consist predominantly of emission bands, it is supposed, following Takakura and others (Takakura [21], Stein and Ney [22], Peterson and Hower [23]), that the radiation is a synchrotron radiation from highly accelerated electrons in the magnetic fields of the activity centers. A support for this is found in the occasional appearance of polarization in the burst radiation.

Although we can assume that many bursts, from their spectra alone, emit thermal or synchrotron radiation exclusively, for the majority of cases we have to assume a combination of the two types of emission mechanisms, with variable amounts of mixing. This is evident from statistical studies where by impulse type, gradual rise and fall type, and even complex type bursts which are clearly recognizable by their behavior in the meter range as Type IV bursts, all exhibit emission bands and monotone intensity distributions. Finally, the fact that excessive X-radiation is generated in both microwave bursts with spectra A and C, seems to be another support for this assumption.

The radiating areas of the microwave bursts are imbedded in the lower corona, just above the chromosphere. As evidence, occasionally one finds a relatively sharp cutoff of the continuous spectrum at 1000 Mc/s. The diameter of the areas was determined by Kundu [24] at 3.2 cm. He found values of between 1 and 2.5′ of arc for bursts associated with flares of importance 2; after the bursts, the radiating areas are about 10 times greater than the flare areas. Contrary to the meter-wavelength bursts of Type II and III, the source of the microwave bursts has no appreciable movement outward.

The electron density within the radiating volume is of the order of $10^9 - 10^{10}$ cm^{-3} (Hachenberg and Wallis [17]). During the maximum phase of the burst, the density is more than 10 times higher than in the undisturbed corona. At the upper boundary, the density drops down steeply as compared to the density of the undisturbed corona. There, the excitation of plasma oscillations occasionally takes place, which give rise to the sharp peaks which are superimposed on burst records between $\lambda = 20$ and $\lambda = 40$ cm (see for instance Fig. 4c). If we assume the continuous spectrum to be of thermal origin, the electron temperature lies between 10^6 and 10^8 °K.

Within the radiating area we have a physical process which is evidently similar for all bursts in the considered frequency range. It consists of the acceleration of electrons which then give off partly thermal emissions and partly magnetic Bremsstrahlung. The great variety of burst forms are not produced by different processes, but come about through variations of this process − perhaps the accelerating process − within the burst volume. There is no doubt that the plasma of the microwave bursts has a certain similarity to the plasma of the coronal condensation.

Concerning Types IV bursts, we can consequently classify all those events which show the appearance of a band emission as Type IV events. However, if we adhere to the original definition of this burst type, which was connected with an expanding particle cloud (continuum in connection with a Type II burst), then we have no definite signs in the centimeter range as to if or when this particle stream started. We can only establish that predominantly the large bursts with especially high radiation power also produce an expanding particle cloud. The classification of bursts of the centimeter-wavelength range as Type IV events therefore is somewhat arbitrary.

REFERENCES

1. J. P. Wild, S. F. Smerd, and A. A. Weiss, *Ann. Rev. Astron. Astrophys.* 1, 291 (1963).
2. O. Hachenberg, *Z. Astrophys.* 46, 67 (1958).
3. T. Takakura, *Publ. Astron. Soc. Japan* 12, 55 (1960).
4. M. R. Kundu and F. T. Haddock, *IRE Trans. Antennas Propagation* AP-9, 82 (1961).
5. A. R. Thompson and A. Maxwell, *Astrophys. J.* 136, 546 (1962).
6. O. Hachenberg and A. Krüger, *J. Atmospheric Terrest. Phys.* 17, 20 (1959).
7. G. A. Harvey, *Astrophys. J.* 139, 16 (1964).
8. O. Hachenberg, Rendiconti della Scuola Internazionale di Fisica "Enrico Fermi" XII, Bologna (1960), p. 217.
9. O. Hachenberg and H. Volland, *Z. Astrophys.* 47, 369 (1959).
10. H. Volland, *J. Atmospheric Terrest. Phys.* 26, 695 (1964).
11. K. A. Anderson and J. R. Winckler, *J. Geophys. Res.* 67, 4103 (1962).
12. A. Krüger, Thesis Berlin (1961); *Z. Astrophys.* 55, 137 (1962).
13. H. W. Dodson, E. R. Hedemann, and A. E. Covington, *Astrophys. J.* 119, 541 (1954).
14. T. Kakinuma and H. Tanaka, *Proc. Res. Inst. Atm. Japan* 8, 39 (1961).
15. A. D. Fokker, *Information Bull. Solar Radio Obs.* No. 11 (1962).
16. O. Hachenberg and A. Krüger, *Z. Astrophys.* 59, 261 (1961).
17. O. Hachenberg and G. Wallis, *Z. Astrophys.* 52, 42 (1961).
18. A. Kawabata, IAU Symposium on the Solar Corona, Cloudcroft (1961).
19. G. B. Gelfreich, *Solnechnye Dannye*, Leningrad No. 5 67 (1962).
20. W. Hirth, Thesis, Berlin (1963).
21. T. Takakura, *Publ. Astron. Soc. Japan* 12, 325 (1960).
22. W. A. Stein and E. P. Ney, *J. Geophys. Res.* 68, 65 (1963).
23. A. M. Peterson and G. L. Hower, *J. Geophys. Res.* 68, 723 (1963).
24. M. R. Kundu, *Ann. Astrophys.* 22, 1 (1959).

Chapter 13

Radio Investigation of the Solar Corona and the Interplanetary Medium

Antony Hewish

Mullard Radio Astronomy Observatory
University of Cambridge, England

SUMMARY OF METHODS OF STUDYING THE INTERPLANETARY MEDIUM

Optical — Zodiacal light . . . . ⟩ $100R_\odot$ . . . . (N, dust)
Radio — Scattering by electron clouds . . $5R_\odot$ to $100R_\odot$ . . (N, H?)
Direct sampling — Space probes . . . . $140R_\odot$ to $200R_\odot$ . . . . (Plasma flux, H)
Indirect methods — Comet tails . . . . (Plasma flux ?)
      — Cosmic rays . . . . (H)
      — Geophysical phenomena . . . . (Plasma streams, etc.)

Zodiacal Light

Sunlight is scattered by its passage through the interplanetary medium. This process gives rise to the well-known phenomenon of zodiacal light, a faint halo centered on the ecliptic which becomes visible, under suitable observing conditions, when the sun is below the horizon. The zodiacal light exhibits substantial linear polarization, and this led early workers (Behr and Siedentopf, [2]) to suppose that free electrons were mainly responsible. An analysis on this basis gave $N \sim 600$ cm^{-3} at a distance of 1 A.U. It was later shown by Blackwell, however, that dust scattering can also give rise to optical polarization.

Blackwell and his co-workers have attempted to separate the dust and electron components of the zodiacal light from spectrum considerations. Scattering from electrons at a high kinetic temperature will show no Fraunhofer lines, due to Doppler broadening, while the light scattered by dust particles will have the normal Fraunhofer lines. By comparing the depths of the lines in the spectrum of the zodiacal light and direct sunlight, it is possible to separate the two components. The best observational data so far obtained (Beggs and Blackwell [1] leads to $N \sim 16$ to 20 electrons/cm^2 at 1 A.U.

Radio Scattering by Electron Clouds

When radio sources are viewed through the interplanetary medium not too far from the sun, irregular scattering takes place, and the angular extent of the sources is apparently increased. This phenomenon gives information about irregular electron clouds at distances up to 0.5 A.U.; this method will be discussed in detail later.

Space Probes

Direct observations of the physical condition in space are now being undertaken by space probes. The most extensive measurements so far available are those of Mariner II launched in August 1962. This vehicle measured a proton flux of approximately $1.2 \times 10^8 \mathrm{cm}^{-2} \sec^{-1}$ in the radial direction with particle velocities in the range 400 to 700 km/sec. Magnetometer data indicated the presence of a significant magnetic field at all times with a value of the order of 5 to 10 gamma and variable in direction. Considerable changes in H and particle flux were observed and some of these could be related to geomagnetic disturbances. A survey of the data has been given by Coleman et al. [8]. Space probe measurements have so far been limited to distances of 0.7 to 1 A.U.

Indirect Methods

Comet Tails

Comets of Type 1, whose tails are predominantly composed of ions, invariably develop long straight tails pointing away from the sun radially. Motions in the tails often indicate accelerations 20 to 200 times stronger than solar gravity, which cannot be explained by radiation pressure alone. Biermann [3] originally suggested that the tails were blown out by a continuous stream of particles emitted by the sun. However, it seems more likely that some type of plasma interaction may occur, but no quantitative theory has been developed.

This behavior of comets has been seen at all stages of the solar cycle suggesting a continuous ejection of plasma from the sun. The close alignment of the tails along a radial direction indicates ejection velocities of several hundred km/sec.

Cosmic Rays

The propagation of solar cosmic rays on their way to the earth is largely governed by the interplanetary magnetic field. Many different types of cosmic ray variation have been studied and one of the most important of these is a sudden increase of intensity associated with a major solar flare. On these occasions the field appeared to be largely radial, but with a significant curvature imposed by solar rotation. More complex variations, which occur more frequently, have been explained by the trapping of particles in magnetic bottles or behind shock fronts. A general discussion has been given by Obayashi [16].

Geophysical Phenomena

It has long been known that phenomena such as auroras and disturbances of the geomagnetic field are associated with solar activity. The precise mechanism for such phenomena is still uncertain, but in general terms they indicate the presence of plasma clouds and streams in the interplanetary medium. The time delay between the occurrence of a visible solar flare and the onset of magnetic disturbances indicates a velocity of 1000 to 2000 km/sec.

CORONAL MODELS

The origin of the interplanetary medium as an extension of the solar atmosphere has received considerable theoretical attention in recent years. Three types of model have been proposed by Chapman, Chamberlain, and Parker. All of the models are preliminary in that idealized coronas are assumed, and there has been much dispute in the publications of Chamberlain and Parker, but Parker's model appears to give the closest agreement with a variety of observational evidence.

Chapman's Model

Chapman bases his simple model on a system in hydrostatic equilibrium. Using the known thermal conductivity of ionized hydrogen and assuming a temperature and density near the sun where such parameters can be measured, he predicts N and T at further distances. In a modification of this model, which makes some attempt to account for solar rotation and a general magnetic field, he derives $N \sim$ 300 cm^{-3} at 1 A.U. (Chapman [6]). Even allowing for a three-fold reduction of this value, in the light of more recent optical data (Ingham [14]), the model is scarcely plausible.

Chamberlain's Model

Chamberlain [5] assumes the corona to be a quiet atmosphere with a base temperature of about 2×10^6 °K. He computes the mass loss due to thermal evaporation of the high-energy tail of the Maxwell distribution and shows that there will be a gentle outflow of material giving $N \propto (r/R)^{-3/2}$. At 1 A.U. the radial velocity is about 18 km/sec and the density $N \sim 30$ cm^{-3}. Magnetic fields are ignored and the corona is taken to be spherical and nonrotating.

Parker's Model

In this model Parker [17] assumes a constant temperature of around 10^6 °K out to a distance of several solar radii. He shows that this gives rise to an outward pressure, and solution of the hydrodynamic equations indicates a large-scale expansion at supersonic velocities. To maintain the constant temperature in the initial stages of the expansion, a heat source in the upper corona is required and it is possible that this might be provided by the dissipation of hydromagnetic waves generated at the photosphere.

At distances beyond about 20 R_0 Parker's theory indicates an outward flow having nearly constant velocity of several hundred km/sec. The precise values of N and the velocity at 1 A.U. depend critically upon the assumed coronal temperature, but reasonable values are obtained for T in the range 0.5 to 3×10^6 °K.

Since the kinetic energy of the solar wind will exceed the magnetic energy density (on the assumption of a 1-G solar magnetic field) at large distances from the sun, the lines of force will tend to be stretched into a co-rotating Archimedes' spiral. The outward motion of the solar wind will eventually lose its momentum by collision with interstellar material at some point in the range 10 to 100 A.U. Since the velocity of flow is almost constant in Parker's model, it follows that the electro density will vary as $(r/R_0)^{-2}$.

SCATTERING OF RADIO WAVES BY IRREGULAR ELECTRON CLOUDS

It is found experimentally that radio waves traversing the interplanetary medium are scattered by random irregularities in the electron density. There is, as yet, no exact theory for this scattering process when the random phase changes introduced by the irregularities exceed one radian. The necessary theory must involve both diffraction concepts and multiple scattering. Some progress has been made by Mercier [15] and Pisareva [18], using numerical methods. However, for the present purpose, an approximate ray analysis will suffice.

Consider a wave traversing an irregular transparent medium as shown in Fig. 1. The emergent wave front will suffer random phase deviations of average magnitude

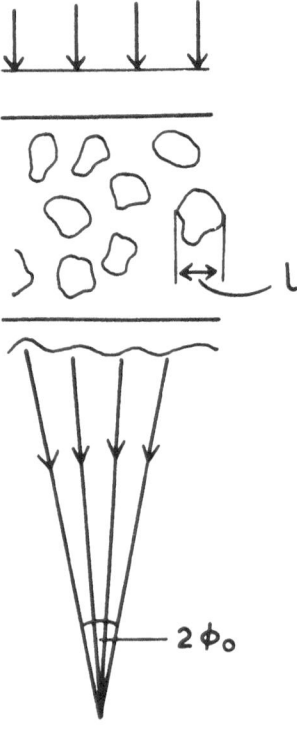

Fig. 1

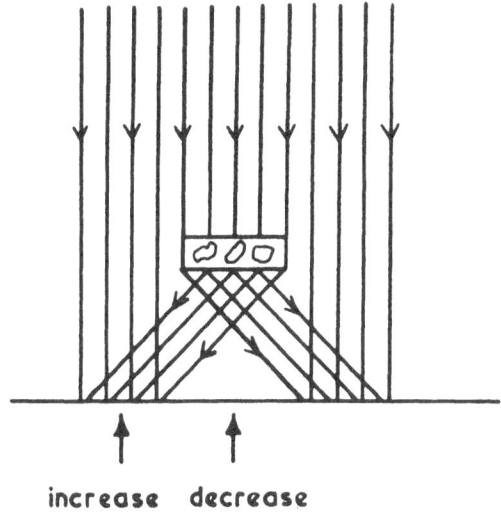

increase   decrease

Fig. 2

$\Delta\mu l\sqrt{n}$, where n is the number of clouds along a line of sight, l the scale of the clouds, and $\Delta\mu$ the variation of refractive index from its mean value. The average angular deviation of a ray ϕ_0 is therefore $4\Delta\mu l\sqrt{n}/l \sim 4\Delta\mu\sqrt{n}$, where $\Delta\mu = 4 \times 10^{-5} N/f^2$ (f in Mc/s, N, electrons/cm^3). Therefore, a point source, viewed through such an irregular medium, appears to have an angular diameter $\sim 2\phi_0$.

To estimate a typical magnitude we put $N = 10^3$, $f = 100$ Mc/s, $n = 100$, and obtain $\phi_0 \sim 0\overset{''}{.}5$. The quantity ϕ_0 will, of course, increase rapidly as the line of sight passes the sun more closely.

A rigorous statistical analysis (see, for example Chernov [7]) shows that no matter how many irregularities are traversed, the scale of the random variations across the emergent wave front is l, provided ϕ_0 is small. This is of importance since it enables an upper limit to be placed on the size of the electron clouds.

Scattering causes no change of the mean intensity of a radio source provided that the scattering region has a sufficiently great extent. For a limited scattering region it is clear that both increases and decreases of intensity may occur as illustrated in Fig. 2. These considerations are important in the case of large scattering close to the sun at meter wavelengths.

When the scattering medium is anisotropic, the largest value of ϕ_0 will occur in a direction corresponding to the minimum value of l. A medium containing filaments aligned in a preferred direction, for example, will give the largest scattering perpendicular to the filaments.

MEASUREMENTS WITH RADIO INTERFEROMETERS

Scattering of the kind discussed in the previous section has been observed for a number of radio sources, in particular the Crab Nebula which passes close to the

sun in June. The usual experimental technique is to measure the angular diameter
of the radio source using an interferometer. If the scattered distribution has a
Gaussian form such that the intensity of the radiation scattered at an angle ϕ is proportional to $\exp(-\phi^2/\phi_0^2)$, then the fringe visibility (the amplitude of the record)
is proportional to $\exp(-\pi^2 d^2 \phi_0^2/\lambda^2)$ for an interferometer of spacing d and wavelength λ. Measurement of the fringe visibility then determines ϕ_0, the angular extent of the scattered rays. The result of a typical experiment on the Crab Nebula
might be as shown in Fig. 3. Sensitive measurements require large spacings and
also long wavelengths since $\phi_0 \propto \lambda^2$.

When radio sources are observed at a small angular separation from the sun,
the scattering is large and ϕ_0 may be measured using a single directional antenna.
Such measurements are difficult since the sun gives intense radio emission which
tends to swamp the radio source under investigation. With interferometers this
difficulty is minimized since the sun, being an extended source, is often resolved
by the interferometer.

Radio interferometers, used simultaneously and having base lines of different
orientation, have been used to study the relative magnitude of the scattering in
different directions.

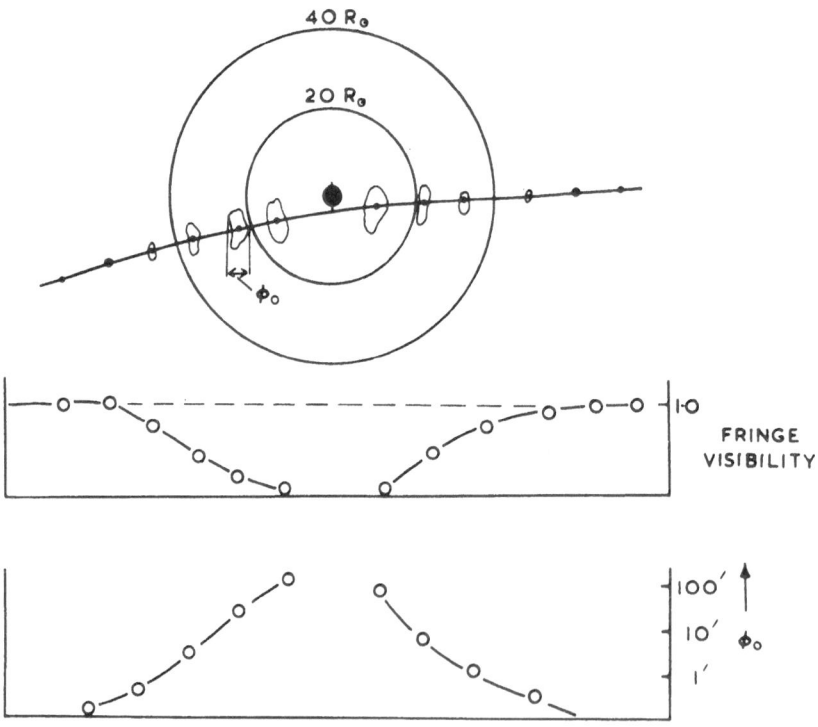

Fig. 3

A SURVEY OF EXPERIMENTAL SYSTEMS

The following list of observations is by no means exhaustive, but is intended to illustrate the range of experiments which have been carried out.

	λ	d
Vitkevitch, 1951	4 m	100 λ
	2.7 m	150 λ
	7.6 m	100 λ
Hewish, 1952	1.9 m	300 λ
	3.7 m	105 λ
	7.9 m	8 λ, 49 λ,
		200 λ, 560 λ,
		900 λ, 1290 λ
	11 m	148 λ
Slee, 1961	3.5 m	2920 λ
Blum and Boischot, 1957	1.9 m	3' fan beam
Ericson, 1959	11.4 m	185 λ
		270 λ
Basu and Castelli, 1962–64	10 cm,	84 ft dish
	25 cm	
Wyndham and Clark, 1963	30 cm	–

THE NATURE OF THE SOLAR CORONA AS DERIVED FROM SCATTERING MEASUREMENTS

Experimental observations give directly the magnitude of ϕ_0 at a variety of wavelengths, and its variation with distance from the sun. The radial variation may be compared to that predicted by different coronal models, while the absolute magnitude of ϕ_0 may be expected to give data about the scale and density of the electron irregularities in heliographic latitude. Observations extended over several years give data concerning the behavior of the outer corona during the sunspot cycle. Investigation of the scattering anisotropy may be expected to throw light on the elongation and direction of alignment of the irregularities, and hence, possibly, of the solar and interplanetary magnetic fields. These topics will now be given brief consideration.

The Size of the Irregularities and Their Origin

A minimum scale of about 10 km can be derived from a more thorough analysis of the scattering phenomenon based on diffraction theory (Hewish [10]). Sizes of this order scarcely seem plausible on physical grounds and the fixing of an upper limit is of greater consequence. An upper limit to the scale may be derived from

Slee's observations which show that the interferometric relation is obeyed for ϕ_0 as small as 6$''$. With reference to Fig. 1, this implies that rays are being received from several irregularities at once, even when ϕ_0 is exceedingly small, and leads to a maximum scale $l \sim 5000$ km at 30 R_0 (Hewish and Wyndham [12]). This scale is remarkably small compared to that of visible rays and streamers. It should, perhaps, be stressed here that this small structure is a basic property of the outer corona at all times since radio scattering is always observed. Occasionally, sudden short-lived enhancements of scattering have been observed (Slee [19, 20] and Vitkevitch [21, 22]) which might be due to disturbances of the outer corona. These are rare events of uncertain interpretation and they will not be considered further in this survey.

The only plausible origin of irregularities having a scale less than 5000 km at great distance from the sun is that they are associated with "frozen in" magnetic fields. In the absence of a magnetic field, irregularities of this scale would rapidly diffuse to form a continuum. In the presence of a field, diffusion is restricted to directions along the lines of force.

The Orientation of the Irregularities

The character of the magnetic field in the outer corona and the interplanetary medium is far from clear. Space probe measurements have given conflicting results while cosmic ray evidence is usually relevant to conditions of solar activity when the quiescent field may be appreciably disturbed. In the presence of a solar wind, the field lines may be largely radial but with a curvature imposed by solar rotation, or the solar wind may carry with it a disordered field. If an ordered radial field exists, then the irregularities are likely to take the form of aligned filaments and the scattering measurements should show evidence of this. Measurements of the scattering anisotropy using interferometers having base lines of different orientation generally show significant, but not large, anisotropy. Three triangularly orientated base lines are required to remove ambiguity and results obtained from such systems (Hewish and Wyndham [12] and Ericson [9]) show that the scattering is usually greatest in a direction perpendicular to the sun's equatorial plane. Some typical results showing the alignment of the filaments are indicated in Fig. 4. These results are in fair agreement with a radial system of filaments, but notable deviations occur and more observations are needed to derive a detailed picture. It should be remembered that the scattering results from irregularities distributed over a considerable depth in the corona and over which uniform conditions may not prevail.

The Variation of Scattering with Heliographic Latitude
and the Sunspot Cycle

Slee's work on 3.5 m, in which many different sources were observed, is important since it gave the first indication of how the irregularities were distributed in heliographic latitude. Over the pole the scattering appeared to be reduced by a considerable factor, compared to the scattering at low latitudes. Quantitative estimates of this effect were made in 1962 (Hewish and Wyndham [12]) by com-

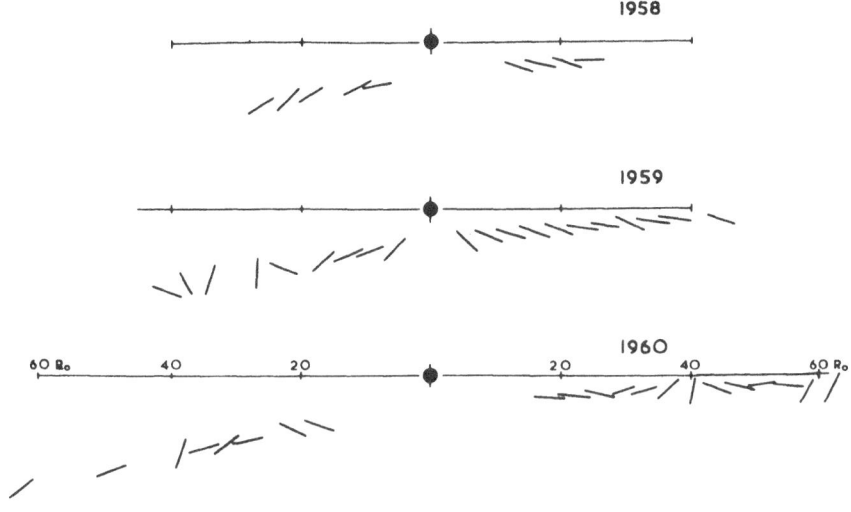

Fig. 4

paring the scattering observed for the Crab Nebula and 3C 123. These results showed that scattering over the pole was reduced by a factor 2.6 : 1 at all radial distances out to $50R_\odot$.

There is a pronounced variation of scattering with the solar cycle (Vitkevitch [21, 22] and Hewish [10, 11]). Extended measurements over the past few years have shown that the increase (by a factor of about 2) at sunspot maximum extends to radial distances of at least $50R$. It had been suggested earlier that the variation might not be significant beyond about $16R$.

The Variation of Scattering with Radial Distance

An accurate determination of the radial variation of scattering demands a series of measurements carried out at the same time, using interferometers of different spacing, in order to cover a wide range of ϕ_0. Observations made in 1960, 1961, and 1962 (Hewish and Wyndham [12]) indicate a rather slow radial variation such that ϕ_0 varies as $(r/R)^{-1.3}$ to $(r/R)^{-1.5}$. This relation corresponds approximately to sunspot minimum conditions. Ericson has derived a variation $(r/R)^{-2}$, but his combination of measurements by different workers shows considerable spread and is not incompatible with the first result. Slee has published observations indicating an even steeper radial variation, but no allowance was made for solar cycle variations and this would cause an appreciable steepening of his radial law. It is also possible that the radial variation becomes steeper at sunspot maximum, but further work is needed to establish this.

The Electron Density in the Irregularities

Since, a priori, the scale of the irregularities may have any value in the range 10 to 5000 km, and the number of such irregularities along the line of sight is un-

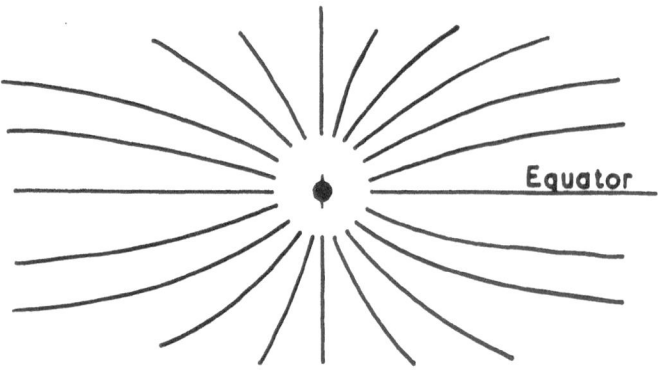

Fig. 5

known, only estimates of the electron density may be made. However, such estimates are not without importance, since prior to space probe measures radio scattering was the only definite evidence for the presence of any electrons in the interplanetary medium. Calculations based on the above scales and assuming a filamentary structure have been carried out (Hewish and Wyndham [12]) for two extreme cases. The least number of irregularities traversed by any line of sight must be greater than one since scattering is always present, while the greatest possible number corresponds to close-packed irregularities. Values for these two extremes are given below:

Widely spaced filaments . . . . $N \sim 7 \times 10^3$ cm^{-3} (independent of l)

Close-packed filaments. . . . . $l \sim 10$ km $\qquad N \sim 15 \qquad$ cm^{-3}

$\qquad\qquad\qquad\qquad\qquad\qquad\quad l \sim 5 \times 10^3$ km $N \sim 3 \times 10^2$ cm^{-3}

Estimates of the mean density due to the irregular component alone may be derived by smearing out the electron content of the irregularities to fill the whole volume. For the case of widely spaced filaments of scale 5000 km, we obtain $N \sim 10$ cm^{-3} at $20R$. On the other hand, if the irregularities are close-packed, we have $N \sim 300$ cm^{-3}; extrapolating this value to 1 A.U. according to the observed radial law gives $N \sim 5$ cm^{-3}, which is a value in good agreement with space-probe figures.

Comparison of the Observations with Coronal Models

To compare the observed radial variation with that predicted by different coronal models it is necessary to make the reasonable assumption that the random variations of density are always proportional to the mean density. From visual data in the lower corona it seems probable, indeed, that the irregularities do not represent a small perturbation but that they constitute the bulk of the coronal material. In this case we have ϕ_0 proportional to N and the observed radial variation of ϕ_0 gives the radial variation of N also. On this basis we have $N \propto (r/R_0)^{-1.3}$

to $(r/R_0)^{-1.5}$. Now Chamberlain's model predicts $N \propto (r/R)^{-1.5}$, while Parker's model gives $N \propto (r/R)^{-2}$. Thus, the observations appear to favor Chamberlain. However, other evidence for a rapid solar wind is strong, and our results can be reconciled with Parker's provided that the wind is not strictly radial, but converges progressively toward the solar equator as shown in Fig. 5. A convergence of this type might arise, for example, from the magnetic pressure of a strong polar field. Only a small degree of convergence is required to reduce the radial variation from an inverse square law to the smaller value actually observed.

Intensity Variations

When sources are observed within about $10R_0$ from the sun, the scattering becomes very large at long wavelengths ($> 1°$) and variations of intensity arise due to the limited extent of the scattering region as mentioned in the section entitled "Scattering of Radio Waves by Irregular Electron Clouds." Measurements at wavelengths of 8 m and above exhibit large decreases (Hewish, Ericson) while at shorter wavelengths both increases and decreases have been reported (Blum and Boischot [4]). Theoretical computations have been carried out by Högbom [13] and Ericson for different coronal models. Both workers derive similar expected variations of the type shown in Fig. 6. Although the decreases predicted by the computations are in fair agreement with observation, provided that a model with radial filaments is adopted, the increase at slightly greater radial distance has not been seen at wavelengths of 8 m and above. On the other hand, at frequencies greater than 169 Mc/s, an increase has been reported by Boischot where one would not be expected. Ericson has suggested that the presence of additional scattering components of small scale may explain some of the discrepancies. It is also possible that more realistic models which take into account enhanced scattering toward the

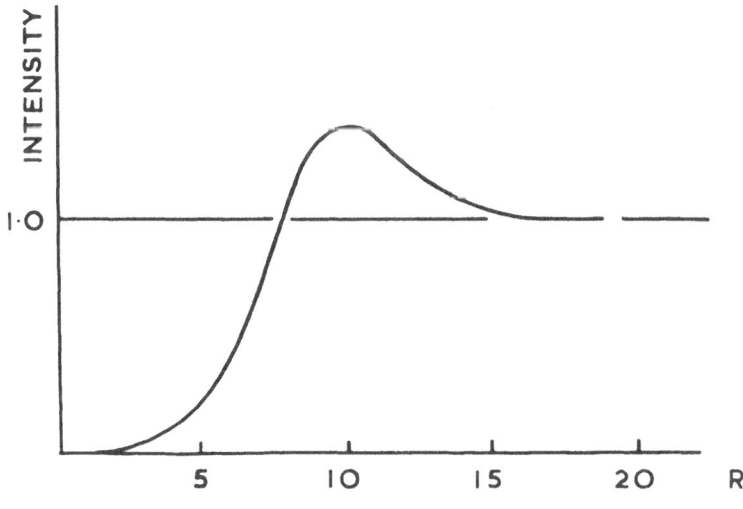

Fig. 6

solar equator might throw light on the difficulties. More work, both theoretical and observational, is clearly needed.

INTERPLANETARY SCINTILLATION

A survey of radio exploration of the interplanetary medium would be incomplete without some mention of an important new technique which holds great promise for the study of the motion and scale of electron clouds in space. It has recently been discovered that "quasi-stellar" radio sources exhibit scintillation phenomena, much akin to ionospheric scintillation, which arises in the interplanetary medium. The phenomenon has not been noticed hitherto since it requires large radio telescopes which operate at meter wavelengths, radio sources of exceptionally small angular diameter, and these sources to be observed within about $90°$ of the sun. The new method makes possible the study of irregularities at very much greater distances than has previously been possible and already they have been detected at a distance of 0.9 A.U. using the large 178 Mc/s reflector at Cambridge (Hewish, Scott, and Wills [23]).

REFERENCES

1. D. W. Beggs and D. E. Blackwell, *Monthly Notices Roy. Astron. Soc.* 127, 329 (1964).
2. A. Behr and H. Siedentopf, *Z. Astrophys.* 54, 200 (1962).
3. L. Biermann, *Observatory* 77, 109 (1957).
4. E. J. Blum and A. Boischot, *Observatory* 77, 205 (1957).
5. J. W. Chamberlain, *Ap. J.* 131, 47 (1960).
6. S. Chapman, *Space Astrophysics* (McGraw-Hill, New York, 1961).
7. L. A. Chernov, *Wave Propagation in a Random Medium* (McGraw-Hill, New York, 1960).
8. P. J. Coleman, L. Davis, E. J. Smith, and C. P. Sonnet, *Science* 138, 1099 (1962).
9. W. C. Ericson, *Ap. J.* 139, 1290 (1964).
10. A. Hewish, *Proc. Roy. Soc., A* 228, 238 (1955).
11. A. Hewish, *Monthly Notices Roy. Astron. Soc.* 118, 534 (1958).
12. A. Hewish and J. D. Wyndham, *Monthly Notices Roy. Astron. Soc.* 126, 469 (1963).
13. J. A. Högbom, *Monthly Notices Roy. Astron. Soc.* 120, 530 (1960).
14. M. F. Ingham, *Monthly Notices Roy. Astron. Soc.* 122, 157 (1961).
15. R. P. Mercier, *Proc. Cambridge Phil. Soc.* 58, 382 (1962).
16. T. Obayashi, *Planetary Space Sci.* 12, 463 (1964).
17. E. N. Parker, *Planetary Space Sci.* 12, 451 (1964).
18. V. V. Pisareva, *Astron. Zh.* 36, 427 (1959).
19. O. B. Slee, *Monthly Notices Roy. Astron. Soc.* 123, 223 (1961).
20. O. B. Slee, *Australian J. Phys.* 12, 134 (1959).
21. V. V. Vitkevitch, *Astron. Zh.* 37, 32 (1960).
22. V. V. Vitkevitch, *Astron. Zh.* 37, 961 (1960).
23. A. Hewish, P. F. Scott, and D. Wills, *Nature* 203, 1214 (1964).

Radar Astronomy of Solar
System Plasmas

Von R. Eshleman

Center for Radar Astronomy
Stanford University and Stanford Research Institute
California, United States of America

ABSTRACT

Harmonic frequency radar echoes from the moon are being used at Stanford to measure the density and dynamics of the ionized cislunar medium by a combination of Doppler frequency (phase path) and Faraday polarization techniques. For example, during the solar eclipse of July 20, 1963, comparisons of the radar data with ionosonde and VLF whistler measurements appear to indicate that at Stanford, where the maximum solar-disk area covered by the moon was 23%, the radar detected a blocking of the solar wind by the moon.

The deep space probes Pioneer A and B, which are due to be launched in the spring and summer of 1965, will carry receivers for radio propagation studies of the interplanetary medium. Harmonic, modulated, circularly polarized signals at 50 and 425 Mc/s will be transmitted from Stanford to the probes, where measurements of the phase path, group path, and polarization will be made. Sensitivities correspond to the ability to measure the average interplanetary volume density to an accuracy of 0.5 cm^{-3} when the probe is at its maximum distance. Changes in average density as small as 0.005 cm^{-3} can be detected. One or both of the spacecraft may be fired on a trajectory such that it will be occulted by the moon, so that measurements on the amplitudes of the two signals, as they are cut off by the limb of the moon, can give a sensitive measure of a possible lunar ionosphere.

Theoretical studies have been made of radio propagation through planetary ionospheres and atmospheres. It appears that a particularly simple and meaningful measurement technique would involve radio-wave propagation at two or more frequencies between the earth and a space probe which is occulted by the planet. As the ray paths pass nearer and nearer to the limb of the planet, refraction in the ionosphere and atmosphere and diffraction at the limb make possible sensitive measures of the electron density profile of the ionosphere and the neutral density profile of the atmosphere.

Continuing radar studies of the sun by the Lincoln Laboratory of MIT show an outward flow of plasma at the reflecting level, and a large amount of Doppler spread due to turbulence. The effective radar cross section of the sun is highly variable from day to day. When space probes which can be sent behind the sun become available, occultation measurements somewhat akin to the planetary studies described above could augment the ground-based radar studies to help define the density, structure, and dynamics of the inner and outer coronal regions of the sun.

INTRODUCTION

The subject of this paper is the study, by radar techniques, of gaseous regions in the solar system. The interplanetary medium, the solar corona, and planetary atmospheres and ionospheres are considered. Past experimental results are reviewed, although much of the discussion is of immediate plans for, and future potentialities of, this method of investigating the solar system.

Most of the regions of interest contain gases which are wholly or partially ionized, so that the radar techniques of study involve propagation of electromagnetic waves in a plasma. However, a consideration of propagation in non-ionized portions of planetary atmospheres is also included since, under certain conditions, both atmospheric and ionospheric perturbations to the radar waves will be important. The atmosphere and ionosphere of the earth are considered here only with respect to their effects on the measurements more distant regions.

Both monostatic and bistatic radar astronomy techniques are discussed. Here these terms are meant to imply that either transmitter and receiver are both on the earth or both on a space probe (monostatic), or one is on the earth and the other on a space probe (bistatic). The term "radar" is used here to include conditions where there may be nearly straight-line propagation from a transmitter to a receiver, as well as circumstances where the propagation path includes reflection from a relatively discrete target.

The subject matter of this chapter is sometimes referred to as "soft-target" radar astronomy to differentiate it from "hard-target" radar studies of the surfaces of the moon, planets, satellites, and asteroids.

FIRST-ORDER THEORY

A first-order set of formulas will be presented to illustrate several of the effects which are amenable to measurement by radar techniques. The aim will be to concentrate more on the techniques as they relate to astronomical studies of the solar system, rather than on the full details of the theory, so that only a simplified set of equations is given.

The refractive index n of a collisionless plasma in the absence of a static magnetic field is real and independent of wave polarization:

$$n = [1 - (f_0^2/f^2)]^{1/2} = \left[1 - \left(\frac{c^2 r_e}{\pi}\right)\frac{N_e}{f^2}\right]^{1/2} \tag{1}$$

where rationalized MKS units are used, and f is the wave frequency, f_0 is the plasma frequency, $c \cong 3 \times 10^8$ m/sec, $r_e = 2.8178 \times 10^{-15}$ m (the classical electron radius), and N_e is the electron volume density. For the case of a uniform and relatively dilute electron gas, where $f \gg f_0$

$$\Delta v = v_\phi - c = (c/n) - c = c - v_g$$

$$= c - cn \cong \frac{c^3 r_e N_e}{2\pi f^2} \text{ [meters/second]} \qquad (2)$$

Thus, Δv is the amount by which the phase velocity exceeds c and also the amount by which c exceeds the group velocity.

If the integrated electron density (columnar electron density) is I electrons/m^2 for a path of length D, then a wave packet after propagating at v_g over the dis-distance D will be displaced by a distance R, behind where it would have been in free space. The distance is $\Delta vD/cm$, or since $N_e D = I$

$$R = \left(\frac{c^2 r_e}{2\pi} \right) \frac{I}{f^2} \qquad \text{[meters]} \qquad (3)$$

The dimensional constant $(c^2 r_e/2\pi)$, which will be encountered in several formulas, has the well-known value of 80.6/2 or 40.3 m^3/sec^2.

Other simple arguments such as the one used above show that after the wave has propagated through the distance D its frequency will be different from the initial frequency if I is changing with time. To first-order, and assuming no relative motion of transmitter and receiver,

$$f_{DE} = \left(\frac{c^2 r_e}{2\pi} \right) \frac{I_t}{cf} \qquad \text{[cycles per second]} \qquad (4)$$

where $I_t \equiv dI/dt$ and f_{DE} is the apparent Doppler frequency, or Doppler "excess," due to the changing density. The Doppler frequency due to path length change would be additive to this effect.

For quasi-longitudinal propagation when there is a weak $(f \gg f_L)$ static magnetic field in the dilute plasma, there will be a Faraday rotation of the plane of polarization of the wave given by

$$\Omega = \left(\frac{c^2 r_e}{2\pi} \right) \frac{f_L I}{cf^2} \qquad \text{[cycles]} \qquad (5)$$

where f_L is the longitudinal gyro frequency. If the plasma is characterized by a collision frequency ν, due to electron-neutral or other types of collisions, then the wave amplitude will vary according to $\exp(-\tau)$, where

$$\tau = \left(\frac{c^2 r_e}{2\pi} \right) \frac{\nu I}{cf^2} \qquad \text{[nepers]} \qquad (6)$$

While the statements above were made for a homogeneous medium, they apply equally well if there are electron density gradients in the direction of propagation, as long as $f \gg f_0$ at all positions, and the gradients are sufficiently gentle to prevent sensible reflections. If both N_e and f_L or ν vary in the longitudinal direction, then $f_L I$ and νI in equations (5) and (6) should be determined by integration.

For transverse gradients in the electron density, the direction of propagation changes away from the gradient. The amount of change in the direction of propagation is

$$\beta = -\left(\frac{c^2 r_e}{2\pi}\right)\frac{I_x}{f^2} \quad \text{[radians]} \tag{7}$$

where $I_x \equiv dI/dx$, and the x-direction is the direction of the maximum gradient in the plane transverse to the direction of propagation.

If in (7) β is a function of x, then the amplitude of the wave will change due to focusing. If the rays do not form caustics and the region of ray bending is localized, then the logarithmic gain G of the signal due to the focusing would be

$$G = -\frac{1}{2}\ln\left(1 + y\,\frac{d\beta}{dx}\right) \cong \left(\frac{c^2 r_e}{2\pi}\right)\frac{y I_{xx}}{2f^2} \quad \text{[nepers]} \tag{8}$$

where $I_{xx} \equiv d^2 I/dx^2$ and y is the distance from the focusing region to the receiver. The extension of this formula to possible two-dimensional focusing is straightforward.

Many phenomena of interest in radar studies of solar-system plasmas are adequately described by the above first-order formulas. There are important exceptions, of course, and these will be mentioned as they are encountered in later sections.

For the case of nonionized planetary atmospheres, the refractive index of the gas is $n = 1 + 10^{-6} N$, where N is the refractivity in the commonly used N-units. For a gas consisting of a mixture of nonpolar molecules,

$$N = \sum_n a_n p_n/T \tag{9}$$

where the a_n are constants for the constituent gases, the p_n are their partial pressures, and T is the absolute temperature. For polar molecules, such as water vapor, there is an extra term varying as p_n/T^2.

Since the refractive index for nonionized gases is greater than unity and not a function of wavelength at the radio wavelengths of interest here, $v_\phi = v_g < c$. Thus, there is equal phase and group retardation and there is ray bending in the direction of the transverse gradient of density. There is little absorption for the atmospheres and wavelengths of interest here, and there are no anisotropic and doubly refractive effects.

The refractivity of an ionized gas corresponding to the conditions of equation (1) can be expressed in N-units as

$$N = - 10^6 \left(\frac{c^2 r_e}{2\pi} \right) \frac{N_e}{f^2} \tag{10}$$

if $f \gg f_0$. Conversely, equations (3), (4), (7), and (8) could be used for a non-ionized gas by replacing I by

$$\pm 10^{-6} \left(\frac{2\pi}{c^2 r_e} \right) NDf^2 \tag{11}$$

where the plus sign should be used in equation (3) and the minus sign in equations (4), (7), and (8).

THE CISLUNAR MEDIUM

Monostatic (earth-based transmitters and receivers) radar echoes from the moon have been used to study the ionized regions between the earth and the moon. Since our principal concern here is in regions beyond the earth's ionosphere and magnetosphere, we will not review the extensive moon-echo results which are based only on Faraday polarization measurements, since from equation (5) these are sensitive only to the near-earth regions, where the magnetic field strength is relatively high. To sense the ionization in the absence of a magnetic field, it is convenient to use methods corresponding to equations (3) or (4). These might then be included with polarization measurements to help make a separate determination of ionization near the earth and ionization in the remaining cislunar medium.

Moon-radar work of this type is being conducted at Stanford (Eshleman et al. [1], Howard et al. [2], and Howard et al. [3]). Since early 1962, about 400 hr of Doppler data corresponding to equation (4) have been obtained. In addition, about 100 hr of this time includes data on polarization rotation [equation (5)] and its rate of change with time. Amplitude data were taken for much of this time so that an analysis of focusing [equation (8)] can be made. A new technique based on pulse compression is just now being initiated in an attempt to measure the group delay [equation (3)].

The radar system consists of two transmitters and two antennas. A 300-kW transmitter at 25 Mc/s is used with an array of 48 log-periodic antenna elements arranged in two 1200-ft north-south rows. A 50-Kw transmitter at 50 Mc/s is used with a steerable 150-ft diameter parabolic antenna. Simultaneous operation of the two systems is possible for about two hours per day, this limit being set by the azimuthal beamwidth of the log-periodic array.

In practice, the use of two frequencies affords self-calibration which often marks the difference between a relatively simple experimental procedure and one which is very difficult or impossible to implement. If the two frequencies are f

and nf with $n > 1$, and Δ is used to indicate the conditions at f minus those at nf, then

$$\frac{\Delta R}{R} = \frac{f_{DD}}{f_{DE}} = \frac{\Delta \Omega}{\Omega} = \frac{\Delta \tau}{\tau} = \frac{\Delta \beta}{\beta} = \frac{\Delta G}{G} = \frac{n^2 - 1}{n^2} \tag{12}$$

In equation (12), f_{DD} is the frequency difference between the received lower frequency and $1/n$ of the received higher frequency. Thus by measuring this frequency, or the difference in delay of the two pulses, or the difference in the received polarization at the two frequencies, etc., the total Doppler excess, pulse delay, Faraday rotation, etc. for the lower frequency can be found from knowledge of n.

Although the main interest lies in regions beyond the ionosphere and magnetosphere, it is now clear that most of the observed effects are due to changes which take place near the earth. Doppler-excess measurements made at different times of day show the expected morning increase and evening decay of ionospheric electron density, together with irregularities in the ionosphere that become more pronounced during magnetic storms. Figure 1 illustrates several of these effects. The top curve shows Doppler excess measurements averaged over 1-min intervals for a 2-hr period during the late morning of March 4, 1962. The bottom curve shows the integral of the top curve, giving the relative value of I as a function of time for the cislunar medium, where most of the change probably occurred in the ionosphere. The fluctuations of Doppler excess, with periods from a few to about 30 min, indicate changes similar to those found in measurements of ionospheric content by Evans and Taylor [4] using the Faraday moon-radar technique, and they are presumably due to large-scale ionospheric irregularities (see, for example, Little and Lawrence [5].

Simultaneous measurements of Doppler excess and Faraday polarization at 25 and 50 Mc/s have been used in an attempt to separate near-earth effects from changes which may take place in the interplanetary medium between the magnetospheric boundary and the moon. In principle, the value of I determined from Doppler measurements includes both regions while I determined from Faraday polarization only includes near-earth ionization, so that the difference between the two values is a measure of effects in the interplanetary medium.

However, in the polarization technique it is necessary to compute from an assumed N_e height profile a weighted average value of f_L along the propagation path in order to compute I from a measurement of Ω. For studies of the ionosphere itself, this problem causes little difficulty since the bulk of the ionospheric ionization of interest exists between narrow height limits where the value of f_L varies very little. However, if we are looking for effects well away from the earth, which may be only a small fraction of the near-earth effects, it then becomes necessary to have relatively precise electron density profile information for both the ionosphere and magnetosphere. The magnetospheric region above several thousand kilometers but below the magnetospheric boundary at 60,000 or more kilometers appears to have an integrated electron content comparable to that

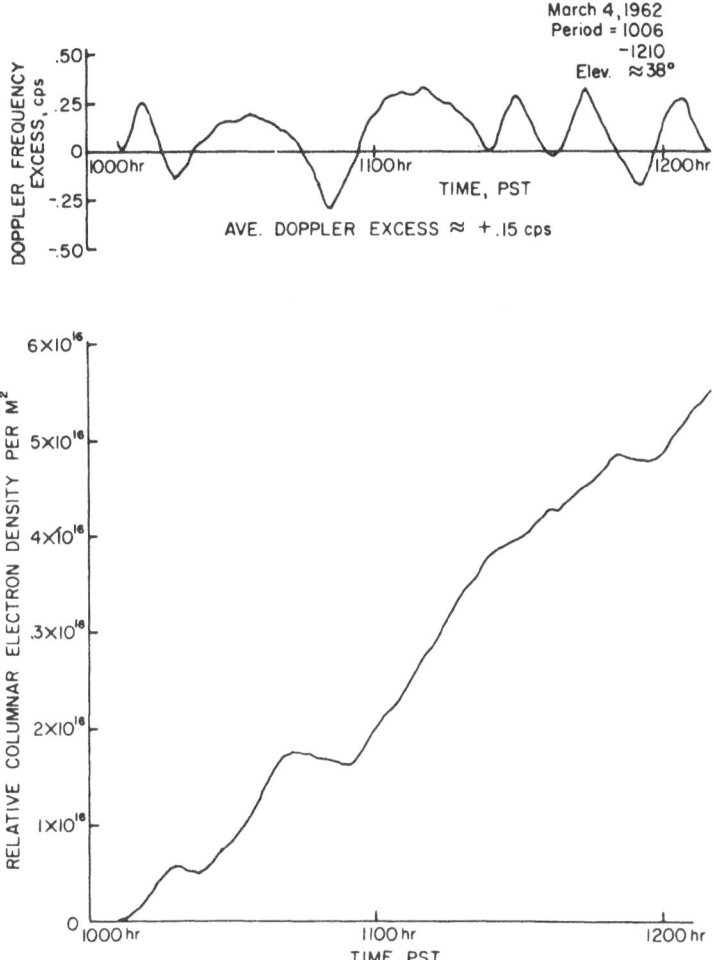

Fig. 1. Smoothed Doppler excess frequency (a) and its integral (b) showing the changing electron density between the earth and the moon.

of the ionosphere (Smith [6]). This magnetospheric content has a much smaller diurnal variation than the ionosphere, although it is markedly affected by magnetic storms.

It now appears that the combined Faraday – Doppler measurements are turning up new effects of interest in terms of the changing electron density profile in and between the ionosphere and magnetosphere, but that the detection of effects beyond the magnetosphere may be relatively infrequent. A major exception to this statement may be the measurements that were taken during the solar eclipse of July 20, 1963.

The Faraday data taken on the eclipse day show the type of ionospheric reduction in content that was expected. This will not be discussed further. Figure

2 shows the difference of the relative integrated density as determined by the Doppler method and as determined by the Faraday method for the earth – moon path on the eclipse and adjacent days. Figures 2a, c, and d show the results for the control days, July 19, 21, and 22. Figure 2b presents a similar curve for the eclipse day; it also includes the optical obscuration curve and a theoretical curve explained below. The electron content difference curve shows an unusually large decrease during the eclipse, reaching a low value soon after maximum obscuration and staying near this value for the remaining 1.5 hr of measurement. The total change during the measurement period was about 2.8×10^{16} electrons/m^2.

It is unfortunate that only about 4 hr of data could be obtained, since it appears that large changes beyond the ionosphere were already occurring at the start of the measurements, and nothing was detected of a possible recovery phase. The change on the eclipse day was greater than on any of the control days, although an important change beyond the ionosphere was observed on the day after the eclipse. The magnetic A_p index for July 21 was 26, as compared with 5, 6, and 13, for July 19, 20, and 22, respectively. The relative intensity and time of the effect noted on July 20 indicate that the change in electron

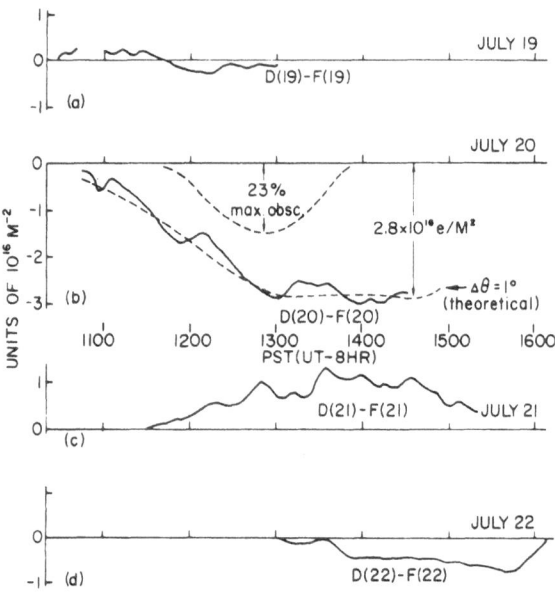

Fig. 2. Changes in electron density beyond the ionosphere measured by combined Faraday and Doppler technique: (a) July 19; (b) eclipse day, July 20. Also shown (dotted line) is the optical obscuration of the sun and the theoretical integrated density in the lee of the moon for a solar-wind dispersion of $\Delta\theta = 1°$; (c) July 21; and (d) July 22.

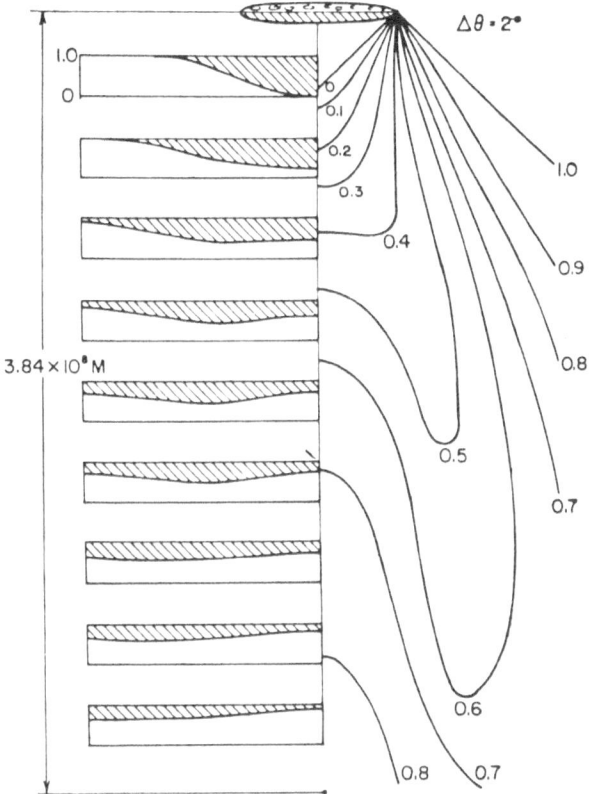

Fig. 3. Calculated electron density in the lee of the moon with a solar-wind dispersion of $\Delta\theta = 2°$. On the left is shown the normalized density as a function of distance from the moon−earth line at various distances behind the moon. On the right is shown the contours of constant electron density in the same region. The horizontal scale is greatly expanded.

density is associated with the eclipse, although the location and cause of the change cannot as yet be specified.

One possible interpretation of the effect observed during the eclipse is that there is a region of decreased solar-wind density in the lee of the moon. That is, the particles in the wind flowing outward from the sun are absorbed or deflected by the moon, causing a decreased density behind it. The amount and extent of this decrease depend on the spread in direction of particle velocities. A large spread will give a small decrease of modest extent, whereas a small spread (all particles coming almost directly from the sun) will give almost complete depletion, extending great distances behind the moon. Figure 3 shows a calculated density distribution in this region for a Gaussian spread in particle velocity direction of $\Delta\theta = 2°$. The integral of this density between the earth and moon would be the quantity measured by the Doppler–Faraday technique.

In Fig. 2b such an integral has been plotted for $\Delta\theta = 1°$, and its shape shows reasonable agreement with the limited eclipse-day data.

One difficulty with the solar-wind interpretation is that the total decrease in density amounts to about 75×10^6 m^{-3} averaged over the cislunar path, which implies a solar-wind density of 10^8 m^{-3} (100 cm^{-3}). This density is nearly an order of magnitude greater than the proton densities measured by the Mariner II space probe (Neugebauer and Snyder [7]) which, however, was not sensitive to low-velocity particles.

A second difficulty is that the motion of the earth and moon about the sun should cause the particle shadow of the moon to be tilted in a direction opposite to this motion. For the maximum density change to occur near the time of visual occultation requires that the solar wind retain sufficient angular momentum from the sun's rotation so that its tangential velocity is approximately equal to the earth's orbital velocity. If this were not so, and the particles moved directly radially from the sun, the change should have occurred about eight hours earlier.

Another possibility is that this density change occurred within the magneto-sphere, as discussed above. Whistler analyses made at Stanford, however, show little magnetospheric change during this time (Carpenter [8]). Also, if the eclipse were responsible for changes in the magnetosphere, the only immediately obvious mechanism would be an effect related to the moon's blocking of the solar wind. However, the shape and time of occurrence of the depression require the same narrow velocity dispersion and tangential velocity component in the solar wind as described above.

Future possibilities for radar studies of the cislunar medium include the use of lunar orbiting spacecraft and transmitters or transponders on the lunar surface. It has been proposed by Stanford that AIMP (anchored interplanetary monitoring platform) spacecraft which will be orbited around the moon transmit at a sub-harmonic of the telemetry frequency. Both frequencies could then be monitored on the earth for Doppler excess and group-path measurements. Such measurements would be more precise than those possible with lunar-radar techniques since reflection from the rough moon limits the accuracy of frequency, time delay, and amplitude measurements. The AIMP measurements would be particularly interesting when the propagation path is along the direction of the solar-wind wake of the earth or moon. If other measurements of the ionospheric and magneto-spheric profile (with topside sounders, a geostationary or highly elliptical earth satellite, or incoherent scatter radar) are made at the same time, it should be possible to obtain information on the interplanetary density near the earth – moon system, and the wake these bodies make in the solar wind.

The proposed radio propagation experiment for these spacecraft includes bi-static radar studies of the lunar surface and occultation measurements of a possible lunar atmosphere or ionosphere.

THE INTERPLANETARY MEDIUM

To keep the near-earth plasma from masking effects of the interplanetary medium it is obvious from the above discussion that propagation paths longer than those to the moon are desirable. Monostatic radar echoes from the planets could be used in principle, but the radar system requirements for such a study would be difficult to achieve. Low frequencies are desired because most of the effects of interest vary as f^{-2}, but because of the increase in cosmic noise and the lessening of antenna gain per unit aperture as frequency is lowered, extremely high transmitter power would be required. With planetary echoes there would also be the difficulty of having ionospheres to contend with at both ends of the path (Priester *et al.* [9]).

The Center for Radar Astronomy at Stanford University and the Stanford Research Institute is preparing a radio propagation experiment for the Pioneer A and B spacecraft. These probes are scheduled to be launched into interplanetary space in 1965. Other experiments are designed for measurements of the magnetic field, local interplanetary plasma, cosmic rays, and micrometeorites. Pioneer program management is by the NASA-Ames Research Center, and the spacecraft are being constructed by the Space Technology Laboratory.

The radio propagation (bistatic radar) experiment involves transmissions from Stanford at 50 and 425 Mc/s and reception in the space probes. Transmitter powers are a continuous 300 kW at 50 Mc/s and 30 kW at 425 Mc/s. Both will be connected to the steerable 150-ft parabolic antenna. Reception in the spacecraft will be by means of coaxial whip antennas and phase-locked receivers. Digital-output circuitry is designed to make Doppler, group-path, and amplitude measurements and encode these onto the telemetry channel for transmission to the NASA deep-space tracking site at Goldstone, California.

Figure 4 shows a sketch of the spacecraft. The antenna for the propagation experiment is shown at the lower right. Figure 5 shows the trajectories of the two spacecraft in the ecliptic plane relative to the earth–sun line.

The design lifetime of the deep-space probes based on the maximum communication range is about 200 days, and the maximum range from the earth to the

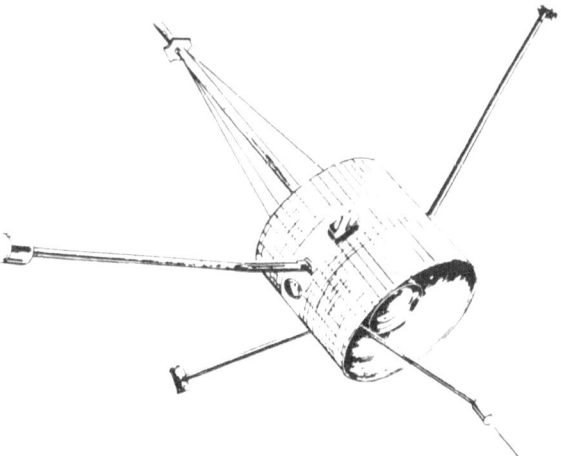

Fig. 4. The pioneer interplanetary spacecraft scheduled
for flight in 1965.

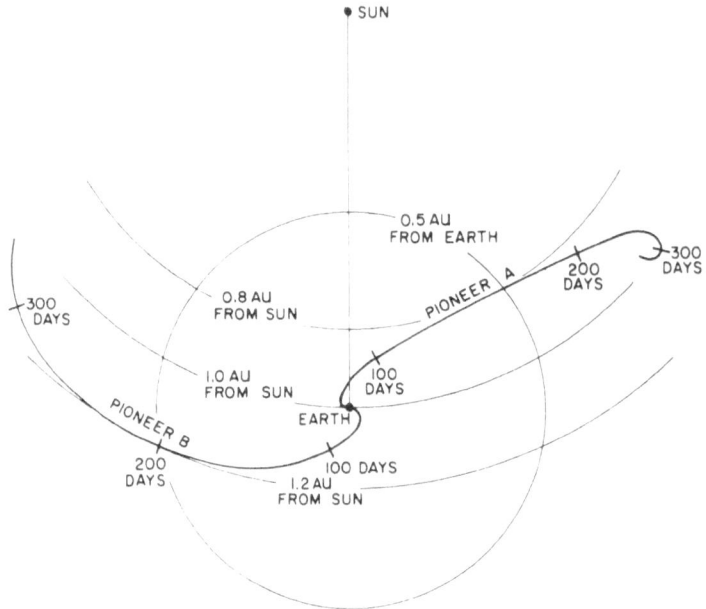

Fig. 5. Nominal trajectories for the interplanetary spacecraft Pioneer
A and B.

two probes would be 0.5 to 0.7 AU. Pioneer A would go to about 0.8 AU from the
sun, while Pioneer B would go out to about 1.2 AU.

For the chosen two frequencies, $n = 8.5$ so that $(n^2 - 1)/n^2$ is nearly unity.
Thus, we can use 50 Mc/s as the frequency in equations (3) and (4) to determine
measurement capabilities. From (4), one cycle change between 50 Mc/s and the
appropriate subharmonic of 425 Mc/s corresponds to a change in I of about
4×10^{14} electrons/m^2. To detect the sense of the change, a bias of 5 c/s will be
used so that there would be 300 beats during 1 min between 50 Mc/s and the 425
Mc/s subharmonic when the medium is not changing. The count would be 301 if
the integrated interplanetary density increased by 4×10^{14} m^{-2}, and 299 if it decreased by a similar amount. At a range of 0.5 AU, this change in the integrated
electron density would correspond to an average volume density change of about
5000 m^{-3}, or only 0.005 electrons/cm^3.

Both frequencies will be sine-wave modulated in phase at the transmitter.
The modulation phase difference will be determined at the spacecraft for the group
delay measurement. Either one of two modulation frequencies, 8.7 or 7.7 kc/s,
will be used. From equation (3) it can be shown that the total electron content I
can be measured to about $\pm 4 \times 10^{16}$ m^{-2}, or an average volume density at 0.5 AU
of 0.5 cm^{-3}, assuming that the modulation phase is measured to $\pm 6°$. By having
two modulating frequencies, any possible ambiguity due to more than one cycle of
modulation phase difference can be evaluated.

Because of the great range of the spacecraft, the value and time changes of I in the ionosphere and magnetosphere will not add much uncertainty in the measurements of the interplanetary medium. It is estimated that these uncertainties due to near-earth ionization will be about the same as the measurement accuracies given above.

Measurements of amplitude are included for several additional experiments. It has been proposed that the spacecraft be fired on a trajectory such that they will be occulted by the moon, as seen from the earth, one or more times during their flight. If there is a lunar ionosphere having a volume density as low as 100 cm^{-3}, it will bend the tangential 50-Mc/s rays more than the 425-Mc/s rays, and this would show up at the spacecraft as a difference in time of signal extinction at the two frequencies. This subject is discussed in greater detail in the next section.

It has been suggested by Lusignan [10] that the relativistic increase in mass of streaming electrons in the solar wind should cause a polarization effect on radio waves propagating through the interplanetary medium. At the spacecraft, polarization changes of the incident wave will be detected as characteristic amplitude changes due to the fact that the antenna and spacecraft will be spinning.

With two probes in space at the same time, it should be possible to monitor the manner in which a solar stream progressively fills the two propagation paths. Plasma instruments and magnetometers on the spacecraft, and studies near the earth, could give more detailed but more localized information about the stream. Such complementary studies should provide important new information about the density, dynamics, and other characteristics of the interplanetary medium.

PLANETARY IONOSPHERES AND ATMOSPHERES*

Ionospheres

Radar probing of planetary ionospheres is possible in principle by monostatic earth-based radar, monostatic probe-based radar (topside sounding), or bistatic radar. In bistatic radar, the surface of the planet might be used to return a signal which penetrates the ionosphere in order to make Doppler, group-path, etc. measurements of ionospheric effects, or a variable low frequency might be used which would be reflected at various levels in the ionosphere itself.

Another bistatic possibility which appears much easier to implement than any of the above (assuming the advent of the space age) requires occultation of a spacecraft by the planet as seen from the earth. Radio waves propagating between the earth and the spacecraft would then pass tangentially through the ionosphere and atmosphere.

Monostatic radar from the earth would be troubled by the earth's ionosphere, and extremely large transmitter powers at relatively low frequencies would be required, as described previously. Similar trouble would be encountered in the bistatic mode involving reflections from the planetary ionosphere. Spacecraft in-

---

*Eshleman *et al.* [11].

strumentation for topside sounding and for using surface-reflected signals would be considerably more complex than is required for the bistatic occultation method.

Figure 6 illustrates the geometry of the occultation problem. It is obvious that such occultation measurements could be accomplished either with a fly-by or an orbiting trajectory. Under the assumption of spherical symmetry of the plasma around the planet, a trajectory that lies outside the sensible ionosphere and magnetosphere, and straight-line propagation between the earth and spacecraft, the integrated electron density as a function of closest approach of the ray to the planet's center a is given by

$$I(a) = 2 \int_0^\infty \frac{N_e(r)r\,dr}{(r^2 - a^2)^{1/2}} \tag{13}$$

In practice, $I(a)$ would be found from $I(t)$ measurements by making use of the measured spacecraft position as a function of time. If a measurable amount of the neutral atmosphere is intercepted, equation (13) could be adjusted to include its effects, as indicated in equations (10) and (11). The problem is also tractable for a nonspherically symmetric ionosphere if the law of the departure from symmetry is known, such as would be the case for a theoretical electron volume density dependence on solar zenith angle. The correction for lack of symmetry is expected to be relatively unimportant, however (Fjeldbo [12]).

The function $I(a)$ in equation (13) can be measured directly from group delay [equation (3)], by integrating the Doppler excess [equation (4)], or doubly integrating gain measurements [equation (8)]. If the magnetic field strength and orientation are known, Faraday polarization measurements [equation (5)] can also be used. On the other hand, polarization measurements might first be used to estimate the near-planet magnetic field strength using knowledge of the plasma density gained in the other measurements. Absorption measurements [equation

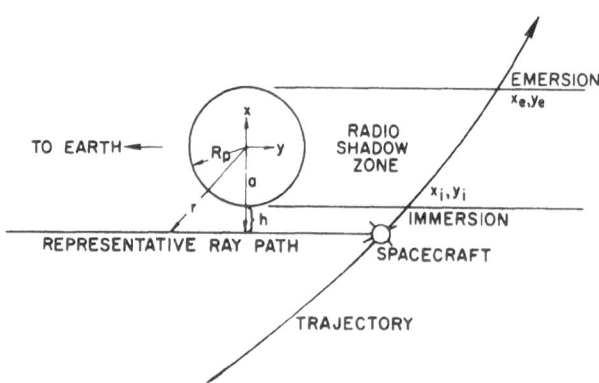

Fig. 6. Simplified geometry of planetary occultation of a fly-by or orbiting space probe.

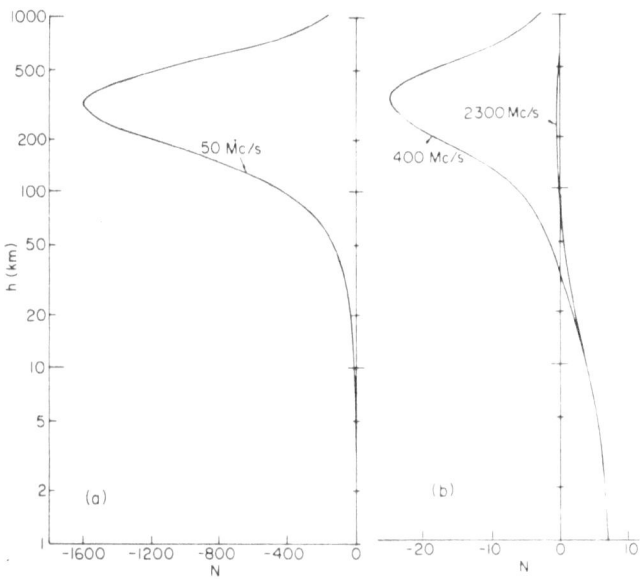

Fig. 7. Refractivity profile in height of assumed Martian ionosphere (peak density 10^{11} m^{-3}) and atmosphere (surface pressure 25 mb) at 50, 400, and 2300 Mc/s (see text).

(6)] and ray bending [equation (7)] could also be measured for additional information and as a check on the other determinations.

Once $I(a)$ is determined, equation (13) can be inverted to solve for the volume density profile of the electrons $N_e(r)$, or the profile of refractivity $N(r)$.
This problem corresponds to the solution of Abel's integral equation, which has
been used extensively in ionospheric physics to determine the true-height electron
density profile from vertical ionosonde records (Kelso [13]).

In practice, corrections will probably be required for higher-order effects
(e.g., ray bending affects phase path) than given in the above simplified formulas.
This can be accomplished either by the use of analytical expressions that include
higher order terms, or by an iterative process involving the determination of a
first-order profile for which correction terms can be computed, etc. (Fjeldbo [12]).

From a consideration of the possible ionospheric measurement methods from
the viewpoint of accuracy and ease of implementation, it appears that a Doppler
excess measurement based on the use of two harmonic frequencies would be preferred. This would give $I_t(a)$, from which $I(a)$ could be found by integration.
Group-path, amplitude, polarization, etc. measurements could be added, if feasible, for extension and check of the Doppler measurements.

Consider, for example, the ionosphere and atmosphere of Mars. By way of
illustration we will assume a Chapman ionosphere with its maximum density at a
height above the surface of 320 km and a scale height of 130 km. Chamberlain's
[14] model of the ionosphere of Mars has a daytime maximum volume density of
10^{11} m^{-3}, and this value is used in this example. For the neutral atmosphere,

we assume an exponential profile of scale height 16 km with a surface pressure of 25 mb, corresponding to the studies of Kaplan, Munch, and Spinrad [15]. The temperature of the atmosphere is assumed to be 250°K.

Figure 7 shows the refractivity profile $N(h)$ for this assumed atmosphere and ionosphere. Figure 7a is for a frequency of 50 Mc/s, while 7b is for 400 and 2300 Mc/s. Note the great change in the scale for N in (a) as compared with (b). The refractivity, of course, varies as f^{-2} for the ionosphere, but is independent of frequency for the neutral atmosphere. (For the atmosphere it was assumed that the average a_n in equation (9) is the same as for the earth's dry atmosphere, and this would be approximately true for the expected constituents in the atmosphere of Mars.)

As expected from equation (12), the ionospheric effect at 50 Mc/s (−1600 N-units maximum) is very large compared to that at 400 mc/s (−25 N-units maximum) so that one could consider that the 400 Mc signal acts as an unperturbed

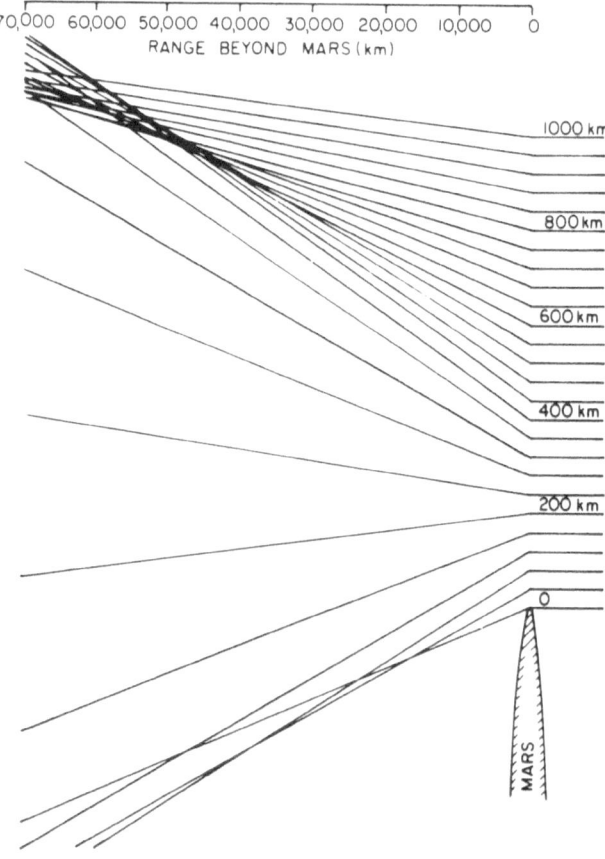

Fig. 8. Asymptotes of 50 Mc/s ray paths near Mars for the assumed ionosphere.

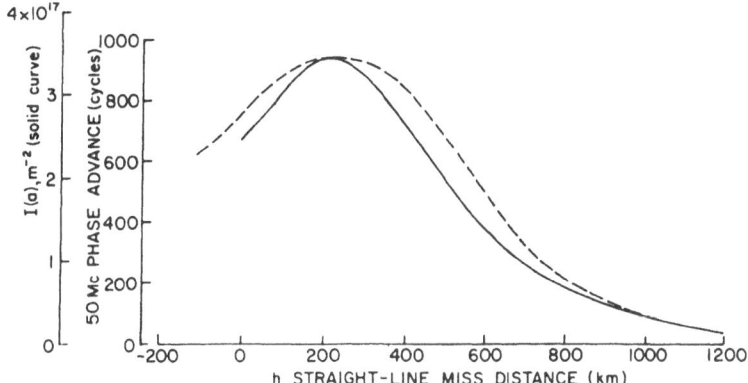

Fig. 9. Phase advance at 50 Mc/s and integrated electron density for the assumed Martian ionosphere. The solid curve represents both quantities to first order. The dashed curve for phase advance includes the correction due to ray bending. It is assumed that the spacecraft is 15,000 km behind Mars.

reference for automatic cancellation of the Doppler effect due to path-length change, thus avoiding the necessity of making very precise measurements of frequency. In regard to this model, the effect of the neutral atmosphere (+7 N-units maximum) is so small that it can be neglected as compared to the ionospheric effect at the lowest frequency.

Figure 8 shows the asymptotes of the ray paths for 50 Mc/s near Mars. The angles of refraction are very small (the maximum here is about 0.5°) so that a greatly compressed horizontal scale is used in the figure. For trajectories beyond about 50,000 km behind the planet, the spacecraft is in areas where the ray paths cross over and form caustics, so that the measurements could not be simply interpreted as discussed above. However, for close-in trajectories, the spacecraft is always within the region where no confusion can arise due to the formation of caustics. From equation (8), the distance y_c to the crossover point of adjacent rays is

$$y_c = \left(\frac{2\pi}{c^2 r_e}\right) \frac{f^2}{I_{xx}} \tag{14}$$

Figure 9 shows the 50-Mc/s phase advance (to be obtained in practice by integrating I_t) and the values of $I(a)$ for this model ionosphere, plotted as a function of $h = a - R_P$, the distance by which an earth-probe straight line would miss the planetary surface. The solid curve results from using the simple formulas given above, while the dashed curve for phase advance includes the correction due to ray bending (Fjeldbo [12]).

Figure 10 shows the 50-Mc/s signal gain due to ionospheric focusing plotted as a function of h. This was computed from the more exact first expression for G in equation (8). Since immersion and emersion of the spacecraft would occur

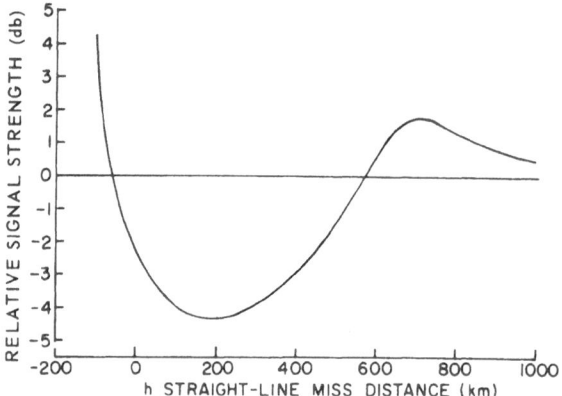

Fig. 10. Focusing gain at 50 Mc/s in the assumed
ionosphere for a spacecraft 15,000 km behind Mars.

at different known values of y (except for a spacecraft in a circular orbit), it
should be possible to separate the y-dependent focusing gain [equation (8)] from
the absorption [equation (6)], which would be independent of the distance to the
spacecraft.

For a fly-by trajectory, dx/dt would be of the order of 1 or 2 km/sec, so
that the occultation measurements over a height range of about 1000 km would be
made in less than about 15 min. Measurement accuracy in this example would
correspond to one part in a thousand at the maximum value of I if the Doppler
excess is counted to an accuracy of one cycle. Changes in the earth's ionosphere
and in the interplanetary medium would set the limiting accuracy due to the environment. It does not appear that such uncertainties could have an appreciable
effect unless there were very unusual and large-scale changes in the interplanetary medium during this short measurement period.

Atmospheres

As seen from the above example, relatively low radio frequencies are desirable for ionospheric measurements. Because of the very large transmitter power
requirements at low frequencies, it appears important that the transmitters be on
the earth and the receivers on the spacecraft. Nevertheless, the first ionospheric
measurements may actually be made using spacecraft transmissions, as a by-
product of the currently planned atmospheric measurements of Mars.

A major engineering constraint to the exploration of the Martian surface is
the uncertainty in the atmospheric parameters which affect entry and soft-landing
of a probe on the planet. Estimates of surface pressure range from less than 10
mb to more than 100 mb. A group (A. J. Kliore, F. D. Drake, D. L. Cain, and G. S.
Levy) from the Jet Propulsion Laboratories (JPL) together with G. Fjeldbo and
the author from Stanford have proposed that the occultation of the S-band telemetry

signals on the Mariner spacecraft which will be sent to Mars in 1964 be
used to make atmospheric measurements.

From Fig. 7b it is seen that an atmosphere of 25 mb surface pressure has a
very slight refractive effect compared with the effect of the ionosphere at 50 Mc/s.
The 2300-Mc/s telemetry signal, on the other hand, is sufficiently high to
cause the atmospheric perturbations to be predominant with the above atmosphere –
ionosphere model.

Since atmospheric refractivity is nondispersive, absolute frequency and amplitude measurements must, in general, be made. That is, the convenient self-
calibration characteristics of dual-frequency ionospheric measurements are not
applicable here. By use of a transponder technique, a ground-based frequency
standard can be used so that extremely small (one part in 10^{11}) effects can be
measured.

The occultation measurement of the atmosphere is, in several respects, an
ideal experiment. It requires no added weight, power, space, or information capacity in the spacecraft, since use is made of the existing telemetry signal.

The profile of atmospheric refractivity depends upon pressure, temperature,
and constituent profiles. Except for a small uncertainty about the effects of unknown constituents, however, refractivity is most closely related to p/T or
density, so that it is a good measure of the probe-entry characteristics of the
atmosphere. There are several different frequency and amplitude measurements
that give independent information on p/T and $T/\overline{m}$ ($\overline{m}$ is the average molecular
mass) at the surface, so that some separation of parameters is possible. Combin-
ing this with other information on temperature and probable constituents makes it
possible to use the radio information to help determine actual constituents, pressure, temperature, and density.

The JPL group plans to make frequency measurements, but in order to find
f_{DE} in equation (4) it will be necessary to know the Doppler effect due to changing spacecraft motion to an accuracy of about one part in 10^5. Nevertheless,
they estimate that the surface pressure can be determined to an accuracy of
better than 5% if the pressure is near 25 mb, and better than 10% if it is near
10 mb.

The Stanford group plans to use the amplitude measurements to find atmospheric defocusing [equation (8)], and, in addition, to study the effects of the
atmosphere on the Fresnel diffraction pattern produced at the limb of Mars. Figure 11 illustrates the difference between vacuum and atmospheric occultation at
80,000 km behind Mars, assuming a frequency of 2300 Mc/s and a surface pressure of 100 mb. The extension of the period of the Fresnel fluctuations (Fresnel "stretch") appears to provide a very sensitive measure of near-surface atmospheric conditions. In addition, the gain   (–8 db in this example) and the time
rate of change of gain give independent information on surface density and scale
height. Finally, this figure also illustrates that there is a difference in time of
occultation for vacuum and atmospheric conditions, and this could provide another
atmospheric measure, if the diameter of Mars is known to within a few kilometers.
It appears more likely, however, that this last measurement plus other atmospheric

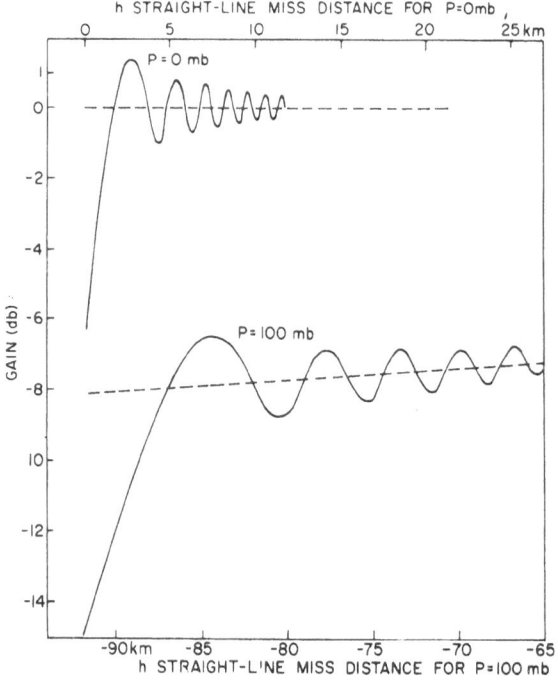

Fig. 11. Fresnel patterns at 2300 Mc/s due to limb diffraction with and without an atmosphere of surface pressure equal to 100 mb, a scale height of 16 km, and a
temperature of 250°K. It is assumed that the spacecraft
is 80,000 km behind Mars. (Note that the top and bottom curves have different abscissas.)

measurements would serve instead to improve our information on the diameter of
Mars.

For an exponential atmosphere, it can be shown from equation (11) and earlier equations that the maximum first-order atmospheric effects depend upon the
surface refractivity N_0, scale height H, and planet-probe distance y:

$$
\begin{array}{ll}
\text{Focusing gain and} \\
\qquad \text{Fresnel stretch} \ldots\ldots\ldots\ldots\ldots\ldots\ldots & \sim N_0 y H^{-3/2} \\
\text{Time or space rate of} \\
\qquad \text{change of focusing gain} \ldots\ldots\ldots\ldots\ldots & \sim N_0 y H^{-5/2} \\
\text{Phase} \ldots\ldots\ldots\ldots\ldots\ldots\ldots\ldots\ldots\ldots\ldots\ldots & \sim N_0 H^{1/2} \\
\text{Frequency} \ldots\ldots\ldots\ldots\ldots\ldots\ldots\ldots\ldots\ldots & \sim N_0 H^{-1/2} \\
\text{Time or space delay} \\
\qquad \text{to signal extinction} \ldots\ldots\ldots\ldots\ldots & \sim N_0 y H^{-1/2}
\end{array}
\qquad (15)
$$

Figures (12) and (13) give computed examples of the profile of frequency and
gain as the spacecraft is immersed or emerges at 20,000 km behind the planet for

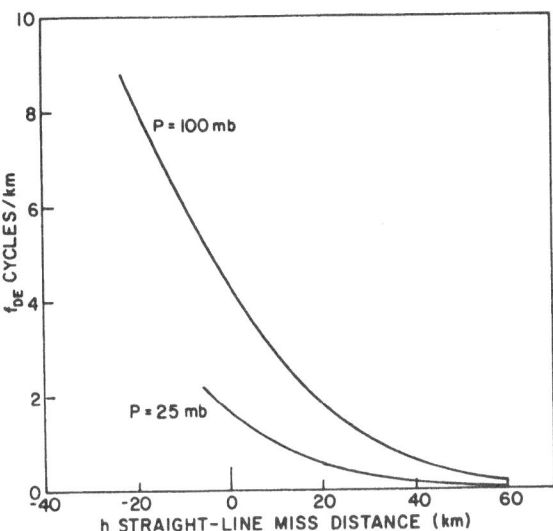

Fig. 12. Changes with time of the Doppler excess frequency at 2300 Mc/s due to the atmosphere for two values of surface pressure and a probe — Mars distance of 20,000 km.

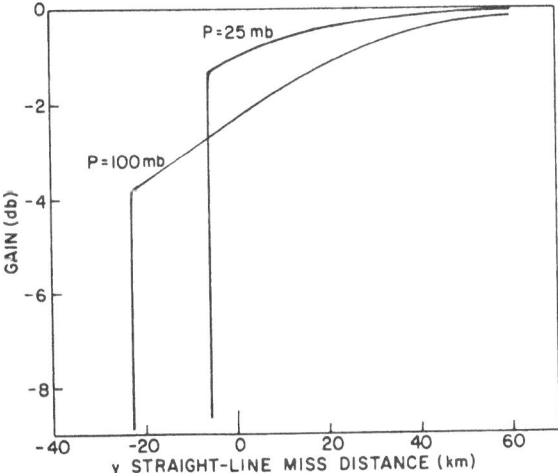

Fig. 13. Changes with time of the average signal strength due to atmospheric focusing for two values of surface pressure and a probe — Mars distance of 20,000 km.

25 and 100 mb surface pressures, assuming $H = 16$ km, $T = 250°$K, and
$f = 2300$ Mc/s. The maximum Fresnel stretch would be 16% and 56% for the two
pressures in this example.

Surface roughness at the limb of the planet is expected to wash out high-order
Fresnel fluctuations, but from radio-astronomical measurements of the occultation
of radio stars by the moon, it is not expected that rough planetary limbs will seriously degrade an atmospheric measurement based on diffraction.

The principal uncertainty in the 2300-Mc/s occultation experiment on Mars
appears to be the possible effect of a dense ionosphere. In the example given
above, it was shown that a Chapman layer with a maximum electron density of
10^{11} m^{-3} had much less of an effect at this frequency than did the atmosphere.
However, the ionospheric model from Chamberlain [14] was based on a surface pressure of 85 mb (2 mb CO_2 and 83 mb N_2), while recent studies (Kaplan et al. [15]) indicate a much lower pressure of 25 ± 10 mb, with 3 to 6 mb of CO_2 and either
N_2 or Ar as the principal remaining constituent. Because of the greater amount
of CO_2 to provide the principal ion O^+, and a lesser amount of N_2 to act through
ion – atom interchange and dissociative recombination to provide an electron loss
mechanism, the new low-pressure atmosphere could have an ionosphere of markedly greater density than in the Chamberlain model. If this increase is not too great,
the S-band experiment could provide both an electron density ionospheric profile
and a neutral density atmospheric profile. On the other hand, a very high ionospheric density or sharp density gradients could mask the atmospheric effect. In
general, it would be advantageous to use several frequencies simultaneously in
such an occultation experiment so that dispersive and nondispersive refraction
could be experimentally separated.

SOLAR RADAR ASTRONOMY

Monostatic radar echoes from the sun have been obtained at 26 and 38 Mc/s.
Bistatic methods similar to those discussed in the two preceding sections can be
used to study the inner and outer solar corona when space probes become available which can be sent behind the sun.

Marginal returns were obtained at Stanford in April 1959 (Eshleman et al. [16])
and again in September 1959 (Barthle [17]) at the lower frequency, using four or
eight rhombic antennas (covering 14 or 28 acres) and a CW transmitter power of
about 40 kW. Antenna gains are estimated at 25 to 28 db. Coding and computer
data processing techniques were used to demonstrate that (1) there was a return
(at a S/N ratio of about −23 db before integration) corresponding to reflection at
about 1.7 solar radii, (2) the Doppler frequency spectrum of the echo must be
wider than the 2-kc/s receiver bandwidth, and (3) the echo strength in this bandwidth corresponded to a radar cross section approximately equal to one to two
times the photospheric disk size. Data from 16-min runs made on several different days had to be combined to produce sufficient signal-to-noise ratio to demonstrate the presence of an echo. Figures 14a, b, and c illustrate the echo indications at 26 Mc/s, and Fig. 14d is a corresponding illustration from early results
with the 38-Mc/s system discussed below.

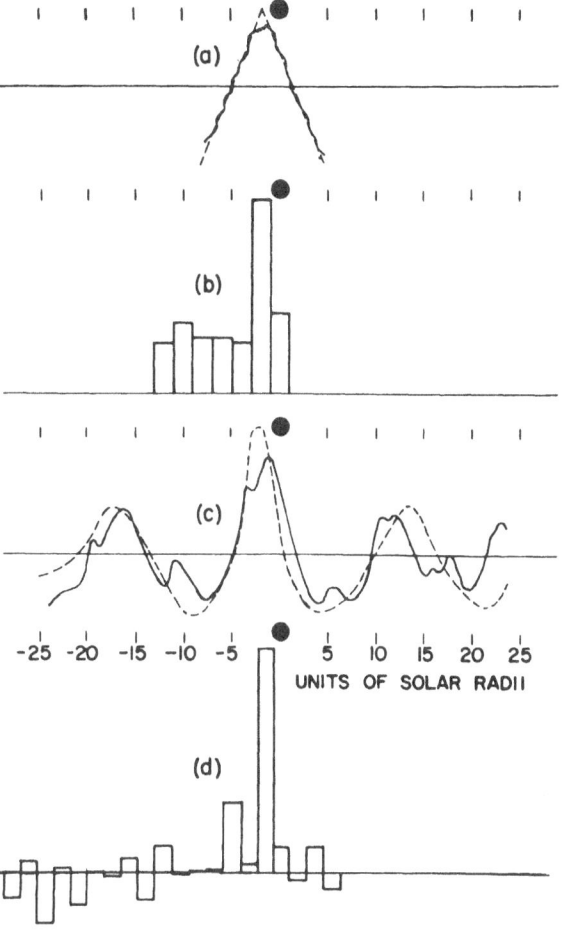

Fig. 14. In (a), the solid curve is the cross correlation of the solar-echo data taken at Stanford in April 1959, and the dashed line is the ideal correlation corresponding to reflection at 1.7 solar radii (Eshleman *et al.* [16]). In (b), the same data are analyzed in terms of the positions of significant peaks in the integrated return, and the largest number of such peaks are at a range position corresponding to solar reflection. Curve (c) shows the ideal (dashed) and measured (solid) correlation for the more complicated code used in the September 1959 tests, the central peak corresponding to the range of the echo (Barthle [17]). In (d), echo energy is plotted as a function of range for the early 38-Mc/s measurements at El Campo, Texas; the graph shows the sum for a number of days of the period April to July 1961 (Abel *et al.* [18]). In each graph the circle represents the size of the photospheric disk relative to the range scale.

At 38 Mc/s, routine radar studies of the sun have been made since April 1961 by the MIT Lincoln Laboratory at a site near El Campo, Texas (Abel *et al.* [18], Chisholm and James [19], and James [20]). Transmitter power is up to 500-kW CW, and the antenna now consists of two crossed-polarized arrays of dipoles, one with a measured gain of 33 db and the other 36 db. One 16-min coded transmission is sent per day, the azimuthal beamwidth being sufficient for only one transmit and receive period. The higher-gain antenna is used for transmission and both antennas are used for reception. Measured Doppler spread of the echo varies between 20 and 70 kc/s, with the center of the spectrum shifted up in frequency by an average of about 4 kc/s. This upward shift represents a net outward flow of plasma at the reflecting level.

Radar cross sections vary from a high of about 16 times the photospheric disk size down to a level that is undetectable by this system. Average values correspond approximately to the size of the photosphere. The energy appears to be returning from a level in the corona corresponding to about 1.5 solar radii, and the spectral and range-spread evidence indicates that this level is rough and rapidly moving (James [20]) with both random and directed velocities. Figure 15 consists of range Doppler spectra for four relatively strong echoes, and these illustrate the wide bandwidth and range spread. There is a decided net positive Doppler shift with the range intervals nearer the earth showing the largest effects. This has been interpreted (Chisholm and James [19] and James [20]) as evidence of the solar wind near its source.

From the 38-Mc/s results (James [20]) it appears that the echo energy in the Stanford experiment should have been spread from 6 to 25 times as wide as the receiver bandwidth of 2 kc/s. Thus the solar radar cross section at the lower frequency would be perhaps an average of 10 times the measured value. This would make it considerably higher than the average radar cross section measured at 38 Mc/s, but still within the measured range of fluctuation of these results. Two factors are expected to contribute to this difference.

Figure 16 shows computed values by Yoh [21] of the expected solar radar cross section as a function of wavelength, assuming spherical symmetry and various electron temperatures and densities. (The *n* values are multiplicative factors to be used with the Allen–Baumbach (Allen [22]) model of the solar corona.) These curves include the effects of ray bending and absorption in this model corona. (The simple formulas given above are not applicable to this problem since complete reflection and high absorption may be encountered.) Note in particular that over the plotted wavelength range, the solar radar cross section increases with wavelength. For a relatively low electron density of a half-million degrees, the 26-Mc/s results might be expected to give a radar cross-section as great as ten times the 38-Mc/s value, due primarily to less absorption above the reflecting level. The theoretical absolute values of radar cross sections under these conditions are less than the measured values, but this could result from the increased cross section of the actual rough target over the theoretical smooth model. The possibility that the electron temperature may be lower than previously suspected is supported by the radar measurements of turbulence in the corona. That is, some of the relatively high estimates of temperature are based on Doppler

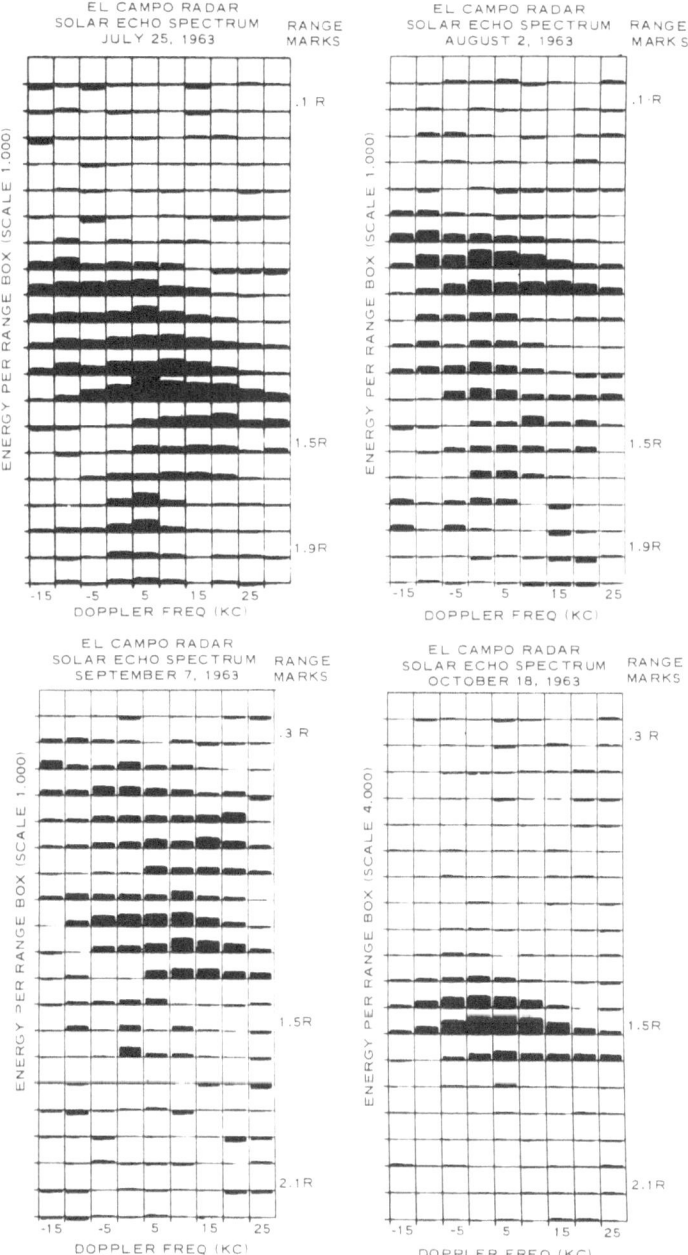

Fig. 15. Range Doppler spectra of four solar radar echoes obtained at the Lincoln Laboratory field site near El Campo, Texas (courtesy of J. James, Lincoln Laboratory). The vertical range scale corresponds to reflection at the indicated number of solar radii, in units of 0.1 R, and the horizontal scale is Doppler frequency intervals in units of 5 kc/s.

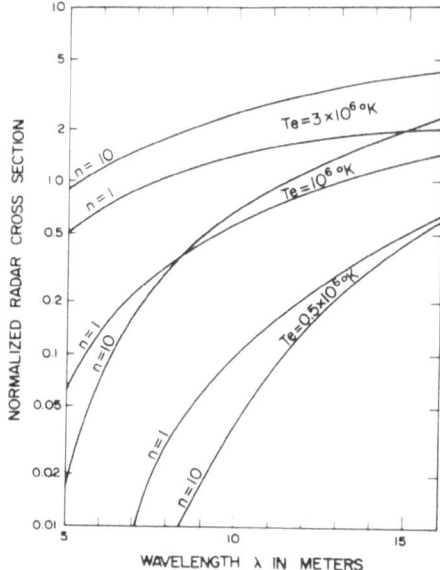

Fig. 16. Theoretical radar cross section of the sun normalized by
the area of the photospheric disk, plotted as a function of wavelength for various electron temperatures and densities (Yoh [21]).

broadening of coronal emission lines assumed to be due to thermal motion of the
electrons. If the electrons have large mass motions in addition to their thermal
motions, then overestimates of electron temperatures would result from this method (James [20] and Billings [23]).

K. L. Bowles [24] of the Jicamarca radar observatory in Peru has conducted
several series of measurements in an attempt to obtain solar echoes at 50 Mc/s.
No echoes were obtained even though the product of transmitter power and antenna
gain is considerably greater than for any other solar radar system. This gives
added evidence for a precipitous reduction of radar cross section with decreasing
wavelength between 12 and 6 m.

The second possible reason for the apparent cross section difference in the
two experiments is the measured change in cross section with average solar
activity. The 38-Mc/s results indicate that the average cross section has decreased steadily from 2.2 in 1961–1962 to 0.6 photospheric areas in 1963–1964
(James [20]). The original 26-Mc/s data were taken in 1959, near the peak of the
sunspot cycle. Since that time, several limited sets of measurements near 26
Mc/s have been made at Stanford, in the summer of 1962 and 1963. No detectable
echoes were obtained even though the transmitter power is greater by 8 to 9 db than
that of the 1959 system. There have been other changes in receiving and coding
techniques, and also in the antenna system. In 1959, rhombic antennas were used
at dawn, while the later experiments employed the log-periodic array near noon.

It does not appear possible at this time to be definite about the effects of
frequency and solar activity from the various bits of evidence at hand. In view
of the limited data taken on the lower frequency in 1959, it would appear that an

understanding of these effects must await further experimentation, at least until the next period of maximum solar activity.

Even with the radar system in Texas, many minutes of signal integration are needed to demonstrate the presence of an echo. In order to obtain frequency and range-spread spectra, the integration of relatively strong returns is required. It is not possible, unfortunately, to measure the variations in echo characteristics from second to second or even from minute to minute.

If a system were built with sufficient sensitivity to obtain reasonable S/N ratios without integration, the dynamic motion, shape, and echoing strength of the solar corona could be studied in considerable detail. Such a system would also be very useful for bistatic solar studies, and for both monostatic and bistatic radar studies of planetary ionospheres and surfaces. Planets and the interplanetary medium over virtually the whole of the solar system could be brought under intensive radar investigation with a combination of deep-space probes and powerful transmitters and large antennas on the earth.

ACKNOWLEDGMENTS

The report given here includes a brief review of some of the recent work at Stanford of my colleagues and students, in particular G. Fjeldbo, O. K. Garriott, H. T. Howard, B. B. Lusignan, F. L. Smith, III, and P. Yoh. Research support includes contracts from the National Aeronautics and Space Administration, the the U. S. Air Force and Navy, and the National Science Foundation.

REFERENCES

1. V. R. Eshleman, P. B. Gallagher, and R. C. Barthle, *J. Geophys. Res.* 65, 3079 (1960).
2. H. T. Howard, P. Yoh, and V. R. Eshleman, *J. Geophys. Res.* 69, 535 (1964).
3. H. T. Howard, B. B. Lusignan, P. Yoh, and V. R. Eshleman, *J. Geophys. Res.* 69, 540 (1964).
4. J. W. Evans and G. N. Taylor, *Proc. Roy. Soc. (London), Ser. A* 263, 189 (1961).
5. C. G. Little and R. S. Lawrence, *J. Res. Nat. Bur. Std.* 64D, 335 (1960)
6. R. L. Smith, Jr., *J. Geophys. Res.* 66, 3209 (1961).
7. M. Neugebauer and C. W. Snyder, *Science* 138, 1095 (1962).
8. D. L. Carpenter, private communication.
9. W. Priester, M. Roemer, and T. Schmidt-Haler, *Nature* 196, 464 (1962)
10. B. B. Lusignan, *J. Geophys. Res.* 68, 5617 (1963).
11. The material in this section is based largely upon material being prepared for several publications by V. R. Eshleman, G. Fjeldbo, O. K. Garriott, and F. L. Smith, III.
12. G. Fjeldbo, dissertation, *Stanford University* (1964).
13. J. M. Kelso, *Radio Ray Propagation in the Ionosphere* (McGraw-Hill, New York, 1964), p. 237.
14. J. W. Chamberlain, *Astrophys. J.* 136, 582 (1962).
15. L. D. Kaplan, G. Munch, and H. Spinrad, *Astrophys. J.* 139, 1 (1964).
16. V. R. Eshleman, R. C. Barthle, and P. B. Gallagher, *Science* 131, 329 (1960).
17. R. C. Barthle, dissertation, *Stanford University* (1960).
18. W. G. Abel, J. H. Chisholm, P. L. Fleck, and J. C. James, *J. Geophys. Res.* 66, 4303 (1961).
19. J. H. Chisholm and J. C. James, *Astrophys. J.* (in press).
20. J. C. James, *Trans. IEEE, PTGME*, Vol. MIL-8, Nos 3 and 4, 210 (1964).
21. P. Yoh, *Stanford University Report No. 2*, contract AF 19(604)-7436 (1961).
22. C. W. Allen, *Monthly Notices Roy. Astron. Soc.* 107, 426 (1947).
23. D. E. Billings, *Astrophys. J.* 137, 592 (1963).
24. K. L. Bowles, private communication.

Chapter 15

The Interpretation of
Thermal Emission from the Moon

Harold Weaver

Radio Astronomy Laboratory
University of California
Berkeley, California, United States of America

INTRODUCTION

After many years of neglect by astronomers, the moon is becoming an object of intensive study. To see the trend one has only to note the increasing number of papers related to lunar physics appearing in journals, and to count the monographs in publishers' lists. Space research has provided much impetus to these studies, but the multiplicity of new observational and theoretical approaches coupled with current emphasis on physical knowledge rather than mere description have also strongly influenced this redirection of interest. In the older literature one finds largely descriptions and maps of craters, maria, and mountains, or detailed mathematical discussions of the dynamics of the earth–moon system. Now one finds discussions of the nature of the lunar surface material, its thermal and electrical properties, radioactivity and the flow of heat through the surface, the effects of bombardment of the lunar surface by solar corpuscular radiation, surface luminescence, and the like. In this discussion we shall deal with infrared and radio observations of the moon — that is, with electromagnetic radiation emitted by the moon itself.

Excluding for the moment radioactive heating and plutonic activity (which we shall discuss briefly in the section entitled "Heat Flow through the Lunar Surface"), the source of this emitted radiation is absorbed sunlight. Our interest, then, will center on (1) the heating of the lunar surface (as a function of position and time) as that surface is irradiated by the sun, and the cooling of the surface during the lunar night; (2) the conduction of surface-absorbed heat into the lunar interior and the resulting lunar subsurface thermal structure; and (3) the energy radiated by the moon from its surface and interior. The amount of solar energy cycled into and out of the lunar material during each lunation, the depth to which the absorbed thermal energy is conducted, the depth of penetration by electromagnetic waves of different frequencies — all these are determined by the nature of the lunar material, by the specific values of the physical parameters describing that material. Thus, in the study of the lunar emission we are, in fact, fundamentally investigating the physical nature of the lunar material to the depth of penetration of the longest wavelength at which we observe. The energy cycling into and out of the lunar material is simply our probe to investigate the properties of the materials through which the flow takes place. We measure radiation, but we are, in fact, studying the properties of the outer skin of the moon.

We may expect to find the nature of the lunar surface quite different from that of the earth. The principal cause of this difference is the moon's lack of an atmosphere and of running water. These are the chief agents of change on the surface of the earth. On the moon, erosion, as we normally understand it on the earth, does not take place. Living things do not exist to cover, change, or transform the surface materials on which they live. The nature of the lunar surface is the direct result of interaction of the moon with its cosmic environment. The moon has no measurable magnetic field; it is constantly swept by the "solar wind," and is bombarded by other high-energy solar and cosmic particles. Meteors rain down upon it constantly. The craters on the moon represent the readily visible results of large meteor strikes over the past few billion years. Whipple estimates that the number, $n(m)$, of impacts per second per square meter of lunar surface by sporadic meteors m grams or greater in mass is given by the equation $n(m) = 10^{-12.5} \, m^{-1}$. This would imply that, statistically, during the last 10^9 years, each square centimeter of the lunar surface has been struck by a meteor weighing between 0.1 and 1.0 g, and each square millimeter of the surface has felt the impact of a meteor weighing between 10^{-2} and 10^{-3} g. Since it travels at an average velocity of 30 km/sec, the average meteor in the mass range 10^{-2} to 10^{-3} g releases 10^{10} ergs of kinetic energy when it strikes. At the point of impact there is the equivalent of a small explosion. Lunar debris (dust) is thrown out from the miniature impact crater. Many of the particles are given velocities sufficient to carry them to considerable distances from the original crater; some particles may escape from the moon. (The lunar escape velocity is only 2.4 km/sec; ejected particles must often escape the moon.)

The surfaces of the explosion fragments produced in the vacuum surrounding the moon are chemically clean. Falling back onto the lunar surface, they probably tend to bond together when they touch, thus forming a very open cellular structure. If there is melting of the lunar material upon meteor impact, differential velocities within the liquid material will draw it out into threads which fall and crisscross the surface. Indeed, structures of these general types are indicated by the character of the sunlight reflected from the moon as well as by the infrared and longer wavelength radiation emitted by the moon, as we shall see in what follows.

To end this section a remark about the point of view adopted in this review may be worthwhile. We shall constantly emphasize the basic physics of the problem under consideration, and we shall try to stress the theoretical aspects of how the parameters of the lunar material are determined. We will make no effort to derive a consistent set of best values of parameters, or to present an extensive list of all values so far determined. While such a discussion would be of very considerable value, it lies outside the scope of this document.

A MODEL OF THE MOON

General Parameters

We consider the moon to have been formed cold (though there could have been subsequent radioactive heating, melting, and solidification), a smooth homogeneous sphere of dielectric material of density ρ (g/cm^3), thermal conductivity K(cal/sec − cm^2 − deg C/cm), heat capacity C(cal/g − deg C), and dielectric constant ϵ^*. The material making up the moon has some conductivity and may additionally be expected to possess various resonances on the molecular scale, particularly at the higher frequencies of radiation with which we shall be concerned, hence ϵ^* is complex. We shall deal with this problem in some detail in a later section of this paper. The synodic period of rotation of the moon (new moon to new moon) is, in seconds, $p = 2\pi/\omega$, where ω specifies the moon's angular rate of rotation; $\omega = 2.46 \times 10^{-6}$ rad/sec. We assume the moon to move in a circular orbit.

The model moon is warmed by radiation from the sun. The energy output of the sun is specified by the solar constant S, which represents the number of ergs/cm^2 − sec delivered by the sun to a surface perpendicular to the line of sight to the sun, the surface being located at the earth's mean distance from the sun. The constant of irradiation for the moon, $S/d^2(t)$, measured in A.U., is the moon's distance from the sun at any instant of time t. Normally, for our model, we shall take $d(t) = 1$, though a precise value is readily calculable. We assume the moon always to be at the same distance from the sun, 1 A.U., hence $S_{moon} = S$.

The pole of rotation of our model moon is so oriented that the subsolar point always lies on the lunar equator. We shall designate position on the moon's surface by the angular coordinates l, longitude, which we will specify from some fixed point, and latitude, ϕ, measured from the lunar equator. Within the body of the moon, position will be designated by spherical coordinates r, l, ϕ, the radius of the moon being r_0.

Differences Between the Model Moon and the Real Moon

The real moon differs from the mathematical model we have just described, though not so much as to make discussion of the model wholly academic. Perhaps the surprising thing is that observations have not progressed to the point where they make such a model inadequate!

Among the differences one might mention are: the moon is not quite spherical; the pole of the moon is tipped $1°32'$ to the ecliptic; the lunar rate of motion is not uniform since the orbit is noncircular; the distance of the moon from the sun varies because the moon orbits around the earth and, further, the earth revolves around the sun in an elliptical orbit. These effects can all be allowed for by making our model more complicated computationally. However, such complications will not lead to any significantly greater understanding of the lunar surface.

A marked oversimplification of this first model is the assumption that the moon is smooth and homogeneous. Clearly, the moon is not smooth (one has only to look at it!) and it does not give the appearance of being everywhere homogeneous in character of material. We shall find, though, that while we shall have to add to our model moon roughness of surface to explain some lunar observations, the clearly oversimplified assumption of smoothness is surprisingly adequate for the discussion and explanation of the great bulk of observations that now exist. (This may only indicate that our observations are not very precise or sophisticated and need to be improved.) We are only beginning to observe that the thermal character of the moon's surface differs from one area to another, that some crater areas behave differently from others, and so forth. Until observations are improved by a considerable factor, we shall probably not have to modify our model by considering inhomogeneities from one area of the moon to another. However, we will have to consider the possibility of inhomogeneity in depth to explain some observations. We shall deal at length with a two-layer model of the moon in a later section of this paper.

The Surface Temperature at a Point on the Moon

Because of the rotation of the (model) moon, each point on the lunar equator is, at some time, the subsolar point. For simplicity of description, we may imagine one point we observed fixed on the moon at l, ϕ, imagine the moon fixed in space, and lastly imagine the sun rotating around the moon. In our model sun−moon system the distance between the sun and the moon remains constant and equal to 1 A.U. As a function of time t, the amount of solar radiation impinging on 1 cm^2 of the lunar surface at position l, ϕ is given by

$$S \cos \phi \sin (2\pi/p)(t - l) \ (\text{ergs/cm}^2 \cdot \text{sec}) \qquad 0 \leq t - l \leq (p/2)$$

$$0 \qquad\qquad\qquad\qquad\qquad\qquad\qquad\qquad\qquad (p/2) \leq t - l \leq p$$

(1)

Figure 1 shows the insolation for (l, ϕ) points $(0°, 0°)$, $(0°, 45°)$, $(0°, 70°)$, $(0°, 80°)$, and $(0°, 90°)$ as a function of time.

The radiant energy impinging on the lunar surface is partly absorbed, partly reflected. Observations indicate that approximately 12% of the solar radiation is reflected, 88% absorbed. Of the 88% absorbed, approximately 99.4% is reradiated directly, while about 0.6% is conducted into the interior of the moon. Since the amount of energy per square centimeter available for absorption is, by equation (1), proportional to $\cos \phi$, we must expect to encounter diminishing lunar surface temperatures as we observe points starting at the equator and progressing toward the pole. The Stefan−Boltzmann law states that emission from a hot surface is proportional to T^4; hence, since absorbed energy available for emission is proportional to $\cos \phi$, we should expect to measure surface temperatures proportional to $\sqrt[4]{\cos \phi}$. Our model moon is thus strongly pole-darkened.

To determine the lunar surface temperature at a point l, ϕ on the lunar surface as a function of time, we employ the equation of heat conduction with appropriate boundary conditions. The temperature on the surface we represent by T; it is a function of l, ϕ, t, and the thermal properties of the lunar material. In writing the equation we may without loss of accuracy replace the spherical moon by a semi-

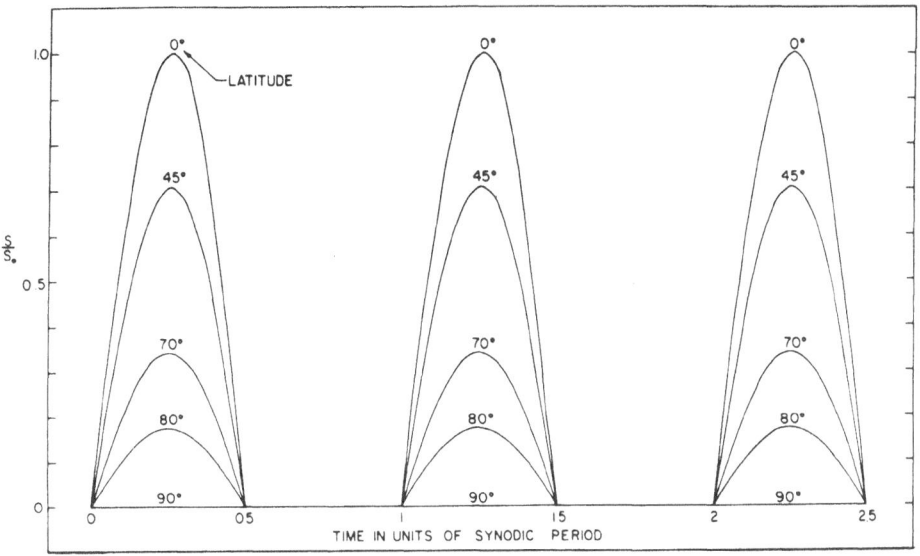

Fig. 1. Insolation on the moon for different latitudes, ϕ, as a function of time, which is given in terms of the synodic period of the moon as unit.

infinite solid with its face forming the tangent plane of the spherical moon at the point l, ϕ. We need consider only radial flow of heat into the moon, since it can be shown that even in a period of the age of the universe, nonradial flow is of no significance in changing the equator-pole temperature gradient.

The equations to solve are

$$\frac{\partial^2 T(x,\, t)}{\partial x^2} - \frac{\rho C}{K} \frac{\partial T(x,\, t)}{\partial t} = 0 \qquad x > 0,\, t > 0 \qquad (2)$$

$$T(x,\, t) = T_0 \qquad x > 0,\, t = 0$$

where x measures distance below the lunar surface and T_0 is the minimum value of the temperature attained at point l, ϕ during a lunation. The boundary conditions to be satisfied at $x = 0$ are

$$K \frac{\partial T(x,\, t)}{\partial x} = E \sigma T^4(x,\, t) - (1 - A) S \cos \phi \cos (2 \pi t / p) \qquad x = 0,\, |t| < \tfrac{1}{4} p$$

$$(3)$$

$$K \frac{\partial T(x,\, t)}{\partial x} = E \sigma T^4(x,\, t) \qquad x = 0,\, \tfrac{1}{4} p < t < \tfrac{3}{4} p$$

In these expressions E designates the emissivity of the lunar material. The boundary conditions state simply that (1) while the area is being irradiated by the sun, the outflow of heat $K[\partial T(x, t)/\partial x]$ is equal to the amount radiated by the moon, $E\sigma T^4(x, t)$, diminished by the amount absorbed from the sun, $(1 - A)S \cos \phi \times \cos (2\pi t/p)$, where the quantity $(1 - A)$ represents the fractional part of the solar radiation absorbed. We start counting time from the instant of local noon at the area under study.

Unfortunately, even at this early stage of our investigation we encounter diffi- culties. Because of the nonlinear boundary conditions — the fourth power of T oc- curs in equations (3) — no simple explicit solution for $T(x, t)$ can be found. The equations have been solved by Wesselink [30] by numerical integration and also by an iterative numerical procedure by Jaeger [7, 8]. The solutions obtained provide a family of curves with the thermal inertia, $(K\rho C)^{1/2}$, which characterizes the lunar surface material, as the parameter distinguishing different members of the family. Figure 2 displays a family of $T(0, t)$ curves drawn for a point on the lunar equator and for a variety of values of $(K\rho C)^{1/2}$. All the curves show some phase lag in that they do not reach maximum temperature at $t = 0$, though only for the curve with $(K\rho C)^{1/2} = 0.05$ is this phase lag visible in the diagram. Figure 3 shows a few observations of the lunar surface temperature cycle for a point on the lunar equator compared with a theoretical $T(0, t)$ curve for which $(K\rho C)^{1/2} = 0.0023$. The fit is reasonable, though the number of observed values is disappointingly small. A sur- prisingly small number of infrared measurements exist to determine the temperature cycle for various points on the lunar surface. Only a few workers are making such observations; many more observers in this field are needed.

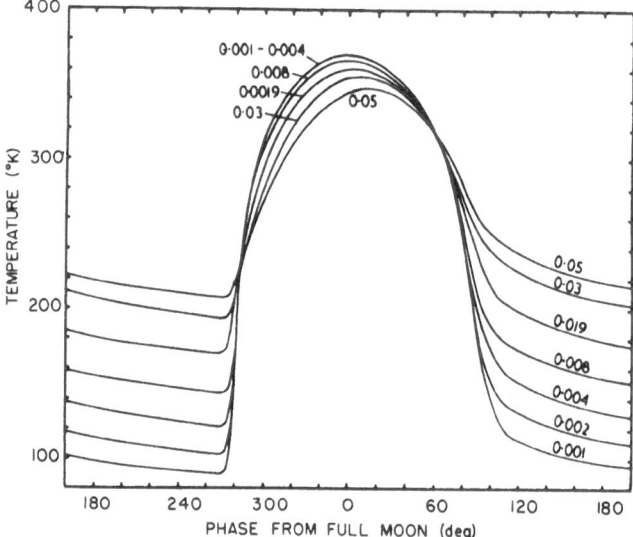

Fig. 2. A family of theoretical curves for the brightness tem- perature of a point on the lunar surface at the equator. The numerical values shown refer to the thermal inertia, $(K\rho C)^{1/2}$, which is the parameter characterizing any member of the family. The data are from Sinton [26].

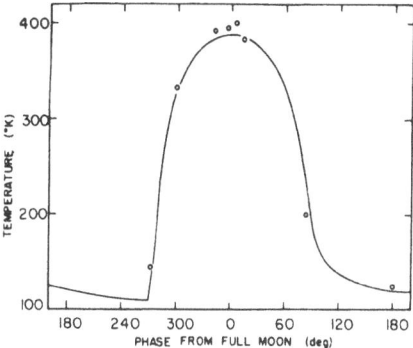

Fig. 3. Observations of the lunar surface brightness temperature fitted by a theoretical curve with $(K\rho C)^{1/2} = 0.0023$. The data are from Sinton [26].

Lunar eclipses are our best source of information on the numerical value of $(K\rho C)^{1/2}$, which is a lunar parameter that will occur many times in our discussions. As the shadow of the earth passes over the moon, the solar radiation is removed and restored in a predictable way and within a short period of time. The temperature at any point on the lunar surface decreases as the solar radiation is removed and increases as the radiation is restored. The thermal response of the surface allows a determination of the value of the thermal inertia, $(K\rho C)^{1/2}$, just as does the slow temperature change during a lunation. A very real advantage of eclipse observations is that they are made within a few hours and observational accuracy is likely to be higher than for observations made during an entire month during which calibrations and instrumental parameters can change. The equations for $T(0, t)$ during an eclipse are the same as those for a lunation; the functions in the boundary conditions change slightly. Specifically, the equations for an eclipse starting at $t = 0$ are

$$\frac{\partial^2 T(x, t)}{\partial x^2} - \frac{1}{\kappa} \frac{\partial T(x, t)}{\partial t} = 0 \qquad x > 0, \, t > 0$$

$$T(x, t) = T_0 \qquad x > 0, \, t = 0 \tag{4}$$

where we have written $(1/\kappa)$ for $\rho C/K$, a second combination of parameters that will frequently recur in our discussions; κ is termed the thermal diffusivity. The boundary conditions to be satisfied are:

$$K \frac{\partial T(x, t)}{\partial x} = E\sigma T^4(x, t) - (1 - A)\sigma T_0^4 f(t) \qquad 0 < t < t_0$$

$$\tag{5}$$

$$K \frac{\partial T(x, t)}{\partial x} = E\sigma T^4(x, t) \qquad t > t_0$$

Here $(1 - A)\sigma T_0^4 f(t)$ describes the way in which radiation is absorbed from the sun as a function of time during the penumbral stages of the eclipse. The time function goes to 0 at t_0, the instant at which the umbral state begins. We do not here discuss the eclipse beyond the time the umbral stage ends. The solution of these equations, which again involve nonlinear boundary conditions, can be derived by numerical methods as described earlier. Figure 4 displays Pettit's [19] observations of the 1939 eclipse, together with a series of solutions of equations (4) and (5) for a variety of values of $(K\rho C)^{1/2}$. From the results displayed in Fig. 4 we see that $(K\rho C)^{1/2} \approx 10^{-3}$, but it is clear that no one of the curves completely fits the observations. During the umbral phase of the eclipse, the observed temperature falls less rapidly than predicted. Jaeger and Harper [9] and also Lettau [15] have pointed out that such an effect would occur if the thermal conductivity increased with depth in the lunar material or if the conductivity increased with temperature.

Variation of the thermal parameters of materials as a function of temperature has long been known and investigated both observationally and theoretically. At the present time, empirical knowledge in this field relevant to lunar problems is increasing at a fairly rapid rate because of the interest of cryogenic engineers in such temperature variations of materials in the general temperature range encountered on the moon. Unfortunately, there now exists no adequate mathematical

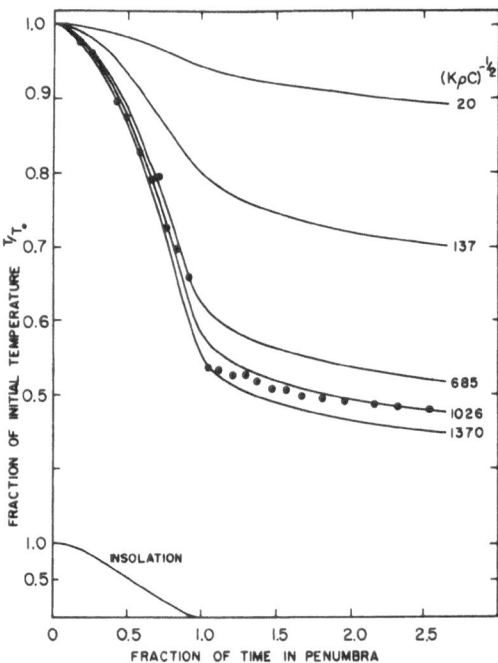

Fig. 4. Pettit's [19] observations of the 1939 lunar eclipse shown with a family of theoretical cooling curves computed by Jaeger [7]. The theoretical curves are for a homogeneous model of the moon.

treatment of the problem of temperature variations of the lunar surface with thermal variations of the parameters of the lunar material included. Only a start on this topic has been made. A thorough discussion making full use of available knowledge in this field is very much needed at the present time.

There have been solutions for the equations under the assumption that conductivity varies with depth. In particular, this alternative to the homogeneous model of the moon we have so far discussed is generally pictured as consisting of a substratum of material, which is the main body of the moon, covered by a thin layer of what, in the literature, has come to be called "dust." The thin surface layer has a much lower conductivity than the material forming the main body of the moon. Such a two-layer model of the moon was first discussed by Piddington and Minnett [21] in conjunction with the interpretation of their observations of lunar temperatures measured at 1.25 cm wavelength. However, before we can adequately discuss such a two-layer model, we must derive the equations for the temperature distribution below the surface of the moon and for the microwave temperature of the moon as a function of time and of the wavelength at which the observations are made. In the next sections we derive these equations.

The Subsurface Temperature Distribution on the Moon (Homogeneous Model)

We consider a homogeneous smooth moon with parameters K, ρ, and C. We know the temperature at a point l, ϕ on the moon's surface as a function of time throughout a lunation. The temperature for point l, ϕ is expressed as a Fourier series:

$$T(0, t) = A_0 + A_1 \cos(\omega t - q_1) + A_2 \cos(2\omega t - q_2) + \cdots \qquad (6)$$

where q_n is the phase associated with term n. The coefficients A_n and phases q_n can, in principle, be determined either empirically from observed temperatures throughout a lunation or by representation of the numerical solution of equations (2) and (3) with the value of the thermal inertia deemed most probable on the basis of eclipse results.

If equation (6) represents the time-varying temperature at the surface of a semi-infinite solid, then, as shown in detail in Section 2.6 of Carslaw and Jaeger [3], the solution for the subsurface temperature distribution is

$$T(x, t) = A_0 + A_1 e^{-x\sqrt{\omega/2\kappa}} \cos\left[\omega t - q_1 - x\left(\frac{\omega}{2\kappa}\right)^{\frac{1}{2}}\right]$$
$$+ A_2 e^{-x\sqrt{(2\omega/2\kappa)}} \cos\left[2\omega t - q_2 - x\left(\frac{2\omega}{2\kappa}\right)^{\frac{1}{2}}\right] + \cdots \qquad (7)$$

In writing equation (7) as the solution, we assume that $x/2\kappa t$ is a small quantity. This implies that we consider the equilibrium solution. The age of the moon is sufficient to assure that our assumption is fulfilled; t is great enough so that all transients in the solution have long since damped out.

Equation (7) represents a series of "thermal waves" penetrating into the lunar material and attenuating with distance penetrated. The wavelength λ_n, associated with term n, is readily found to be

$$\lambda_n = 2\pi \sqrt{\frac{2\kappa}{n\omega}} = \frac{\lambda_1}{\sqrt{n}} \tag{8}$$

where, specifically, λ_1, the wavelength of the fundamental, is $2\pi\sqrt{2\kappa/\omega}$. The velocity of propagation λ_n/p_n is given by the expression $\sqrt{2\kappa n\omega}$. For each component (that is, for each term n), there is a phase lag $x \cdot \sqrt{n\omega/2\kappa}$. As a wave of wavelength λ_n penetrates into the lunar material, its amplitude diminishes as

$$e^{-x\sqrt{(n\omega/2\kappa)}} = e^{-2\pi(x/\lambda_n)} \tag{9}$$

A thermal wave is damped to a value e^{-1} of its initial value in a distance

$$L_{T,n} = \sqrt{\frac{2\kappa}{n\omega}} = \frac{1}{\sqrt{n}} L_{T,1} \tag{10}$$

where the subscript T denotes thermal attenuation. This characteristic distance for thermal waves will appear frequently in discussions that follow. In terms of $L_{T,1}$, equation (7) becomes

$$T(x, t) = A_0 + A_1 e^{-x/L_{T,1}} \cos(\omega t - q_1 - x/L_{T,1})$$
$$+ A_2 e^{-x/L_{T,2}} \cos(2\omega t - q_2 - x/L_{T,2}) + \ldots \tag{11}$$

In penetrating the surface a depth of one wavelength, each component wave is damped by the factor $e^{-2\pi} = 1.87 \times 10^{-4}$, a very strong attenuation. Since the harmonics have shorter wavelengths than the fundamental, and since each is attenuated by the same amount, $e^{-2\pi}$, in traveling a distance of one wavelength, it is clear that the harmonics damp out faster than the fundamental as we go to greater depths below the surface. Whatever the form of the temperature variation on the surface, the temperature variation rapidly becomes sinusoidal as we go to increasing depths below the surface. Further, since attenuation is high, $e^{-2\pi}$ per wavelength penetration, the amplitude of the temperature variation, whatever may be its form, rapidly decreases with depth. These remarks are well illustrated by the case displayed in Fig. 5, which is adapted from a diagram in Carslaw and Jaeger [3]. On the surface, the temperature variation is a square wave. At subsurface depths $x/L_{T,1} = 0.5, 1.0, 1.5, 2.0$. The variation rapidly loses its square-wave character as shown in the figure.

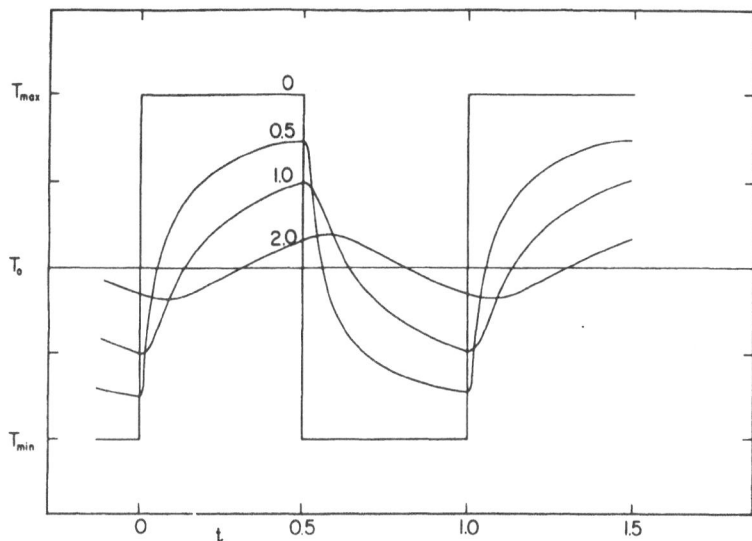

Fig. 5. A square-wave temperature fluctuation is present on the surface of a semi-infinite solid. At subsurface depths, $X/L_{T,1} = 0.5$, 1.0, 1.5, 2.0, the temperature fluctuations will be as shown. Higher frequencies damp out very quickly, only a near-sinusoidal variation remains. The illustration is adapted from one by Carslaw and Jaeger [3].

The Microwave Temperature of the Moon

Electromagnetic radiation in the radio range penetrates the material of the moon, hence the microwave temperature we observe represents a weighted mean value of the subsurface temperature distribution to the greatest depth penetrated by radiation of the wavelength at which we are observing. We may derive the required weighted mean temperature from equation (11) for $T(x, t)$.

In the frequency range $\nu \pm (\Delta\nu)/2$ we observe the radiation emanating from a specific point having coordinates l, ϕ on the lunar surface. Our line of sight makes an angle θ (see Fig. 6) with the normal to the lunar surface at the point l, ϕ; θ is directly related to the fractional distance of the point observed from the center of the lunar disk. The radiation we see emanating from the point l, ϕ traverses a trajectory of different angle when it is below the surface of the moon. Below the lunar surface it traverses a path that makes an angle χ with the normal to the surface. The two angles involved are related through Snell's law of refraction:

$$\frac{\sin \theta}{\sin \chi} = n_\nu \qquad (12)$$

where n_ν is the index of refraction of the lunar material for radiation of frequency ν. From any point such as P, which is located a distance x below the surface, radiation flows along the trajectory indicated to the telescope. The emissivity

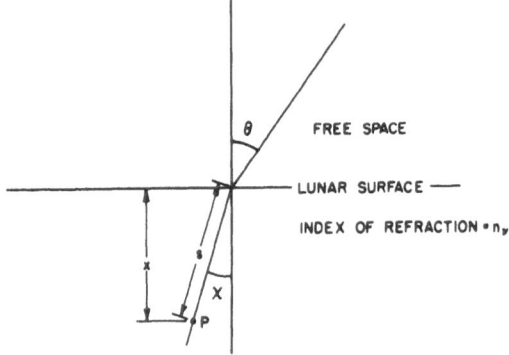

Fig. 6. The relations between the angles θ and χ.

j_ν per unit solid angle and per square centimeter for the frequency range of inter-est at point P is, by Kirchhoff's law

$$j_\nu = n_\nu^2 \rho k_\nu B_\nu(T) \Delta\nu \tag{13}$$

In this expression k_ν represents the mass absorption coefficient for radiation of frequency ν, ρ the density of lunar material, and $B_\nu(T)$ the Planck function at frequency ν.

The contribution from P to the intensity of radiation arriving on the inner side of the lunar surface (which we designate by 0^-) in a column 1 cm^2 in cross section making an angle χ with the normal is

$$dI_\nu(0^-, \chi, t, s)\, ds\, \Delta\nu = n_\nu^2 \rho k_\nu B_\nu(T, t)\, e^{-k_\nu \rho s}\, ds\, \Delta\nu \tag{14}$$

where we have now added the time t as an additional variable, since the tempera-ture at point P is changing with time in accordance with equation (11). In equa-tion (14), s represents the path distance from P to the surface. For the total in-tensity at 0^- from all points P below the surface we integrate over s to find

$$I_\nu(0^-, \chi, t) = n_\nu^2 \rho k_\nu \int_0^\infty B_\nu(T, t)\, e^{-\rho k_\nu s}\, ds$$

With the transformation of variable $s = x \sec \chi$, we have

$$I_\nu(0^-, \chi, t) = n_\nu^2 \rho k_\nu \int_0^\infty B_\nu(T, t)\, e^{-\rho k_\nu x \sec \chi}\, dx \sec \chi \tag{15}$$

In the radio wavelength range the Rayleigh–Jeans law suffices for $B_\nu(T)$, hence

$$I_\nu(0^-, \chi, t) = \frac{2 k n_\nu^2 \rho k_\nu}{\lambda_0^2} \int_0^\infty T(x, t) e^{-\rho k_\nu x \sec \chi} dx \sec \chi$$

where k represents the Stefan–Boltzmann constant and λ_0 is the vacuum wavelength corresponding to frequency ν.

The function $T(x, t)$ is given by equation (11), thus

$$I_\nu(0^-, \chi, t) = \frac{2 k n_\nu^2 \rho k_\nu}{\lambda_0^2} \int_0^\infty [A_0 + A_1 e^{-x/L_{T,1}} \cos(\omega t - q_1 - x/L_{T,1})$$

$$+ A_2 e^{-\sqrt{2} x/L_{T,1}} \cos(2\omega t - q_2 - \sqrt{2} x/L_{T,1})$$

$$+ \ldots] e^{-\rho k_\nu x \sec \chi} dx \sec \chi \qquad (16)$$

We integrate equation (16) term by term as follows:

Term 0:

$$A_0 \int_0^\infty \rho k_\nu e^{-\rho k_\nu x \sec \chi} dx \sec \chi = A_0$$

Term 1:

$$A_1 \int_0^\infty \rho k_\nu e^{-x/L_{T,1}} \cos(\omega t - q_1 - x/L_{T,1}) e^{-\rho k_\nu x \sec \chi} dx \sec \chi$$

We now define $L_{R,\nu}$ (in analogy to $L_{T,n}$) as a characteristic length in which radiation of frequency ν is attenuated by the factor e^{-1}. Thus, $e^{-\rho k_\nu L_{R,\nu}} = e^{-1}$, or

$$L_{R,\nu} = \frac{1}{\rho k_\nu} \qquad (17)$$

In terms of this definition we continue with Term 1, defining

$$\frac{x \sec \chi}{L_{R,\nu}} = y,$$

$$A_1 \int_0^\infty e^{-y[(L_{R,\nu}/L_{T,1})(1/\sec \chi)]} \cos\{\omega t - q_1 - y \cdot [(L_{R,\nu}/L_{T,1}) \cdot (1/\sec \chi)]\} e^{-y} dy$$

If we now set

$$\frac{L_{R,\nu}}{L_{T,1}} \cdot \frac{1}{\sec \chi} = \delta_1(\chi)$$

we find

$$A_1 \int_0^\infty e^{-y\,\delta_1(\chi)} \cos\left[\omega t - q_1 - y\,\delta_1(\chi)\right] e^{-y}\,dy$$

$$= A_1 \cos\left(\omega t - q_1\right) \int_0^\infty e^{-[\delta_1(\chi)+1]y} \cos y\,\delta_1(\chi)\,dy$$

$$+ A_1 \sin\left(\omega t - q_1\right) \int_0^\infty e^{-[\delta_1(\chi)+1]y} \sin y\,\delta_1(\chi)\,dy$$

These are standard Laplace transforms; therefore we find at once for Term 1

$$\frac{A_1}{1 + 2\delta_1(\chi) + 2\delta_1^2(\chi)} \left\{ [\delta_1(\chi) + 1] \cos\left(\omega t - q_1\right) + \delta_1(\chi) \sin\left(\omega t - q_1\right) \right\}$$

from which we readily derive

$$\frac{A_1}{[1 + 2\delta_1(\chi) + 2\delta_1^2(\chi)]^{\frac{1}{2}}} \cos\left[\omega t - q_1 - \psi_1(\chi)\right] \qquad (18)$$

where

$$\psi_1(\chi) = \text{arc tan} \frac{\delta_1(\chi)}{\delta_1(\chi) + 1} \qquad (19)$$

Terms 2, 3, 4, ... are completely analogous to Term 1; hence we can write with no further effort

$$I_\nu(0^-, \chi, t) = \frac{2kn\nu^2}{\lambda_0^2} \left\{ A_0 + \frac{A_1}{[1 + 2\delta_1(\chi) + 2\delta_1^2(\chi)]^{\frac{1}{2}}} \cos\left[\omega t - q_1 - \psi_1(\chi)\right] \right.$$

$$\left. + \frac{A_2}{[1 + 2\delta_2(\chi) + 2\delta_2^2(\chi)]^{\frac{1}{2}}} \cos\left[2\omega t - q_2 - \psi_2(\chi)\right] + \ldots \right\} \qquad (20)$$

where

$$\delta_n(\chi) = \frac{L_{R,\nu}}{L_{T,n}} \cdot \frac{1}{\sec \chi} = \sqrt{n}\,\delta_1(\chi) \qquad (21)$$

and

$$\psi_n(\chi) = \text{arc tan} \frac{\delta_n(\chi)}{\delta_n(\chi) + 1} = \text{arc tan} \frac{\delta_1(\chi)}{\delta_1(\chi) + 1/\sqrt{n}} \qquad (22)$$

Kopal [11] has pointed out that equations (20) and (22) are very slightly inaccurate owing to the fact that sec χ (or cos θ as we shall later derive) cannot be taken out of the integral. Kopal finds that where we have written $(1 + 2\delta_n + 2\delta_n^2)^{\frac{1}{2}}$ we should have $[(1 + \delta_n)^2 + (\delta_n - d_n)^2]^{\frac{1}{2}}$, and for ψ_n = arc tan $\delta_n/(\delta_n + 1)$ we should have arc tan $(\delta_n - d_n)/(\delta_n + 1)$. The quantity d_n is equal to $n\omega/K\rho c$, where c represents the velocity of light. With the parameters of the lunar surface material we readily compute that $d_n \sim n/c$, that is, $\sim 10^{-10}$. We shall neglect this vanishingly small correction.

$I_\nu(0^-, \chi, t)$ specifies, for time t, the intensity in a pencil of radiation on the inside of the lunar surface at position l, ϕ, making an angle χ with the normal to the surface. The angle χ is a nonobservable. The angle we can observe is θ, where θ and χ are related through Snell's law, equation (12). We can transform $I_\nu(0^-, \chi, t)$ to $I_\nu(0^-, \theta, t)$, provided we know the index of refraction (which may be complex) of the lunar material. Additionally, we wish to transform $I_\nu(0^-, \theta, t)$ to $I_\nu(0^+, \theta, t)$, that is, to the intensity in a pencil of radiation outside the lunar surface. This latter transformation requires that we know the transmissivity of the lunar surface, a quantity given by the Fresnel equations which also depend upon the index of refraction. We now digress briefly to discuss the problems of refraction and reflection at the lunar surface.

Refraction and Reflection at the Lunar Surface

Arguing from analogy with the earth, we may assume that the lunar surface material will have a small conductivity σ of the order 10^{-2} to 10^{-5} mhos/m, and a relative dielectric constant ϵ_ν/ϵ_0 (ϵ_0 is the dielectric constant for free space) of the order 2 to 4. Using the test for a dielectric that $\sigma/\omega\epsilon \ll 1$, where ω represents the angular frequency $2\pi\nu$, we may conservatively estimate that as far as reflection and refraction are concerned, the earth (and by analogy the moon) will act as a dielectric for $\nu > 10^3$ Mc/s and possibly even for $\nu > 1$ Mc/s. This is fortunate, since under such circumstances we may use Snell's law in its familiar form [equation (12)] and the Fresnel equations in their simplest forms, omitting absorption terms arising from conduction or from molecular or crystal lattice resonances. If $\sigma/\omega\epsilon$ is not $\ll 1$, we must use the general form of the laws of refraction and reflection. These are very cumbersome. It is therefore of importance to ascertain observationally the value of $\sigma/\omega\epsilon$ for lunar material. We start this discussion, then, by considering some of the relations between absorption parameters arising from the fundamental properties of materials and the absorption coefficient k_ν introduced in equation (13).

The quantity $\sigma/\omega\epsilon$ is generally termed the loss tangent and is written as

$$\tan \Delta = \frac{\sigma}{\omega\epsilon} \tag{23}$$

It is a quantity found tabulated for many dielectric materials. It may be expected to vary with frequency; its value and variation serve to characterize the particular material for which it was determined in the same general way that an optical spectrum serves to identify a substance.

For an electromagnetic field propagating through material having electromagnetic parameters μ, σ_ν, and ϵ_ν, the E-vector satisfies the wave equation

$$\nabla^2 E = \mu \epsilon_\nu \ddot{E} + \mu \sigma_\nu \dot{E} \tag{24}$$

The term involving $\dot{E}$ implies attenuation of the wave as it is propagated; such attenuation, the result of Joule heating, will exist as long as $\sigma \neq 0$. If the field is strictly monochromatic and of angular frequency ω, equation (24) becomes

$$\nabla^2 E + k*^2 E = 0 \tag{25}$$

where

$$k*^2 = \omega^2 \mu [\epsilon_\nu + i(\sigma_\nu/\omega)] = \omega^2 \mu \epsilon_\nu{}^* \tag{26}$$

The material may be thought of as possessing a complex dielectric constant

$$\epsilon_\nu{}^* = \epsilon_\nu + i(\sigma_\nu/\omega) \tag{27}$$

In this expression we must look upon σ_ν as a generalized dielectric conductivity which may arise from conductivity involving migrating charge carriers or may refer to other loss-causing processes on the molecular scale such as molecular rotation, lattice vibration, and the like.

In terms of the generalized Maxwell relation, we may write for the complex velocity of the wave in the material

$$v_\nu{}^* = 1/\sqrt{\mu \epsilon_\nu{}^*} \tag{28}$$

In terms of the complex index of refraction

$$n_\nu{}^* = \frac{c}{v_\nu{}^*} = c\sqrt{\mu \epsilon_\nu{}^*} = c(k_\nu{}^*/\omega) \tag{29}$$

If we now write

$$n_\nu{}^* = n_\nu(1 + i\kappa_\nu) = n_\nu + i\kappa_{0,\nu} \tag{30}$$

we find upon squaring, making use of equation (29), and equating real and variable parts, respectively,

$$n_\nu{}^2(1 - \kappa_\nu{}^2) = c^2 \mu \epsilon_\nu$$
$$\tag{31}$$
$$n_\nu{}^2 \kappa_\nu = c^2 \mu \sigma_\nu/2\omega$$

We solve these for n_ν and $n_\nu \kappa_\nu$ to find

$$n_\nu = c \left[\frac{\mu \epsilon_\nu}{2} \left(\sqrt{1 + \frac{\sigma_\nu^2}{\omega^2 \epsilon_\nu^2}} + 1 \right) \right]^{1/2}$$

$$n_\nu \kappa_\nu = c \left[\frac{\mu \epsilon_\nu}{2} \left(\sqrt{1 + \frac{\sigma_\nu^2}{\omega^2 \epsilon_\nu^2}} - 1 \right) \right]^{1/2}$$

(32)

Adopting the plane, time-harmonic wave solution of equation (25), we find

$$E = E_0 e^{i\left[k_\nu^* (\mathbf{r} \cdot \mathbf{s}) - \omega t \right]}$$

(33)

or, employing the relation $k_\nu^* = (\omega/c) n_\nu^*$ and taking only the real part of the solution for E, we have

$$E = E_0 e^{-(\omega/c) n_\nu \kappa_\nu \mathbf{r} \cdot \mathbf{s}} \cos \left[(\omega/c) n_\nu \mathbf{r} \cdot \mathbf{s} - \omega t \right]$$

(34)

The wave thus attenuates in amplitude as

$$e^{-(\omega/c) n_\nu \kappa_\nu \mathbf{r} \cdot \mathbf{s}}$$

as it penetrates the material. It attenuates in power as

$$e^{-2(\omega/c) n_\nu \kappa_\nu \mathbf{r} \cdot \mathbf{s}}$$

(35)

We may write the coefficient of the exponent in the power attenuation law as

$$2 (\omega/c) n_\nu \kappa_\nu = 2 (2\pi/\lambda_0) \kappa_{0,\nu} = (4\pi/\lambda_0) \kappa_{0,\nu}$$

(36)

The latter expression is the equivalent of ρk_ν which appears in equation (13); thus

$$k_\nu = (4\pi/\lambda_0) (\kappa_{0,\nu}/\rho)$$

(37)

and, in accordance with equation (17),

$$L_{R,\nu} = \lambda_0 / 4\pi \kappa_{0,\nu}$$

(38)

If $\tan \Delta \ll 1$, then, from equation (32) for $n_\nu \kappa_\nu$, we find

$$\sigma_\nu / \omega \epsilon_\nu = \tan \Delta \simeq 2 \kappa_\nu$$

(39)

and from equations (38) and (39)

$$L_{R,\nu} = (1/\sigma_\nu) \sqrt{(\epsilon_\nu/\mu)}$$

(40)

If we take $\mu = \mu_0$, we find

$$L_{R,\nu} = \sqrt{\epsilon_\nu/\epsilon_0}/377\,\sigma_\nu \tag{41}$$

For dry ground $\sigma_\nu \sim 10^{-4}$ mhos/m; $\epsilon_\nu/\epsilon_0 \sim 2$, therefore $L_{R,\nu} \sim 37$ m if ν is in an appropriate frequency range at which the surface material can be treated as dielectric.

In the section entitled "Absorption of Radiation in the Lunar Material," where we shall deal in greater detail with the absorption of radiation in the lunar material, we shall see that for the moon, $\tan \Delta \sim 10^{-2}$; hence, we can, in fact, treat the lunar material as a dielectric as we have here assumed. The reflectivity of the surface is thus governed by Fresnel's equations in their simple form without absorption terms. Snell's law is

$$\frac{\sin \theta}{\sin \chi} = n_\nu \tag{42}$$

or by Maxwell's relation in which $v = 1/\sqrt{\mu\epsilon}$,

$$\frac{\sin \theta}{\sin \chi} = n_\nu = \sqrt{\frac{\mu\epsilon\nu}{\mu_0\epsilon_0}} = \sqrt{\frac{\epsilon_\nu}{\epsilon_0}} \tag{43}$$

since we take $\mu = \mu_0$, the permeability in free space.

To specify the Fresnel equations for reflectivity in the two polarizations, we first define a plane by the normal to the lunar surface and the pencil of radiation emerging from the surface at angle θ. This plane will pass through the center of the lunar disk. We define parallel polarization p as that which has its E-vector in the plane just defined and the perpendicular to the pencil of radiation. Perpendicular polarization s is that which has its E-vector perpendicular to the plane just described. The reflection coefficients of the lunar material at the point l, ϕ for these two polarizations are given by the Fresnel equations

$$R_p = \left[\frac{\tan (\chi - \theta)}{\tan (\chi + \theta)} \right]^2$$
$$R_s = \left[\frac{\sin (\chi - \theta)}{\sin (\chi + \theta)} \right]^2 \tag{44}$$

Derivation of $T_b(\nu, \theta, t)$ from $I_\nu(0^-, \chi, t)$

We return now to consideration of equations (16) and (17), which we rewrite in the forms

$$I_\nu(0^-, \chi, t) = \frac{2 k n_\nu^2}{\lambda_0^2} \left\{ A_0 + \frac{A_1}{a_1(\chi)} \cos \left[\omega t - q_1 - \psi_1(\chi) \right] \right.$$

$$\left. + \frac{A_2}{a_2(\chi)} \cos \left[2\omega t - q_2 - \psi_2(\chi) \right] + \cdots \right\}$$

where $a_n(\chi)$, the attenuation factor, is given by

$$a_n(\chi) = [1 + 2\delta_n(\chi) + 2\delta_n^2(\chi)]^{\frac{1}{2}} \tag{45}$$

and, in terms of $\kappa_{0,\nu}$,

$$\delta_n(\chi) = \sqrt{n} \sqrt{\frac{\omega}{2}} \frac{C}{\sqrt{K\rho C}} \frac{\lambda_0 \rho}{4\pi \kappa_{0,\nu}} \frac{1}{\sec \chi} = \sqrt{n} \delta_1(\chi) \tag{46}$$

$$\psi_n(\chi) = \arctan \frac{\delta_n(\chi)}{\delta_n(\chi) + 1} = \arctan \frac{\delta_1(\chi)}{\delta_1(\chi) + 1/\sqrt{n}} \tag{47}$$

Using Snell's law, we transform from $\chi \longrightarrow \theta$, to derive

$$\delta_n(\theta) = \sqrt{n} \sqrt{\frac{\omega}{2}} \frac{C}{\sqrt{K\rho C}} \frac{\lambda_0 \rho}{4\pi \kappa_{0,\nu}} \frac{\sqrt{n_\nu^2 - \sin^2 \theta}}{n_\nu} \tag{48}$$

and

$$\psi_n(\theta) = \arctan \frac{\delta_1(\theta)}{\delta_1(\theta) + 1/\sqrt{n}} \tag{49}$$

We can, of course, substitute ϵ_ν/ϵ_0 for n_ν^2 if that is desirable. Equation (48), we note, contains, beside the density parameter ρ, the thermal parameters K and C and both radiation parameters n_ν and $\kappa_{0,\nu}$.

A pencil of radiation that passes through a series of media 1, 2, ... having indices of refraction $n_1, n_2, \ldots$ will have intensities $I_1, I_2, \ldots$ within the media. In accordance with Kirchhoff's law, the quantity I_i/n_i^2 is a constant provided there are no reflection losses at the interfaces. In the case of radiation traversing the lunar surface, there are reflection losses. The reflection coefficients are specified by Fresnel's equations [equations (44)]. To write Fresnel's equations, we consider a point l, ϕ on the lunar surface. The point is located a fraction $\sin \theta$ of the distance from the center of the lunar disk toward the limb of the moon, where θ is the angle between the normal to the lunar surface at the point l, ϕ and the normal at the center of the lunar disk. If $R_p(\theta)$ and $R_s(\theta)$ denote the fractional losses, through reflection, of intensity in the pencil in the p and s directions of polarization, then the fractional amounts of radiation escaping through the lunar surface are given by $1 - R_p(\theta)$ and $1 - R_s(\theta)$. The total intensity of radiation escaping to free space through the surface of the moon is then given by

$$I_\nu(0^+, \theta, t) = \tfrac{1}{2}[1 - R_p(\theta) + 1 - R_s(\theta)] \cdot (1/n_\nu{}^2) I_\nu(0^-, \theta, t) \qquad (50)$$

where the symbol 0^+ denotes the fact that we discuss radiation on the outside surface of the moon. Since radio telescopes are normally calibrated in terms of temperature rather than intensity, we can, for convenience, write

$$I_\nu(0^+, \theta, t) = \frac{2 k T_{b,\nu}(0^+, \theta, t)}{\lambda_0{}^2} \qquad (51)$$

where T_b represents the brightness temperature characterizing the lunar radiation emerging from l, ϕ. Thus, finally

$$T_b(\nu, \theta, t) = \left\{1 - \frac{R_p(\theta) + R_s(\theta)}{2}\right\}\left\{A_0 + \frac{A_1}{a_1(\theta)} \cos[\omega t - q_1 - \psi_1(\theta)]\right.$$
$$\left. + \frac{A_2}{a_2(\theta)} \cos[2\omega t - q_2 - \psi_2(\theta)] + \cdots\right\} \qquad (52)$$

where we have dropped the index 0^+, since from now on we shall be discussing radiation outside the surface of the moon. We have added frequency ν more prominently to T_b to indicate that our values refer to a specific frequency and frequency range related to observations.

Equation (52) represents the solution to the problem of relating microwave temperatures to the (infrared) surface temperature of the homogeneous, smooth moon that we have adopted as the model for discussion. We call special attention to several features of equation (52).

1. The factor involving $R_p(\theta)$ and $R_s(\theta)$ multiplies the entire Fourier series. This will have an important consequence in our discussion of the brightness temperature $T_b(\nu, \theta, t)$, and will be crucial in our discussion of polarization of lunar radiation.
2. The constant term A_0 is, at all frequencies, the same in the equation for the surface temperature, the temperature as a function of depth, and the microwave temperature. A_0 represents the constant temperature we should expect to find throughout the moon at the l, ϕ point at which we are observing, if only solar radiation is responsible for heating the moon.
3. Comparing corresponding terms in the surface temperature equation and the microwave temperature equation for a given point on the moon, we note that, except for the factor containing $R_p(\theta)$ and $R_s(\theta)$, the coefficients of the variable terms are in the ratio $1 : a_n(\theta)$, where the amplitude attenuation factor is a rather involved function of the thermal and optical parameters of the lunar material and of the angle the line of sight makes with the normal to the surface.

4. Comparing corresponding terms in the surface temperature and microwave temperature equations for a given point on the moon, we note that there is a phase lag $\psi_n(\theta)$ in the microwave equation. Like the amplitude attenuation factor $a_n(\theta)$, $\psi(\theta)$ depends upon both the properties of the surface material and angle the line of sight makes with the normal to the surface: $a_n(\theta)$ and $\psi(\theta)$ are directly related quantities.

The effects mentioned in Items (3) and (4) occur, of course, because the microwave temperature is a weighted mean of the subsurface temperature as mentioned in the section entitled "The Microwave Temperature of the Moon," and at each depth there are attenuation and phase effects as the thermal waves penetrate the lunar material.

We now deal with some of the properties of equation (52) in greater detail.

Amplitude Attenuation and Phase Effects in the Microwave Temperature Equation

The quantity $\delta_n(\theta)$ in the amplitude attenuation factor

$$a_n(\theta) = [1 + 2\delta_n(\theta) + 2\delta_n^2(\theta)]^{\frac{1}{2}}$$

is defined in equation (48). As we see from equation (48), for any frequency ν, we may write

$$\delta_n(\theta) = \delta_n(0)\frac{\sqrt{n_\nu^2 - \sin^2\theta}}{n_\nu} \tag{53}$$

and the attenuation factor can thus be written

$$a_n^2(\theta) = 1 + 2\delta_n(0)\frac{\sqrt{n_\nu^2 - \sin^2\theta}}{n_\nu} + 2\delta_n^2(0)\frac{n_\nu^2 - \sin^2\theta}{n_\nu^2} \tag{54}$$

The phase angle, as a function of $\delta_n(0)$, n_ν, and θ is

$$\psi_n(\theta) = \text{arc tan }\frac{\delta_n(0)}{\delta_n(0) + n_\nu/\sqrt{n_\nu^2 - \sin^2\theta}} \tag{55}$$

We examine now the amplitude attenuation factor $a_n(\theta)$, the phase lag $\psi_n(\theta)$, and the value of $\delta_n(\theta)$ for term n (we take $n = 1$ for this illustration) as a function of θ, along, let us say, the lunar equator. We shall take $\delta_1(0) = 2$, $n_\nu = 1.26$ (which is equivalent to $\epsilon_\nu/\epsilon_0 = 1.6$), values characteristic of the moon at a frequency of 30 Gc/sec. The results of our calculations are shown in Fig. 7. There is a very significant variation of a_1 and ψ_1 with θ. The variation in a_1 is the lunar counterpart of the familiar limb darkening of the sun.

However, the value of $a_1(\theta)/a_1(0)$ plotted in Fig. 7 is not what we should expect to observe. Equation (52) indicates that the brightness temperature contains

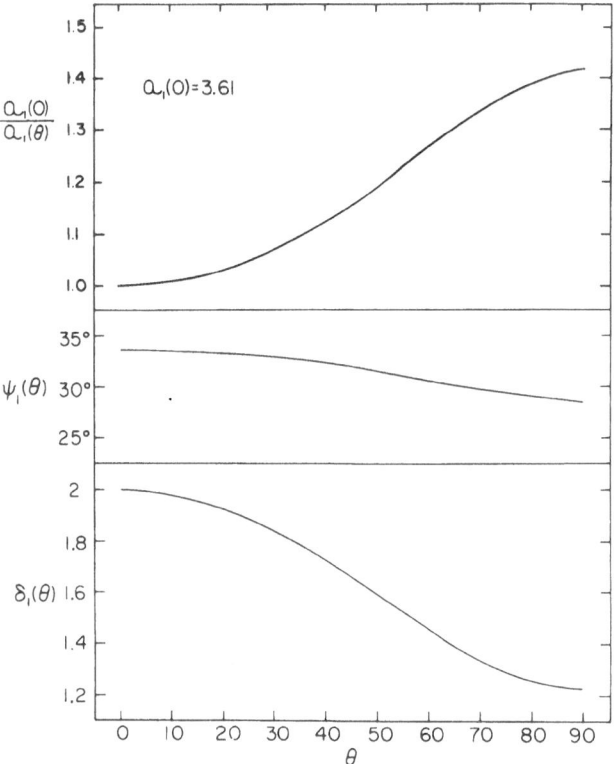

Fig. 7. Calculated values of the ratio of amplitude attenua-
tion factors $a_1(0)/a_1(\theta)$, the phase lag, $\psi_1(\theta)$, and $\delta_1(\theta)$ for
lunar material for which $\epsilon_\nu/\epsilon_0 = 1.6$.

the multiplicative factor

$$1 - \frac{R_p(\theta) + R_s(\theta)}{2}$$

which specifies the fraction of the total radiation escaping into space. We must
multiply $A_1/a_1(\theta)$ by the appropriate value of this multiplicative factor to derive
the amplitude we should expect to find in the $n = 1$ term in the brightness temperature. However, we make our observations with a real antenna. An antenna
does not accept all the radiation implied by the factor $\{1 - [R_p(\theta) + R_s(\theta)]/2\}$.
Normally, an antenna accepts only one mode of polarization. In a practical case,
then, we must use a multiplicative factor that relates to the direction of polarization our antenna accepts. For illustration, we assume the horn accepts p polarization. We make a series of observations and then turn the horn so it accepts s
polarization and make a second series of observations. In Fig. 8 we illustrate
the predicted amplitude attenuation factors ($n = 1$), including the appropriate

multiplicative factors, in the brightness temperature for the two modes of polarization p and s, for a series of points lying along the lunar equator. These two cases represent the extremes. Other directions of polarization will give results that lie between them.

We can eliminate the effect of polarization from our results and, with some additional effort, regain the desired $a_n(\theta)$ value even if we do not know the value of $1 - R_p(\theta)$ or $1 - R_s(\theta)$. For a specific mode of polarization, say, p, we will in fact derive from observations the values [refer to equation (52)] of each of the series of terms

$$[1 - R_p(\theta)] A_0$$

$$[1 - R_p(\theta)] A_1/a_1(\theta)$$

$$[1 - R_p(\theta)] A_2/a_2(\theta)$$

If we form the ratio

$$\frac{[1 - R_p(\theta)] A_0}{[1 - R_p(\theta)] A_1/a_1(\theta)} = \frac{[1 - R_p(\theta)]}{[1 - R_p(\theta)]} \frac{A_0}{A_1} a_1(\theta) \tag{56}$$

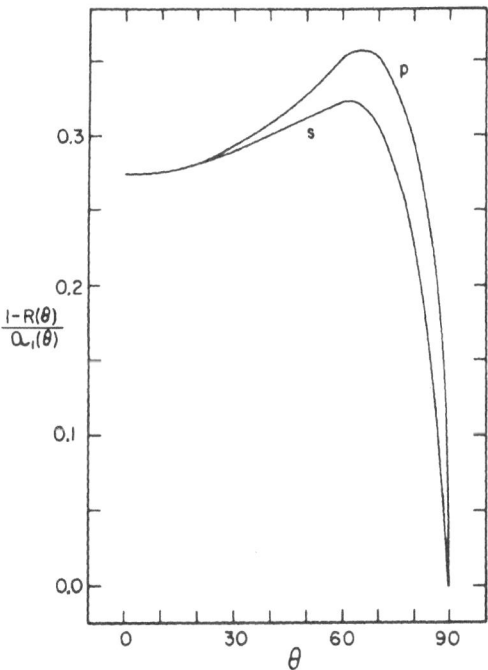

Fig. 8. Calculated values of $[1 - R_p(\theta)]/a_1(\theta)$ and $[1 - R_s(\theta)]/a_1(\theta)$ for lunar material in which $\epsilon\nu/\epsilon_0 = 1.6$.

and compare this with the value A_0/A_1 derived from the surface-temperature curve, we find

$$\frac{[1 - R_p(\theta)]/[1 - R_p(\theta)] \cdot (A_0/A_1) \cdot a_1(\theta)}{A_0/A_1} = a_1(\theta) \tag{57}$$

which is, of course, precisely the quantity we started with and now desire. We can then form $a_1(\theta)/a_1(0)$, which is what is plotted in Fig. 7. Taking the series of ratios indicated in equation (57), that is, using the observationally determined A_0 term as well as the observationally determined amplitude of the fundamental to find $a_1(\theta)$, does increase the observational difficulties. The procedure proposed demands that the zero point of the observational temperature system be precisely determined as well as the scale of the system and adds to the already difficult problem of antenna calibration.

Observation of the variation of the amplitude attenuation factor $a_n(\theta)$ provides us with a means of deriving values of n_ν and $\delta_n(0)$. Assume that we have chosen $m + 1$ points along the lunar equator, $\theta = 0, \theta_1, \ldots, \theta_m$, and for these we have derived from observations the surface temperature curve and the microwave temperature curve as some specified wavelength. We represent the two curves by Fourier series. From a pair of series for a θ point we then, by ratios like equation (57), determine for term n the amplitude attenuation factor $a_n(\theta)$. In principle, we then have for any θ_i value amplitude attenuation factors $a_1(\theta_i)$, $a_2(\theta_i)$, $\ldots$, $a_n(\theta_i)$. (In practice, we will normally be able to find from observations only $a_1(\theta_i)$, but we here consider the general case.) Noting that $\delta_n(\theta) = \sqrt{n}\,\delta_1(\theta)$, we then write

$$1 + 2\delta_1(0) \qquad\qquad\qquad + 2\delta_1^2(0) \qquad\qquad\qquad = a_1^2(0)$$

$$1 + 2\delta_1(0)\frac{\sqrt{n_\nu^2 - \sin^2\theta_1}}{n_\nu} \qquad + 2\delta_1^2(0)\frac{n_\nu^2 - \sin^2\theta_1}{n_\nu^2} \qquad = a_1^2(\theta_1)$$

$$\cdot \qquad\qquad\qquad\qquad\qquad\qquad\qquad \cdot$$
$$\cdot \qquad\qquad\qquad\qquad\qquad\qquad\qquad \cdot$$
$$\cdot \qquad\qquad\qquad\qquad\qquad\qquad\qquad \cdot$$

$$1 + 2\delta_1(0)\frac{\sqrt{n_\nu^2 - \sin^2\theta_m}}{n_\nu} \qquad + 2\delta_1^2(0)\frac{n_\nu^2 - \sin^2\theta_m}{n_\nu^2} \qquad = a_1^2(\theta_m)$$

$$1 + 2\sqrt{2}\,\delta_1(0) \qquad\qquad\qquad + 2 \cdot 2\delta_1^2(0) \qquad\qquad\qquad = a_2^2(0)$$

$$1 + 2\sqrt{2}\,\delta_1(0)\frac{\sqrt{n_\nu^2 - \sin^2\theta_1}}{n_\nu} \qquad + 2 \cdot 2\delta_1^2(0)\frac{n_\nu^2 - \sin^2\theta_1}{n_\nu^2} \qquad = a_2^2(\theta_1)$$

$$\cdot \qquad\qquad\qquad\qquad\qquad\qquad\qquad \cdot$$
$$\cdot \qquad\qquad\qquad\qquad\qquad\qquad\qquad \cdot$$
$$\cdot \qquad\qquad\qquad\qquad\qquad\qquad\qquad \cdot$$

$$1 + 2\sqrt{2}\,\delta_1(0)\frac{\sqrt{n_\nu^2 - \sin^2\theta_m}}{n_\nu} \qquad + 2 \cdot 2\delta_1^2(0)\frac{n_\nu^2 - \sin^2\theta_m}{n_\nu^2} \qquad = a_2^2(\theta_m) \tag{58}$$

(more)

.
.
.

$$1 + 2\sqrt{n}\,\delta_1(0) \qquad\qquad\qquad\qquad + 2n\delta_1^2(0) \qquad\qquad\qquad\qquad = a_n^2(0)$$

$$1 + 2\sqrt{n}\,\delta_1(0)\frac{\sqrt{n_\nu^2 - \sin^2\theta_1}}{n_\nu} + 2n\delta_1^2(0)\frac{n_\nu^2 - \sin^2\theta_1}{n_\nu^2} = a_n^2(\theta_1)$$

.
.
.

$$1 + 2\sqrt{n}\,\delta_1(0)\frac{\sqrt{n_\nu^2 - \sin^2\theta_m}}{n_\nu} + 2n\delta_1^2(0)\frac{n_\nu^2 - \sin^2\theta_m}{n_\nu^2} = a_n^2(\theta_m) \qquad (58)$$

from which we can find $\delta_1(0)$ and n_ν.

In principle, we might also determine and study the phases $\psi_n(\theta)$ to find $\delta_1(0)$ and n_ν, but, as a practical matter, $\psi_n(\theta)$ is less well determined than $a_n(\theta)$ and, moreover, has a smaller range of variation as θ goes from 0 to $\pi/2$.

The amplitude attenuation factor $a_n(\theta)$ and the phase lag $\psi_n(\theta)$ are related in a one to one manner since each is a function of $\delta_n(\theta)$. In Fig. 9, we plot for the fundamental $n = 1$, and for the center of the moon $\theta = 0$, the phase ψ_1 as a function of a_1. As $\delta_1 \longrightarrow \infty$, $a_1 \longrightarrow \infty$ and $\psi_1 \longrightarrow 45°$.

The a_1, ψ_1 plot in Fig. 9 provides a means of determining the value of δ_1 directly from the observed values of a_1, ψ_1: we have only to determine a_1, ψ_1 observationally, plot the corresponding point in Fig. 9 and read off δ_1. However, when such an attempt was made by Piddington and Minnett [21] from the first ex-

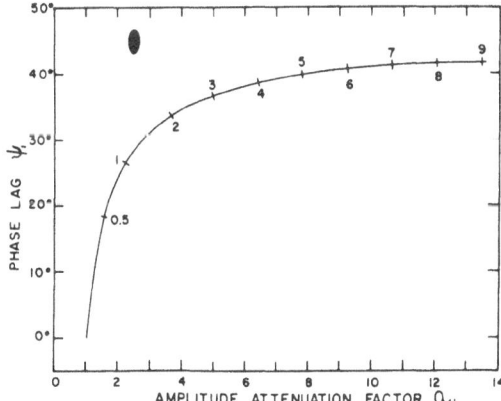

Fig. 9. The relation between amplitude attenuation factor, $a_1(0)$, and phase lag, $\psi_1(0)$, with associated values of $\delta_1(0)$ indicated on the curve. The large point represents the $a_1(0)$, $\psi_1(0)$ pair found observationally by Piddington and Minnett [21] for wavelength 1.25 cm.

tensive series of microwave measurements of lunar radiation (wavelength 1.25 cm), they found that their observed a_1, ψ_1 point ($a_1 = 2.58$, $\psi = 45°$) was above the theoretical a_1, ψ_1 curve by an amount which they considered far greater than their probable errors. Their observational point is plotted in Fig. 9 with their estimate of observational uncertainty indicated by the size of the point. Piddington and Minnett stated that their observations are incompatible with the predictions from a homogeneous model of the moon. They considered their results indicative of lunar surface material of such nature that δ was a function of depth below the surface. However, they did not attempt a general solution in which $\delta = \delta(x)$; rather, they considered a model of the moon in which a layer of very poorly-conducting material ("dust") covers a solid (or at least more solid) uniform material of higher conductivity. A similar two-layer model had been suggested by Jaeger and Harper from their investigation of eclipse observations made in the infrared. These considerations of a "dust-layer" as a mathematically tractable case of a variable δ, have thus led to the "two-layer" model of the moon, which we next consider.

A TWO-LAYER MODEL OF THE MOON

We generalize the model of the moon as indicated in Fig. 10. This two-layer model consists of a layer of material of thickness r and conductivity K_d, density ρ_d, and specific heat C_d. The corresponding parameters of the material in the body of the moon are K, ρ, and C. The thin surface layer is of such a nature that the infrared measurements of the lunar surface temperature indicate the temperature at the level S_0, the top of the thin layer. However, microwaves penetrate the layer. As far as microwave temperatures are concerned, surface S plays the role of the surface of the moon. In this discussion, x will designate distance below S.

Amplitude Attenuation

We assume that the time variation at a point on the surface S is specified by the Fourier series

$$T(0, t) = T_0 + T_1 \cos(\omega t - q_1) + T_2 \cos(2\omega t - q_2) + \ldots \tag{59}$$

In the material below the point on S, at depth x and time t, the temperature is then given by

$$T(x, t) = T_0 + T_1 e^{-b_1 x} \cos(\omega t - q_1 - b_1 x)$$
$$+ T_2 e^{-b_2 x} \cos(2\omega t - q_2 - b_2 x) + \ldots \tag{60}$$

as we found before in the section entitled "The Subsurface Temperature Distribution on the Moon." In equation (60), $b_n = \sqrt{(\omega/2)(n/\kappa)}$ as previously in equation (7).

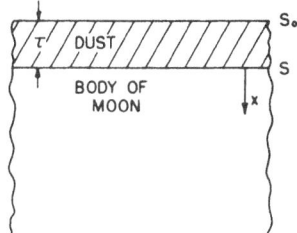

Fig. 10. A two-layer model of the moon.

At any depth x in this semi-infinite solid, the thermal flux is given by

$$F(x) = -K \frac{\partial T(x, t)}{\partial x} \qquad (61)$$

or, with the value of $\partial T(x, t)/\partial x$ we derive from equation (60)

$$F(x) = K T_1 b_1 e^{-b_1 x}[\cos(\omega t - q_1 - b_1 x) - \sin(\omega t - q_1 - b_1 x)]$$
$$+ K T_2 \sqrt{2} b_1 e^{-\sqrt{2} b_1 x}[\cos(2\omega t - q_2 - \sqrt{2} b_1 x)$$
$$- \sin(2 \omega t - q_2 - \sqrt{2} b_1 x)] + \cdots$$

or

$$F(x) = K T_1 b_1 e^{-b_1 x} \sqrt{2} \cos\left(\omega t - q_1 - b_1 x + \frac{\pi}{4}\right)$$
$$+ K T_2 \sqrt{2} b_1 e^{-\sqrt{2} b_1 x} \sqrt{2} \cos\left(2\omega t - q_2 - \sqrt{2} b_1 x + \frac{\pi}{4}\right) + \cdots \qquad (62)$$

At $x = 0$,

$$F(0) = K T_1 b_1 \sqrt{2} \cos\left(\omega t - q_1 + \frac{\pi}{4}\right)$$
$$+ K T_2 b_1 2 \cos\left(2\omega t - q_2 + \frac{\pi}{4}\right) + \cdots \qquad (63)$$

At surface S_0, the outermost surface of the layer of "dust," the temperature is represented by $T_s(t)$:

$$T_s(t) = T(0, t) + \Delta T$$

where ΔT denotes the temperature drop that takes place through the dust layer.

$$\Delta T = \frac{\tau F(0)}{K_d}$$

hence

$$T_s(t) = T(0, t) + \frac{\tau F(0)}{K_d}$$

or

$$T_s(t) = T_0 + T_1\left[\cos(\omega t - q_1) + \frac{K\tau b_1 \sqrt{2}}{K_d} \cos\left(\omega t - q_1 + \frac{\pi}{4}\right)\right]$$

$$+ T_2\left[\cos(2\omega t - q_2) + \frac{K\tau b_1 2}{K_d} \cos\left(2\omega t - q_2 + \frac{\pi}{4}\right)\right] + \dots \quad (64)$$

We now put

$$\delta_{sn} = \frac{K\tau b_1 \sqrt{n}}{K_d} = \frac{\tau \sqrt{n}}{K_d} \sqrt{\frac{\omega}{2}} \sqrt{K\rho C} = \delta_{s1} \sqrt{n} \quad (65)$$

with the aid of which we readily derive from equation (64)

$$T_s(t) = T_0 + T_1(1 + 2\delta_{s1} + 2\delta_{s1}{}^2)^{\frac{1}{2}} \cos(\omega t - q_1 + \psi_{s1})$$

$$+ T_2(1 + 2\delta_{s2} + 2\delta_{s2}{}^2)^{\frac{1}{2}} \cos(2\omega t - q_2 + \psi_{s2}) + \dots \quad (66)$$

where δ_{sn} is defined by equation (65) and

$$\psi_{sn} = \arctan \frac{\delta_{sn}}{\delta_{sn} + 1} = \frac{\delta_{s1}}{\delta_{s1} + 1/\sqrt{n}} \quad (67)$$

If for convenience we now write for the amplitude amplification factor for the dust layer

$$a_{sn} = (1 + 2\delta_{sn} + 2\delta_{sn}{}^2)^{\frac{1}{2}} \quad (68)$$

we find

$$T_s(t) = T_0 + a_{s1} T_1 \cos(\omega t - q_1 + \psi_{s1})$$

$$+ a_{s2} T_2 \cos(2\omega t - q_2 + \psi_{s2}) + \dots \quad (69)$$

Equation (69) specifies the time-varying temperature at S_0, the top of the dust layer, in terms of the temperature $T(0, t)$ at the interface between the dust layer and the underlying material. If, term for term, we compare this expression with equation (59) for $T(0, t)$ we find:

1. In the expression for the temperature at S_0, term n leads the corresponding term in the expression for the temperature at S by the angle ψ_{sn}, and

2. The amplitude of term n in the equation for the temperature at S_0 is greater than the amplitude of the corresponding term in the expression for temperature at surface S, the amplitudes being in the ratios $a_{sn} : 1$.

In terms of the temperature at surface S, the microwave temperature is, in analogy with the corresponding equation derived previously [see equation (52)]:

$$T_b(\nu, \theta, t) = \left[1 - \frac{R_p + R_s}{2} \right]$$

$$\times \left[T_0 + \frac{T_1}{a_1} \cos(\omega t - q_1 - \psi_1) + \frac{T_2}{a_2} \cos(2\omega t - q_2 - \psi_2) + \dots \right] \qquad (70)$$

If, term for term, we now compare the expression for the microwave temperature, equation (70), with the expression for the surface temperature, equation (69), we find that:

1. Compared to term n in the surface temperature expression term n in the microwave brightness temperature lags by the phase angle $\psi_{sn} + \psi_n$, which can exceed $45°$, theoretically reaching $90°$.
2. Compared to the amplitude of term n in the equation for the surface temperature, the amplitude of term n in the expression for the microwave brightness temperature is decreased in the ratio:

$$1 : \frac{1 - (R_p + R_s)/2}{a_{sn} a_n}$$

Observational Isolation of Effects Caused by a Dust Layer on the Moon

If there is on the moon a surface layer of poor conducting material overlaying a layer of greater conductivity, the phase lag of each term in the expression for the microwave brightness temperature is increased over what it would be for a homogeneous moon, and the amplitude of term n in the microwave brightness temperature is attenuated by a factor $[1 - (R_p + R_s)/2]/a_{sn}a_n$. If we take ratios as in equation (57), we can always eliminate the effects of the multiplicative factor and find the product $a_{sn}a_n$. However, from observations one derives only the resultant phase lag of the microwave brightness temperature, $\psi_n + \psi_{sn}$, and the product of the amplitude factors $a_n a_{sn}$. Our task is to find how the observed resultant phase lag is to be apportioned between ψ_n and ψ_{sn}, and how the observed product of the amplitude factors is to be divided between a_n and a_{sn}. The problem is most easily solved graphically.

Since we have a product of two quantities for one coordinate and a sum for the other, it will be expedient to plot logarithmically on the axis along which we graph the product, and numerically on the axis along which we graph the sum. With such a scheme we shall be able to make the desired apportionment directly. We plot ψ_n

as a function of log a_n in quadrant 1 and ψ_{sn} as a function of log a_{sn} in quadrant 3 as shown in Fig. 11.

A synthetic fully worked example will provide insight into the process of using this graph to make the apportionment between dust layer and underlying material. Let us say that $\delta_{sn} = 0.50$ and $\delta_n = 4.0$. Then we readily compute that

$$a_{sn} = 1.581 \qquad\qquad a_n = 6.403$$

$$\psi_{sn} = 18\overset{\circ}{.}43 \qquad\qquad \psi_n = 38\overset{\circ}{.}66$$

Observationally, we would be able to determine only that $a_{sn} \times a_n = 10.123$ and that $\psi_{sn} + \psi_n = 57\overset{\circ}{.}09$.

We now mark, on transparent paper, the origin of our coordinate system and a point at log 10.123 (= 1.0053) and 57°09. We may slide the transparent paper freely along the two axes of our underlying graph (Fig. 11) taking care to avoid rotation. We will find that the two plotted "observed" points will fit on the underlying graph only at

$$\log a_{sn} = 0.199 \qquad\qquad \log a_n = 0.806$$

$$\psi_{sn} = 18\overset{\circ}{.}4 \qquad\qquad \psi_n = 38\overset{\circ}{.}7$$

$$(a_{sn} = 1.58) \qquad\qquad (a_n = 6.40)$$

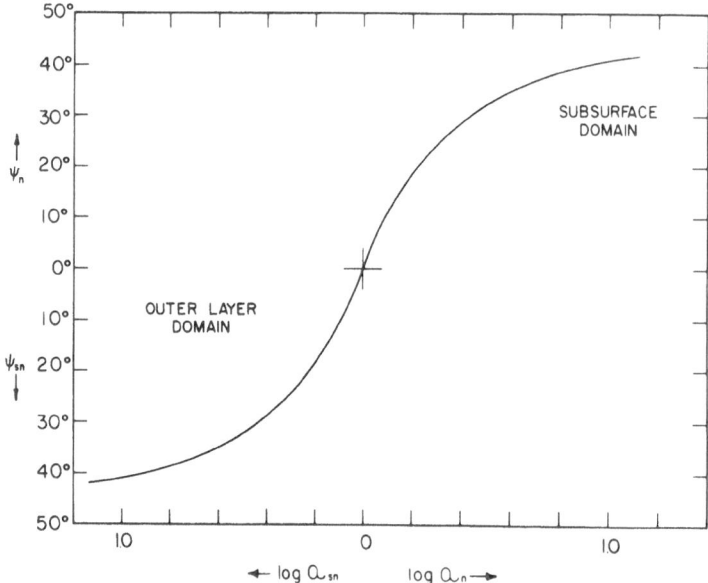

Fig. 11. The log $a_1(0)$, $\phi_1(0)$ relation joined with the log $a_{s1}(0)$, $\phi_{s1}(0)$ relation.

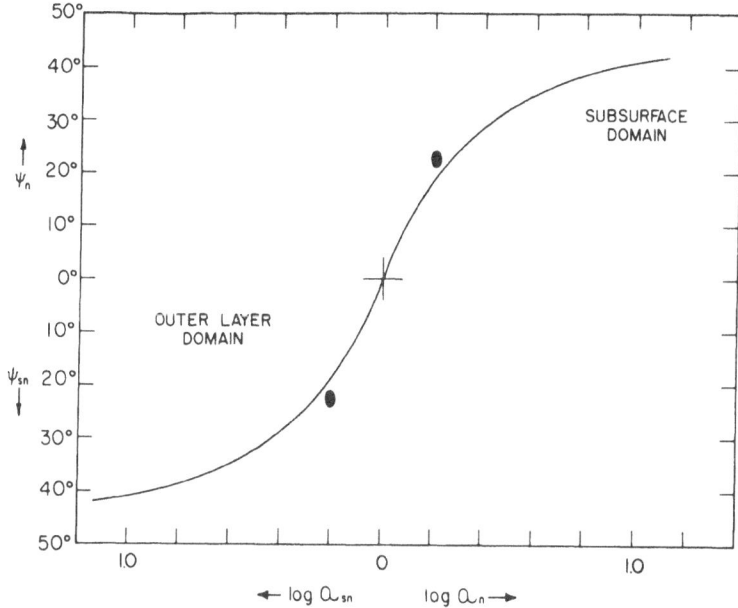

Fig. 12. Solutions for a_1, a_{s1}, ϕ_1, ϕ_{s1}, δ_1, and δ_{s1} for Piddington
and Minnett's [21] microwave results.

Within reading accuracy on the graph we recover the original values from the "observed" quantities; we make the division into the effects of the dust layer and the
body of the moon as required.

It is clear that this graphical method of apportionment will provide immediate
insight into inadequacies of the observations. In most instances, no precise fit
between observations and graph will be found. No observations are absolutely precise. It will then be necessary to make some compromise fit, the exact nature of
which will depend upon the relative accuracies of the observations in the two coordinates' phase angle and amplitude attenuation.

Interpretation of Piddington and Minnett's Results on the Basis of a Two-Layer Model of the Moon

We consider now the microwave observations ($\lambda = 1.25$ cm) of Piddington and
Minnett [21], who found an amplitude attenuation $a_1 = 2.58$ (log $2.58 = 0.412$) and
a phase lag ψ_1 of 45°. Plotting the points $(0, 0)$ and $(0.412, 45)$ on transparent
paper and sliding it as indicated in the previous section, we find no exact fit. The
residuals are, in fact, quite large, rather far outside Piddington and Minnett's estimate of their uncertainty. There appears to be no solution better than that originally suggested by Piddington and Minnett and illustrated in Fig. 12. We find
$\psi_{s1} = \psi_1 = 22°.5$, $a_{s1} = a_1 = 1.606$, $\delta_{s1} = \delta_1 = 0.71$. (Piddington and Minnett
used only the fundamental, $n = 1$, in their analyses, hence no additional values

of ψ, a, or δ can be derived.) With the observed value of δ_{s1} we find from equation (65) that

$$\frac{\tau}{K_d} = 640 \times \frac{1}{\sqrt{K\rho C}} \tag{71}$$

The microwave results thus provide a relation between the thickness of the dust layer τ, its conductivity K_d, and the thermal inertia of the underlying material. For an assumed value of K_d we may then, for various values of $\sqrt{K\rho C}$ determine the thickness of the dust layer as did Piddington and Minnett. The relation can, however, be more effectively utilized in the discussion of eclipse observations as was done by Jaeger and Harper [9].

The Use of Eclipse Observations to Evaluate τ and $\sqrt{K\rho C}$

In the section entitled "The Surface Temperature at a Point on the Moon," we discussed Jaeger [7, 8] and Wesselink's [30] interpretation of Pettit's observations of the 1939 eclipse. Figure 4 shows how the eclipse data are fitted by a homogeneous model of the moon with $1/\sqrt{K\rho C} = 1030$. As we pointed out in the above-mentioned section, the fit is not entirely satisfactory; the predicted cooling rate for the umbral stage is greater than the rate observed. The disagreement did, in fact, lead to consideration of a two-layer model of the moon. Jaeger and Harper [9] adopted the results of Piddington and Minnett's microwave observations as given in equation (71). (They used 610 in place of 640.) They assumed $K_d = 2.8 \times 10^{-6}$, and hence that

$$\tau = \frac{1.71 \times 10^{-3}}{\sqrt{K\rho C}} \tag{72}$$

Compatible values of τ and $\sqrt{K\rho C}$ from equation (72) are shown in Table I. Using these compatible values as parameters, Jaeger and Harper computed eclipse curves by numerical procedures. The theoretical results are compared with observations in Fig. 13; the curves are identified by the value of the parameter $\sqrt{K\rho C}$. The curve with $\tau = 0.17$ cm, $\sqrt{K\rho C} = 1.0 \times 10^{-2}$, provides the best fit to the data. However, considered from the point of view of representing the

TABLE I

Compatible Values of τ and $\sqrt{K\rho C}$
Computed from Equation (72)

Model	τ, cm	$\sqrt{K\rho C}$
II	0.24	7.15×10^{-3}
III	0.17	1.0×10^{-2}
IV	0.12	1.43×10^{-2}
V	0.05	3.33×10^{-2}

observations, this curve is scarcely more satisfactory than the one derived for the homogeneous model with $\sqrt{K\rho C} = 9.71 \times 10^{-4}$. The predicted rate of cooling for the homogeneous model is too fast; that for the two-layer model is too slow. To satisfy the observations we might take some average of the two theoretical curves: it could be partly two-layer, partly deep homogeneous fluff, or it could be mostly two-layer with a small admixture of essentially bare rock with a value of $\sqrt{K\rho C} \sim 0.05$.

Clearly, the case for or against either model is not proved by these observations. More data are required. We therefore consider in the next section evidence from other microwave observations.

THE USE OF MICROWAVE OBSERVATIONS TO TEST THE HOMOGENEOUS AND TWO-LAYER MOON MODELS

Angular Resolution in Radio Observations

The theoretical developments made in the preceding sections all refer to a point on the moon's surface. Unfortunately, observations in the radio wavelength range of, say, 4 mm to 20 cm or more, all refer to large or relatively large areas of the moon. The largest antenna now in use for high-frequency radio measurements is the 22-m dish at the Lebedev Physics Institute in the USSR. The angular resolution of this instrument is quite high; the beam diameter for wavelengths of 3.2 cm, 2 cm, and 8 mm is, between half-power points, $6'.3$, $4'$, and $2'$, respectively. However, in the great majority of observations made in the past — observations on which many of the currently quoted results have been based — the beam has encompassed all or at least a very large part of the moon at one time. Thus, the beam of the antenna used by Piddington and Minnet for the 1.25-cm observations quoted earlier was approximately $22'$ between half-power points. Such poor angular resolution causes details to be blurred; it becomes difficult to take the blurring into account in any theoretical treatment; there results a decrease in the certainty with which the numerical values of important physical parameters can be derived.

Adequate interpretation of the results obtained with an antenna of small angular resolution poses a very difficult problem. In principle, we could, for families of n_ν, δ_n values construct mathematical models of the theoretical brightness temperature over the lunar disk, taking all effects fully into account. We might then numerically integrate these models with a specified antenna pattern, a pattern empirically determined for the specific antenna used for the lunar observations. A series of such integrations of theoretical models could be made as a function of time throughout a lunation. We could then determine for series with one n_ν value an a_1, ψ_1 curve (or a group of a_n, ψ_n curves) which would then serve for the interpretation of the observations of the moon. Such a calculation would be enormously time-consuming and difficult even on a modern computer, but it could be done.

In such an integration there would be one question in regard to input data that would be very troublesome to settle in a satisfactory manner at the present time: the moon's latitude distribution of brightness temperature. Simple theory indicates

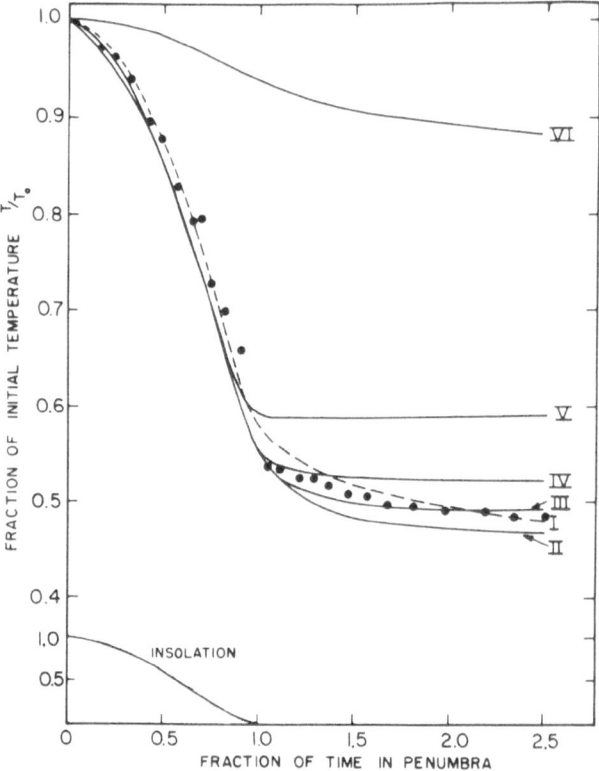

Fig. 13. Jaeger and Harper's calculated eclipse cooling curves for a two-layer model of the moon. Curve I, dotted, is for a homogeneous model of the moon with $(K\rho C)^{-\frac{1}{2}} = 1030$; curve VI is for a homogeneous model with $(K\rho C)^{-\frac{1}{2}} = 20$. Curves II – V are for two-layer models identified more specifically in Table I.

that since insolation will vary as cos ϕ, temperature will vary as $(\cos \phi)^{\frac{1}{4}}$ in accord with the Stefan–Boltzmann law. Observation indicates that T varies more nearly as $(\cos \phi)^{\frac{1}{2}}$ or $a + b(\cos \phi)^{\frac{1}{2}}$, this departure from simple theory occurring presumably because of surface roughness and shadowing. Troitskii [27] has pointed out that the observationally determined constant term in the brightness temperature equation depends upon the angular resolution of the antenna beam and the form of the pole darkening law; this dependence has been used to verify the $(\cos \phi)^{\frac{1}{2}}$ law, but the test is not a critical one. Unfortunately, no determination of the pole darkening law is completely convincing at the present time. Uncertainty in the theoretical temperature distribution over the lunar disk will enter directly in making the results of an integration uncertain and will cause doubt as to the relevance of our models to the real moon. The seemingly simple task of determining the distribution of brightness temperature in the direction of the poles remains a problem of importance and high priority at the present time. We return to this question again in the section entitled "Roughness on the Lunar Surface."

Troitskii [27] has made the most extensive investigation of the effect of antenna beam on observations of the lunar brightness temperature. He has derived, particularly for wide-beam antennas in which the entire moon is encompassed, expressions for the constant temperature term and the amplitude of the fundamental temperature variation. Troitskii has shown, for example, that by integration the amplitude of the temperature variation is reduced to approximately 60% of its value at the center of the lunar disk. However, the exact percentage must depend rather heavily on the precise weighting of the outermost parts of the lunar disk [see Fig. 8]. Troitskii also showed that the phase was rather little affected by integration as one might expect from examination of Fig. 7.

Troitskii's 1954 paper is a classic in the field; its appearance marked a great advance in our understanding of lunar phenomena. However, in the integrations performed, many approximations are made — for example, the variation of the amplitude attenuation factor with θ is not included — hence it is not possible to make detailed use of many of these results in studies of the precision we should be achieving today. It is unfortunate that many of our deductions regarding properties of the moon are based upon integrated disk observations and therefore carry with them great uncertainty. High angular resolution studies made with large precision paraboloids or large interferometers are greatly needed at the present time.

Some Observational Traditions

It has become customary in the analysis of microwave lunar observations to use only the $n = 0$ and 1 terms in the series representing the observations. Thus, traditionally,

$$T_b(\nu, t) = \left[1 - \frac{R_p + R_s}{2}\right]\left[A_0 + \frac{A_1}{(1 + 2\delta_1 + 2\delta_1^2)^{\frac{1}{2}}}\cos(\omega t - q_1 - \psi_1)\right]$$

The reason for this truncation of the series is that to date observations have been relatively inaccurate and made with small antennas, hence the already reduced amplitude A_1/a_1 is generally further reduced by integration over the lunar disk.

Our theoretical treatment readily permits us to compute what amplitudes and phase lags we might expect near the center of the moon for observations at wavelength 1 cm, for which $\delta_1 \sim 2$. We use Sinton's [26] values of the amplitudes of the terms with $n = 1, 2, 3$ in the Fourier series expansion of the surface temperature to estimate the amplitudes of the corresponding terms in the microwave brightness temperature variation. We find, for a homogeneous model of the moon,

Fundamental: expected temperature amplitude = $41°$; phase lag = $33°.7$

First harmonic: expected temperature amplitude = $7°$; phase lag = $36°.5$

Second harmonic: expected temperature amplitude = $5°$; phase lag = $37°.8$

Detection of terms with amplitudes of 7° and 5° will be difficult even if a beam of high angular resolution is used. Use of beams of low angular resolution will make the task even more difficult. There is small wonder that in earlier studies the fundamental has been adequate to represent the observations. It is encouraging to note, however, that recent radio observations at 4 mm by Kisliakov and Salomonovich [10] have shown up to the second harmonic.

In establishing the variation of the lunar radiation as a function of phase, we encounter one of the most difficult problems in radio astronomy: the measurement of a temperature on an absolute scale. Moreover, when we test the adequacy of the homogeneous model to represent the microwave observations, we will want to test the consistency of the amplitude attenuation factors and their associated phase lags for a variety of wavelengths. Thus, we are asking for consistency in temperature calibration for a variety of wavelengths. An examination of Table II, which lists published results on lunar temperature measurements collected by Mayer [17], will convince one of the lack of satisfactory calibration among currently available data. A completely new attack on this problem must be made if we are to make progress in understanding the nature of the lunar surface.

Lack of consistent and adequate reduction procedures also contributes to the scatter exhibited in Table II. It is impossible to tell in many instances how the observations were reduced and what corrections have been applied.

A Test of the Homogeneous Model Using Microwave Data

To test for the adequacy of the homogeneous model we plot the observed phase lag ψ_1 as a function of the observed amplitude attenuation parameter a_1. If the plotted points define the theoretical a_1, ψ_1 curve shown in Fig. 9, the homogeneous model adequately represents the observations. If the plotted a_1, ψ_1 points lie consistently above the theoretical a_1, ψ_1 curve indicating that for a given a_1 value the phase lag is larger then predicted by the homogeneous model, we must reject the homogeneous model and adopt some other model. Presumably we would then consider as our next alternative the two-layer model, the only other model now fully worked out mathematically. Failure of the homogeneous model should not be taken as automatically indicating success of the two-layer model; quite a different model, not yet fully discussed, might, in fact, be preferable to both.

To make the desired test we shall use the observational data tabulated by Troitskii [28]. With Troitskii, we shall consider the ratios $A_0/(A_1/a_1)$ which we will then divide by A_0/A_1 derived from surface temperature observations. Troitskii [27] states that this type of comparison is free from many errors of calibration and is more reliable than the ratio $A_1/(A_1/a_1)$. Since in establishing an observational temperature system both a zero point and a scale must be set up, there is no assurance that errors in one compensate errors in the other. In fact, it might be simpler to establish an accurate scale than a correct zero point, hence, better to compare amplitudes than quantities involving constant terms. For these reasons we do not find Troitskii's proposition entirely convincing, but for reasons enumerated in the section entitled "Amplitude Attenuation and Phase Effects in the Microwave Temperature Equation," we do take the same ratio.

TABLE II

Radio Observations of the Moon

Wave-length, cm	Apparent blackbody temperature,* $^{\circ}$K	Uncertainty	Observer
0.43	T_c = 170 to 290	25% (p.e.)	Coates (1959)
0.75	T_d = 125 to 175	....	Mitchell and Whitehurst (1958)
0.80	T_c = 197 − 32 cos $[\omega t - 40(\pm 5)^{\circ}]$	10%	Salomonovich (1958)
0.86	T_c = 150 (single observation)	40%	Hagen (1949)
0.86	T_c = 145 to 225	....	Gibson (1958)
1.25	T_d = 270 (single observation)[†]	....	Dicke and Beringer (1946)
1.25	T_c = 249 − 52 cos $(\omega t - 45^{\circ})$[‡]	5%	Piddington and Minnett (1949)
1.63	T_c = 224 − 36 cos $(\omega t - 40^{\circ})$	$\pm$10−15%	Zelinskaya, Troitskii, and Fedoseev (1959)
2.20	T_c = 200 (constant to $\pm 10^{\circ}$K)	5%	Grebenkemper (1958)
3.15	T_c = 195 − 12 ($\pm$5) cos $[\omega t - 44(\pm 15)^{\circ}]$	$\pm 25^{\circ}$K (p.e.)	Mayer, McCullough, and Sloanaker (unpublished)
3.20	T_d = 183 (constant to $\pm 9^{\circ}$K)	$\pm 13^{\circ}$K	Zelinskaya and Troitskii (1956)
3.20	T_d = 170 (constant to $\pm 12^{\circ}$K)	$\pm$20%	Troitskii and Zelinskaya (1955)
3.20	T_d = 133 (constant to $\pm 10^{\circ}$K)	$\pm 20^{\circ}$K	Kaidenovsky, Turusbekov, and Khaikin (1956)
10.0	T_d = 130 (single observation)	20%	Kaidenovsky, Turusbekov, and Khaikin (1956)
10.0	T_c = 315 (constant to $\pm 50^{\circ}$K)	$\pm 50^{\circ}$K	Akabane (1955)
10.3	T_d = 207 (single observation)	$\pm 27^{\circ}$K	Sloanaker
21.0	T_c = 250 (constant to $\pm 5^{\circ}$K)	12%	Mezger and Strassl (1959)
22.0	T_d = 270 (single observation)	$\pm 60^{\circ}$K	Westerhout (1958)
32.0	T_d = 246 (constant to $\pm$5%)	$\pm 40^{\circ}$K	Ko
33.0	T_d = 220 (constant to $\pm 9^{\cup}$K mean dev.)	$\pm 33^{\circ}$K	Denisse and LeRoux
75.0	T_d = 160 (single observation)	....	Seeger, Westerhout, and van de Hulst (1956)
75.0	T_d = 185 (constant to 10%)	....	Seeger, Westerhout, and Conway (1957)

* T_c is the radio brightness temperature at the center of the disk; T_d is the radio brightness temperature averaged over the disk.

[†] Corrected by Piddington and Minnett (1949).

[‡] Corrected for an emissivity of 0.9 by the observers.

Corresponding values of ψ_1 and a_1 derived from the tabulation prepared by Troitskii for a variety of wavelengths are listed in Table III. In computing the values listed we have taken A_0/A_1 = 1.5. The results are plotted in Fig. 14. Though the values for the longest wavelengths lie generally above the curve, we

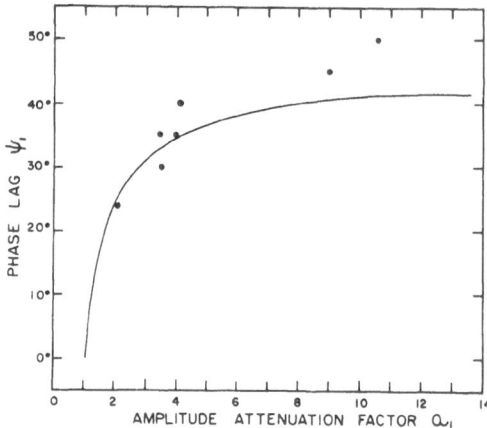

Fig. 14. Test of the two-layer model of the
moon by means of microwave observations. If
the homogeneous model serves to represent the
microwave data, the points lie on the curve; if
a two-layer model is more appropriate, the
points will lie systematically above the curve.

must conclude, with Troitskii, that the data listed by him as the most reliable
indicate that the homogeneous model of the moon appears to represent the observations reasonably adequately at the present time. Clearly, however, the test
we have performed is not a highly discriminating one. More work is required and
further investigation of other models is required, particularly in representation of
eclipse data. We shall touch upon this topic again in the section entitled "New
Models of the Moon."

POLARIZATION OF LUNAR RADIATION

An important feature of equation (52) is the factor $[1 - (R_p + R_s)/2]$, which
stands outside the series. At every point on our smooth homogeneous model moon
except the center $\theta = 0$, and the limb $\theta = \pi/2$, R_p differs from R_s, hence the
emitted lunar radiation is polarized. Since the Fresnel coefficients are outside

TABLE III
Amplitude Attenuation Factors and Phase Lags

λ, cm	a_1	ψ_1, deg	λ, cm	a_1	ψ_1, deg
3.2	10.5	50	0.86	3.4	35
3.2	9.0	45	0.8	4.1	40
1.63	4.15	40	0.8	3.5	30
1.25	4.0	35	0.4	2.1	24

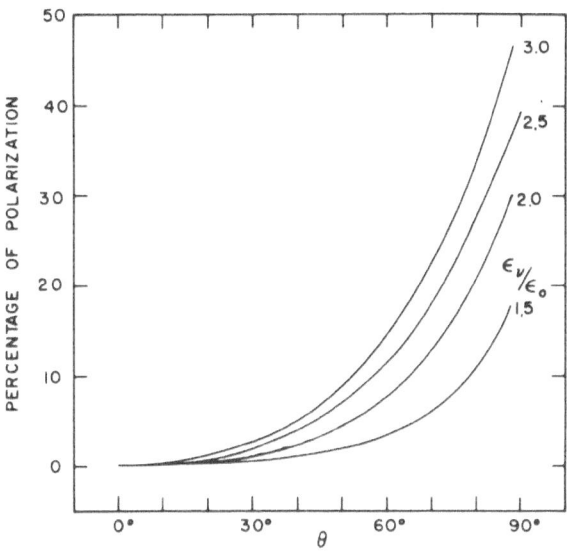

Fig. 15. Polarization as a function of θ for different
values of ϵ_ν/ϵ_0.

the integral, the percentage of polarization at any point is independent of phase
and provides an extremely important means of estimating the dielectric constant
of the lunar material. The polarization effect is not a small one. We show in
Fig. 15, as a function of apparent radius of the moon, the percentage of polarization expected for the cases ϵ_ν/ϵ_0 = 1.5, 2.0, 2.5, and 3.0. We assume that the antenna used has infinite resolving power. If an antenna of finite angular resolving
power is used, some integration of the curves in Fig. 15 will result.

To illustrate further the role of polarization, we consider a hypothetical
smooth moon made of dielectric material of uniform temperature and having a dielectric constant ϵ_ν/ϵ_0 = 3.0. In Fig. 16 we show (a) the predicted distribution

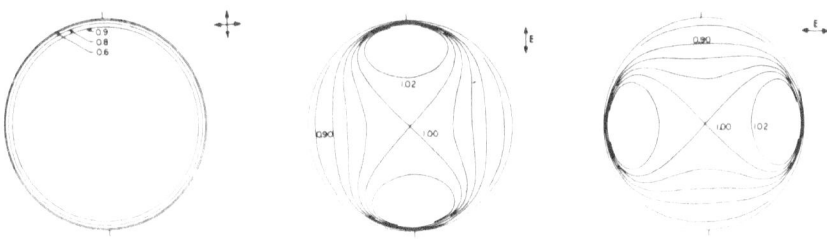

Fig. 16. Calculated brightness temperatures for an isothermal smooth moon for (a)
an antenna accepting both modes of polarization and for antennas accepting (b) the
vertical mode and (c) the horizontal mode of polarization; ϵ_ν/ϵ_0 = 3.0.

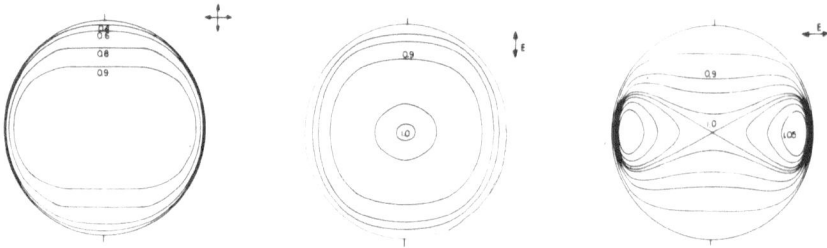

Fig. 17. Calculated brightness temperatures for a moon with pole darkening according to the law $T \propto \cos^{1/2} \phi$ for (a) an antenna accepting both modes of polarization
and for antennas accepting (b) the vertical mode and (c) the horizontal mode of
polarization; $\epsilon_\nu/\epsilon_0 = 3.0$.

of brightness temperature of the isothermal moon if we could observe it with a
hypothetical antenna that accepted both modes of polarization, (b) the predicted
brightness temperature if the antenna accepted radiation with its E-vector vertical,
and (c) the predicted brightness temperature if the antenna accepted radiation with
its E-vector horizontal.

A more realistic model of the moon will have a more complicated distribution
of brightness temperature. We assume our model moon, for which $\epsilon_\nu/\epsilon_0 = 3.0$, to
be pole-darkened in accord with the customary form of the law $T \propto \cos^{1/2} \phi$. In
Fig. 17 we show the predicted brightness temperature (a) for a hypothetical antenna
accepting both modes of polarization, (b) for an antenna accepting radiation with its
E-vector vertical, and (c) for an antenna accepting radiation with its E-vector horizontal.

The diagrams presented are drawn for an antenna of infinite resolving power;
for an antenna of finite resolving power, the pictures must be somewhat blurred or
smeared out. For an antenna of finite resolving power there may also be some
change in measured polarization as a function of lunar phase. There is no change
in polarization at any one point on the moon's surface with phase, but each incre-
ment of area within the beam has a different polarization; these areas are differently weighted as first one, then another, is at the higher temperature during a
lunation.

It is clear from these diagrams that in polarization we have a very powerful
means of investigating the lunar surface. We may determine percentage of polarization as a function of radius, or we may observe the distribution of brightness
temperature with different orientations of the E-vector.

DETERMINATION OF THERMAL AND ELECTRICAL PARAMETERS
OF THE MOON

In this section we consider some of the methods of evaluating the principal
thermal and electromagnetic parameters of the lunar surface material. No effort
will be made to produce an exhaustive list of all parameters or all possible ways
of evaluating them. Rather, our purpose is to call attention to the physical nature

of the more frequently encountered parameters and to indicate the types of observations needed for their evaluation. In the process of deriving precise parameter values, nothing is more important than good observing techniques, precise instrumental calibrations, and proper data reduction procedures; however, consideration of these problems lies outside the scope of this paper.

The Thermal Inertia $(K\rho C)^{1/2}$

Three types of observational data provide practical information on the value of $(K\rho C)^{1/2}$: (1) infrared observations of the temperature variation during a lunar eclipse, (2) infrared observations of the surface temperature of the moon during a lunation, and (3) variation of microwave temperature during a lunation. Of these, the eclipse measurements appear to provide the most critical set of observations to be fitted, though eventually the results throughout a lunation may be of equal accuracy. Because of observational difficulties of the type discussed earlier, the microwave measurements would appear to be the least trustworthy at the present time.

To discuss the eclipse results, we solve equations (4) with the boundary conditions (5); $(K\rho C)^{1/2}$ is a parameter in the solution. The proper value of the thermal inertia is the one which provides the best fit between the observations and the theoretical curve. Jaeger and Harper [9] found $(K\rho C)^{-1/2}$ between 1370 and 1030. It now appears that these values may be in need of revision because of errors in the temperature scale used for the observations. With a rescaling of the temperatures, the best value of $(K\rho C)^{-1/2}$ decreases to possibly 600 or less. This entire analysis needs to be repeated with an improved temperature scale, or, preferably, with new eclipse observations.

The measurements during a lunation are discussed in the same general way as the eclipse data. However, during the dark part of a lunation the temperature becomes so low ($\sim 120°K$) that observational difficulties arise. There are very few observed values in this part of the curve to which one can fit a theoretical function. Wesselink [30] fitted a curve to the temperature data for a lunation with $(K\rho C)^{-1/2} = 920$, but Jaeger [7, 8] states that because of the large uncertainty in the value of the temperature for lunar midnight, the value of $(K\rho C)^{-1/2}$ may lie anywhere in the range 200 to 1000. For the curve fitted to Sinton's observations in Fig. 3, the value $(K\rho C)^{-1/2} - 435$ has been used.

Krotikov and Troitskii [12] have combined surface temperature theoretical data and radio observations in an interesting way to find the value of $(K\rho C)^{-1/2}$. By numerical methods Krotikov and Troitskii obtain a family of solutions for the temperature variation during a lunation, with $(K\rho C)^{-1/2}$ as the parameter of the family. They represent each member of the family by the Fourier series, equation (6), obtaining the coefficients $A_0, A_1, \ldots, A_n$ as functions of $(K\rho C)^{-1/2}$. They plot $A_0[(K\rho C)^{-1/2}]$, $A_0[(K\rho C)^{-1/2}]/A_1[(K\rho C)^{-1/2}]$, and $T_m[(K\rho C)^{-1/2}]$, where T_m is the surface temperature at lunar midnight. Krotikov and Troitskii then analyze the microwave data to determine the best values of A_0, A_0/A_1, T_m, for a variety of wavelengths and extrapolate to zero wavelength. Unfortunately, the radio data involve averaging over a substantial area of the lunar disk because of finite antenna resolving power; this reduces the reliability of the final results since from the space-averaged observed values, values of A_0, T_m, and A_0/A_1 characteristic of the center of the moon must be inferred. This is difficult. By combining radio observations with surface temperature theory, Krotikov and Troitskii find $(K\rho C)^{-\frac{1}{2}}$ = 350 ± 75. This appears to be at variance with eclipse results. Additional work, both theoretical and observational, is required here.

The Value of the Dielectric Constant of the Lunar Material

The dielectric constant of the lunar surface material appears in the discussion when microwave brightness temperatures are discussed. Analyzing equation (52), which relates the brightness temperature in the microwave range to the surface temperature of the moon, we note that the relative dielectric constant $\epsilon\nu/\epsilon_0 = n_\nu^2$ is involved in the equation in three ways: (1) in the multiplicative factor $[1 - (R_p + R_s)/2]$, (2) in the amplitude attenuation factor $a_n(\theta)$, and (3) in the phase $\psi_n(\theta)$. There are very many ways in which these can be utilized and combined to evaluate $\epsilon\nu/\epsilon_0$:

Probably the simplest way to derive a value of $\epsilon\nu/\epsilon_0$ is to compare (at the center of the lunar disk) the constant temperature term A_0 as derived from infrared measurements, and the constant temperature term $[1 - R(0)]A_0$ as determined from radio observations in the shorter-wavelength range. The ratio of the two gives $R(0)$ from which n_ν or $\sqrt{\epsilon\nu/\epsilon_0}$ can be immediately found. Mayer [17] by this method estimated $\epsilon\nu/\epsilon_0 \approx 3.5$. This appears to be very high compared to more recently determined values. Using the same principles, but combining a theoretical surface temperature with radio observations, Troitskii [28] found $\epsilon\nu/\epsilon_0 = 1.5 \pm 0.3$.

At the limb of the lunar disk emissivity is lower than at the center of the disk; the multiplicative factor $\{1 - [R_p(\theta) + R_s(\theta)]/2\} \longrightarrow 0$ as $\theta \longrightarrow \pi/2$. Salomonovich [23] used the 22-m Lebedev telescope to scan the moon at wavelength 8 mm; from the decrease in emissivity toward the limb he found $1 \leq \epsilon\nu/\epsilon_0 \leq 2$. Cudaback [5], from scan-beam scans at a wavelength of 9 cm found $\epsilon\nu/\epsilon_0 = 1.1$ to 1.3. In employing this particular approach involving measurement of change of emissivity, one should, if possible, scan along the lunar equator to avoid confusion with the pole darkening law. Unfortunately, in practice, such an equatorial scan is not always possible.

A method that is more promising than either of the above procedures utilizes the polarization of lunar radiation to derive the dielectric constant. Cudaback [6], observing at a wavelength of 3.75 cm, has employed this method to find $\epsilon\nu/\epsilon_0 \sim 1.5$. As angular resolution becomes higher, measurement of polarization will undoubtedly become one of the most useful methods for investigation of the dielectric constant and other properties of the lunar surface.

A method that is also of potentially great value as angular resolution grows larger, is suggested by equations (58), through which one can determine both $\delta_1(0)$ and $\epsilon\nu/\epsilon_0$. These equations, it will be recalled, specify the change in the amplitude attenuation factors, $a_n(\theta)$, as a function of θ. For best results one should have infrared measurements corresponding to the same positions on the moon at which the microwave measurements are made so that highly accurate values of $a_n(\theta)$ can be found.

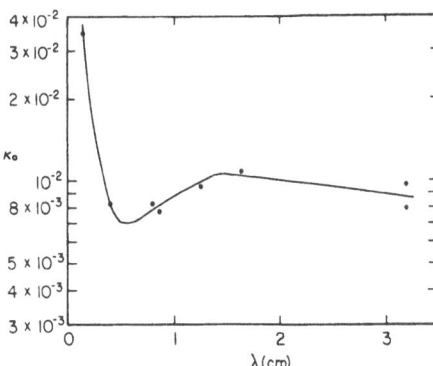

Fig. 18. The extinction index, $\kappa_{0,\nu}$, computed from observations of the moon, shown as a function of wavelength.

Absorption of Radiation in the Lunar Material

The second electromagnetic parameter of the lunar surface material, $\kappa_{0,\nu}$ or σ_ν or $\tan\Delta_\nu$, whichever way one wishes to specify it, can be derived from $\delta_n(0)$ which, in turn, we can find from solution of equations (58) or from the position of observed a_n, ψ_n points on the theoretical curve. Of the three descriptive parameters mentioned above, we shall use the extinction index, $\kappa_{0,\nu}$, in this discussion. It can, of course, be converted to any of the other parameters at once.

From equation (48)

$$\kappa_{0,\nu} = \sqrt{\frac{\omega}{2}} \frac{C}{\sqrt{K\rho C}} \frac{\lambda_0\rho}{4\pi\delta_1(0)} \tag{73}$$

To provide an example of the determination of $\kappa_{0,\nu}$ we have, for the series of points shown in Fig. 14, read off the values of $\delta_1(0)$ for each wavelength. In reading these off we have taken that value of $\delta_1(0)$ which minimizes the distance of the observed point from the theoretical curve. For other lunar parameters in equation (73) we take $\sqrt{K\rho C} = 10^{-3}$, $C = 2 \times 10^{-1}$, $\rho = 1$. The derived values of $\kappa_{0,\nu}$ are plotted as a function of wavelength in Fig. 18.

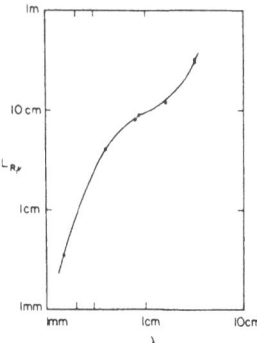

Fig. 19. The penetration depth, $L_{R,\nu}$, corresponding to the values of extinction index shown in Fig. 18, as a function of wavelength.

The values of $\kappa_{0,\nu}$ displayed in Fig. 18 are systematically uncertain because of uncertainties in the numerical values of $\sqrt{K\rho C}$, C, and ρ. The value
of C has a rather small uncertainty, but the assumed value of $\sqrt{K\rho C}$ we have
taken is a minimum, while that of ρ is probably a maximum. The derived values
of $\kappa_{0,\nu}$ are thus probably maximum ones; they may be smaller by a factor of 4.

Over the wavelength range covered by these results, $\kappa_{0,\nu} \ll 1$; clearly then,
the lunar material acts as a good dielectric, and we are justified in employing the
refraction and reflection formulas for a lossless dielectric, at least for wavelengths longer than 1 mm.

Using the $\kappa_{0,\nu}$ values plotted in Fig. 18, we can calculate the depth of penetration for radiation, $L_{R,\nu}$, from equation (38): $L_{R,\nu} = \lambda_0/4\pi\kappa_{0,\nu}$. The resultant
values are plotted in Fig. 19. We note that if the $\kappa_{0,\nu}$ values are systematically
reduced by a factor of 4, the penetration depths are increased by the same factor.

Curves of the type shown in Fig. 18, especially if the index of refraction can
be determined and plotted on the same diagram, can, in principle, make possible
the identification of the material in the upper crust of the moon. The features in
the κ_0 plot are analogous to optical spectral features; they are characteristic of
specific substances (or types of bonds within molecules) as are spectral lines or
bands. They are, of course, broad; precise identification, particularly of mixtures,
may be impossible.

Unfortunately, the observational data on which Fig. 18 is based are few in
number and not strongly determined. However, their trend is not what would be expected from conduction alone; they do, in fact, appear to indicate what we might
expect from a molecular or crystal lattice resonance. With increasing accuracy of
observations, it is probable that determinations of $\kappa_{0,\nu}$ will provide important information on the composition and condition of the lunar material.

HEAT FLOW THROUGH THE LUNAR SURFACE

A number of authors have considered the question of a thermal gradient in the
moon arising from radioactivity or remaining from an earlier thermal condition. A
test for a thermal gradient in the deeper layers of the moon is simple. Since radiation of long wavelengths penetrates the lunar material deeper than that of short
wavelengths, we look for an increase in the constant component of the lunar radiation with wavelength at, say, the moon's center. Alternatively, we might choose a
relatively long wavelength (> 20 cm if possible) and look along the lunar equator for
limb-darkening. Both of these effects would, if found, indicate within the moon a
temperature increasing with depth below the surface. Krotikov and Troitskii [13]
have presented observational data (Fig. 20) which indicate that just such an effect
is observed.

In this discussion we will largely follow Troitskii's development of the theory,
though we will consider somewhat different aspects of the problem than those he
treated.

We postulate that there exists a thermal gradient in the body of the moon; the
temperature, as a function of depth x below the lunar surface, is given by the linear
equation

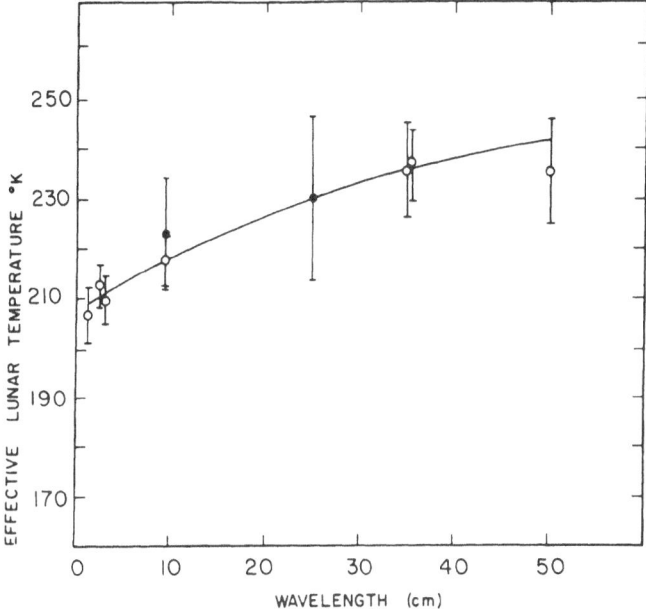

Fig. 20. Wavelength dependence of the effective lunar temperature as shown by Krotikov and Troitskii [13] (open circles). The
two filled circles show the results of well-calibrated measurements made by Castelli (AFCRL-64-313, April 1964) and indicate
the uncertainty of the slope of the curve.

$$T(x) = T_0 + \frac{\partial T}{\partial x} x \tag{74}$$

where $\partial T/\partial x$ represents the average temperature gradient to at least one or two
times the penetration depth of radiation of wavelength λ, at which we observe. In
accord with the methods used previously in the section entitled "The Microwave
Temperature of the Sun," we write for the brightness temperature of wavelength λ
escaping the surface at point l, ϕ

$$T_\lambda(l, \phi) = \left(1 - \frac{R_p + R_s}{2}\right) \int_0^\infty T(x) e^{-k_\nu \rho x \sec \chi} k_\nu \rho \sec \chi \, dx$$

$$= \left(1 - \frac{R_p + R_s}{2}\right)\left(T_0 + \frac{1}{k_\nu \rho \sec \chi} \frac{\partial T}{\partial x}\right) \tag{75}$$

or, since by Snell's law $\sin \theta = n_\nu \sin \chi$

$$T_\lambda(l, \phi) = \left(1 + \frac{R_p + R_s}{2}\right)\left(T_0 + \frac{\sqrt{n_\nu^2 - \sin^2\theta}}{k_\nu \rho n_\nu} \frac{\partial T}{\partial x}\right) \tag{76}$$

We shall so frequently encounter the multiplicative factor $[1 - (R_p + R_s)/2]$ that we abbreviate it $F(n_\nu, \theta)$, the variables being inserted to call sharp attention to the fact that it is a function of n_ν (or $\sqrt{\epsilon_\nu/\epsilon_0}$) and the angle of view θ. Some interpretation must be made in using this Fresnel factor $F(n_\nu, \theta)$, for emissivity. We wrote it symbolically as $1 - (R_p + R_s)/2$, which specifies the total fraction of radiation escaping the surface. However, since we use a real antenna in our observations and only one mode of polarization is accepted, we must interpret $F(n_\nu, \theta)$ to mean the appropriate direction of polarization for our particular observation. Thus, rewriting equation (76) with $F(n_\nu, \theta)$, we find

$$T_\lambda(l, \phi) = F(n_\nu, \theta)\left(T_0 + \frac{\sqrt{n_\nu^2 - \sin^2\theta}}{k_\nu \rho\, n_\nu} \frac{\partial T}{\partial x}\right) \tag{77}$$

If, in accord with equation (17), we make use of the concept of penetration depth $L_{R,\nu}$, for radiation of frequency ν,

$$T_\lambda(l, \phi) = F(n_\nu, \theta)\left(T_0 + L_{R,\nu} \frac{\sqrt{n_\nu^2 - \sin^2\theta}}{n_\nu} \frac{\partial T}{\partial x}\right) \tag{78}$$

At the center of the moon, $\theta = 0$, the brightness temperature is

$$T_\lambda(0, 0) = F(n_\nu, 0)\left(T_0 + L_{R,\nu} \frac{\partial T}{\partial x}\right)$$

that is, it is greater than the surface temperature T_0 by an amount that depends upon the product of the radiation penetration depth and the thermal gradient. On the lunar equator and on the limb of the moon, the brightness temperature is

$$T_\lambda\left(\frac{\pi}{2}, 0\right) = F\left(n_\nu, \frac{\pi}{2}\right)\left[T_0 + L_{R,\nu} \frac{\sqrt{n_\nu^2 - 1}}{n_\nu} \frac{\partial T}{\partial x}\right]$$

For the lunar material $n_\nu^2 \sim 1.6$, hence along the lunar equator there will be a decrease in brightness temperature by approximately $0.39\, L_{R,\nu}(\partial T/\partial x)\, F(n_\nu, \pi/2)$, as we go from center to limb. Unfortunately, this limb-darkening will not be very large even if we observe at a relatively long wavelength of 30 cm, but, if measured with care, it will permit us to determine the thermal gradient.

As an example we compute the limb-darkening effects to be expected and show how they might be used to derive the value of $\partial T/\partial x$ from observations at one wavelength, say, 30 cm. For this example we take from Fig. 20 $L_{R,\nu}(\partial T/\partial x)$ $= 25°\text{K}$ for $\lambda = 30$ cm. With this value and $n_\nu = \sqrt{1.6}$, we compute the limb-darkening factor $(\sqrt{n_\nu^2 - \sin^2\theta}/n_\nu)\, L_{R,\nu}(\partial T/\partial x)$ as a function of θ; this factor is exhibited in Fig. 21a. We then adopt $T_0 = 210°\text{K}$, compute the Fresnel factor $F(n_\nu, \theta)$ for both p and s polarizations and derive $T_\lambda(\theta, 0)$; this is plotted in Fig. 21b. Clearly the Fresnel emissivity factor is the dominant one, but if one does measure both p and s polarizations along the equator, one can determine the

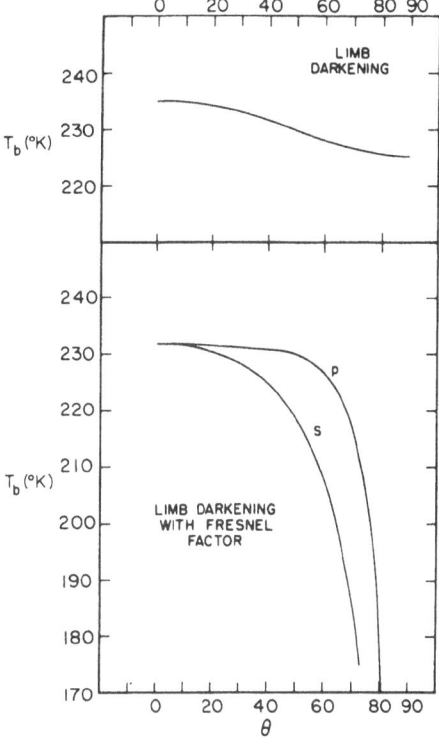

Fig. 21. Limb-darkening computed for observations made at a wavelength of 30 cm.

quantity $L_{R,\nu}(\partial T/\partial x)$. Effectively, one determines the value of n_ν (or ϵ_ν/ϵ_0) from the polarization; combining n_ν with the observations, one determines the limb-darkening function as shown in Fig. 21a, and from the limb-darkening function solves for $L_{R,\nu}(\partial T/\partial x)$. The measurement will be a difficult one; we are trying to determine only a few degrees of temperature variation between the center of the disk and θ about 70°, which is near the practical limit of measurement in θ. If λ is as small as 10 cm, the center to $\theta = 70°$ limb-darkening is of the order of 3°K.

Troitskii has suggested another approach in which observations at only the center of the moon are utilized, but these must be made in at least two wave-lengths. If we make such observations in two wavelengths, λ_1 and λ_2, we have

$$T_{\lambda_1}(0,0) = F(n_{\nu 1}, 0)\left(T_0 + L_{R,\nu 1}\frac{\partial T}{\partial x}\right)$$

$$T_{\lambda_2}(0,0) = F(n_{\nu 2}, 0)\left(T_0 + L_{R,\nu 2}\frac{\partial T}{\partial x}\right)$$

(79)

To a very high degree of approximation we may place $F(n_{\nu 1}, 0) = F(n_{\nu 2}, 0) = F(n_\nu, 0)$ and write

$$T_{\lambda_1}(0,0) - T_{\lambda_2}(0,0) = F(n_\nu,)\frac{\partial T}{\partial x}(L_{R,\nu 1} - L_{R,\nu 2}) \tag{80}$$

If we make use of Troitskii's empirical observation that $\delta_1 = 2\lambda$ [this follows at once from equation (48) if $\kappa_0 \sim$ constant as we see it is for $\lambda > 4$ mm], where λ is measured in cm, and combine this with equation (21) that $L_{T,1}\delta_1 = L_{R,\nu}$, we have from equation (80)

$$\frac{\partial T}{\partial x} = \frac{T_{\lambda_1}(0,0) - T_{\lambda_2}(0,0)}{\lambda_1 - \lambda_2}\frac{1}{2F(n_\nu,0)L_{T,1}} \tag{81}$$

Or, if we make use of equation (10) for $L_{T,1}$,

$$\frac{\partial T}{\partial x} = \frac{T_{\lambda_1}(0,0) - T_{\lambda_2}(0,0)}{\lambda_1 - \lambda_2}\frac{1}{2F(n_\nu,0)}\sqrt{\frac{\omega}{2}}\frac{\sqrt{K\rho C}}{K} \tag{82}$$

All quantities on the right-hand side of equation (82) are known from observation or theory, hence $(\partial T/\partial x)$ can be computed. From the curve for $T_\lambda(0,0)$ (Fig. 20) we find the average value of $\Delta T/\Delta \lambda$ to be 0.8 deg/cm. Combining this with numerical values of the other parameters favored by Krotikov and Troitskii, we calculate $\partial T/\partial x = 1.6$ deg/m.

The heat flow out from the moon, Q, in cal/cm^2-sec, is given by $K(\partial T/\partial x)$ or, in terms of equation (82),

$$Q = \frac{T_{\lambda_1}(0,0) - T_{\lambda_2}(0,0)}{\lambda_1 - \lambda_2}\frac{1}{2F(n_\nu,0)}\sqrt{\frac{\omega}{2}}\sqrt{K\rho C} \tag{83}$$

Krotikov and Troitskii compute

$$Q = 1.3 \times 10^{-6} \text{ cal/cm}^2\text{-sec}$$

or, from the entire moon, 1.6×10^{19} cal/year. Presumably this heat is radiogenic in origin. We compute for the specific heat flow of the moon 2.2×10^{-7} cal/g. This value is large — 4 to 6 times greater than laboratory values reported for chondrites, and 6 to 7 times greater than the value observed for the earth; such a large value requires that there be a high concentration of radioactive materials in the lunar material. The value of the temperature of the lunar interior depends upon the distribution of radioactive materials throughout the moon. If, with Krotikov and Troitskii, we assume the lunar crust to consist of granite, the most radioactive terrestrial rock, which releases 7×10^{-6} cal/year, we find a granite layer 20 km thick accounts for the lunar specific heat flow. If, as in the earth, there is a mixture of granite and basalt, the layer must be about 50 km thick. If there is no further heat generation below the under surface of this layer, the temperature there

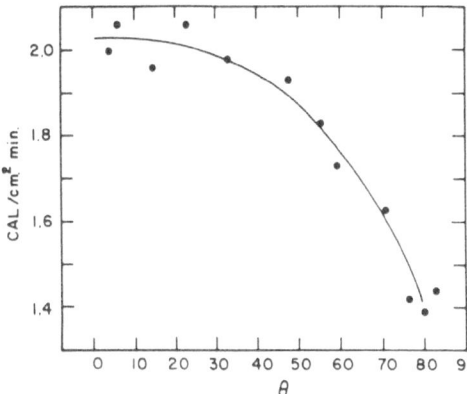

Fig. 22. The measured brightness of the subsolar point as a function of the angle θ of the subsolar point. The data have been taken from Sinton [26].

and at greater depths will be approximately 1000°K. Before strong conclusions are based on these calculations, however, further observational work in both the infrared and microwave regions is called for.

ROUGHNESS ON THE LUNAR SURFACE

The moon obviously has a very rough surface, which on the large scale is everywhere visible to the eye. On the small scale, we have very little direct information on what the surface of the moon is like. Observationally, our resolving power does not permit us to see lunar objects smaller than the range of possibly 0.1 mile in diameter. Roughness is a dominant factor in the brightness distribution over the moon and very greatly complicates the simple smooth-sphere theory developed in the preceding sections.

Figure 22 provides an example of the effect of roughness on emitted energy as measured in the infrared. The data have been taken from the review by Sinton [26]. The diagram displays the measured thermal energy from the subsolar point as a function of θ, which is equivalent to the longitude of the subsolar point measured from the center of the lunar disk as zero. If the moon were smooth, there would be no brightness variation of the subsolar point with θ. Surface roughness, which causes shadows and obstructs our view of illuminated areas of the lunar surface, provides the most direct and feasible explanation of the effect shown.

A second illustration of the effects of roughness is shown in Fig. 23, taken from the paper by Pettit and Nicholson [20]. The figure displays measured infrared brightness as a function of sin θ for the instant of time at which the subsolar point coincides with the center of the lunar disk. The observed brightness distribution strongly departs from that predicted for a smooth moon.

When the subsolar point is at the center of the disk, the limb region of the moon is brighter than predicted on the basis of a smooth moon; when the subsolar point is at the limb of the moon, the radiation from the subsolar point is much less than predicted from a smooth-model moon. These observations can be explained qualitatively if we imagine a large number of obstructions (craters, mountains,

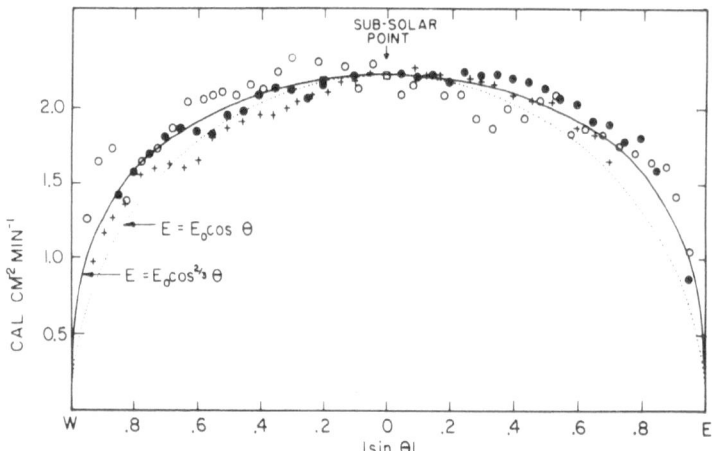

Fig. 23. Measured infrared brightness along the lunar equator for the case in which the subsolar point is at the center of the lunar disk. The data are from Pettitt and Nicholson [20].

mounds, ...) which we idealize as cylinders of all sizes protruding from the surface of the moon. For such objects located at the limb of the moon, with the sun directly in their zenith, we see the unilluminated sides. Moreover, when so viewed from the side they stand in front of illuminated areas; our view is obstructed by them. Under such conditions we receive less energy from the limb region of the moon than we would if the moon were smooth. When the subsolar point is at the center of the moon, the cylindrical protuberances in the limb regions of the moon are quite fully illuminated; they are warmed; they radiate more energy than we would predict for a smooth moon.

It is mathematically simple to construct a statistical model of the lunar surface from which protrude N such cylindrical bodies of average diameter $2r$ and average height h. These upright cylinders are distributed randomly over the surface of the moon. One can then make predictions of the infrared brightness of the lunar surface for various configurations of observer and subsolar point. Unfortunately, and perhaps surprisingly, the observations needed to test adequately such a model are lacking. They very much need to be made.

Pettit and Nicholson represented the observed brightness distribution shown in Fig. 23 by the empirical relation

$$E(\theta) = E(0) \cos^{2/3} \theta \tag{84}$$

There is no theoretical basis for such a cosine power law. The cylindrical-protuberance model of the lunar surface mentioned just previously predicts an expression of quite different mathematical form, but which nevertheless shows much the same general behavior as $\cos^{2/3} \theta$.

For the time of full moon, when we assume the Pettit and Nicholson expression, equation (84), to hold, we can readily predict the observable infrared brightness temperature along any locus of constant latitude. We have that $\cos \theta = \cos l \cos \phi$, where the angle coordinates l, ϕ specify longitude and latitude of a point on the lunar surface. By transformation of equation (84) we find that

$$E(l, \phi) = \cos^{2/3} \phi \cos^{2/3} l \qquad (85)$$

From this expression one might infer that the pole darkening law $P_E(\phi)$ (where the subscript E refers to energy pole darkening as contrasted to $P_T(\phi)$ which signifies the temperature pole darkening law) is given by $\cos^{2/3} \phi$ is, of course, at variance with the commonly accepted, observationally determined, microwave law $P_T = \cos^{1/2} \phi$ or its equivalent (since in the microwave region $T \propto E$).

$$P_E(\phi) = \cos^{1/2} \phi \qquad (86)$$

The difference between equation (85), which we might term the "instantaneous" infrared pole darkening law, and equation (86), the currently used microwave pole darkening laws emphasizes an important feature of the microwave law. The pole darkening law, as it has been understood in what has gone before, involves an integration over both longitude and time; essentially it provides, as a function of latitude ϕ, a measure of the average thermal energy content per gram of lunar material above a depth of $\sim L_{R,\nu}$ in the line of sight.

Geometrical effects of the lunar topography, shadowing, and obscuration all strongly influence the integrated amount of insolation on the lunar material at any particular point and hence influence the thermal energy constant of the lunar material in different regions of the moon. The numerical value of the exponent of $\cos \phi$ in the observed microwave darkening law compared to the numerical exponent which theory predicts for the smooth model indicates that there is rather strong shadowing and obscuration. There are substantial regions of the moon in which shadowing must reduce the insolation markedly below the predicted smooth-moon value. It is not impossible that, because of the nature of the shadow-forming obstructions, different forms of the pole darkening law will be found for infrared and millimeter waves and for centimeter waves. An interesting problem for the future will be the search for such differences. It will also be of interest to establish the mathematical form of the pole darkening law more securely than is now the case.

Since the law depends upon chance geometrical forms and configurations as well as the absorption coefficient of the lunar material (which may change from one area to another), there is no compelling reason to expect a simple cosine power law. It may be considerably more complicated in form and it is essentially a certainty that there will be localized departures from any general expression set up to represent the entire disk.

A second effect of surface roughness will be found in the measured percentage of polarization of lunar radiation. The degree of polarization of emitted radiation will be influenced by the distribution of slopes on the lunar surface. We imagine the surface of the moon to be made up of flat plates which are large compared to the wavelength of radiation in which we make our observations. At any point l, ϕ

on the lunar surface, the normals of the flat plates making up the lunar surface at point l, ϕ are tipped with respect to the l, ϕ zenith direction. Clearly, such tipping will affect the resultant average emissivity and polarization of radiation from point l, ϕ.

To illustrate the effect such a cracked and tilted surface has on polarization, we imagine the region of the lunar equator to be made up of flat plates whose normals point (a) at the zenith or (b) $10°$ from the zenith. [In case (b) the normals are uniformly distributed in the $10°$ circle around the zenith.] Half the area of the moon at any point on the equator is in condition (a); half in condition (b). This model is, of course, artificial, but it will serve to illustrate the effect of roughness. We assume that the material making up this model moon has a dielectric constant $\epsilon\nu/\epsilon_0$ of 3.0. In Fig. 24 we show curves of the predicted percentage of polarization for a smooth moon composed of material having a dielectric constant of (a) 1.5, (b) 2.0, and (c) 3.0. The predicted curve for the rough moon ($\epsilon\nu/\epsilon_0$ = 3.0) is shown as a dotted line. There is rather little difference between the predicted percentage of polarization for the rough moon and the smooth moon of the same dielectric constant except in the central portion of the moon and at the extreme limb where the ratio of the two computed percentages differs significantly from unity. In the central region of the moon where, on the basis of area there is some hope to detect the difference,

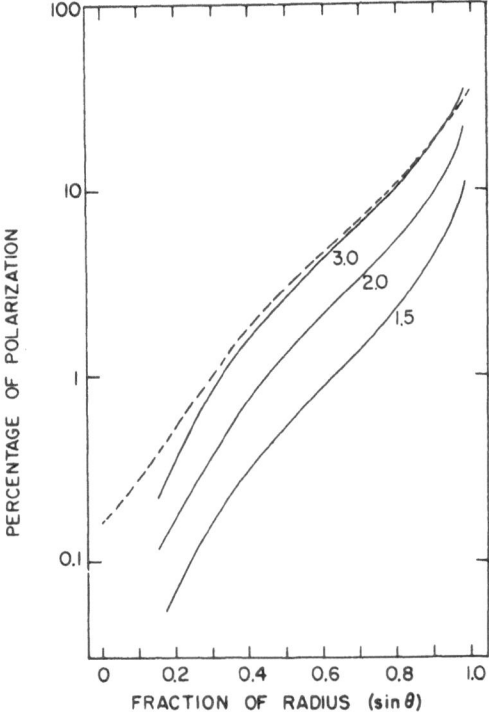

Fig. 24. Percentage of polarization computed for a rough model and a smooth model of the moon.

the percentage of polarization is very small (<1 or 2%), hence detection of the difference is observationally very difficult because of accuracy requirements. In the observable limb region of the moon the curves are quite close together; one would make only a slight error in predicting the dielectric constant from the polarization measurements ignoring the roughness altogether, unless that roughness were rather considerably greater than that of the model we have just studied.

Polarization measurements of high precision and high angular resolution performed at a variety of wavelengths could, in principle, provide information on the scale of roughness, and may provide important information in the future.

The examples given in these calculations indicate some aspects of lunar radiation measurements in which surface roughness plays an important role. A task of high priority is the formulation of a theory of roughness which will satisfy the various boundary conditions that can now be specified. Such a theory should permit us to achieve a better understanding of the nature of the lunar surface, and to derive a consistent set of numerical estimates of the parameters of the lunar surface material.

VARIATIONS ON THE LUNAR SURFACE

In the preceding sections the point of view has been implicit that the moon is uniform in all of its aspects — that one portion of the lunar surface is like every other portion. Optically, this is clearly not so. There are various types of regions, maria, highland areas, craters having special properties such as rays and the like, all easily identified as being quite different. Different areas have different albedos; the range appears to be from approximately 5 to 18%. There are now clear demonstrations that different parts of the moon behave thermally in different ways also. Such different thermal behavior may occur simply because of the landforms involved (Sinton, [25]). Valleys which cannot radiate into space through a complete hemisphere are, under insolation, warmer than adjacent peaks which can radiate to a full hemisphere. On the other hand, if the solar radiation illuminating a valley is at a large angle to the vertical, the valley becomes shadowed and cool while the peaks receive sunshine and are heated. These are merely topographical effects. More important for our consideration here, there appear to be thermally anomalous areas which must be materially quite different in character from surrounding regions.

Localized Thermal Variations on the Lunar Surface

From observations made during the March 13, 1960, lunar eclipse in the 11 μ range, Shorthill, Borough, and Conley [24] found that rayed craters cooled less rapidly than their surrounding regions. In particular, they found Tycho to be 40°K warmer than its surrounding area an hour after it entered the penumbra. The findings for Tycho were later confirmed by infrared measurements made by Sinton [25]. The measurements were later greatly extended by Saari and Shorthill [22], who demonstrated that all five of the rayed craters they examined during the eclipse of September 5, 1960, exhibited anomalous cooling. The effect observed — that the crater remained warmer than its surroundings during the eclipse — was most pronounced for Tycho and was progressively less for Aristarchus, Copernicus, Proclus, and Kepler. Murray and Wildey [18], from their infrared measurements in the 8 to 14 μ range, have identified various regions of the moon which remain anomalously warm during the lunar nighttime; they found the anomalous regions clustered around the craters Tycho and Copernicus. Observations made by Coates [4] at a wavelength of 4.3 mm show that variations in radio brightness over the lunar disk are related to surface features. The maria heat up and cool off more rapidly than surrounding regions. Mare Imbrium is an exception to this statement; Coates found that it always remained cooler than its surroundings. Cudaback [6], observing at a wavelength of 3.75 cm, has found an area of anomalous polarization related, possibly, to some of these same craters found to be anomalous in the infrared.

To illustrate the nature of the effect observed for a rayed crater during eclipse, we exhibit in Fig. 25, adapted from data given by Saari and Shorthill [22], eclipse curves for Aristarchus and its environs.

Possible Explanation of the Observed Anomalies

Various explanations of these thermal anomalies can be suggested. Subsurface thermal sources, possibly volcanic in nature, offer an easy but unlikely explanation. Power requirements to maintain such thermal differences are large; observations in the radio range, though less numerous and of smaller angular resolution than one would like, do not indicate the existence of such subsurface sources.

The higher-than-normal temperature (at least for Tycho) appears to vanish as the lunar night wears on; it therefore appears to be the effect of simple heating

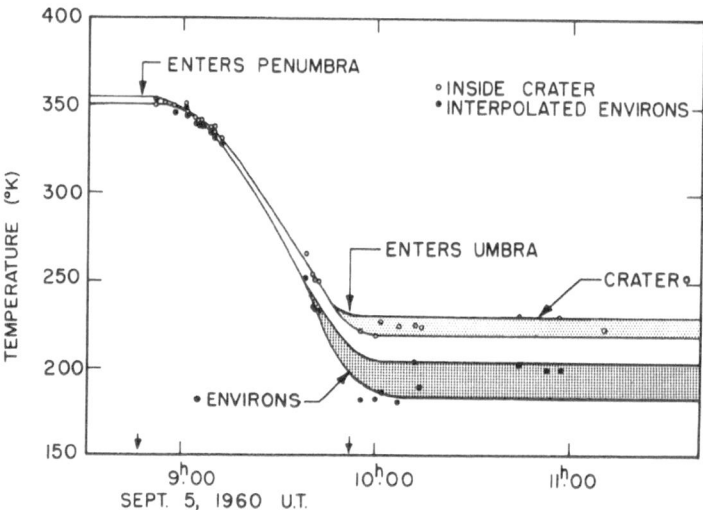

Fig. 25. Eclipse cooling curves for Aristarchus and its surroundings as observed in the infrared by Saari and Shorthill [22].

and cooling, not a continuously existing anomaly. The higher temperature of Tycho observed during lunar night does not appear to result simply from the crater's greater absorption of energy throughout the day. Tycho does, in fact, appear slightly cooler throughout the day, an effect which appears to be related to its albedo, which is measured slightly higher than that of the surrounding area.

The thermal anomalies do not appear to be a result of anomalous emissivities. Unreasonably large and unlikely differences in the emissivities of the crater floor and the surrounding area (in the ratio of approximately 1 to 2) are required to account for the observed temperature differences. Also, the daytime temperatures are not consistent with such differences in emissivity, which therefore do not provide a reasonable explanation of the effect.

The most plausible explanation of the thermal anomalies is found in variations of the thermal conductivity of the lunar surface. The rayed craters are believed to be the youngest ones visible; they represent the results of relatively recent meteor impact. Within a crater and for some distance around the crater, certain areas, possibly secondary impact areas, are presumed to be less densely covered than is normal by the dust that seems to be found everywhere on the moon. In a thinly dusted area — for example, within a crater — the thermal conductivity is higher because the surface is not thickly covered with a highly insulating material. Heat, cycled into the moon during periods of insolation, is more readily cycled out (the area therefore appears warmer) during lunar nighttime or during an eclipse. For such an area of high thermal conductivity, the temperature would be slightly less than normal during insolation periods; more than an average amount of thermal energy is conducted into the moon. However, it is important to note, as pointed out by Murray and Wildey, that the crater floors and the surrounding thermally anomalous areas are not completely free of dust or the dustlike material that covers the moon. The crater floors and surrounding areas do not show anomalous optical polarization or optical brightness-phase anomalies when compared with the general surface of the moon (van Diggelen [29] and Wildey and Pohn [31]). All areas of the moon appear to be covered with a dust layer, though the layer may vary in thickness from one area to another. The most plausible explanation of the thermal anomalies observed for the rayed craters thus leads us directly back to consideration of the two-layer model of the lunar surface, which we now proceed to test as an explanation of the observed cooling curves.

The Two-Layer Model of the Moon Reconsidered

In Fig. 26a are plotted observed eclipse cooling curves for Aristarchus and its surrounding areas. In the same figure are plotted theoretical cooling curves giving predicted fraction of brightness temperature, T/T_0, for smooth homogeneous models of the moon for a variety of values of $(K \rho C)^{-\frac{1}{2}}$. These data are adapted from the paper by Saari and Shorthill [22].

The cooling curves predicted for a homogeneous model of the moon bear very little resemblance to the observed curves; the homogeneous model of the moon with constant parameters clearly does not fit the observations.

In Fig. 26b, which is also adapted from the paper by Saari and Shorthill [22], predicted fractional brightness temperature curves for four models of a two-layer

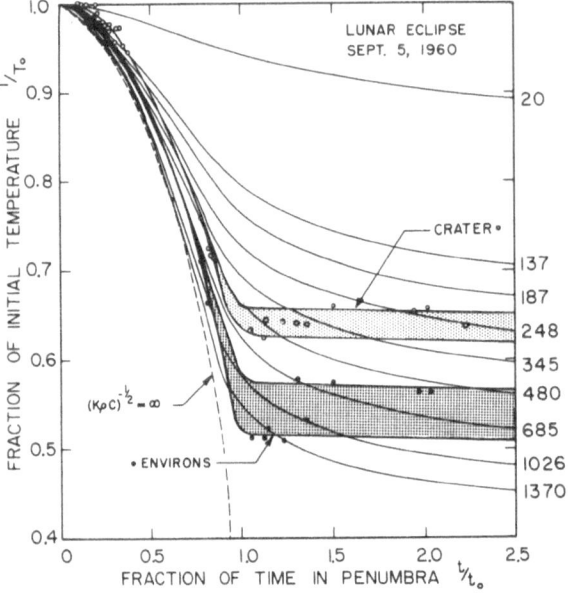

Fig. 26a. Observed and theoretical cooling curves for
Aristarchus and its environs. A homogeneous model of the
moon has been utilized for the theoretical calculations.
Data taken from Saari and Shorthill [22].

moon are plotted. Saari and Shorthill matched the layer and the substratum by
means of Jaeger and Harper's [9] equation [our equation (71)].

$$\frac{\tau}{K_d} = 610 \frac{1}{\sqrt{K\rho C}}$$

[Equation (71) has the coefficient 640, which differs from Jaeger and Harper's
value 610 for reasons mentioned earlier.] Clearly, predicted curves from the two-
layer model provide a much more satisfactory fit to the eclipse data (both for the
floor of Aristarchus and for the environs of the crater) than do the curves predicted on the basis of the homogeneous model. Though no specific two-layer
model computed by Saari and Shorthill adequately fits the eclipse data for the
umbral period, we can readily see that if we assume that for the underlying
stratum $(K\rho C)^{-1/2} \sim 100$ to 140, τ, the thickness of the dust layer in the crater
is ~ 0.01 mm, whereas outside the crater, $\tau > 0.1$ mm. (Sinton [25] uses the
thickness of the dust layer in the crater to estimate the crater age, but we will
not deal with this interesting possibility in this discussion.)

We are thus faced with a dilemma: the infrared observations during eclipse
(or during a lunation, see Murray and Wildey, [18]) are not compatible with a
homogeneous model of the moon which has constant thermal parameters. The ob-

servations are much more satisfactorily represented by a two-layer model of the moon. However, in the section entitled "A Test of the Homogeneous Model Using Microwave Data," we concluded, with Troitskii, that a homogeneous model of the moon appeared to be adequate to represent the microwave data. It is much to be regretted that high-angular-resolution microwave observations do not now exist to permit a direct comparison for microwave and infrared data for the anomalous crater areas and for other areas on the moon as well. Until such direct comparisons are made, it is not likely that we shall be able to resolve the present disagreement between these two types of data in regard to the nature of the lunar model. Microwave observations of the type we require to resolve our dilemma do lie within our present capabilities but are not available; it is greatly to be hoped that precision, high-angular-resolution microwave studies of the moon will become available in the near future.

NEW MODELS OF THE MOON

In view of the disagreement between the microwave and infrared results, we must give thought to other possible lunar models. In particular, we note that while the two-layer model of the moon provides a more satisfactory fit to the eclipse observations than does the homogeneous model, the fit afforded by the two-layer

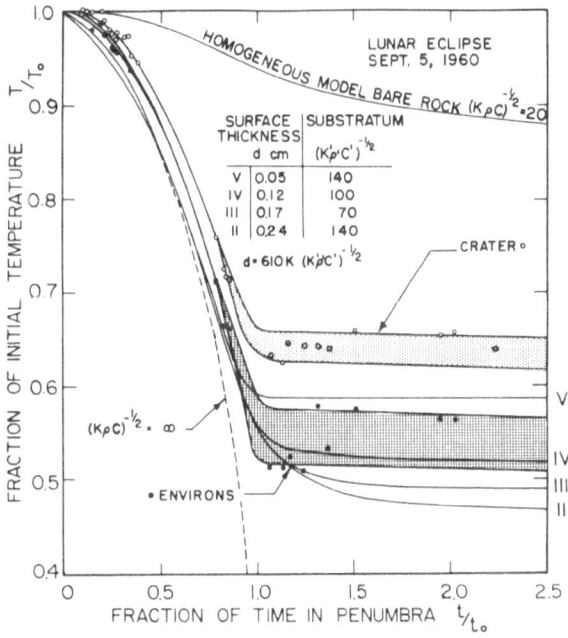

Fig. 26b. Observed and theoretical cooling curves for Aristarchus and its environs. A two-layer model of the moon has been utilized for the theoretical calculations. Data taken from Saari and Shorthill [22].

model is not as satisfactory as might be desired. During the penumbral period of the eclipse when the brightness temperature is rapidly decreasing, the observed points for both the crater floor and for the environs of the crater lie above the theoretically predicted curve, especially in the upper portion of the curve. The observed points appear to approach the predicted curve more closely as the temperature decreases to the value reached in the umbral stage of the eclipse. This type of discrepancy may be expected if some of the properties of the lunar material change with temperature. Future models of the moon must undoubtedly include variations of the thermal parameters with temperature (see Buettner [1]).

In the lunar models discussed in the preceding sections, we have considered only transport of heat by conduction. It is clear from a variety of recent laboratory investigations that heat transport by radiation must also be considered. For the porous outer skin of the moon the process of heat transport can become quite complicated, involving both conduction and radiation processes. Heat transport under conditions relatively similar to those one finds on the moon have recently received much attention in various laboratories and have been reviewed in the monograph edited by Blau and Fischer [2] and in papers that will be found referenced there.

Resistance to the conduction of heat through porous media can be traced to two basic sources: (1) the intrinsic physical properties of the individual particles comprising the porous solid (here we refer to such properties as conductivity, crystal structure, grain size, and impurities, which control the transport of heat through a given particle) and (2) contact or constriction resistance which imposes a thermal resistance owing to the small contact area of particles and to acoustic mismatch. Significant variation of transport rate with temperature variation may be expected.

Radiative transport of thermal energy through the material depends upon radiative exchange between particles and, near the surface, the direct escape of radiation through interstices. The transport rate will depend very much upon the nature of the particle, particle size, particle density, and the nature of scattering and absorbing centers in the material. The existence of infrared absorption bands in the particulate material will strongly affect the transport process. Admixture of metallic powders even in small percentages can change the heat transport properties of the material markedly. The temperature sensitivity of the process is fairly large: in the crudest theory, the radiation part of the heat transport may be expected to vary as T^3.

On the basis of laboratory experiments with powders in vacuum and in the general thermal range of interest for lunar investigation, one might expect changes in effective thermal conductivity by a factor of two over the lunar temperature range.

In regard to the interpretation of infrared observations, two questions in particular need further investigation. Silicates show variations in emissivity in the 8 to 14 μ range, a point discussed by Pettit and Nicholson [20]. Lyons and Burns [16] showed that as a consequence of such wavelength fluctuations, the emissivity of silicates is a function of temperature. Further analysis of this problem, particularly as it relates to porous or powdered silicate and nonsilicate materials is needed. Observed thermal variations over the lunar surface may, to some extent, be traceable to the effects of local variations of this nature in lunar material.

The partial transparency of finely divided silicates (Launer [14]) in some portions of the 8 to 14 μ region of the spectrum implies that in some areas of the moon there may be infrared radiation from below the surface of the dust or porous layer. In view of the high thermal gradient that must exist in the uppermost portion of the lunar surface, such penetration by infrared radiation could show itself in an interesting and significant way in different areas of the lunar surface. Narrow-band photometry in the infrared may, in the future, provide extremely interesting insights into the small-scale structure of the lunar surface.

We have started to learn a fair amount about the lunar surface from studies of the radiation emitted by it. Many new avenues of investigation are open to us. For further progress our needs appear to be fourfold: We need continuing excellent infrared observations as they are being made at several laboratories, but these observations should be extended to more wavelengths and should be made in narrow wavelength bands. The observations should be densely packed in time throughout a lunation. We need vastly improved observations in the radio range — improved as to accuracy of calibration and improved as to angular resolution. We need more realistic models of the moon in which we make full use of all relevant knowledge of the properties of materials (particularly porous materials) at low temperatures and in vacuum conditions. Lastly, we need a continuing series of laboratory investigations of those aspects of solid state physics which have a direct bearing on the lunar problem.

ACKNOWLEDGMENTS

I should like to express my thanks to Dr. David D. Cudaback of the Radio Astronomy Laboratory who several years ago first excited my interest in problems of the nature of the lunar surface. This paper has, in part, been prepared under ONR Contract No. 222(66), and under NASA Grant No. NsG 225-62(S-2).

REFERENCES

1. K. J. K. Buettner, *The Moon's First Decimeter* (Memorandum Rand Corp., No. RM-3263 — JPL, September 1962).
2. H. H. Blau and H. Fischer, eds., *Radiative Transfer from Solid Materials* (Macmillan, New York, 1962).
3. H. S. Carslaw and J. C. Jaeger, *Conduction of Heat in Solids*, second ed. (Oxford University Press, 1959).
4. R. J. Coates, *Ap. J.* 133, 723 (1961).
5. D. D. Cudaback, thesis, University of California, Berkeley (1962); see also R. Bracewell, *Astron. J.* 67, 786 (1962).
6. D. D. Cudaback, Meeting Am. Astron. Soc. (Flagstaff, July 1964).
7. J. C. Jaeger, *Australian J. Phys.* 6, 10 (1953).
8. J. C. Jaeger, *Proc. Cambridge Phil. Soc.* A, 49, 355 (1953).
9. J. C. Jaeger and A. F. A. Harper, *Nature* 166, 1026 (1950).
10. A. G. Kisliakov and A. E. Salomonovich, *Izv. Vysshikh. Uchebn. Zavedenii Radiofiz.* 6, 431 (1963).
11. Z. Kopal, *Radiative Transport of Heat in Lunar and Planetary Interiors*, Boeing Scientific Research Laboratories Report D1-82-0328 (January 1964).

12. V. D. Krotikov and V. S. Troitskii, *Soviet Astron.-AJ* 7, 119 (1963).
13. V. D. Krotikov and V. S. Troitskii, *Soviet Astron.-AJ* 7, 822 (1964).
14. P. Launer, *Am. Mineralogist* 37 (1952).
15. H. Lettau, *Geofis. Pura Appl.* 19, 1 (1951).
16. R. Lyons and E. A. Burns, *Nature* 196, 463 (1962).
17. C. H. Mayer, in: *Planets and Satellites*, G. P. Kuiper and B. M. Middlehurst, eds. (Univ. of Chicago Press, 1961), Chap. 12.
18. B. C. Murray and R. L. Wildey, *Ap. J.* 139, 734 (1964).
19. E. Pettit, *Ap. J.* 91, 408 (1940).
20. E. Pettit and S. B. Nicholson, *Ap. J.* 71, 102 (1930).
21. J. H. Piddington and H. C. Minnett, *Australian J. Sci. Res.* 2, 63 (1949).
22. J. M. Saari and R. W. Shorthill, *Icarus* 2, 115 (1963).
23. A. E. Salomonovich, *Astron. Zh.* 39, 79 (1962).
24. R. W. Shorthill, H. C. Borough, and J. M. Conley, *P.A.S.P.* 72, 481 (1960).
25. W. M. Sinton, *Lowell Obs. Bull.* 5, 25 (1960).
26. W. M. Sinton, in: *Physics and Astronomy of the Moon*, Z. Kopal, ed. (Academic Press, New York, 1962), Chap. 11.
27. V. S. Troitskii, *Astron. Zh.* 31, 511 (1954).
28. V. S. Troitskii, in: *The Moon*, Z. Kopal and Z. Mikhailov, eds. (Academic Press, New York, 1962).
29. J. van Diggelen, *Rech. Astron. Obs. Utrecht* 14, 114 (1960).
30. A. J. Wesselink, *Bull. Astron. Inst. Neth.* 10, 351 (1948).
31. R. L. Wildey and H. Pohn, 1963 (in press).

Chapter 16

Lunar Radar Reflections

G. Pettengill

Arecibo Ionospheric Observatory*
Arecibo, Puerto Rico

INTRODUCTION

In the 18 years since the first detection of radio echoes from the moon, radar systems have undergone tremendous improvements. Measurements have now been made at dozens of laboratories which together span a frequency spectrum of from about 30 Mc/s up to 37 Gc/s. Taken together with the many years of careful optical observation and the more recent infrared and millimeter wave thermal measurements, the radar data can add significantly to an understanding of the lunar surface.

Because of his control over the emitted waveform, the radar astronomer has an inherent measurement advantage as compared to experimenters who rely on thermal emission or solar irradiation as energy sources. First of all he knows the carrier frequency of his transmitted power so that only a small bandwidth region need be considered for detection. In addition, the coherence of his emission enables very precise measurements to be made of both the frequency displacement (Doppler shift) and dispersion of the returned signal.

If the waveform is marked in some way, as by amplitude or phase pulsing, it is possible to measure the echo delay with high accuracy and to isolate the returns from various distance zones on the target. Finally, since the radio energy sent out has a unique and known polarization, it is possible to study some aspects of the propagating medium as well as the roughness of the reflecting surface from the altered properties of the echo.

Optical techniques for studying the moon offer unparalleled angular resolution, but, at least from the surface of the earth, are limited to detecting the presence of objects no smaller than several hundred meters in size. On a statistical basis it is possible to infer the degree of surface smoothness on a scale of a few microns from the scattering laws observed. In the range of structure sizes from a few millimeters up to several hundred meters, however, optical techniques can offer little information. What little data there are have been obtained from observing shadows cast on the lunar surface. It is just this range of sizes, however, that is of greatest interest in the design of spacecraft to land and move about on the moon, and it is this domain that radar can most effectively explore.

_____
* Operated by Cornell University with the support of ARPA under contract with AFOSR.

355

HISTORY

The first radio echoes from the moon were obtained in 1946 by Bay [1] and de Witt and Stodola [6]. These early measurements consisted of little more than bare detection, although the latter group noted severe fading with echoes totally absent on certain days. Later Kerr and Shain [20] recognized that there were two sources of fading, one having an ionospheric origin and a second, faster component attributable to the moon itself. Murray and Hargreaves [25] showed that the long-period fading was due to the effect of Faraday rotation in the earth's ionosphere of linearly polarized waves. Fricker et al. [13] showed that this long-period fading could be eliminated by using circularly polarized transmissions.

In a paper published in 1948, Grieg et al. [14] suggested that multiple reflection points on the lunar surface could cause interference fading as the relative positions of these points changed with the libration of the moon. By making careful measurements of the fading autocorrelation function of lunar echoes at transit over several months, Evans [7] was able to prove that the short-period fading was indeed associated with changes in the lunar libration; an even more careful series of observations was reported later by Evans et al. [9].

Evans [7] and Trexler [33], both working at wavelengths near 2 m discovered that, in contrast to the optical situation, the moon exhibits a very strong "highlight" near the center when illuminated by radio waves. This nearly specular echo gave the first hint that the moon might be relatively smooth at radio wavelengths. Somewhat later this result was verified for wavelengths near 10 cm by Yaplee et al. [37] and Hey and Hughes [18]. All these observers agreed that about 50% of the reflected energy appeared to originate from a region at the center of the visible disk of about $1/10$ the lunar radius.

Improvements in radar sensitivities enabled Leadabrand et al. [22] and Pettengill [27] to observe echo power continuously from all ranges out to the lunar limb. Quantitative measurements pursued by Evans at 2.5 m [7], Trexler at 1.99 m [33], Pettengill at 68 cm [27], Hughes at 10 cm [19], Pettengill and Henry at 68 cm [28], Evans and Pettengill at 3.6 cm, 68 cm and 7.84 m [11], and Lynn et al. at 8.6 mm [23] have extended measurements of the lunar surface scattering laws over a wavelength range of almost 10 octaves. Polarization measurements have been reported by Blevis and Chapman [2], Victor et al. [35], and Evans and Pettengill [12].

Theoretical attempts to explain the observed scattering laws have been made principally by Hargreaves [16], Daniels [3-5], Hagfors [15], Muhleman [24], and Winter [36].

More recently, digital processing methods have been applied to the received lunar echo data in order to obtain spectral analyses for each of many range intervals. Through the use of these techniques, Pettengill [27], Pettengill and Henry [29], and Thompson et al. [32] have localized regions of unusually intense reflectivity on the lunar surface which appear to correlate with the rayed craters.

ECHO STRENGTH

Radar Equation

Basic to any experiment involving radar is the radar equation, which relates the strength of the echo received to the properties of the target, its distance, and the observing equipment used. The radar equation can be stated in many ways, but for purposes of understanding, the simplest seems to be to write it as the product of a "path loss" containing astronomical parameters beyond the experimenter's control, and a term containing the equipment characteristics. The path loss can be written as

$$\sigma / (4\pi R^2)^2 \tag{1}$$

where σ represents the radar scattering cross section of the target and R the distance between the observer and his target. The expression yields the attenuation in signal undergone by radio energy transmitted from an isotropic antenna (gain equal to unity) and received in an antenna of unit capture area.

Supplying the radar parameters then yields for the actual received power:

$$P_r = P_t G_t A_r \text{ (path loss)} \tag{2}$$

where P_t is the transmitted power; G_t, the transmitting antenna gain; and A_r, the receiving antenna aperture. A_r may be related to G_r, the gain of the receiving antenna, through the expression

$$A_r = (\lambda^2/4\pi) \ G_r \tag{3}$$

where λ is the operating wavelength.

The range to the moon changes only slightly over the course of a month, averaging to 3.844×10^8 m (equivalent to a round-trip echo delay of 2.56 sec). As will be discussed later, the total lunar cross section for radio wave scattering is largely independent of frequency. Expressed as a fraction of the geometric cross section (the area of a lunar great circle), the mean value of the measurements is quoted as close to 0.07. Thus, the mean value of the path loss is approximately $1.9 \times 10^{-25} \ m^{-2}$ or $-247.2 \ db/m^2$. The monthly variation about this mean caused by the variation in range is ± 1 db. The total lunar cross section does not appear to vary significantly as changes in libration vary the aspect presented to earth.

Lunar Cross Section

Table I lists all known measurements of the total lunar radar cross section, showing the frequency at which the value was determined, and the author. In some cases, revision by the author (privately communicated) has changed the originally published value slightly; the modified value is listed. Figure 1 plots the data of Table I and shows how little the total radar cross section varies with wavelength over 3 orders of magnitude.

TABLE I

Author	Year	Wavelength, cm	$\sigma/\pi a^2$	Estimated error, db
Lynn et al.	1963	0.86	0.07	±1
Kobrin	1963*	3.0	0.07	±1
Morrow et al.	1963*	3.6	0.07	±1.5
Evans and Pettengill	1963a	3.6	0.04	±3
Kobrin	1963*	10.0	0.07	±1
Hughes	1963*	10.0	0.05	±3
Victor et al.	1961	12.5	0.022	±3
Aarons	1959†	33.5	0.09	±3
Blevis and Chapman	1960	61.0	0.05	±3
Fricker et al.	1960	73.0	0.074	±1
Leadabrand	1959†	75.0	0.10	±3
Trexler	1958	100.0	0.07	±4
Aarons	1959†	149.0	0.07	±3
Trexler	1958	150.0	0.08	±4
Webb	1959†	199.0	0.05	±3
Evans	1957	250.0	0.10	±3
Evans et al.	1959	300.0	0.10	±3
Evans and Ingalls	1962	784.0	0.06	±5

* Revised value — privately communicated.
† Reported by Senior and Siegel (1959, 1960).

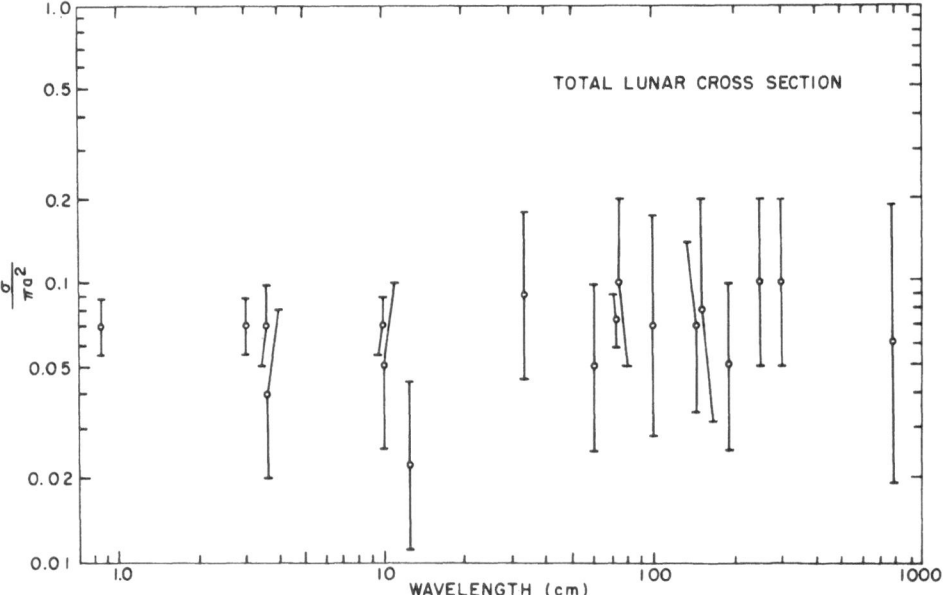

Fig. 1

In order to relate the scattering cross section of the moon to some of the properties of its surface, it is useful to write an expression separating the geometrical effects of the surface configuration from the electrical properties of the surface material:

$$\sigma = g \rho \pi a^2 \tag{4}$$

The geometrical cross section or intercept area is given by πa^2, the electrical reflectivity by ρ, and the directivity of the surface configuration by the factor g.

For a smooth, perfectly conducting sphere of dimensions which are large compared to the wavelength, the incident energy will be reflected equally into every element of solid angle of the scattered wave [26]. A smooth dielectric surface will not scatter isotropically, of course, since the reflectivity will vary with the angle of the incident energy to the local surface. Since we are here concerned only with the backscattered energy, however, we may hypothesize a fictitious surface which maintains the reflectivity at normal incidence over all angles of incidence without altering the measured results. It follows, therefore, from the definition for radar cross section contained implicitly in equation (1) that $g = 1$ for any smooth spherical surface. The geometry of a roughened or undulating surface, however, may alter the effectiveness of the integrated backscattering.

We know the dimensions of the moon with fair precision, but to determine g and thus isolate the reflectivity ρ from the measured cross section requires a knowledge of the scattering behavior of the moon in all directions. From a measuring base located only on the surface of the earth, this is impossible. Fortunately, however, for many reasonable, naturally occurring types of surfaces, this factor may be calculated.

The Lambert law, which describes the scattering from a surface so covered with irregularities that the emergent energy "forgets" the state of the incident, is given by

$$\sigma \sim \cos \phi \cos \phi' \tag{5}$$

where ϕ and ϕ' are the incident and emergent angles to the normal. By integration it may be shown (Grieg et al. [14]) that in this case $g = 8/3$. For a perfectly smooth surface, $g = 1$. For a surface which is smooth (in terms of the wavelength employed) but undulating, Hagfors (unpublished result) has shown that $g = 1 + 2a^2$, where a is the mean gradient of the undulating slopes (assumed here to have a Gaussian distribution).

At all wavelengths greater than a few centimeters, the bulk of the returned energy appears to be scattered by a smooth undulating surface characterized by values for $a < 0.3$ (see following section). Therefore, g may be estimated with sufficient precision in most cases to determine ρ from the measured radar cross section.

Since most of the returned energy has been scattered from a smooth surface at normal incidence, the reflectivity may be related to the surface dielectric constant k through

$$\rho = \left(\frac{\sqrt{k} - 1}{\sqrt{k} + 1} \right)^2 \tag{6}$$

The assumption, presumably justified, has been made here that the surface has near zero conductivity and unity permeability. Inserting $g = 1.1$ and $\sigma/\pi a^2 = 0.07$ in equations (4) and (6) yields $k = 2.8$. Since almost all solid terrestrial materials have a dielectric constant greater than 5, this lower value must imply a degree of porosity in the lunar surface. It is known that at optical wavelengths the surface is highly porous, and it is not surprising to find a similar indication from the radar data. The precise value of porosity indicated depends somewhat on the grain shape assumed, but is probably close to that for sand. (Dry sand has a measured dielectric constant of about 3.)

The low value of dielectric constant obtained from the radar data is somewhat larger than the values deduced from the passive radiometric observations. Soviet workers (Troitsky, [34], Salomonovich and Losovsky, [31], and Krotikov and Troitsky, [21]) arrive at values less than 2. A recent value reported by American observers (Heiles and Drake [17], was 2.1 ± 0.2. However, a difficulty with the radiometric observations is that the results are sensitive to the surface roughness and require a correction (to greater values) which is difficult to make. Therefore, it is not felt that the disagreement is necessarily significant.

ANGULAR SCATTERING LAW

If sufficient power is available, it is possible to receive echoes from all portions of the visible lunar disk, (Leadabrand et al. [22] and Pettengill, [27]). As is shown in Fig. 2, there is a direct relation between the range depth from

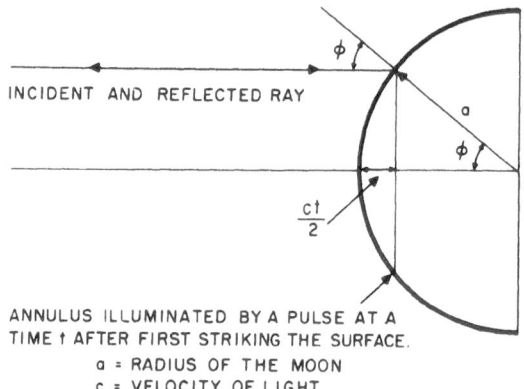

Fig. 2. The relation between the range delay t and the angle of incidence and reflection ϕ of the radio waves.

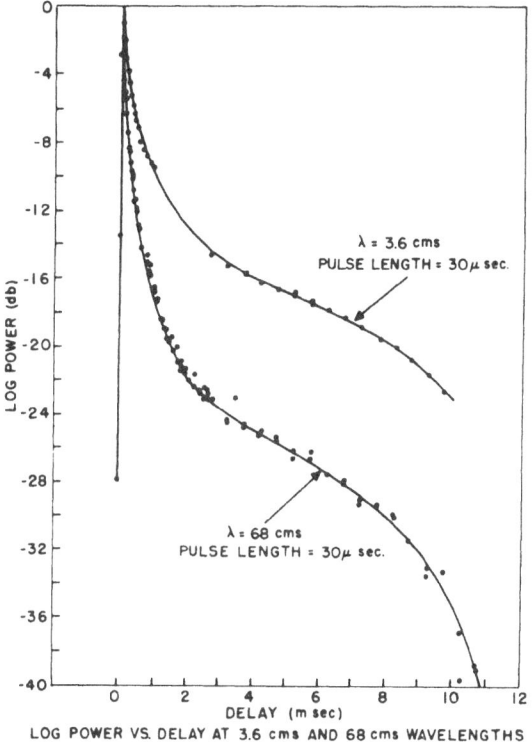

LOG POWER VS. DELAY AT 3.6 cms AND 68 cms WAVELENGTHS

Fig. 3. The average echo power reflected by the
moon $\bar{P}(t)$ as a function of delay measured with
respect to the point closest to the radar. The 68-
cm results were obtained using the same methods
of averaging as were employed to obtain the 3.6-
cm results. The curves have been normalized at
zero delay.

which an echo is received and the angle of incidence and emergence, ϕ, of the
scattered ray to the mean surface.

$$\tfrac{1}{2}\ ct\ =\ a(1\ -\ \cos\ \phi) \tag{7}$$

where c is the velocity of light, t is the echo delay measured from the front edge
of the moon, and a is the lunar radius.

Figure 3 shows the results of the two radar experiments carried out at MIT
Lincoln Laboratory (Evans and Pettengill, [12]) at wavelengths of 3.6 and 68 cm.
In these experiments a pulse width of 30 μsec was employed — enough to resolve
all but the fine details of the initial spike. Both curves show the same rapid
initial decay followed by a more gradual descent until the limb is reached at 11.6
msec. Strikingly different, however, is the level at which the leveling-off occurs.
From Fig. 1, we saw that the total integrated cross section was not sensibly dif-

ferent at 3.6 and 68 cm. (In Fig. 3, both curves have been normalized to 0 db at their peaks.) Therefore, it is clear that at 3.6 cm a trade-off of power has occurred in favor of the greater delays as compared with the delay scattering law valid at 68 cm.

Figure 4 shows the 68-cm data plotted as a function of the log cosine of the angle of incidence to the surface. Also shown is the scattering behavior to be expected were the surface to scatter according to the Lambert ($\cos^2\phi$) or Lommel–Seeliger ($\cos\phi$) laws. In the tail of the distribution corresponding to scattering at very oblique angles of incidence (where, presumably, the scattering is controlled by surface irregularities of the order of a wavelength), the actual behavior appears to lie midway between the two laws illustrated.

In the initial portion of the angular spectrum, it is clear that the decay is significantly less rapid at 3.6 cm than at 68 cm (see Fig. 3). The initial decay is controlled by the number of large flat surfaces found sufficiently inclined at increasing ranges to present a normal face to the incident energy. Therefore, one must conclude that as the scale of the examined region decreases (by shortening

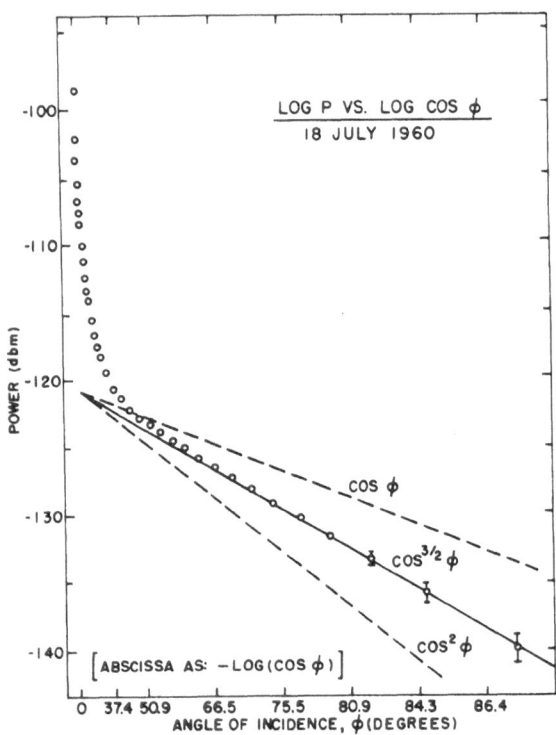

Fig. 4. The angular power spectrum $\bar{P}(\phi)$ obtained from the echo power distribution. The Lambert ($\cos^2\phi$) and Lommel–Seeliger or uniformly bright behavior ($\cos\phi$) are shown for comparison.

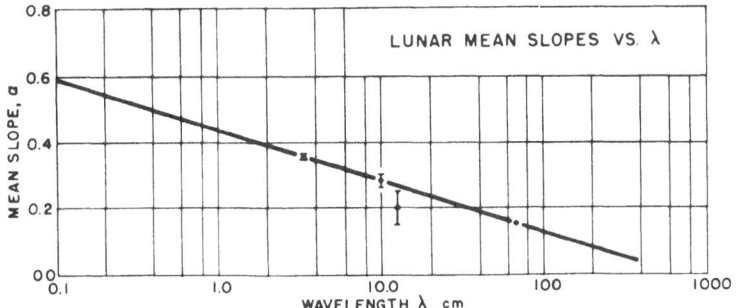

Fig. 5. The mean surface slope α vs. wavelength according to a model of the surface by Muhleman (1964). Each point has been obtained by adjusting the value of α in the theory until the published angular scatter laws $P(\phi)$ were matched at each wavelength (after Muhleman 1964).

the probing wavelength), the mean slopes of the lunar surface increase — much in the same way that the surface slopes of a choppy sea would behave.

Theoretical expressions for the relationship between various distributions of surface slopes and the measured scattering laws have been sought principally by Daniels [3–5], Hagfors [15], and Muhleman [24]. Muhleman has achieved a good fit to the observed data at a wide range of angles and wavelengths and finds that the mean slope must be adjusted with frequency as shown in Fig. 5.

Thus, at 68 cm the mean slope is found to be 0.145 or about 1 in 7, while at 3.6 cm it is 0.35 or about 1 in 3. These values depend slightly on whether one chooses a Gaussian or exponential distribution for the slopes; the experimental data tend to favor an exponential distribution.

DEPOLARIZATION

Since the emitted waveform is completely polarized in a known way, it is possible to measure the distortion in polarization suffered on scattering from the lunar surface. In order to obviate the effects of Faraday rotation of a linearly polarized wave in the earth's ionosphere, circularly polarized transmitted radiation is generally used. On reflection at normal incidence to a smooth area, right-circular incident polarization is converted to left-circular reflected polarization. However, if the surface is rough, there will be a large fraction of scattered power possessing no definite polarization. Thus, the degree of depolarization is a measure of surface roughness. Figure 6 shows the degree to which the returned energy is polarized as a function of scattering depth at 68 cm wavelength. It may be seen that in the region where the returned signal is largely obtained by reflection from large smooth surfaces, the degree of polarization is high as expected. At greater ranges, however, where the energy has been backscattered at nearly grazing angles, the polarization has been substantially reduced.

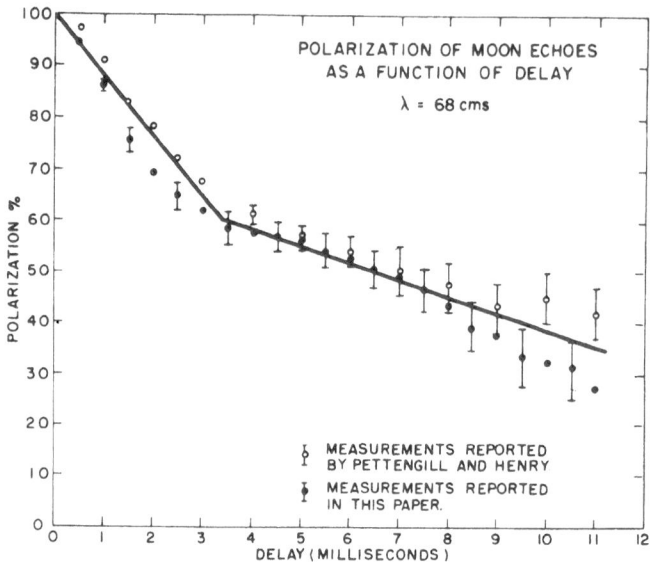

Fig. 6. The percentage polarization of moon echoes as a function
of range delay observed at 68 cm wavelength.

DISTRIBUTION OF REFLECTIVITY

It was shown earlier how the mean surface reflectivity and configuration
could be determined from radar pulse measurements from an antenna which illuminated the entire visible lunar disk. However, if sufficient angular resolution exists, it is useful to explore the distribution in reflectivity around a contour of
constant range depth. Evans [8], Lynn et al. [23], and Thompson et al. [32] have
explored this technique. Only Thompson has succeeded in using the angular resolution of his beam ($\frac{1}{6}$ deg) to successfuly isolate regions of unusually intense
reflectivity by this means.

A much more powerful method for exploring the distribution of reflectivity over
the lunar surface in minute detail has been exploited by Pettengill [27] and Pettengill and Henry [29]. In this technique, advantage is taken of the frequency coordinate to localize the origin of reflected power. Figure 7 shows how the small apparent rotation of the moon with respect to the radar results in a differential frequency shift across the surface. If a frequency spectrum is made of the returned
power, components at different frequencies may be identified with strips lying
parallel to the apparent axis of rotation. By taking spectra at a selected range,
the origin of the reflected component may be assigned to a pair of arbitrarily small
regions lying at the intersections of the appropriate range and frequency contours.
The appearance of such spectra may be judged from Fig. 8, which shows how the
method was used by Pettengill and Henry [29] to isolate a region of unusually intense reflectivity associated with the crater Tycho. A single measurement of re-

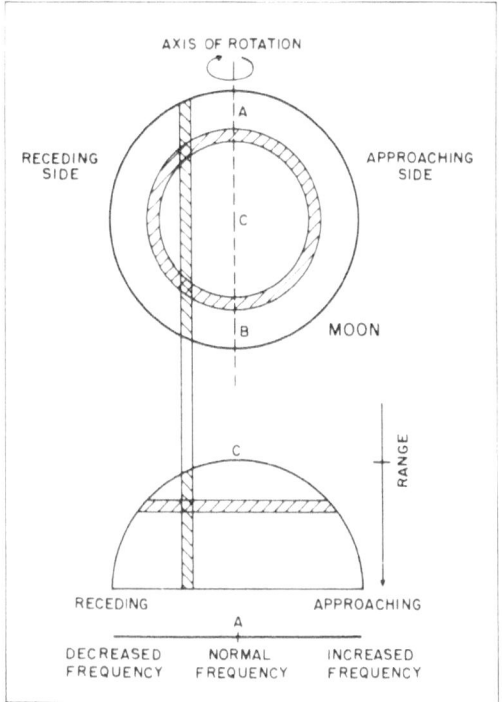

Fig. 7. The contours of constant delay τ and constant Doppler shift f are shown for a hypothetical pair of values. The shaded regions indicate the two areas having the same frequency and range coordinates.

flected power yields only a single value, of course, and since many individual reflecting centers are interfering within the resolution cell, represents a poor estimate of the mean reflectivity. Thus the results shown in Fig. 8 are the mean of 20 individually determined spectra and have a consequent root-mean-square fluctuation of about 22%.

Despite these fluctuations, however, an echo significantly stronger than its surroundings may be seen at a range and frequency which correspond to the position of Tycho on the lunar surface. In this case, the echo was sufficiently strong to permit unambiguous assignment to one of the pair of ambiguous regions of intersection shown in Fig. 7 through changes in its relative frequency position with time. Several modern radar systems have sufficiently narrow beamwidths to remove even this remaining ambiguity (except in cases where the ambiguous regions are not sufficiently separated). Using this method of range-frequency radar mapping, selected regions of the lunar surface are being investigated at the Arecibo Observatory to a resolution of about 10 km.

The anomalously strong reflectivity exhibited by Tycho (some 10 times greater than its surroundings) appears to be caused by its relatively rougher and denser

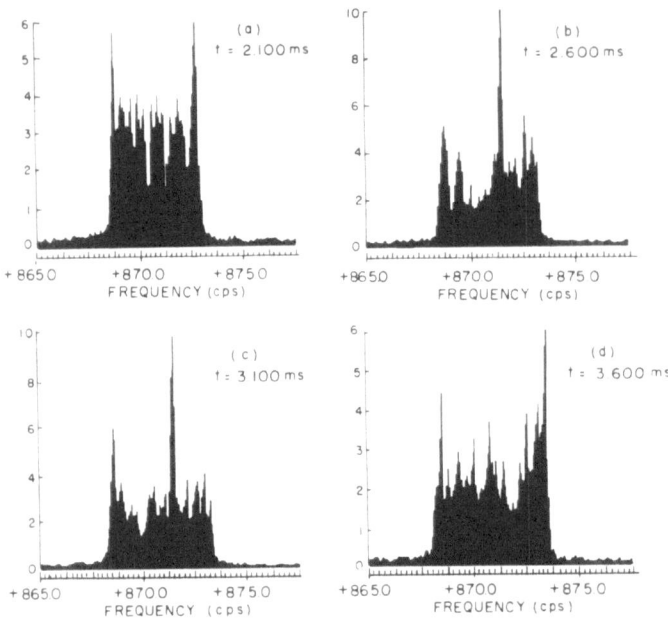

Fig. 8. Lunar echo power spectra at 4 intervals of range, taken 19h
00m 19s U.T., June 20, 1961. The central peaks in (b) and (c) correspond to the position of the crater Tycho. Peaks at the extremes
of the spectra result from the large common area of the intersecting
range and Doppler contours (Fig. 7) and probably do not represent
individual surface features.

surface. As remarked earlier, the bulk of the lunar surface appears to be relatively smooth at radio wavelengths and porous. If, however, the surface is disrupted
by a meteor impact and a shattered, compressed surface exposed, the scattering at
oblique angles of viewing could easily be expected to increase by a factor of 10.
Pettengill and Henry [29] have also investigated the depolarization of the enhanced scattering and found it to be characteristic of a rough surface. Observation over the limited range of scattering angles permitted by librational motions
discloses no sharp angular dependence, thus making less likely an accidental enhancement caused by a suitably placed steep slope.

Beyond the radar measurements, Saari and Shorthill [30] have found infrared
temperature anomalies associated with Tycho (and the other rayed craters). They
explain the slow cooling of these craters under eclipse conditions as caused by a
relatively high thermal conductivity of the surface material. It seems, therefore,
that a likely explanation for the radar anomalies is that a relatively recent meteor
impact has disturbed the mantle of smooth, porous surface accretion creating a
rough relatively dense surface in the vicinity of the crater. Recent radar observations by Thompson et al. [32] have identified more than five other rayed craters
by their enhanced radar scattering.

SUMMARY

The radar cross section of the moon has been measured over a frequency range of about 10 octaves, stretching from 7.84 m to 8.6 mm. Within the accuracy of measurement, which in only a few cases exceeds a factor of 2, there is no significant frequency dependence. The mean of all measurements yields a value of 0.07 of the geometrical (disk) cross section.

The angular scattering law has also been determined over a wide range of frequencies, although the best measurements appear to lie in the region from 70 to 0.86 cm. There is a marked frequency dependence both in the mean surface slopes and in the fraction of surface which appears to scatter roughly. At the longer wavelengths the mean surface slope is about 1 in 7, while in the millimeter region the corresponding value approaches 1 in 2. The division of power between the quasi-specular peak corresponding to normal reflection from large smooth areas and the largely nondirectional scattering from rough areas varies from 4 to 1 (favoring quasi-specular) at 68 cm to 1 to 6 (favoring rough) at 0.86 cm.

The measurements described above may be used to deduce a surface dielectric constant of about 3. For most likely surface materials this requires a bulk density of about one-half that of siliceous rocks typical of the earth's surface. This value is larger than that deduced for the visible lunar surface and undoubtedly corresponds to conditions a few meters below.

As yet the measurements of depolarization of the scattered wave have not been extended over a wide frequency range nor have they been used quantitatively to argue the precise form of scattering. Qualitatively, however, they support with considerable force the implication derived from the angular scattering law that the echo power at greater ranges is returned chiefly from wavelength-sized irregularities in the surface.

Perhaps the most exciting recent development in the study of the moon by radar is the detailed mapping of surface reflectivity using the range-Doppler technique of analysis. With the extra coordinate afforded by frequency measurement, resolution of surface areas only 10 by 20 km in size has been possible. In particular, this method has established that the relatively young, rayed craters possess unusually high radar reflectivity as compared to most areas of the lunar surface lying at the same range.

For the future, further measurements would be of value if they: (1) extended the radar observations to the limits in frequency permitted by atmospheric and ionospheric penetration, providing better measurement of the angular scattering, particularly at low frequencies, and more accurate values of cross section at all frequencies; (2) clarified, either theoretically or through model measurements, the quantitative relationship between types of rough surfaces and the depolarization of the scattered wave.

REFERENCES

1. Z. Bay, *Hung. Phys. Acta* 1, 1 (1946).
2. B. C. Blevis and J. H. Chapman, "Characteristics of 488 Mc/s Radio Signals Reflected from the Moon," *J. Res. Natl. Bur. Std.* 64D, 331–334 (1960).

3. F. B. Daniels, "Radar Determination of the Scattering Properties of the Moon," *Nature* 187, 399 (1960).
4. F. B. Daniels, "A Theory of Radar Reflection from the Moon and Planets," *J. Geophys. Res.* 66, 1781–1788 (1961).
5. F. B. Daniels, "Radar Determination of the Root-Mean-Square Slope of the Lunar Surface," *J. Geophys. Res.* 68, 449–453 (1963).
6. J. M. de Witt, Jr. and E. K. Stodola, "Detection of Radio Signals Reflected from the Moon," *Proc. IRE* 37, 229–242 (1949).
7. J. V. Evans, "The Scattering of Radio Waves by the Moon," *Proc. Phys. Soc. London* B 70, 1105–1112 (1957).
8. J. V. Evans, "Radio Echo Observations of the Moon at 3.6 cm Wavelength," MIT Lincoln Lab. Tech. Rept. 256, ASTIA No. DDC 274669 (1962).
9. J. V. Evans, S. Evans, and J. H. Thompson, "The Rapid Fading of Moon Echoes at 100 Mc/s," *Paris Symp. Radio Astron.* (Stanford Univ. Press, Stanford, 1959), p. 8, R. N. Bracewell, ed.
10. J. V. Evans and R. P. Ingalls, "Radio Echo Studies of the Moon at 7.84-m Wavelength," MIT Lincoln Lab. Tech. Rept. 288, ASTIA No. DDC 294008 (1962).
11. J. V. Evans and G. H. Pettengill, "The Radar Cross Section of the Moon," *J. Geophys. Res.* 68, 5098–5099 (1963).
12. J. V. Evans and G. H. Pettengill, "The Scattering Behavior of the Moon at Wavelengths of 3.6, 68, and 784 cm," *J. Geophys. Res.* 68, 423–447 (1963).
13. S. J. Fricker, R. P. Ingalls, W. C. Mason, M. L. Stone, and D. W. Swift, "Computation and Measurement of the Fading Rate of Moon-Reflected UHF Signals," *J. Res. Natl. Bur. Std.* 64D, 455–465 (1960).
14. D. D. Grieg, S. Metzger, and R. Waer, "Consideration of Moon-Relay Communication," *Proc. IRE* 36, 652–663 (1948).
15. T. Hagfors, "Some Properties of Radio Waves Reflected from the Moon and Their Relation to the Lunar Surface," *J. Geophys. Res.* 66, 777–785 (1961).
16. J. K. Hargreaves, "Radio Observations of the Lunar Surface," *Proc. Phys. Soc. (London)* B 73, 536–537 (1959).
17. C. E. Heiles and F. D. Drake, "The Polarization and Intensity of Thermal Radiation from a Planetary Surface," *Icarus* 2, 281–292 (1963).
18. J. S. Hey and V. A. Hughes, "Radar Observations of the Moon at 10 cm Wavelength," *Paris Symp. Radio Astron.* (Stanford Univ. Press, Stanford, 1959), p. 13, R. N. Bracewell, ed.
19. V. A. Hughes, "Radio Wave Scattering from the Lunar Surface," *Proc. Phys. Soc.* 78, 988–997 (1961).
20. F. J. Kerr and C. A. Shain, "Moon Echoes and Transmission through the Ionosphere," *Proc. IRE* 39, 230–242 (1951).
21. V. D. Krotikov and V. S. Troitsky, "The Emissivity of the Moon at Centimeter Wavelengths," *Astron. Zh.* 39, 1089–1093 (1962).
22. R. L. Leadabrand, R. B. Dyce, A. Fredriksen, R. I. Presnell, and R. C. Barthle, "Evidence that the Moon is a Rough Scatterer at Radio Frequencies," *J. Geophys. Res.* 65, 3071–3078 (1960).
23. V. L. Lynn, M. D. Sohigian, and E. A. Crocker, "Radar Observations of the Moon at 8.6 mm Wavelength," *J. Geophys. Res.* 69, 781–783 (1964).
24. D. O. Muhleman, "Radar Scattering from Venus and the Moon," *Astron. J.* 69, 34–41 (1964).
25. W. A. S. Murray and J. K. Hargreaves, "Lunar Radio Echoes and the Faraday Effect in the Ionosphere," *Nature* 173, 944–947 (1954).
26. K. A. Norton and A. C. Omberg, *Proc. IRE* 35, 4 (1947).
27. G. H. Pettengill, "Measurements of Lunar Reflectivity Using the Millstone Radar," *Proc. IRE* 48, 933–934 (1960).
28. G. H. Pettengill and J. C. Henry, "Radio Measurements of the Lunar Surface," The Moon, *Intern. Astron. Union Symp.*, 14th, Z. Kopal and Z. K. Mikhailov, eds. (Academic Press, London, 1962), p. 519.
29. G. H. Pettengill and J. C. Henry, "Enhancement of Radar Reflectivity Associated with the Lunar Crater Tycho," *J. Geophys. Res.* 67, 4881–4885 (1962).
30. J. M. Saari and R. W. Shorthill, "Isotherms of Crater Regions on the Illuminated and Eclipsed Moon," *Icarus* 2, 115–136 (1963).

31. A. E. Salomonovich and B. Y. Losovsky, "Radio Brightness Distribution of the Lunar Disk at 0.8 cm," *Astron. Zh.* **39**, 1074–1082 (1962).

32. T. W. Thompson, R. B. Dyce, and A. D. Sanchez, "Measurements of Radar Reflectivity of the Moon in Range and Angle at 430 Mc/s," paper presented at URSI (Washington, D.C., April 1964).

33. J. H. Trexler, "Lunar Radio Echoes," *Proc. IRE* **46**, 286–292 (1958).

34. V. S. Troitsky, "Radio Emission of the Moon, Its Physical State and the Nature of Its Surface," The Moon, *Intern. Astron. Union Symp., 14th*, Z. Kopal and Z. Mikhailov, eds. (Academic Press, London, 1962), p. 475.

35. W. K. Victor, R. Stevens, and S. W. Golomb, "Radar Exploration of Venus," Jet Propulsion Lab. Tech. Rept. No. 32-132 (1961).

36. D. F. Winter, "A Theory of Radar Reflections from a Rough Moon," *J. Res. Natl. Bur. Std.* **66D**, 215–226 (1962).

37. B. E. Yaplee, R. H. Bruton, K. J. Graig, and N. G. Roman, "Radar Echoes from the Moon at a Wavelength of 10 cm," *Proc. IRE* **46**, 293–297 (1958).

Chapter 17

Some Problems of Planetary Radio Astronomy

Harold Weaver

Radio Astronomy Laboratory
University of California
Berkeley, California, United States of America

GOALS

Three major goals serve to orient infrared and radio wavelength investigations of the bodies of the solar system.

1. We wish to know the composition and physical properties of the surfaces of the planets, satellites, asteroids, and comets of the solar system, and to understand the interactions between those surfaces and the cosmic and atmospheric environments in which they are located. Under this goal we extend to the more distant members of the solar system studies of the type discussed earlier in the chapter on the interpretation of thermal emission from the moon (Chapter 15).

2. We wish to know the composition and nature of the gaseous envelopes that surround some of the planets and satellites and the comets of the solar system and, eventually, to know the origin and evolution of those envelopes. We wish to understand the physics of the interactions between the gaseous envelopes and the cosmic environment in which they are located.

3. We wish to determine the properties of magnetic fields that surround some of the bodies of the solar system and to understand interactions between those fields and the cosmic environment.

Investigation of at least some of the problems implied by these goals will lead us into the study of the sun itself since the sun is the principal source of the "cosmic environment" mentioned in the paragraphs above. The sun supplies the radiant energy that constantly falls upon and heats the bodies in the system; it is the source of the flares that occasionally irradiate them with ultraviolet and shorter wavelength radiations, the solar winds that sweep over them, and many of the high-energy particles that bombard them. However, we shall not specifically include the study of the physics of the sun in this discussion; our purpose here is to focus attention on the bodies moving around the sun.

TABLE I

Angular Diameters of Planets and Selected Objects in the Solar System

Object	Range of diameter (sec of arc)	Object	Range of diameter (sec of arc)
Mercury	4".7 to 12".9	Neptune	2".2 to 2".4
Venus	9".9 to 64".0	Pluto	0".1
Mars	3".5 to 25".1	Jupiter III	1".1 to 1".7
Jupiter	30".5 to 49".8	Jupiter IV	1".0 to 1".6
Saturn	14".7 to 20".5	Titan	0".6 to 0".8
Uranus	3".4 to 4".2	Ceres	0".3 to 1".3

SURFACE STUDIES

There are available to us, at least in principle, all of the techniques discussed in the chapter dealing with the interpretation of thermal radiation from the moon, to study the surface properties of the bodies of the solar system. The great difficulty in practice in the study of the planets is, of course, that they are very much smaller in angular size than the moon. Therefore, we face a severe problem in angular resolution when we observe the planets. Table I provides data on the range of angular sizes of the planets as well as some asteroids and satellites. For comparison purposes it may be well to keep in mind that the moon is approximately 30' of arc in diameter. An added complication for some objects may be that at some frequencies and at some times the surface may be obscured by the planetary or satellite atmosphere.

In order to observe the planets with angular resolution adequate to determine surface detail — resolution of the general order of 1" of arc — we require an antenna far larger than any now in existence; the antenna must be capable of operating over a wide variety of wavelengths down to a few centimeters, preferably a few millimeters; it must be capable of measuring polarization. A cross-type antenna consisting of a substantial number of paraboloids — perhaps 32 — of fairly large size — possibly 50 or 60 ft in diameter — in each arm and extending over a distance of possibly 6 km would provide resolution of the sort required. Construction of such a giant antenna is a project for the distant future.

An important start in the investigation of the brightness distribution across the disks of the planets can be made, however, with a two-element interferometer of the type now in use at the Owens Valley Observatory of the California Institute of Technology. At a specified frequency and for a specified mode of polarization, the fringe visibility, as a function of the separation of the two antennas, is determined by the diameter and brightness distribution of the planetary disk. As was shown in the discussion of thermal emission from the moon, the apparent brightness distribution is a strong function of the properties of the surface material, principally the dielectric constant. In Fig. 1 are shown fringe-visibility curves computed by

Heiles and Drake [31] for Venus as a function of the dielectric constant of the planet's surface material. Shown on the set of theoretical curves are a number of fringe visibility observations made at the Owens Valley Observatory. A high dielectric constant for Venus is indicated.

Lacking an instrument of high angular resolution which permits examination of surface details, we can still make useful observations of radiation from the entire disk of the planet. Such whole-disk temperature measurements have, in fact, provided nearly all existing microwave planetary data. They provide an immediate check on the agreement (or lack of agreement) between infrared or theoretical temperatures and indicate the existence of any unusual conditions on the planetary surface or in the atmosphere or the magnetosphere.

Whole-disk temperatures can provide additional detailed information on, for example, the direction of rotation of an inner planet. Let us observe an inner planet in direct motion in its orbit, following the planet from elongation to elongation through inferior conjunction. We can then derive the phase function for the planet. The phase function may be either symmetrical or asymmetrical about inferior conjunction. If the relation is symmetrical, rotation and revolution of the planet are synchronous. If the phase relation exhibits a minimum before conjunction, the direction of rotation is direct; if the phase relation exhibits a minimum after conjunction, the direction of rotation is retrograde. The reason is readily apparent from inspection of Fig. 2 in the chapter dealing with the interpretation of thermal radiation from the moon.

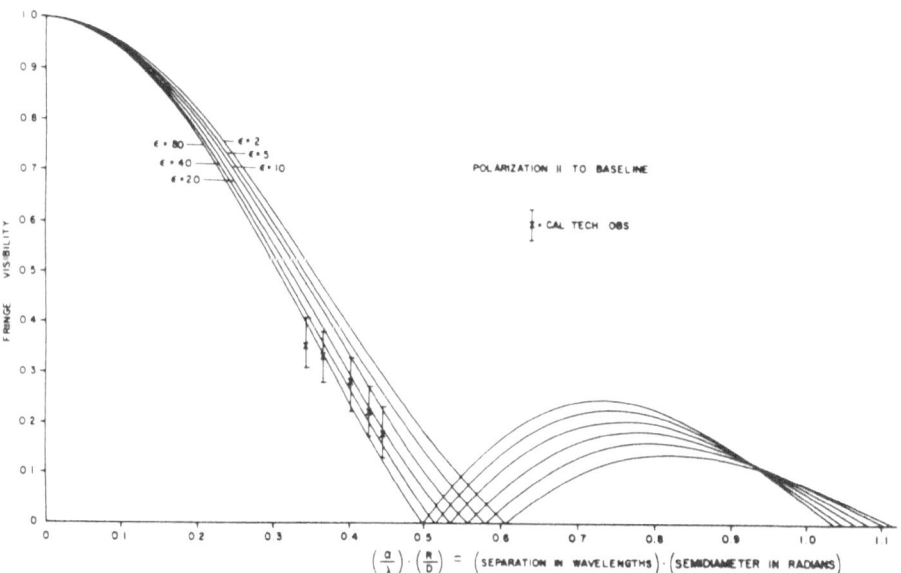

Fig. 1. Fringe visibility curves of Venus for different values of the dielectric constant, ϵ, for the case in which polarization is parallel to the baseline. Several observations of Venus made at the Owens Valley Observatory of the California Institute of Technology are shown. The diagram is taken from Heiles and Drake [31].

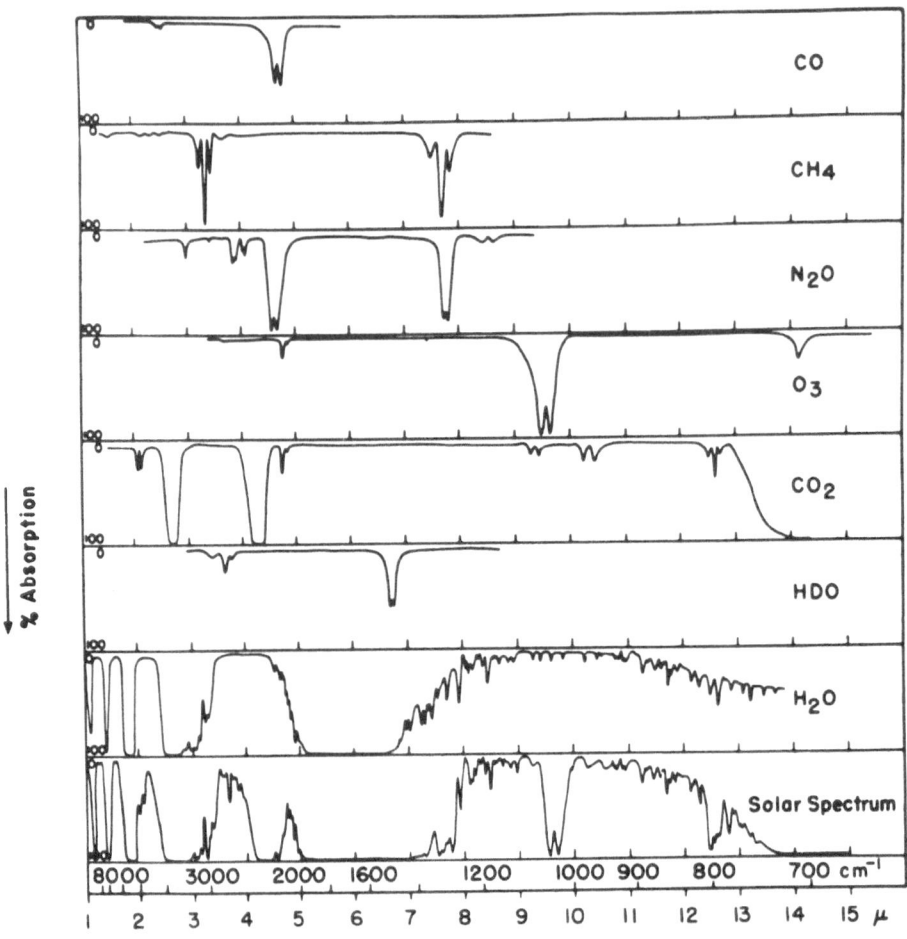

Fig. 2. Spectral features arising from various molecules that might be expected in planetary atmospheres. The diagram is from Howard, Burch, and Williams [33].

SOME APPROACHES TO THE STUDY OF
PLANETARY ATMOSPHERES

Spectroscopic observations — and this term is taken here in its broadest
sense; it includes the case in which the spectrum from, say, 4 mm to 40 cm may
be shown with chromatic resolution, $\lambda/\Delta\lambda$, of 10, as well as the case in which
a very short range of spectra for which $\lambda/\Delta\lambda = 10^5$ and in which highly resolved
lines are exhibited — permit us to make four principal types of atmospheric investigations.

1. Identification of atmospheric gases.
2. Measurement of abundances of atmospheric gases.

TABLE II

Composition of Planetary Atmospheres
(From Urey, 1961)

Planet	Substance	Detected	Amount cm atm (NTP)	Basis of estimate	Remarks
Mercury					Probably fluorescing free radicals and ions produce haze.
Venus	CO_2	Yes	10^5	Spectroscopic	Much below the cloud layer.
	H_2O	Yes?		Spectroscopic	Though once identification seemed certain, new observations make identification very questionable.
	N_2	Yes? ?	?	Spectroscopic	N_2 and N_2^+ bands in the night sky; observation very doubtful.
	CO	Yes? ?	<100	Spectroscopic	CO^+ bands in night sky and absorption bands in day sky. Limit fixed by near infrared bands; observation very doubtful.
Earth	N_2		624,000		
	O_2		167,400		
	H_2O		800 to 22,000		
	Ar		7450		
	CO_2		260		
	Ne		14.6		
	He		4.2		
	CH_4		1.2		
	Kr		0.91		
	CO		0.05 to 0.8		
	SO_2		1		
	H_2		0.4		
	N_2O		0.4		
	O_3		0.25		
	Xe		0.07		
	NO_2		0.0004 to 0.02		
	Rn		5×10^{-14}		
	NO		trace		
Mars	CO_2	Yes	3600	Spectroscopic	Precise amount now subject to question.
	H_2O	Yes	$\sim 10\mu$ precipitable H_2O	Spectroscopic	Polar caps consist of ice. Spectrum of water measured.
	N_2	No	1.8×10^5	Total pressure measured	N_2 is accepted as the most likely noncondensable constituent.
Jupiter	CH_4	Yes	1.5×10^4	Spectroscopic	
	NH_3	Yes	700	Spectroscopic	

TABLE II (Continued)

Planet	Sub-stance	Detected	Amount cm atm (NTP)	Basis of estimate	Remarks
Jupiter (Cont'd)	H_2	No	2.7×10^7		Assumed to be present in
	He	No	5.6×10^6	Density of	solar proportions relative
	N_2	No	4×10^3	the planet	to methane on the basis of
	Ne	No	1.7×10^4		calculations by Marcus.
Saturn	CH_4	Yes	35,000	Spectroscopic	
	NH_3	Yes	200	Spectroscopic	
	H_2	No	6.3×10^7		Assumed to be present in
	He	No	1.3×10^7	Density of	solar proportions relative
	N_2	No	9.5×10^3	the planet	to methane.
	Ne	No	2.7×10^4		
Uranus	CH_4*	Yes	2.2×10^5	Spectroscopic	He and H_2 are assumed to
	H_2	Yes	9×10^6	Calculated on	be effective molecules in
	He	No	2.7×10^7	Herzburg's assumption.	producing transitions of H_2.
	H_2	Yes	4.2×10^6		N_2 and H_2 are assumed to be effective molecules in
	He	No	8.6×10^5	Calculated	producing transitions of
	N_2	No	4.2×10^6		H_2. Solar proportions of He and H_2 assumed.
Neptune	CH_4*	Yes	3.7×10^5	Spectroscopic	
				Spectroscopic	Assumed to be effective
	H_2	Yes			molecules in producing
	N_2	No	Larger than in Uranus		transitions of H_2. Solar
	He	No			proportions of H_2 and He assumed.
Titan	CH_4	Yes	2×10^4	Spectroscopic	
	He	No	- - -		Assumed high diffusion layer for the preservation of H_2.

*Methane has not been considered as the molecule inducing the hydrogen transitions, but it should contribute to this effect.

3. Determination of atmospheric structure on the basis of individual bands or lines.

4. Examination of high-altitude atmospheric emission.
All of these approaches are applicable in the infrared and longward ranges of the spectrum.

A number of excellent reviews of these fields exist. We note particularly volumes edited by Kuiper [45] and Ratcliffe [64], and a monograph by Chamberlain [12]. Goldberg [28] has given an excellent account of the telluric spectrum; Urey [80] has reviewed the atmospheres of the planets, as has Rea [65]. Various papers in the Liege Symposium Volume [46] will be found especially pertinent.

Identifications

In Table II we have listed known constituents of the atmospheres of the planets. In the infrared and millimeter range, absorptions arise from vibration – rotation or pure rotation transitions. The molecular bands in the infrared over the spectral region 1 to 20 μ arise from vibrational states of the molecules. Each vibrational state has a rotational level "fine structure" (i.e., with respect to the basic vibrational energy involved) hence the associated infrared absorption extends over an appreciable spectral region and takes the form of an extended band. A simple criterion for an active vibrational transition can be stated in terms of a classical model, though the details of the separations in energy of the components of a band must, of course, be obtained by quantum mechanics. If the classical mode of oscillation associated with the given vibrational state has associated with it an induced electric dipole moment, that is, if the mode of oscillation generates asymmetry in the molecular charge distribution, the state will be active. Correspondingly, the fine structure arising from rotation of the induced dipole is active. On the other hand, if the mode of oscillation does not result in periodic asymmetrical changes in charge distribution, no induced dipole moment is generated, and the state is inactive. Homonuclear molecules such as N_2 and O_2 have no active infrared levels associated with the simple basic vibrational mode, and a linear symmetrical molecule such as CO_2 has no band associated with the symmetrical mode of oscillation.

The far infrared and millimeter wave regions are associated with purely rotational states of molecules. The classical criterion for spectrally active states is that the molecule possess a permanent dipole moment associated with its vibrational ground state; rotation is equivalent to oscillating dipole moments in the space-fixed reference system and, hence, gives rise to electromagnetic activity. Homonuclear molecules such as N_2 and O_2, which, by virtue of their symmetry have no dipole moments, are inactive in the far infrared and millimeter regions.

However, some rotational transitions and vibration–rotational transitions may become active because of higher order effects. O_2, for example, has weak forbidden bands at 1.06 and 1.27 μ, and much stronger ones at 0.762 and 0.688 μ, arising from forbidden magnetic dipole radiation.

The inversion spectrum of ammonia (NH_3), an axially symmetrical pyramidal molecule, is an example of a fine-structured rotational spectrum. In addition to the active rotational transitions associated with the presence of the permanent dipole moment of the equilibrium pyramidal configuration, the states are split by inversion of the pyramid which produces a 180° change in the direction of the dipole moment in the frame of reference of the molecule.

Minor atmospheric constituents provide most of the bands observable in the infrared and longward spectral region. Therefore, the spectroscopist interested in planetary atmospheres should consider the entire spectral range from ultraviolet to microwave his proper province if he wishes to make a maximum contribution. Since we discuss only one portion of the spectrum, these remarks are correspondingly incomplete and fragmentary.

Figure 2 taken from the well-known publication by Howard, Burch, and Williams [33] exhibits infrared spectra for several molecules that might be expected

TABLE III

Possible Molecular Lines in Planetary Atmospheres

We indicate whether Kuiper or Barrett has suggested the compound. Under the column headed "found, suspected to be likely or possible," the planets Venus, Mars, Jupiter, Saturn, Uranus, and Neptune are indicated by initial. A capital letter indicates that the compound has been found, a small letter that the compound has been mentioned as a particular possibility.

Discovery of many of the molecules will be difficult since some decompose photochemically, others polymerize and drop out of the atmosphere. The discussion by Kuiper is very helpful on these points. We have omitted a few inert gases. When information is available we indicate the existence of an observed microwave spectrum (or its absence) and the strength (or presence or absence) of an observed infrared spectrum.

Gas	Suggested by Kuiper (K) or Barrett (B)	Found, suspected to be likely, or possible	Microwave resonance spectrum	Infrared spectrum
H_2	KB	E, j, s, u, n	No	s
He	KB	E, j, s, u, n	No	s
N_2	KB	v, E, m, j, s, u, n	No	No
O_2	KB	v, E, m	Yes	vs
OH	B	E	Yes	vs
O_3	KB	v, E, m	Yes	s
Ne	KB	v, E, m, j, s, u, n	No	No
A	KB	v, E, m, j, s, u, n	No	No
H_2O	KB	v, E, M	Yes	vs
CO_2	K			Yes
HF	K			
HCl	K			s
SO_2	KB	E	Yes	s
SO_3	K			s
H_2S	KB	j, s	Yes	vs
NO	KB	E	Yes	vs
NO_2	KB	v, E, m	Yes	vs
N_2O	KB	v, E, m	Yes	vs
N_2O_3	K			
NOF	K			
NOCl	K			
NH_3	K	E, J, S	Yes	vs
H_2O_2	B		Yes	vs
N_3H	K	j, s		s - vs
HNO_3	K			s
PH_3	K	j, s		s
CO	KB	v, E, m	Yes	vs
CS	B		Yes	Yes
CO_2	KB	V, E, M	No	vs
C_3O_2	K			vs
COS	KB		Yes	s
CS_2	KB		No	s
HCN	KB		Yes	vs
HCNO	B		Yes	Yes
CH_2N_2	K			
CH_2O	KB	E	Yes	vs
CH_2O_2	B		Yes	vs

TABLE III (Continued)

Gas	Suggested by Kuiper (K) or Barrett (B)	Found, suspected to be likely, or possible	Microwave resonance spectrum	Infrared spectrum
CH_4	KB	v, E, m, J, S, U, N	No	vs
CNSH	B		Yes	
C_2N_2	KB	j, s	No	s
C_2H_2	KB		No	s
C_2H_2O	KB		Yes	Yes
C_2H_4	K			s - vs
C_3HN	B		Yes	Yes
SiH_4	K	j, s, u, n		Yes

in planetary atmospheres. A more inclusive list of possible molecules in planetary atmospheres is given in Table III. These have been taken from Kuiper [45] and Barrett [4]. Following Barrett, we have not listed compounds containing more than five atoms.

Identification of many (if not all) of these compounds in planetary atmospheres will be difficult because the same compounds in the earth's atmosphere, or strong terrestrial lines from water, will obscure them. Balloon-borne and space-probe experiments have obvious advantages for studying such spectral components. From the surface of the earth, infrared observations must generally be confined to wavelength ranges in which there are atmospheric windows, mainly in the 3.5-μ region, and in the range 8 to 14 μ. Between 14 and 24 μ there is a region of limited transparency. At longer wavelengths, from 24 μ to approximately 1000 μ, the earth's atmosphere is made almost completely opaque by the pure rotational spectrum of water. In the millimeter range, the atmosphere is partially transparent from wavelengths of a few millimeters to about 10 cm. At longer wavelengths, until we reach the wavelength of the ionospheric cutoff, it is almost completely transparent. The principal sources of absorption in the millimeter range is a rotational band of H_2O at 1.35 cm and a series of O_2 lines at 5 and 2.5 mm as shown in Fig. 3. Great advantage is to be gained in attaining minimal interference from the water vapor by going to high mountains in dry regions or, with the present state of technology, by making observations from balloons at altitudes of 80,000 ft or more. Alternatively, one can, in some instances, make optical or infrared observations during the times at which the planet has a large radial velocity with respect to the earth, and thus use the planetary Doppler shift to carry a planetary line out from behind its terrestrial counterpart. Large telescopes and spectrographs of very high spectral resolution are required for such observations. Detailed investigations from satellites or planetary probes are highly desirable, but much instrumental development is required before adequate infrared or microwave equipments are ready to fly.

Barrett [4] has discussed microwave spectroscopy of planetary atmospheres. He points out that planetary radiation belts or dense ionospheres will not hinder observations at short centimeter or millimeter wavelengths. On the other hand, clouds in a planetary atmosphere may exert a strong influence on the planetary

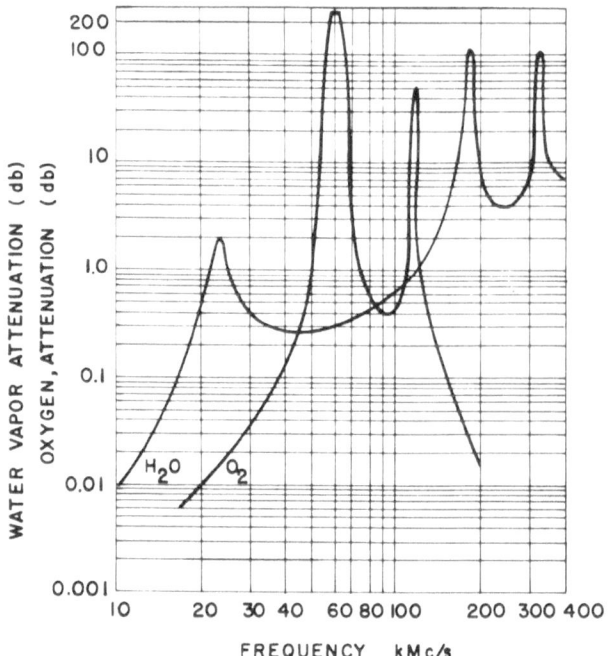

Fig. 3. Absorption bands of water vapor and oxygen.

spectrum, generally at shorter wavelengths rather than longer ones. Of particular interest would be clouds of variable density and of material having sufficient absorption and scattering strengths so that one might find variability in the planetary spectrum. Considerable meteorological information might be gained from radio observations of a planet having such clouds.

It is not likely that microwave lines will be especially useful in identifying molecular species in planetary atmospheres. In Table III we note the occurrence of infrared bands as well as microwave lines or bands for the molecules in Barrett's list. All the molecules have strong or very strong bands in the infrared. It is likely that identification will be simpler in the infrared than in the microwave range.

Abundances of Atmospheric Gases

Abundance determination for an atmospheric constituent involves measurement of the equivalent widths or total absorptions of lines or bands arising from the constituent. The relation between the equivalent width of an absorption line and the abundance and physical state of the absorbing gas is given by the curve of growth, which has found frequent use in stellar research. The curve of growth relates the equivalent width of a line to the line absorption coefficient and the number of atoms in the path. Many investigations of the curve of growth exist, but these refer mostly to the case for a homogeneous atmosphere at constant temperature and pressure. The earth's atmosphere or a planetary atmosphere is not homogeneous; pressure, temperature, and density of the absorbing gas all vary with altitude. An atmospheric model is required; this complicates the discussion, but, in some instances, leads to new atmospheric information. Curves of growth techniques as applied to the atmosphere of the earth are discussed in the article by Goldberg [28].

Occasionally for the earth and quite generally for the other planets, equivalent widths of entire bands rather than individual lines within a band must be used in the discussion. High angular resolution on the planetary surface is desired, since it is convenient to have the spectrum refer to a specific path through the atmosphere. However, in many instances the entire planetary disk must be included in the observation; this degrades the quality of the data. It is very difficult to devise an appropriate weighting function allowing for contributions to the spectrum from various parts of the disk, from each of which there is a different slant path through the atmosphere.

For interpretation of entire bands, the observer will find indispensable the empirical curves of growth for important atmospheric constituents published by Howard, Burch, and Williams [33], Burch and Williams [9, 10], and Howard, Burch, and Williams [34, 35]. Recent discussions of slant depth transmissions through a nonhomogeneous atmosphere will also be helpful; see, for example, Howard and Garing [32].

Atmospheric Structure

Astronomers have long used the shapes of spectral lines and limb darkening to determine the atmospheric structure of the sun and the stars. The techniques developed are applicable to the planets also. High angular resolution is requisite for such investigations. For the line-profile observation one must look along a single path through the atmosphere; for the limb darkening one might preferably use such a pencil beam or, possibly, but less desirably, a fan beam which can be swept across the planetary disk.

The problem of determining the radiative-equilibrium law of darkening and vertical thermal structure for a nongray, finite, plane–parallel atmosphere heated from below by uniform, isotropic radiation has been solved by King [42]; the theory has been applied to Venus by Sinton [72].

Kaplan [39, 40] and others have suggested the use of artificial satellites to observe the 15-μ CO_2 band profile to determine the large-scale temperature structure of the earth's atmosphere; Lilley [47] and Lilley and Meeks [48] have proposed that the 5-mm blend of O_2 lines be used to derive the high-altitude temperature structure; Barrett [4] and Barrett and Chung [5] have discussed how H_2O microwave bands can be used to determine atmospheric structure. Barrett [4] has provided a number of diagrams that show sensitivity of line profile to the vertical distribution of H_2O. He points out, though, that a unique determination of the molecular distribution can be made only if there is available an accurate temperature profile throughout the atmosphere. All these proposals are applications of the same physical principles. The specific observational data required for such applications are measurements of intensity (or temperature) at several different specified wavelengths, λ_i, within the spectral band. The specified λ_i's are so chosen that the atmospheric opacity varies considerably from one λ_i to another. As a result, at any λ_i the contributions of different layers of the atmosphere to the measured intensity of radiation or temperature of λ_i vary markedly. Inversion of the radiative transfer equations that obtain at the selected λ_i's will then permit determination of the temperature of the atmosphere as a function of pressure.

The principle of using band profiles for determination of atmospheric structure is difficult to apply in practice because observations of very high accuracy are required. The results are quite sensitive to small observational errors; high angular resolving power is required; a single path through the atmosphere is desirable. It is likely that the best results could be obtained by application of infrared and microwave results simultaneously, since the several observations would provide various restraints that would provide a well-determined solution.

Goldberg [28] has discussed means of determining the vertical distribution of a molecular species by fitting a curve of growth. Very high spectral resolution is required. No one of the methods discussed here for planetary atmospheric studies holds immediate promise because of the high angular resolution and observational accuracy that is required.

Spinrad [75, 76] has observed that the NH_3 lines in the spectrum of Jupiter have tilts (which are caused by planetary rotation) which differ significantly from those of the atmospherically scattered solar lines. A similar observation for the CH_4 lines in Saturn has been made by Münch and Spinrad [54]. These observations, which clearly provide us with atmospheric structure data, are not yet interpretable physically.

Airglow and Auroras

We have gained considerable knowledge of the upper atmosphere of the earth from studies of the airglow and auroras. It is highly likely that similar phenomena exist on the planets with atmospheres, and that new data on the planets can be gained from observing them. Because of lack of significant phase variations in the outer planets, Venus may be the only object observable for airglow and auroras. For this planet the lack of a magnetic field will be of considerable interest in conjunction with these atmospheric effects.

On the earth, the near infrared region of the nightglow is dominated by OH; we may expect (Chamberlain and Smith [13]) strong bands in the 1 to 5-μ region. However, from the surface of the earth, no OH beyond 2.5 μ has been detected because of the strong thermal radiation which arises from the lower atmosphere. The spectrum differs greatly from that of a blackbody, and changes little from day to night. The most prominent emission features in the nightglow arise from CO_2, O_3, and H_2O in the region shortward of 15 μ. Longward of 15 μ, N_2O is a dominant feature (Sloan, Shaw, and Williams [73] and Burch and Shaw [8]). It is likely that for infrared studies of the nightglow, observations from high-flying balloons would be very much worthwhile. Many new bands will undoubtedly be discovered. Swings [78] suggests that [OI] at 63 and 147 μ may be seen.

Kozyrev [43, 44] and Newkirk [57] have reported seeing airglow spectra on Venus. Later, in 1961, Weinberg and Newkirk [86] were unable to repeat Newkirk's experiment with improved apparatus and concluded that the existence of emission on the dark side of Venus is uncertain. The question of planetary airglow requires reexamination.

A dayglow exists on the earth; it should contain spectral features arising from a variety of atoms and molecules found in the airglow. Brandt [7] predicted that some infrared lines of [OI] should be strong in the dayglow. Lines of [OI] at 6300.3 and 6363.8 Å were first reported in the dayglow by Noxon and Goody [59] and Noxon [58]. Observations of the dayglow [OI] lines extending over some time suggest that at least two physical processes are involved in their production. Barth [1] has observed NO in the dayglow.

There may be some interest in looking for the dayglow in other planets. In a spectral region very different from that under consideration in this report, Urey and Brewer [81] suggest that there is appreciable fluorescence in planetary atmospheres.

PLANETARY MAGNETIC FIELDS

The discovery of low-frequency nonthermal radiation from the planet Jupiter and its interpretation in terms of a Jovian Van Allen belt is one of the major planetary discoveries of radio astronomy. Through studies of such nonthermal radiation, we can investigate the magnetosphere of the planet and the interactions between that magnetosphere and the energetic solar particles that strike it. Only a few years ago the direct study of planetary magnetic fields appeared to be quite beyond the realm of possibility for an earth-bound astronomer. Now such studies are well within his range. It is even possible that we shall learn more about some aspects of Van Allen belts by studying radio radiation from Jupiter than is possible by investigating the earth's belt, since we can see the Jovian belt as an entity.

BRIEF REVIEWS OF OBSERVATIONS OF THE PLANETS

Radio astronomy has provided us with much new and exciting information about the planets. This has been accomplished in a very short period of time through very great efforts on the part of radio astronomers. It must be pointed out, however, that only a start in planetary radio astronomy has been made. Only two planets, Venus and Jupiter, have been observed in any spectral detail, and even those have not been thoroughly studied. In particular, Jupiter has been very little observed in the high-frequency portion of the spectrum, and even Venus has not been observed in that range with the accuracy and precision that are desirable. Observations of Venus at wavelengths longward of 10 cm are particularly lacking. Aside from Venus and Jupiter, the planets have not been observed much, and a great deal remains to be done.

Mercury

Howard, Barrett, and Haddock [36] have observed Mercury in the X-band range at wavelengths of 3.75 and 3.45 cm. They report a mean disk temperature of $350°K$. Under several simplifying assumptions — namely, that the planet is smooth, uniform, synchronously rotating, has no atmosphere, and exhibits limb darkening proportional to $\cos^{1/4} \theta$, where θ is measured from the subsolar point — Howard, Barrett, and Haddock have, from the observed mean disk temperature of $350°K$, observed for the temperature of the subsolar point the value of $1050 \pm 350°K$. This is a high value compared to the infrared temperature of the subsolar point ($613°K$). The high subsolar temperature, which results from a high observed mean disk temperature, has been interpreted by Field [23] as arising from an unexpectedly high temperature for the unilluminated side of the disk. Walker [83], assuming that the radioactive heat generation in Mercury is equal to that of the chondritic meteorites, finds, for a thermal steady state, that the un-illuminated side of Mercury would be at a temperature of $25°K$ if no other heat source were present. If thermal conductivity from the illuminated side of the planet is included in the calculation, the temperature is $28°K$. Polarimetric observations of the sunlit side of Mercury can be interpreted in terms of a thin atmosphere of surface pressure of the order of 1 mm of Hg. Radiogenic argon is presumed to be the principal constituent of the atmosphere. If the argon is not to freeze out on the dark side of the planet, the temperature of the coldest spot must be greater than $56°K$. A large increase in radioactivity over that assumed by Walker could account for such a temperature. Alternatively, atmospheric circulation of approximately 10 m/sec with temperature differences of the order of $10°K$ would also suffice. However, the observed near agreement between insolation and the observed infrared emissions at the subsolar point suggests that less than 3% of the total heat is carried by convection to the dark side of the planet. If the heat carried away by convection heats the nonilluminated part of the planet, the dark side temperature is less than $250°K$. However, before excessive speculation on the disk side temperature of Mercury is based on these results, it might be well to note that the X-band temperature value depends upon the 49 drift curves of Mercury made when the planet was in the range 19 to $28°$ from the sun. The mean antenna temperature derived from the 49 observations is $0.05 \pm 0.01°K$. Clearly more observations are needed.

Venus

Venus has received a large share of the observing time devoted to planets by radio astronomers. The reason for such attention is clear: at conjunction the planet has a large angular diameter, about $1'$ of arc; it is a hot object readily observed with a conventional receiver; most of all, Venus possesses a strange microwave spectrum: at a wavelength of 4 mm the radiation received is characteristic of a temperature of approximately $300°K$, while at 4 cm or more the radiation is characteristic of a temperature of approximately $600°K$. A diagram showing the spectrum derived from the observations reported to date is shown in Fig. 4. The observational data are listed in Table IV.

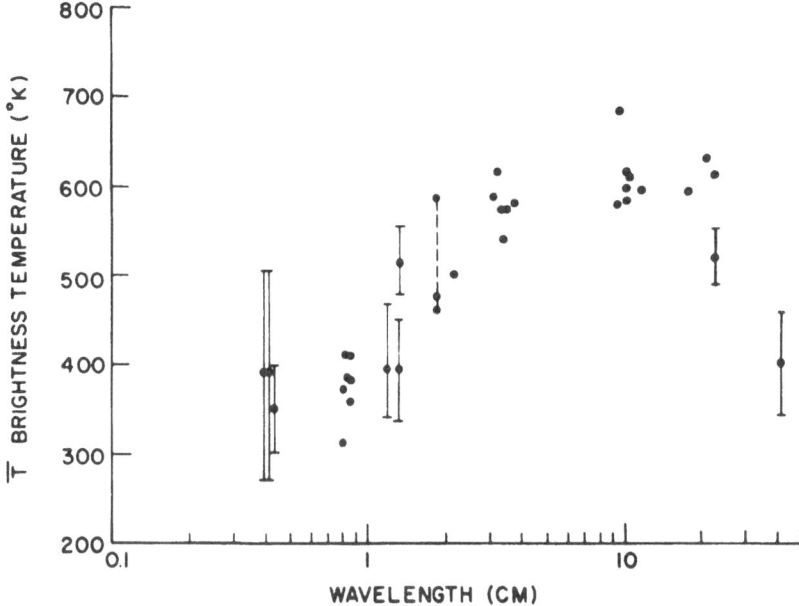

Fig. 4. Observed spectrum of Venus. The diagram is taken from the paper by Barrett and Staelin [6].

TABLE IV

Observations of Venus in the Radio Range
(Adapted from Barrett and Staelin [6])

Wavelength (cm)	T_b (°K)	Range of phase angle* (deg)	Reference
0.4	390 ± 120[†]	143 to 260	a, b
0.4	390 ± 120	216 to 270	c
0.43	350 + 50, −30	113 to 262	d
0.8	375 ± 75[†]	143 to 260	a, e
0.8	315 ± 70[†]	216 to 271	f
0.835	383 ± 60	130 to 165	g
0.85	380 + 72, −34	126	h
0.86	360[†]	174 to 230	i
0.86	410 + 30, −20	167 to 222	j
0.86	410 ± 160	180	k
1.18	395 + 75, −55	126	l
1.35	393 ± 100	**	m
	400 ± 100	**	m
	396 ± 100	**	m
1.35	520 ± 40	∼165	n

TABLE IV (Continued)

Wavelength (cm)	T_b ($^\circ$K)	Range of phase angle* (deg)	Reference
1.90	480 ± 25	**	m
	590 ± 30	**	m
	460 ± 25	**	m
2.07	500 ± 70	285 to 299	o
3.15	621†	~120 to 304	p
3.15	595 ± 55	106 to 186	q, r
3.3	542 ± 85	211 to 281	a, s
3.37	575 ± 58	275	t
3.4	575 ± 60	212 to 241	u
3.75	585‡	101 to 238	v
9.4	580 ± 160	214	q, r
9.6	690 ± 104	143 to 260	a
10.0	622‡	100 to 285	w, x
10.0	610 ± 55	14 to 37	y
10.2	600 ± 65	214 to 245	u
10.7	580 ± 60	149 to 220	z
12.5	600 ± 200	203 to 230	aa
18.0	596 ± 100	121	z
21.0	630 ± 200	165	bb
21.0	616 ± 100	107 to 111	z
21.4	528 ± 33	218	x
40.0	400 ± 60	218	x

* Phase angle is defined here as the sun−Venus−earth angle measured from the sun−Venus line in the direction of Venus' orbital motion. Thus angles less than 180° are prior to inferior conjunction and angles greater than 180° are after, except for minor effects of the noncoplanar orbits of Venus and the earth.

† These observations have been shown to be phase-dependent over the range of phase angles indicated.

‡ No estimate of error given which includes the usually large uncertainty in antenna calibration.

** The temperatures are the peak temperatures of the high-resolution scans by the Mariner II spacecraft. The concept of phase is not applicable in the context of the other entries in the table.

[a] V. P. Bibinova, A. G. Kislyokov, A. D. Kuzmin, A. E. Salomonovich, and I. V. Shavlovsky, *Mem. Soc. Roy. Sci. Liege,* 5^{ieme} Ser. 7, 331 (1962).

[b] A. G. Kislyakov, A. D. Kuzmin, and A. E. Salomonovich, *Astron. Zh.* 39, 410 (1962) [English translation: *Soviet Astron. - AJ* 6, 328 (1962)].

[c] A. D. Kuzmin and A. E. Salomonovich, *Izv. Radio Phys.* 4, 573 (1961).

[d] C. R. Grant, H. H. Corbett, and J. E. Gibson, *Astrophys. J.* 137, 620 (1963).

[e] A. D. Kuzmin and A. E. Salomonovich, *Astron. Zh.* 39, 660 (1962) [English translation: *Soviet Astron. - AJ* 6, 518 (1963)].

[f] A. D. Kuzmin and A. E. Salomonovich, *Astron. Zh.* 37, 297 (1960).

[g] D. D. Thornton and W. J. Welch, *Astrophys. J.* 69, 71 (1964).

[h] V. L. Lynn, M. L. Meeks, and M. D. Sohigian, *Astrophys. J.* 139, 409 (1964).

[i] J. Copeland and W. C. Tyler, *Astrophys. J.* 139, 409 (1964).

[j] J. E. Gibson, *Astrophys. J.* 137, 611 (1963).

[k] J. E. Gibson and R. J. McEwan, *Paris Symp. Radio Astron.* (Stanford Univ. Press, Stanford, Calif., 1959), p. 50, R. N. Bracewell, ed.

[l] D. H. Staelin, A. H. Barrett, and B. R. Kusse, *Astrophys. J.* 69, 69 (1964).

[m] F. T. Barath, A. H. Barrett, J. Copeland, D. E. Jones, and A. E. Lilley, *Astrophys. J.* 69, 49 (1964).

The basic explanation of the form of the spectral energy distribution is that the low temperature in the infrared and millimeter regions refers to the top of the cloud deck of the perpetually cloud-covered Venus. The totally unexpected high temperature observed in the centimeter region of the spectrum refers to (1) the surface of the planet which is not visible in the optical, infrared, and millimeter regions of the spectrum and is, indeed, at a temperature of approximately 600°K, or (2) an ionosphere, or (3) some physical process that acts as a microwave generator in the atmosphere. Reviews of these three types of models of Venus have been provided by a number of authors: Kellogg and Sagan [41], Sagan [68, 69], Sagan and Kellogg [70], Roberts [66], and Barrett and Staelin [6]. The models are treated here in barest detail.

The Greenhouse Model

In this model it is assumed that the observed high temperature truly exists at the surface of the planet; the generally flat temperature-wavelength variation longward of 3 cm (omitting the 40-cm point which may be inaccurate) is taken as indicative of blackbody radiation. The high temperature of the surface is maintained by an efficient greenhouse effect. Solar radiation in the visible and near infrared ranges differs through the atmosphere and heats the surface of the planet. The surface of the planet emits predominately in the longer infrared region to which, it is presumed, the atmosphere is essentially opaque. The outflow of radiation is effectively retarded; the surface of the planet and the lower atmosphere heat up; an equilibrium condition at high temperature is finally reached. Various plausible atmospheric mixtures supplying the necessary infrared opacity and involving nitrogen, carbon dioxide, and water vapor have been studied (Sagan [68, 69], Barrett [3]). It now appears that the presence of water vapor on Venus is very doubtful, hence high pressure carbon dioxide – nitrogen models have been investigated by Barrett and Staelin [6].

---

[n] J. E. Gibson and H. H. Corbett, *Astrophys. J.* **68**, 74 (1963).

[o] T. P. McCullough and J. W. Boland, *Astrophys. J.* **69**, 68 (1964).

[p] C. H. Mayer, T. P. McCullough, and R. M. Sloanaker, *Mem. Soc. Roy. Sci. Liege*, 5ieme Ser. 7, 357 (1962).

[q] C. H. Mayer, T. P. McCullough, and R. M. Sloanaker, *Astrophys. J.* **127**, 1 (1958).

[r] C. H. Mayer, T. P. McCullough, and R. M. Sloanaker, *Proc. IRE* **46**, 260 (1956).

[s] V. P. Bibinova, A. D. Kuzmin, A. E. Salomonovich, and I. V. Savlovsky, *Astron. Zh.* 40, 154 (1963 [English Translation: *Soviet Astron.-AJ* 7, 116 (1963)].

[t] L. E. Alsop, E. A. Giordmaine, C. H. Mayer, and C. H. Townes, *Paris Symp. Radio Astron.* (Stanford Univ. Press, Stanford, Calif., 1959), p. 69, R. N. Bracewell, ed.

[u] C. H. Mayer, T. P. McCullough, and R. M. Sloanaker, Paper presented to the Thirteenth General Assembly, URSI (London, September 5–15, 1960).

[v] F. D. Drake, quoted by C. H. Mayer, "Planets and Satellites," *The Solar System*, G. P. Kuiper and B. M. Middlehurst, eds. (The Univ. of Chicago Press, Chicago, Ill., 1961), Vol. VIII, Chap. 12.

[w] F. D. Drake, *Publ. Natl. Rad. Astron. Obs.* 1(11), 165 (1962).

[x] F. D. Drake, *Astrophys. J.* **69**, 62 (1964).

[y] F. D. Drake, *Nature* 195, 3 (1962).

[z] B. G. Clark and C. L. Spencer, *Astrophys. J.* **69**, 591 (1964).

[aa] C. T. Stelzried and D. Schuster, Jet Propulsion Lab. Tech. Rept. No. 32-132 (1961), W. K. Victor, R. Stevens, and S. W. Golomb, eds., pp. 74–76.

[bb] A. E. Lilley, *Astrophys. J.* **66**, 290 (1961).

Assuming that the surface of Venus is at a temperature of 700°K, various authors have tried to explain the observed radio range spectrum by means of absorption and scattering in clouds alone (water clouds, Deirmendjian [17], and other types of clouds, Barrett and Staelin [6]).

The Aeolospheric Model

Öpik [61] considered the greenhouse effect to be inadequate to produce the observed temperature and has proposed an aeolospheric model of the atmosphere of Venus in which the blanketing is due to dust. Friction of the dust blowing over the surface of the planet provides the source of energy for the high temperature. Öpik suggests a dust composed of calcium and magnesium carbonates; the atmosphere is primarily CO_2 and N_2; no water need be present. Important calculations of the expected microwave spectrum of Venus for aeolospheric models have been presented by Barrett and Staelin [6].

The Ionospheric Model

D. E. Jones [38] has suggested the existence of an ionosphere around Venus. The large concentration of free electrons in the ionosphere will, through free – free transitions, provide a source of radiation which we interpret in terms of a high temperature. In the ionospheric model the temperature at the surface is not unduly high; the apparent high temperature in the observed wavelength interval above 3 cm is an ionospheric radiation effect.

An Electrical Discharge Model

Tolbert and Straiton [79] have proposed a model in which there are fluctuations of electrical charges on particles in the atmosphere of Venus. Water droplets would, at the same time they produced the discharge, attenuate it by absorption. Tolbert and Straiton suggest crystalline materials as the atmospheric particles.

A Plasma Model

The magnetic field of Venus is very weak. Scarf [71] assumes that the solar proton wind strikes Venus and interacts with the ionosphere of the planet. A plasma instability is then presumed to grow, with consequent radiation at high harmonics of the plasma frequency.

Observational results rule against some of the models fairly clearly; no model comes through the observational test without some difficulty. Jastrow and Rasool [37] have pointed out that the atmospheric opacity predicted by Sagan [68, 69] is inadequate to produce the observed temperature on the basis of the greenhouse model. The failure of Gibson and Corbett [25], working at 1.25 cm, and Spinrad [76, 77], working in the infrared, to observe water vapor is also contrary to the models used by Sagan [68, 69] and Barrett [3]. On the other hand, Ohring [60] has pointed out that these models did not include the infrared opacity of the clouds. Inclusion of the effects of the clouds in the radiative transfer problem suggests that the greenhouse effect with only CO_2 may be adequate to account for the observed results. Barrett and Staelin [6] have reexamined Barrett's high pressure

$CO_2 - N_2$ model which has no water in the atmosphere, and have tried to represent the observed spectrum. A very high lapse rate must be used; moreover, atmospheric surface pressures of 200 atm are required for the model to represent the observations even approximately. On the other hand, the observations of Venus made from Mariner II show limb darkening for the planet. Such an effect is what might be expected for a planet with positive temperature gradient toward the surface as would be the case for the greenhouse model or the aeolospheric model. Observations made by Spinrad [75, 76] support the assumption that Venus has a hot surface. Spinrad observed the CO_2 band at 7820 Å and made use of line shape and intensity distribution within the band to determine pressure and temperature. He found that pressure and temperature vary markedly from observation to observation, that is, from day to day, as though one were looking to different depths as atmospheric conditions vary with time. Pressure and temperature were correlated positively; the maximum conditions found by Spinrad were a pressure of 6 atm and a temperature of $440°K$.

If, in the aeolospheric model, small particles are assumed to be present (as suggested by Öpik [61]) so that only absorption takes place, very high density of dust must be assumed (100 gm/m^3 at the ground level for quartz dust), and even then the predicted temperature-wavelength relation does not closely resemble the observed microwave curve. Lower densities suffice for other substances, but these do not permit a satisfactory fit to the observed spectrum. Barrett and Staelin [6] have demonstrated that particles of larger size (up to 0.6 mm in diameter), for which scattering as well as absorption must be considered and for which ground-level densities of the order of 10 gm/m^3 must be postulated, permit a better (but still not satisfactory) representation of the observed spectral curve. One can include with the dust a $CO_2 - N_2$ atmosphere of about 20 atm pressure and obtain a relatively good fit to the observations or, alternatively, add selective absorption to the properties of the dust. However, both of these approaches are purely *ad hoc* and do not provide a convincing model.

Results derived from the aeolospheric model are in disagreement with the phase effect observed by Drake [18-20] or Haddock and Dickel [30], who find that the sunlit side of the planet is hotter than the terminator region. Öpik [62] states that the hottest part of the planet should be along the terminator.

The ionosphere model is rejected by 68-cm radar observations (Muhleman [53]); the observed electron density is far below what is required by the ionospheric model. No way to produce the very high electron densities required by the model is known. Additionally, the limb darkening found by Barath, Barrett, Copeland, Jones, and Lilley [2] from Mariner II observations is contrary to predictions of the ionospheric model.

The amount of rain or dust required to maintain the emission in the wavelength range longward of 3 cm is large, but not impossibly so. However, the discharging particles would, presumably, produce limb brightening rather than limb darkening as observed. Also the high temperature derived from the CO_2 bands tends to support a real temperature at the surface rather than a discharge phenomenon.

Drake [18, 19] has pointed out that the time constancy of the microwave spectrum of Venus is evidence against any model depending upon the interaction of solar corpuscular emission and the atmosphere of Venus. If solar corpuscular emission were directly involved in production of the microwave spectrum, the spectrum should change with the level of solar activity. Such variation does not appear to be present.

At the present time it appears that a greenhouse model of some sort in which a surface of the planet is at an elevated temperature provides the most likely explanation of the microwave spectrum of Venus.

A number of observers have determined the phase function for Venus. There is complete agreement that the effect is such as to indicate retrograde rotation of the planet. The period of rotation is approximately 250 days.

Mars

The planet Mars has so far been little investigated in the radio range of wavelengths. A number of infrared studies of Mars have been made; enough, in fact, to permit some generalizations in regard to climatic conditions on the planet (de Vaucouleurs [82]). For the infrared average disk temperature of Mars, Menzel, Coblentz, and Lampland [52] find the range to be 237 to 254° K. Only three radio observations of Mars have been reported. Mayer, McCullough, and Sloanaker [50, 51] averaged 70 observations at wavelength 3.15 cm to obtain an apparent blackbody disk temperature of 218 ±50° K. Later, Giordmaine, Alsop, Townes, and Mayer [27] used a maser at 3.14-cm wavelength to derive an apparent blackbody disk temperature of 211 ±20° K. Drake [20], working at 10-cm wavelength, found the apparent blackbody disk temperature to be 177 ±17° K. Drake attributes the lower microwave temperature to the effect of surface emissivity changing with wavelength, the smoother surface at the longer wavelengths resulting in smaller emissivitives in the region near the limbs.

Much work remains to be done on the planet Mars, both in the radio wavelength range and in the infrared. If we suitably modify and apply methods developed for the moon, we should be able to learn much about the nature of the Martian surface material.

Jupiter

Observed in the radio range, the planet Jupiter, like Venus, provides new insights into the physics of the planetary system: Venus into the properties of planetary atmospheres, and Jupiter into properties of planetary magnetospheres.

The spectrum of Jupiter in the centimetric and decimetric ranges is shown in Fig. 5. The spectrum in the decametric range is highly variable and cannot be properly illustrated on the diagram. The spectrum may be divided into three wavelength regions on the basis of the form of the spectrum and/or the variability of the radiation.

1. In the centimetric range and shortward, the radiation is of thermal origin; the radiant flux is consistent with what would be expected from a blackbody hav-

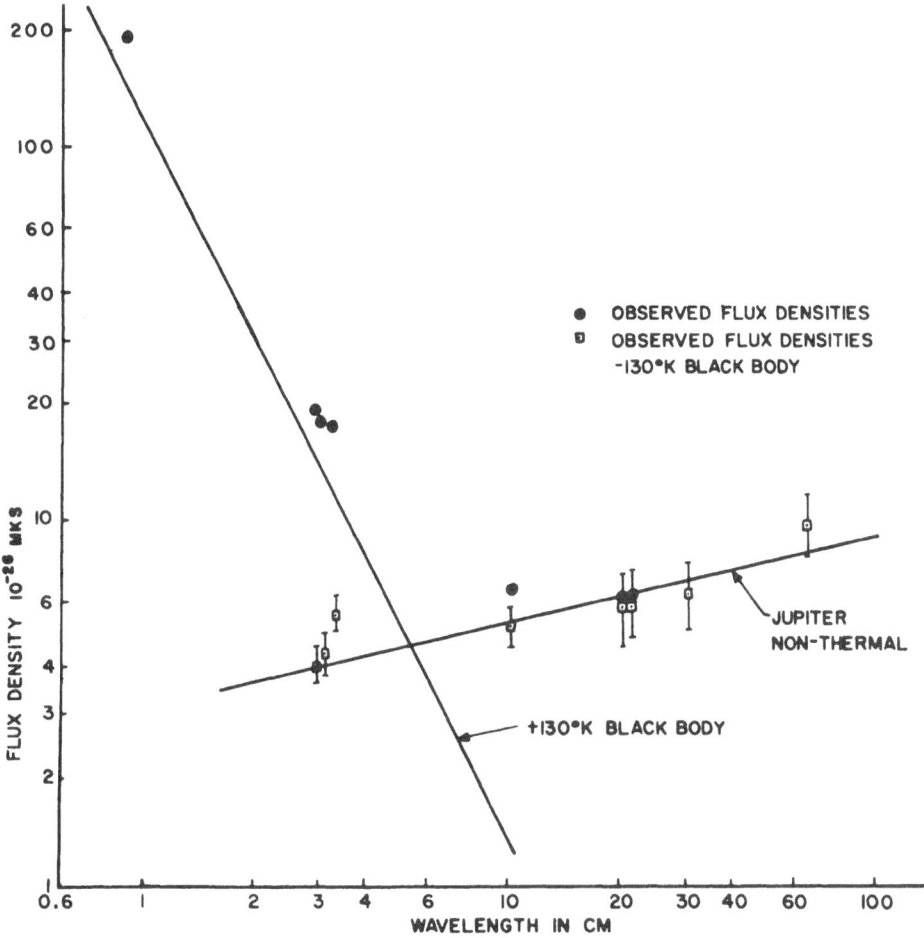

Fig. 5. The spectrum of Jupiter in the centimetric and decimetric ranges.

ing a temperature of $130°$ K, the infrared value found by Menzel, Coblentz, and Lampland [52].

2. Throughout the decimetric range, the flux is constant and far in excess of what is to be expected from a blackbody at a temperature of $130°$ K; it is non-thermal radiation arising, as we shall see, from the Van Allen belts of Jupiter.

3. In the decametric range, the emission, clearly nonthermal in origin, occurs in sporadic very short bursts, which may continue for minutes or hours. The frequency range in which these bursts are observed is restricted, the extreme limits being 4.8 to 43 Mc/s. The decametric radiation appears to originate in a few restricted areas on or near the surface of Jupiter. The radiation originating in a specific area is generally characterized by a certain general range of frequencies, frequency direction of drift, and a fairly consistent polarization.

Jupiter, observed accidentally in the decameter range during a period of high
activity, was the first planet to be observed in the radio range. It has attracted
much attention from radio astronomers since its first discovery as a radio source
in 1954 – 1955, and has been the subject of many reviews and summarizing articles
(Mayer [49], Burke [11], Gallet [24], Newburn [56], Wildt, Smith, Salpeter, and
Cameron [88], Roberts [66], Haddock [29], Smith and Carr [74], and Warwick
[85]). Many references to earlier papers will be found in these articles. In view
of the large number of reviews readily available, because of the great complexity
of the phenomena that have now been observed, and because of the detailed nature
of some of the model calculations necessary to reproduce the observations, we
shall deal only briefly with the observations and theories of the phenomena observed.

It was early discovered that the decametric bursts did not originate uniformly
from all longitudes of the planet. The majority of the bursts were observed when
a certain narrow range of longitudes, of the order of $40°$ wide at the half-frequency
point, faced the earth. (This range of longitudes may be termed the "main source.")
Additionally, there are two other longitude ranges from which decametric radiation
also preferentially emanates. (These are called the "early source" and the "late
source.") The three sources are well separated in longitude; the main source is
substantially more active than the others. The longitude dependence for all decametric data for a period of 12 years is shown in Fig. 6; the three sources appear
as the three lobes of the figure. Long-base-line interferometry indicates that the
sources are small, with angular extents $\leq 15''$.

The decametric radiation emanating from a restricted area on Jupiter's surface must additionally be confined within a narrow cone of the order of 10 to $20°$
between half-power points and beamed at the earth. That this must be so can be
seen from the following argument. Radiation, originating in one small area of the
planetary surface, but not so confined and beamed, but rather emitted over a hemisphere, would not appear to come from a narrow longitude range facing the earth
as is observed to be the case. It would be seen whenever the emitting area was
in the hemisphere facing the earth.

The decametric emission occurs in bursts which last from one to a few seconds; there is very likely fine structure within such bursts. Bursts that persist
for a few seconds often drift rapidly in frequency, 10 Mc/s in a characteristic
period of 10 sec. Events that last minutes or hours are composed of bursts or
groups of bursts. These longer-lived events may drift slowly in frequency, possibly 10 Mc/s in 10 min. Drifts may take place in either direction, from low
frequency to high frequency or high frequency to low frequency. Warwick [85]
notes that the directions of the slower drifts during events are correlated with
longitude. In the early source the frequency drift is from low frequency to high
frequency, while in the main and late sources the drift is in the opposite sense.

Higher frequencies in the decametric range, 30 to 40 Mc/s, are restricted to
certain longitudes. Radiation of frequency 4.8 Mc/s appears to be present at all
times, but varies in intensity as a function of the Jovian longitude of the sub-
earth point. The higher frequencies occur more nearly at the longitudes of the
sources; the distribution lobes (Fig. 6) become more directive the higher the fre-

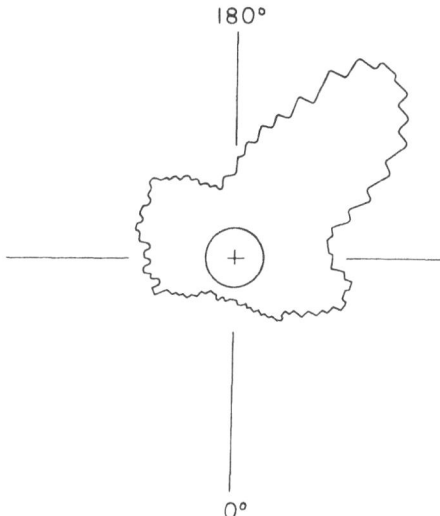

180°

0°

Fig. 6. The longitude distribution of
decametric radiation from Jupiter. (The
diagram is from Wildt, Smith, Salpeter,
and Cameron [88]).

quency of radiation. Certain ranges of frequencies appear to be characteristic of certain longitudes.

The fact that specifiable frequency ranges and a certain direction of frequency drift are characteristic of a certain longitude range implies that, in spite of the bursty and variable nature of the decametric radiation, each of the sources has a permanent (but not always exactly reproduced) dynamic spectrum. This significant observation has been stressed by Warwick [85] and is important in the explanation of the decametric radiation.

The decametric radiation is polarized. In the middle- to high-frequency range, 20 to 30 Mc/s, polarization is essentially always right-handed circular or elliptical. Below 20 Mc/s changing polarizations occur, especially in the third source. At 10 Mc/s the right-handed sense still predominates except in the early source which is left-handed.

Bandwidths of ½ Mc/s are seen in some fine structures in the dynamic spectrum. Longer duration bursts or burst groups generally possess wider bandwidths up to 5 to 10 Mc/s. The general correlation between decametric emission and solar activity is negative.

The nonthermal radiation in the decimetric range is very different in observational character and behavior from the nonthermal radiation in the decametric range. The radiation in the decimetric range does not appear to vary in intrinsic intensity; if it does so vary, the rate of variation must be very slow. There is no agreement between observers on the question of variation of intensity of the decimetric radiation with solar activity. The spectrum of the decimetric radiation is flat between the frequency limits of the observations, 3200 and 178 Mc/s.

The emitting region for the decimetric radiation is considerably larger than the visible disk of Jupiter. Radhakrishnan and Roberts [63], from their observations with the two-element interferometer at the Owens Valley Observatory of the

California Institute of Technology, deduced that the volume emitting the decimetric radiation is, in the equatorial direction, 3.3 times the polar diameter if the
radiation has a Gaussian distribution of intensity, or 3.7 times the polar diameter
if the intensity distribution is made up of two symmetrical Gaussian curves. The
polar extent of the radio source of the decimeter radiation is of the same order as
the polar diameter of the visible disk of Jupiter.

The decimetric radiation is linearly polarized, 30% at frequency 960 Mc/s,
decreasing toward higher frequencies; polarization appears to be greatest at the
edge of the emitting volume. The direction of polarization rocks through an angle
of approximately 20° in each rotation of Jupiter; the period of rocking is very
precise.

There is a small periodic variation in the intensity of the decimetric radiation;
this appears to result from occultation of the emitting volume by the solid disk of
the planet. The variation of intensity, the periodic change of direction of polarization, and the degree of polarization are shown in Fig. 7.

It is now generally agreed (in view of the form and size of the emitting volume
and the linear polarization of the radiation) that in the decimetric range of radiation we are examining the Van Allen belt of Jupiter as originally suggested by
Drake and Hvatum [21]. Models of Jupiter's Van Allen belts computed by Chang
[14] indicate that all observational aspects of the radiation can be accounted for,
though a definitive model, which will be quite complicated, has not been derived.

Briefly, we may say that the observed rocking of the plane of polarization
indicates that the magnetic axis of Jupiter is inclined at an angle of about 8° to
the axis of rotation. Occultation of parts of the emitting volume by the solid disk
of the planet will take place as the planet and its tipped Van Allen belt rotate.
To construct a model that permits prediction of the observed relations between,
and time rates of variation of, the total decimeter radiation intensity, the degree
of polarization, and the direction of polarization, it is necessary to postulate an
inner and an outer Van Allen belt. For the outer belt the electrons move in rather
flat helices; in the inner belt a broader, less restricted distribution of pitch angles
is predicted. For such a model one predicts that the degree of polarization will
increase from the center to the edge of the emitting volume. Such a center-to-edge
increase in degree of polarization is observed.

There is no generally accepted model or theory of the decametric bursts as
there is for the decimetric nonthermal radiation from Jupiter. The model proposed
by Warwick [84] provides the most complete and satisfactory explanation of the
decametric radiation; it is directly related to the model of the Van Allen belts by
which we have accounted for the decimetric radiation. Warwick assumes that the
decametric bursts are produced when energetic electrons are precipitated from the
belts by magnetic disturbances. The fast electrons spiral down around the lines
of force and generate Čerenkov radiation as they enter the ionosphere. Emission
occurs at frequencies near the gyro frequency and is directed toward the planet.
The radiation is reflected from the denser layers of the ionosphere or from the
planetary surface. It is important to note that only for suitable orientations of
the magnetic field and the ionospheric layers or the surface of the planet will this
reflected radiation strike the earth. The tipped and eccentric position of the di-

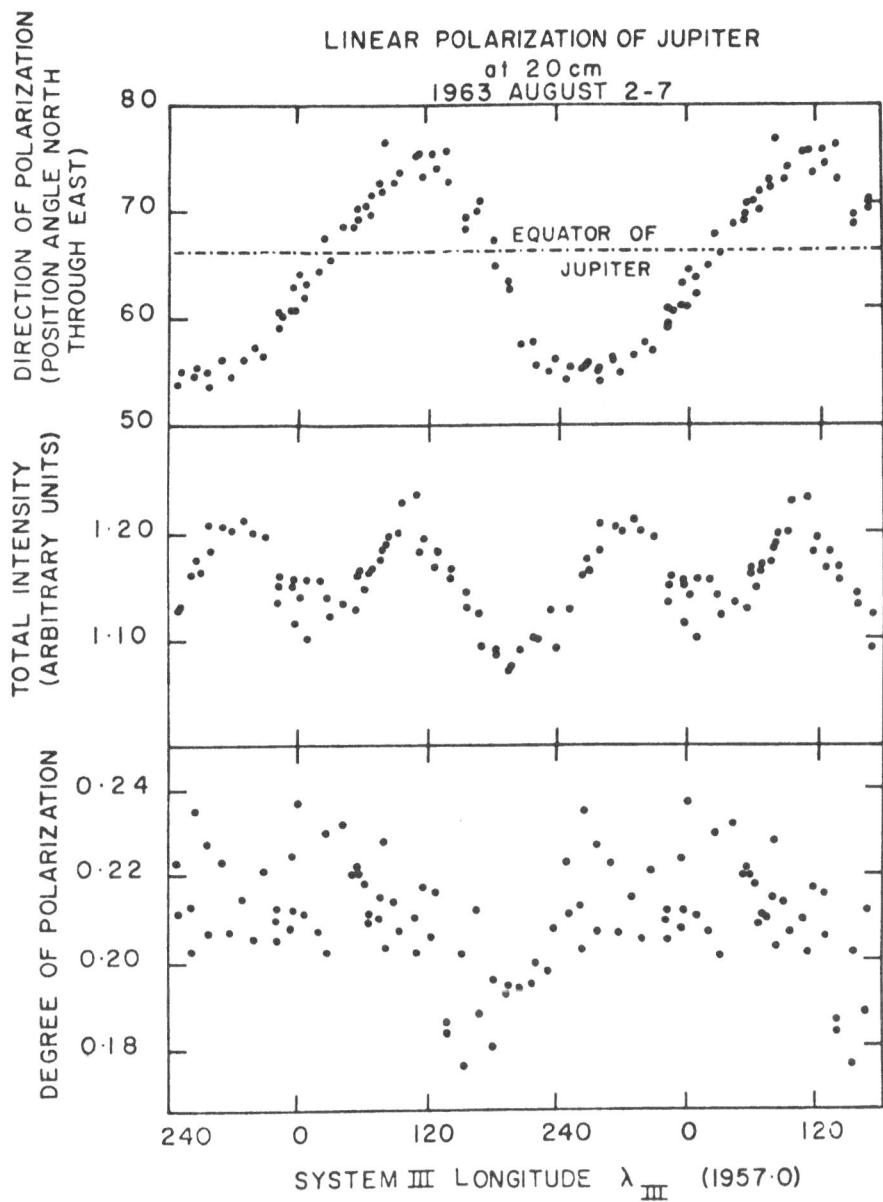

Fig. 7. The total intensity, polarization, and direction of polarization of decimetric radiation shown as a function of the longitude of the central meridian of Jupiter. The diagram is from the paper by Warwick [85].

pole field of Jupiter and the consequently displaced radiation belts account for the longitude locations of the three principal Jovian decametric sources. The physical process by which radiation is generated and the reflection properties of the ionosphere (or surface) account for the narrow emission distribution in the sources. The dynamic spectrum predicted from the model closely agrees with the permanent dynamic spectrum described by Warwick.

The principal features of the Warwick model are illustrated in Fig. 8. The tilt of the magnetic axis is in agreement with the value derived from the decimetric radiation. The center of the dipole is displaced from the center of Jupiter by about 0.7 radius from the equatorial plane; it lies behind the plane of the paper by about 0.1 radius. The Warwick model appears to account for all of the major observational features of the decametric radiation; it predicts that the northern pole of Jupiter must be a magnetic north-seeking pole, which is opposite to what we find on the earth. It is to be hoped that for this interesting model, which brings together so many features of the decimetric and decametric radiation, more observational tests of many types can be made in the future.

Few centimetric and shorter wavelength observations of Jupiter exist. As more observations are made, it is highly likely that interesting discoveries will be made. The Jovian atmosphere contains NH_3, CH_4, and other materials so far undetected. Since a fundamental NH_3 line falls in the spectral range 9 to 13 μ, the 130° K infrared temperature derived by Menzel, Coblentz, and Lampland [52] from measurements made in the 8- to 14-μ window may not have general applicability. The measured temperature refers to some weighted mean level in the atmosphere in which the optical depth in the 9- to 13-μ region of the spectrum becomes large. A rather different temperature may be found in a different wavelength interval.

Gallet has suggested a very complicated Jovian atmosphere containing separate layers of high-altitude ammonia ice-crystal clouds (below which ammonia rainstorms may take place) and lower level water clouds below which water rainstorms may occur. Whatever the precise structure of the atmosphere, it is likely that there is a complicated infrared, millimeter, and centimeter spectrum of the planet. We will therefore see to quite different depths and hence will observe quite different temperatures in different wavelengths.

Giordmaine [26] has interpreted the 3-cm observations of Jupiter in terms of emission from the pressure broadened NH_3 line at wavelength 1.28 cm; Welch and Thornton [87] have made a similar interpretation for their observation at wavelength 8.35 mm. Field [22] and Giordmaine [26] have calculated the probable wavelength variation in Jupiter's thermal radiation in the centimeter range in the spectrum. Much more work, both theoretical and observational, needs to be done in this subject. The few observations now available in the centimeter range combined with our current knowledge of the atmosphere of Jupiter point clearly to the fact that the infrared, millimeter, and centimeter wavelength ranges of the Jupiter atmosphere constitute a rich and interesting field for research.

Saturn

Few observations of Saturn exist; those that do are contradictory. Menzel, Coblentz, and Lampland [52] report an infrared temperature of 128° K for Saturn, but Murray and Wildey [55] report that they were unable to detect the planet in the 8 to 13 μ range, and therefore assign an upper limit of 105° K. The theoretically computed temperature lies below this limit.

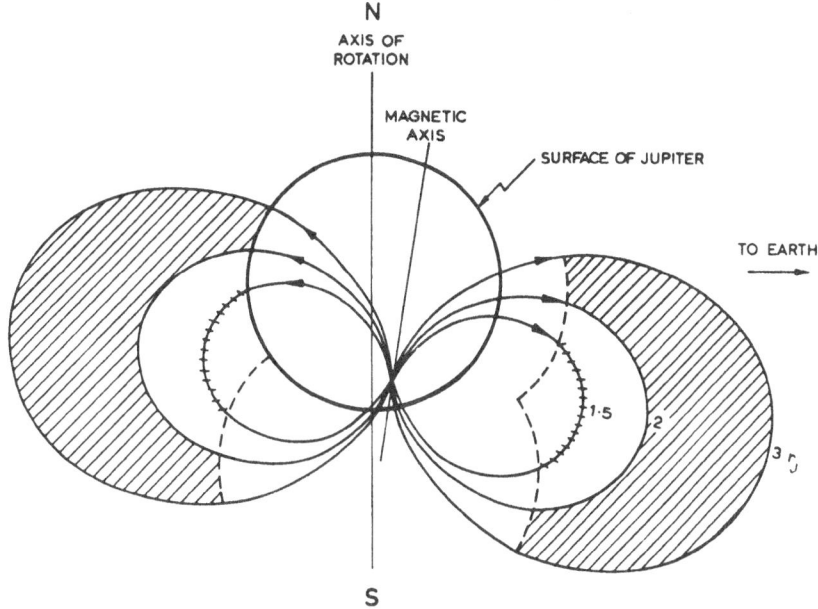

Fig. 8. The Warwick model of Jupiter showing the axis of rotation, the tilted and displaced magnetic axis, and the decentered Van Allen belts. The diagram is from the paper by Warwick [85].

Saturn was first reliably measured in the radio range at 3.45 cm by Cook, Cross, Bair, and Arnold [15], who derived the disk temperature of $106 \pm 21°$K. Observing at 10 cm, Drake [18, 19] found a temperature of $196 \pm 44°$K, and suggested that nonthermal radiation might be present.

Rose, Bologna, and Sloanaker [67] observed at 9.4 cm with an antenna for which they could change the plane of polarization accepted. They found black-body disk temperatures of $140°$K with the electric vector oriented parallel to the equator of Saturn, and $213°$K for the case with the electric vector perpendicular to the equator. This indicates strong linear polarization. However, Cooper, Beard, and Davies [16] find that the polarization is only 6% at 10 cm. Clearly, more observations are needed before any firm conclusions can be drawn in regard to Saturn.

ACKNOWLEDGMENTS

Parts of this review are closely related to and updated from a paper by H. Weaver and S. Silver entitled "Planetary Research in the Millimeter and Infrared Region of the Spectrum" and presented at the XIVth General Assembly of the International Scientific Radio Union, Tokyo, Japan, September 10, 1963. The excellent discussion of Jupiter by J. W. Warwick [85] has been of great help in preparing the review of Jupiter.

This document was prepared in large part under Contract Nonr 222 (66).

REFERENCES

1. C. A. Barth, *J. Geophys. Res.* 69, 3301 (1964).
2. F. T. Barath, A. H. Barrett, J. Copeland, D. E. Jones, and A. E. Lilley, *Science* 139, 908 (1963).
3. A. H. Barrett, *Astrophys. J.* 133, 281 (1961).
4. A. H. Barrett, in: Liege Symp. Eleventh Intern. Colloq. on Astrophys., Liege, Belgium, *Mem. Soc. Roy. Sci. Liege* 5^{ieme} Ser. 7, 197 (1962).
5. A. H. Barrett and V. K. Chung, *J. Geophys. Res.* 67, 4259 (1962).
6. A. H. Barrett and D. H. Staelin, *Space Sci. Rev.* 3, 109 (1964).
7. J. C. Brandt, *Astrophys, J.* 130, 228 (1959).
8. D. E. Burch and J. H. Shaw, *J. Opt. Soc. Am.* 47, 227 (1957).
9. D. E. Burch and D. Williams, *Appl. Opt.* 1, 473 (1962).
10. D. E. Burch and D. Williams, *Appl. Opt.* 1, 587 (1962).
11. B. F. Burke, in: *Planets and Satellites*, G. P. Kuiper and B. M. Middlehurst, eds., *The Solar System, Vol. III* (Univ. of Chicago Press, Chicago, Ill., 1961), p. 473.
12. J. W. Chamberlain, *Physics of the Aurora and Airglow* (Academic Press, New York, 1961).
13. J. W. Chamberlain and C. A. Smith, *J. Geophys. Res.* 64, 611 (1959).
14. D. B. Chang, *Synchrotron Radiation as the Source of Polarized Decimeter Radiation from Jupiter*, Ph.D. thesis, Calif. Inst. of Tech. (1962); see also D. B. Chang and L. Davis, Jr., *Astrophys. J.* 136, 567 (1962).
15. J. J. Cook, L. G. Cross, M. E. Bair, and C. B. Arnold, *Nature* 188, 393 (1960).
16. Cooper, Beard, and Davies, quoted in C. H. Mayer, *Radio Astronomical Studies of the Planets and the Moon* (Draft Reports, Comm. 40, T.A.U., 1964).
17. D. Deirmendjian, *Icarus* 3, 109 (1964).
18. F. D. Drake, *Nature* 195, 893 (1962).
19. F. D. Drake, *Publ. Natl. Radio Astron. Obs.* 1, 165 (1962).
20. F. D. Drake, quoted in D. G. Rea and W. J. Welch, *Space Sci. Rev.* 2, 611 (1963).
21. F. D. Drake and S. Hvatum, *Astron. J.* 64, 329 (1959).
22. G. Field, *J. Geophys. Res.* 64, 1169 (1959).
23. G. B. Field, *Astron. J.* 67, 575 (1962).
24. R. M. Gallet, in: *Planets and Satellites*, G. P. Kuiper and B. M. Middlehurst, eds., *The Solar System, Vol. III* (Univ. of Chicago Press, Chicago, Ill., 1961), p. 434.
25. J. E. Gibson and Corbett, *Sky Telescope* (1963), p. 1.
26. J. A. Giordmaine, *Proc. Natl. Acad. Sci.* 46, 267 (1960).
27. J. A. Giordmaine, L. E. Alsop, C. H. Townes, and C. H. Mayer, *Astron. J.* 64, 332 (1959).
28. L. Goldberg, in: *The Earth as a Planet*, G. P. Kuiper, ed., *The Solar System, Vol. II* (Univ. of Chicago Press, Chicago Ill., 1954), p. 434.
29. F. T. Haddock, *Radio Emission and Radar of the Moon and Planets: Recent Progress (1960–1963)*, Presented to the Fourteenth Gen. Assembly, URSI, Tokyo, September 1963.
30. F. T. Haddock and J. R. Dickel, *Trans. Am. Geophys. Union* 44, 886 (1963).
31. C. E. Heiles and F. D. Drake, *Icarus* 2, 281 (1963).
32. J. N. Howard and J. S. Garing, *Infrared Phys.* 2, 155 (1962).
33. J. N. Howard, D. L. Burch, and D. Williams, *Geophys. Res. Papers* (U.S.) 40, Rept. AFCRL-TR-55-213 (1955).
34. J. N. Howard, D. E. Burch, and D. Williams, *J. Opt. Soc. Am.* 46, 237 (1956).
35. J. N. Howard, D. E. Burch, and D. Williams, *J. Opt. Soc. Am.* 46, 242 (1956).
36. W. E. Howard, A. H. Barrett, and F. T. Haddock, *Astron. J.* 66, 287 (1961).
37. R. Jastrow and S. I. Rasool, in: *Space Res., Proc. Intern. Space Sci. Symp. 3rd COSPAR* (Washington, D.C., 1962) W. Priester, ed. (North Holland Publ. Co., Amsterdam, 1963).
38. D. E. Jones, *Planetary Space Sci.* 5, 166 (1961).
39. L. D. Kaplan, *J. Opt. Soc. Am.* 49, 1004 (1959).
40. L. D. Kaplan, *J. Quant. Spectry. Radiative Transfer* 1, (2) (1961).
41. W. W. Kellogg and C. Sagan, *Publ. Natl. Acad. Sci.* 944, (Natl. Res. Council, Washington, D.C., 1961).
42. J. I. F. King, *Astrophys. J.* 124, 272 (1956).
43. N. A. Kozyrev, *Izv. Krymsk. Astrofiz. Observ.* 12, 169 (1954).

44. N. A. Kozyrev, *Izv. Krymsk. Astrofiz. Observ.* 12, 177 (1954).
45. G. P. Kuiper, ed. *The Atmospheres of the Earth and Planets* (Univ. of Chicago Press, Chicago, Ill., 1952), revised ed.
46. *Liege Symp.*, Eleventh Intern. Colloq. on Astrophys., Liege, July 9–11, 1962, *Mem. Soc. Roy. Sci. Liege*, 5^{ieme} Ser. 7, fascicule unique.
47. A. E. Lilley, in: *Space Age Astronomy* (Proc. Symp. Calif. Inst. Technol., Pasadena, Calif., 1961), A. J. Deutsch and W. B. Klemperer, eds. (Academic Press, New York, 1962), p. 253.
48. A. E. Lilley and M. L. Meeks, *J. Geophys. Res.* 68, 1683 (1963).
49. C. H. Mayer, in: *Planets and Satellites* G. P. Kuiper and B. M. Middlehurst, eds., *The Solar System, Vol. III* (Univ. of Chicago Press, Chicago, Ill., 1961), p. 442.
50. C. H. Mayer, T. P. McCullough, and R. M. Sloanaker, *Astrophys. J.* 127, 1 (1958).
51. C. H. Mayer, T. P. McCullough, and R. M. Sloanaker, *Proc. IRE* 46, 260 (1958).
52. D. H. Menzel, W. W. Coblentz, and C. O. Lampland, *Astrophys. J.* 63, 177 (1926).
53. D. O. Muhleman, *Icarus* 1, 401 (1963).
54. G. Münch and H. Spinrad, *Liege Symp.* (1962), p. 541.
55. B. C. Murray and R. L. Wildey, *Astrophys. J.* 137, 692 (1963).
56. R. L. Newburn, Jr., in: *Advances in Space Science and Technology, Vol. 3*, F. I. Ordway, ed. (Academic Press, New York, 1961), p. 196.
57. G. Newkirk, *Planetary Space Sci.* 1, 32 (1959).
58. J. F. Noxon, *J. Geophys. Res.* 69, 3245 (1964).
59. J. F. Noxon and R. M. Goody, *J. Atmospheric Sci.* 19, 342 (1962).
60. G. Ohring, and O. Coté, *Geophys. Corp. Am. Tech. Rept. No. 63-6-N* (1963).
61. E. J. Öpik, *J. Geophys. Res.* 66, 2807 (1961).
62. E. J. Öpik, *Irish Astron. J.* 6, 59 (1963).
63. V. Radhakrishnan and J. A. Roberts, *Phys. Rev. Letters* 4, 493 (1960).
64. J. A. Ratcliffe, ed., *Physics of the Upper Atmosphere* (Academic Press, New York, 1960).
65. D. G. Rea, *Space Sci. Rev.* 1, 159 (1962).
66. J. A. Roberts, *Planetary and Space Sci.* 11, 221 (1963).
67. W. K. Rose, J. M. Bologna, and R. M. Sloanaker, *Phys. Rev. Letters* 10, 123 (1963).
68. C. Sagan, *Icarus* 1, 151 (1962).
69. C. Sagan, in: *Space Age Astronomy*, A. J. Deutsch and W. B. Klemperer, eds. (Academic Press, New York, 1962), p. 425.
70. C. Sagan and W. W. Kellogg, in: *Ann. Rev. Astron. Astrophys.* Leo Goldberg, ed. (Annual Reviews, Inc., Palo Alto, Calif., 1963), p. 235.
71. F. L. Scarf, *J. Geophys. Res.* 68, 141 (1963).
72. W. M. Sinton, in: *Planets and Satellites*, G. P. Kuiper and B. M. Middlehurst, eds., *The Solar System, Vol. III* (Univ. of Chicago Press, Chicago, Ill., 1961), p. 429.
73. R. Sloan, J. H. Shaw, and D. Williams, *J. Opt. Soc. Am.* 45, 455 (1955).
74. A. G. Smith and T. D. Carr, *Radio Exploration of the Planetary System* (D. Van Nostrand Co., Inc., Princeton, N. J., 1964).
75. H. Spinrad, *Icarus* 1, 266 (1962).
76. H. Spinrad, *Publ. Astron. Soc. Pacific* 74, 187 (1962).
77. H. Spinrad, *Astrophys. J.* 136, 311 (1962). See also H. Spinrad and L. M. Trafton, *Icarus* 2, 19 (1963).
78. P. Swings, in: *Space Age Astronomy*, A. J. Deutsch and W. B. Klemperer, eds. (Academic Press, New York, 1962), p. 9.
79. C. W. Tolbert and A. W. Straiton, *J. Geophys. Res.* 67, 1741 (1962).
80. H. Urey, *Handbuch der Physik*, Vol. 52, S. Flügge, ed. (*Astrophysics III: The Solar System*) (Springer, Berlin, 1959).
81. H. Urey and A. W. Brewer, *Proc. Roy. Soc. (London)* A241, 37 (1957).
82. G. de Vaucouleurs, *Physics of the Planet Mars* (Macmillan, New York, 1954).
83. J. C. J. Walker, *Astrophys. J.* 133, 274 (1961).
84. J. W. Warwick, *Astrophys. J.* 137, 41 (1963).
85. J. W. Warwick, in: *Ann. Rev. Astron. Astrophys.* Leo Goldberg, ed. (Annual Reviews, Inc., Palo Alto, Calif., 1964), p. 1.
86. J. L. Weinberg and G. Newkirk, *Planetary Space Sci.* 5, 163 (1961).
87. W. J. Welch and D. D. Thornton, *Space Sci. Lab. Ser. No. 4*, Issue No. 25 (Univ. of Calif., Berkeley, Calif., 1963).
88. R. Wildt, H. J. Smith, E. E. Salpeter, and A. G. W. Cameron, *Phys. Today* 16, 19 (1963).

Chapter 18

Planetary Radar Astronomy

Gordon Pettengill

Arecibo Ionospheric Observatory*
Arecibo, Puerto Rico

INTRODUCTION

The extension to the planets of radar techniques in use against the moon presents a severe technical challenge. The closest and most easily detected object beyond the moon is Venus, but even under the most ideal conditions this planet returns an echo some 5 million times weaker than does the moon. Nevertheless, there now exist a number of facilities with sufficient sensitivity to observe not only Venus, but Mars, Mercury, and perhaps Jupiter.

Far less is known about the planets than about the moon and, therefore, there is potentially more value in applying radar methods to the study of planetary surfaces. The techniques are generally the same, but because of the weaker echo signal strengths, more compromises with data resolution must be made. It is only in the past year or so that radar information on planetary surfaces is becoming available.

Since the distance to the planets is greater than 100 times the distance of the moon, even relatively inaccurate range measurements represent a very useful fractional precision in total time delay. Because range and Doppler measurements explore a coordinate which can be determined only poorly by optical methods, they nicely complement the accurate angular positions already available.

Of particular importance is the extension to the planets of the range-Doppler mapping techniques discussed in connection with the moon. Since the resolution inherent in these methods depends primarily only on the signal-to-noise available, and not on angular resolution, they may yield far more information on surface details of distant planets than visual observation. In the case of Venus and Jupiter, the greater penetrating power of radio wavelengths also permits investigation of surface layers inaccessible to optical observers. An immediate and obvious application of this capability is to the problem of determining rotation rates where these are poorly known because of distance or obscuration.

---

* Operated by Cornell University with the support of ARPA under contract with AFOSR.

HISTORY

The first serious discussion of the possibilities and technical difficulties of using radar to explore the planets was given by Kerr [7]. It was not until 1958, however, that sufficient radar capability became available at the Lincoln Laboratory of Massachusetts Institute of Technology (MIT) to justify an attempt to observe Venus. In their report of this attempt (Price *et al.* [19]), success was claimed on the basis of two observations made on February 10 and 12, 1958. However, the signals were quite weak, barely exceeding 3 standard deviations of the accompanying noise and were obtained only after extensive analysis on a digital computer. On the basis of the agreement in Doppler shift with the expected value and the self-consistency of the values obtained on the two days, it was felt that the results were genuine. A radar cross section for the planet of approximately πa^2 (where a is the planetary radius) was obtained. However, measurements made at subsequent close approaches of the planet have produced a different value for the orbital dimension, and a very much lower value for the cross section than the earlier work. Therefore, it must be assumed that for some reason a spurious response was present which gave rise to the early results.

At the following conjunction in 1959, using much the same equipment, Price and Pettengill [20] failed to observe echoes from Venus. Evans and Taylor [3], working at the Jodrell Bank Experimental Station of the University of Manchester, England, reported a successful result in agreement with the 1958 measurement. However, their signal was also weak, and they did not exclude the possibility that it might have been caused by noise alone.

By the inferior conjunction of spring 1961, sufficient capability had become available to permit attempts to observe Venus by groups not only at the Lincoln Laboratory (Pettengill *et al.* [17]) and Jodrell Bank (Thomson *et al.* [22]), but also at the Jet Propulsion Laboratory (JPL) of the California Institute of Technology (Victor and Stevens [23]), the Radio Corporation of America (Maron *et al.* [13]) and in the Soviet Union (Kotelnikov *et al.* [9]). The 1961 measurements were in reasonably good agreement concerning the values of the astronomical unit which were obtained (shown in Table I), and in the cross sections – approximately $0.1 \pi a^2$. The JPL and MIT results required that the planet rotate very slowly on its axis, perhaps as slowly as its orbital rate around the sun, while the Soviets reported observations which led to a rotational period of 10 or 11 days.

The inferior conjunction of fall 1962 was again observed by the Soviets (Kotelnikov *et al.* [11]), and JPL (Goldstein [4] and Carpenter [1]). By measuring the way in which the bandwidth of the returned signal varied with the date of observation, JPL was able to establish that the planet possessed a retrograde rotation of about 250 days. The Soviets were able to reproduce their results of 1961 from which they had deduced a rotation of about 10 days, but could not satisfactorily explain them; a 300-day rotation was claimed.

In addition, in 1962, two new groups reported observations of radio echoes from Venus. Working at 50 Mc/s, Klemperer *et al.* [8] reported a cross section of approximately $0.2 \pi a^2$. In common with the groups working at higher frequencies, he also noted that the Venusian surface slopes were significantly smoother than the moon. A series of measurements at 38 Mc/s were also reported by James and Ingalls [6]. They found a mean cross section of about 0.15 times the geometrical, but with substantial variation from day to day.

In June 1962, radar detection of Mercury was reported by the Soviets (Kotelnikov et al. [10]), using much the same equipment and techniques as were used shortly afterward to observe Venus. The signal was not strong; however, detection seems reasonably certain. A fractional cross section of about 0.06 was obtained, which is substantially less than for Venus, but comparable with the value obtained for the moon. In 1963, JPL also reported detection of Mercury with no severe disagreement with the earlier Soviet result.

Radar measurements of Mars have been reported during the opposition of early 1963 by both JPL (Goldstein and Gillmore [5]) and the USSR (Kotelnikov et al. [12]). Both groups of workers find a fractional cross section which varies according to the region of Mars under observation from 0.02 to over 0.10. Both are in agreement that Mars possesses a surface which at radar wavelengths is significantly flatter than Venus. Because of the rapid rotation of Mars, and the consequent Doppler broadening of the returned signal, it was not possible to obtain an accurate value for the astronomical unit with the continuous wave systems used.

In the fall of 1963, Jupiter was observed by both the Soviets and JPL. Since Jupiter possesses a very deep and presumably absorbing atmosphere, it might be expected that signals would not be returned at frequencies high enough to penetrate the ionosphere. Nevertheless, both groups of workers have reported success, although few details have been released.

Currently, efforts are underway to observe Venus during the 1964 inferior conjunction by groups at JPL, the Arecibo Ionospheric Observatory of Cornell University, MIT, Jodrell Bank, and in the USSR. Although unpublished, the results of these measurements appear to confirm the earlier conclusions that Venus is smoother than the moon and is rotating with a retrograde period of about 250 days.

TECHNIQUES AND RELATIVE DETECTABILITIES

As was brought out in the introduction, the planet most easily detected by radar is Venus, which at inferior conjunction still returns an echo 5 million times (or 67 db) weaker than the moon. This enormous gap in level was spanned in less than 15 years as radar systems and techniques were improved. While it seems unlikely that the next 15 years will see a corresponding further improvement, it is interesting to calculate which targets would be uncovered with an additional level of performance. Figure 1 plots the relative signal strength to be expected from the various planets and their moons under the assumption that they scatter like Venus. Although this assumption is now known not to be strictly true for the inner planets, and is likely to be considerably in error for the giant planets, it still serves as an approximate guide. Along the abscissa is shown the echo delay for each target. The dates shown for Mars and Mercury mark the value for the corresponding closest approach when this occurs in the months shown. Perhaps the most immediate and obvious conclusion is how large a fraction of the solar system

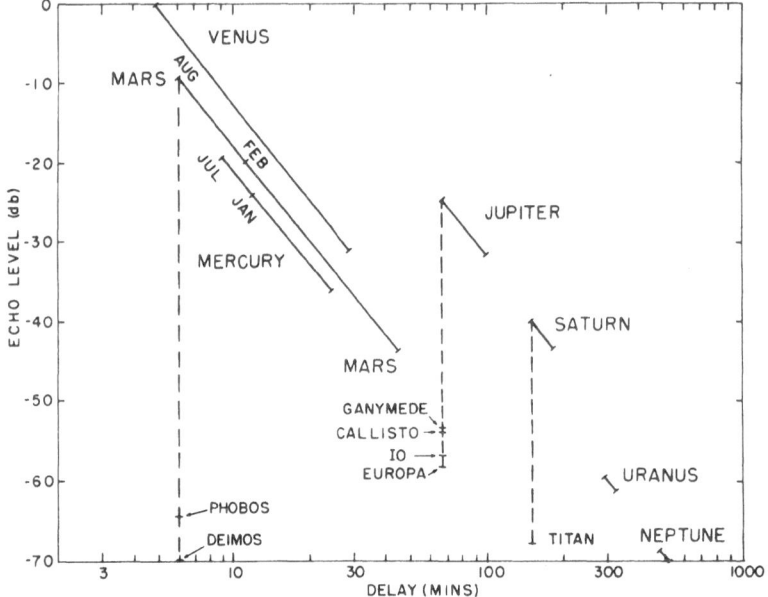

Fig. 1. Relative detectability and echo delay for the planets and their moons by radar, assuming equal surface reflectivity. Months listed by tick marks for Mars and Mercury show highest detectability when closest approach occurs on that date.

can be covered with a further improvement equivalent to that necessary to achieve the first Venus echoes.

Most of the increase in performance which has made radar astronomy possible has stemmed from the use of very large antennas. Since the gain of the antenna enters twice in the radar equation (see "Lunar Radar Reflections," chap. 16) and is proportional to the area, performance depends on the fourth power of antenna diameter. Currently, the largest antenna in regular use for the radar study of the planets is the 1000-ft-diameter reflector of the Arecibo Ionospheric Observatory (AIO) in Puerto Rico. Average transmitted power is important; most of the facilities now in operation use between 100 and 150 kw. Finally, the excess noise temperature of the receiving system is important since it determines the weakest signal which can be observed. Perhaps the most striking example of how low this quantity may be reduced is given by JPL (Goldstein [4]), who has achieved a value of only 37°K or roughly one-ninth the ambient temperature of the operating environment. By combining all these factors, MIT has reached −20 db, JPL and the Soviets approximately −25 db, and the AIO −35 db below the bare ability to detect Venus at closest approach.

In their attack on the problem of planetary detection, the various groups appear to divide into two camps. The older, represented by MIT, the early Jodrell work (Thomson et al. [22]). Jicamarca (Klemperer et al. [8]) and the AIO send out pulses of energy having peak powers in the neighborhood of 1 MW. Pulse widths range from 100 μsec up to many msec at repetition rates such that the average power is in the vicinity of 100 kw. More recently, JPL, the Soviets, El Campo (James and Ingalls [6]), and Jodrell Bank have employed a CW transmission, with various forms of phase and frequency coding. In the past several years both camps have employed increasingly sophisticated modulation and detection schemes which have had the effect of leveling the relative advantages and disadvantages of the two techniques, making them both depend only on the average power employed. Suffice it to say that the major groups in the field now have a signal delay measurement accuracy of about 50 μsec where signal-to-noise permits, and a Doppler frequency shift measurement accuracy approaching 0.1 c/s. Systems now in existence are adequate for detecting Venus and Mercury at almost all points in their orbits and Mars and perhaps Jupiter at close approach.

THE PLANETARY ORBITS

The orbits of the planets have been studied seriously by astronomers for many hundreds of years since the time of Tycho Brahe. However, the only tool available has been the telescope, and thus only angular measurements of position as a function of time could be made. By observing over long periods of time, it was possible to deduce the relative motions and distances between the planets with high accuracy. It was possible to specify roughly the relative distances to an accuracy of about one part in a million. More difficult, however, was the task of determining an absolute calibration of the scale of distance. The two obvious methods which have been used may be called the parallax or surveying method, where angular measurements at two ends of a known base line are made, and the dynamic method where a measured period of revolution permits the orbital distance to be calculated from the gravitational constant, provided the mass of the center of attraction is known. Both methods suffer seriously, however, in the case of the planets. Parallax measurements are difficult because of the great distance to the planets as compared to the length of available base lines which are accurately known in terms of terrestrial units of distance. Dynamic measurements are limited by the lack of independent methods for establishing the mass of the sun.

In practice, both parallax and dynamic measurements have been made on minor planets which have approached earth closely enough to be severely perturbed. Since the approach may be relatively close (a few times the distance of the moon for Eros), accurate triangulation may be employed using the terrestrial diameter as base line. Similarly, since the mass of the earth is relatively well known from lunar measurements, the perturbation observed may be used to calculate a distance. Both methods have been applied, chiefly, to the minor planet Eros during close approach to the earth, and have yielded values for the orbital distance scale. The unit commonly used to express planetary distances is called the astronomical unit and as defined is approximately equal to the mean distance of the earth around the sun. Measurements of the type just described have succeeded in establishing the size of this unit (which is approximately 149,600,000 km) only to an accuracy of about one part in a thousand.

Thus, prior to the advent of radar, astronomy was in the position of knowing the relative positions of the planets to an accuracy some one thousand times better than their absolute.

However, with the use of planetary radar measurements it became possible to measure the echo delay over long distances to an accuracy of several parts in a hundred million. Through the value for the velocity of light, known to an accuracy of about one part in a million, it is possible to relate these delay measurements to a distance expressed in terrestrial units such as kilometers and thus to calibrate the astronomical unit. Table I lists the results of observations made of Venus by several groups during the inferior conjunctions of 1961 and 1962. The errors claimed by the authors are shown; where these errors are small, i.e., of the order of 500 km, they reflect inaccuracies in our present knowledge of the details of the planetary orbits. Where they are larger, they generally result from measurement noise. It may be seen that for the most part they are quite self-consistent. Further refinement will require adjustment of many of the orbital elements of Venus and earth in addition to the A.U. Only in the past year have measurements with sufficient accuracy and spread over a sufficiently long time interval become available to make a solution for these adjustments worthwhile. It is interesting to note that since range measurements include the planetary radius in their reduction, it will be possible also to solve for improved values of this quantity.

Measurements of the other planets have not yet reached sufficient signal strength to permit detailed conclusions concerning orbital accuracy, although it is possible to say that the values of the A.U. obtained do not differ significantly from the values which best fit the Venus observations.

PLANETARY SURFACES

In a preceding lecture concerning lunar radar echoes it was shown how the intensity of the target echo could be related, at least roughly, to the type of surface material comprising the reflecting surface. Similarly for the planets, one

TABLE I

Radar Determination of the Astronomical Unit
(Venus Data Only)

Conjunction	Reference	A.U. (km)
1961	Pettengill et al. [17]	149,597,850 ±400
1961	Muhleman et al. [16]	149,598,500 ±500
1961	Thomson et al. [22]	149,600,000 ±5000
1961	Maron et al. [13]	149,596,000 ±900
1961	Kotelnikov et al. [9]	149,599,300 ±2000
1962	Muhleman [14]	149,598,757 ±670
1962	Kotelnikov et al. [11]	149,597,900 ±500
1962	Ponsonby et al. [18]	149,596,600 ±900

can gain some idea of the surface conditions from the strength of the return.

For Venus, Victor and Stevens [23], Pettengill *et al.* [17], Carpenter [1], Kotelnikov *et al.* [11], James and Ingalls [6], and Klemperer *et al.* [8] have found values for the fractional cross section lying between 0.1 and 0.2 with the higher values generally associated with the longer wavelengths. All the measurements indicate a significantly higher reflectivity than for the moon. If one applies the same analysis used in the case of the moon, one finds a surface dielectric constant of between 4 and 5. Such a value is quite consistent with a surface composed of dry rocks not unlike many rocks of the earth's surface. The presence of significant amounts of liquid water seems to be ruled out. The low value and relative independence of frequency of the reflectivity also make a dense Venusian ionosphere appear unlikely as a source of the reflections. A variation in intensity reported by some workers (James and Ingalls [6]) raises the possibility that differentiation of the surface may exist which affects the signal intensity as various portions of the planet rotate into the central region of the planetary disk.

Several workers (Smith [21], Kotelnikov [11], and Muhleman [15]) have reported angular scattering measurements of Venus. All are in agreement that the echo possesses the same qualitative features as do lunar echoes, namely, that the bulk of the return appears to be related to coherent reflection from flat areas of the surface near the center of the disk. However, in the case of Venus, the surface appears much flatter than for the moon. Muhleman [15] has applied an analysis similar to that developed for the moon and derives a mean surface slope parameter some three times smaller than for the moon. In Fig. 2 is plotted some unpublished results obtained during the 1964 conjunction at the Arecibo Ionospheric Observatory. For comparison, the lunar scattering law obtained under similar conditions is also plotted. Again it will be noted that the initial decay of power with range is less than for the moon, indicating a smoother planetary surface. Also of interest is the presence of a rough scattering component at greater delays, not departing too greatly from the lunar law.

For Mercury and Mars, the available information is much more primitive. The Soviets (Kotelnikov [10]), JPL (Muhleman [14]), and the AIO (Dyce *et al.* [2]) are in good agreement that the surface reflectivity is 0.06, very close to the value obtained for the moon. The surface appears smooth, although the mean slope has not been established. For Mars, there appears to be some variability which is associated with planetary longitude. Surface reflectivities between 0.02 and 0.10 have been reported by JPL (Goldstein and Gillmore [5]) and by the Soviets (Kotelnikov *et al.* [12]).

PLANETARY ROTATION

In the earlier discussion of lunar radar reflections, the importance of frequency spectra as a tool to determine the distribution of reflectivity over the surface was brought out. In reducing the measurements to actual positions on the surface of the moon, it was necessary to use the known rates of libration of the lunar surface with respect to the observer. However, by turning the process around, it would have been possible to obtain the apparent rotation of the moon

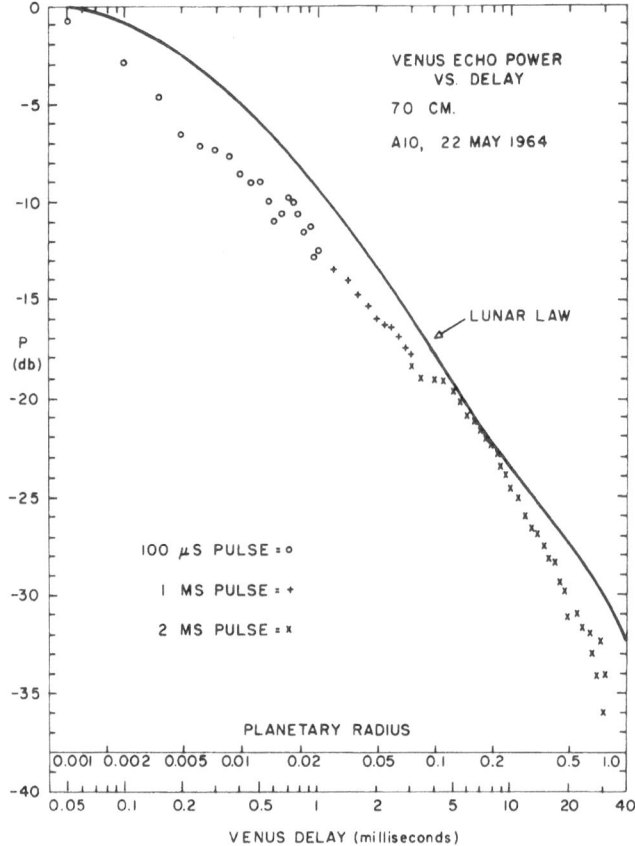

Fig. 2. Venus echo power *vs.* delay. Data taken at the
Arecibo Ionospheric Observatory.

from the radar measurements to a high degree of accuracy. For Venus, where the
rotation is essentially unknown from visual measurements because of the extensive cloud cover, extreme interest attaches to radar measurements.

If only frequency spectrum measurements are available, it is possible to
deduce the apparent instantaneous angular velocity from the maximum extent of
the spectrum, provided sufficient signal-to-noise ratio is available to ensure that
one is actually observing echo power from the limbs of the planet; and, of course,
one must assume that the radius of the planet is known. Working within these
limitations, Carpenter [1] has calculated that the rotation of Venus has a sidereal
period of approximately 266 days in the retrograde sense with an orientation approximately perpendicular to the plane of its orbit. The earlier Soviet results
(Kotelnikov *et al.* [9]) that the rotational period was as rapid as 11 days appear
to be discredited.

An even more powerful attack is possible if one can obtain frequency spectra
at a selected and known range. Reference to Figs. 7 and 8 of the lunar radar lec-

ture will show that the spectra in such a case are sharply defined and avoid the necessity of assuming that echo power is being returned from the actual planetary limb.

Using this technique, Goldstein [4] has obtained from the 1962 observations a retrograde rotation of 248 days, again with an axis orientation very close to perpendicular to the orbital plane of Venus. Unpublished data from the 1964 conjunction taken at the Arecibo Ionospheric Observatory confirm this latter measurement, and yield a value for the sidereal rotation of 247 ± 5 days retrograde and with an axis orientation approximately $6°$ off perpendicular to the orbital plane. Twelve measurements, reduced to the equivalent limb-to-limb Doppler spread are shown in Fig. 3, together with the curve which is the least-mean-squares fit to the data. The three parameters which yield the best fit curve are also shown. From the extreme accuracy with which the 12 observed points may be made to fit a theoretical curve of three degrees of freedom, it must be concluded with no reservation that we are observing, at least at 70 cm, the actual surface of a solid, rotating body. However, we must not rule out the possibility that there are also weaker reflections from a variable and turbulent atmosphere which can give rise to a much broader spectral component.

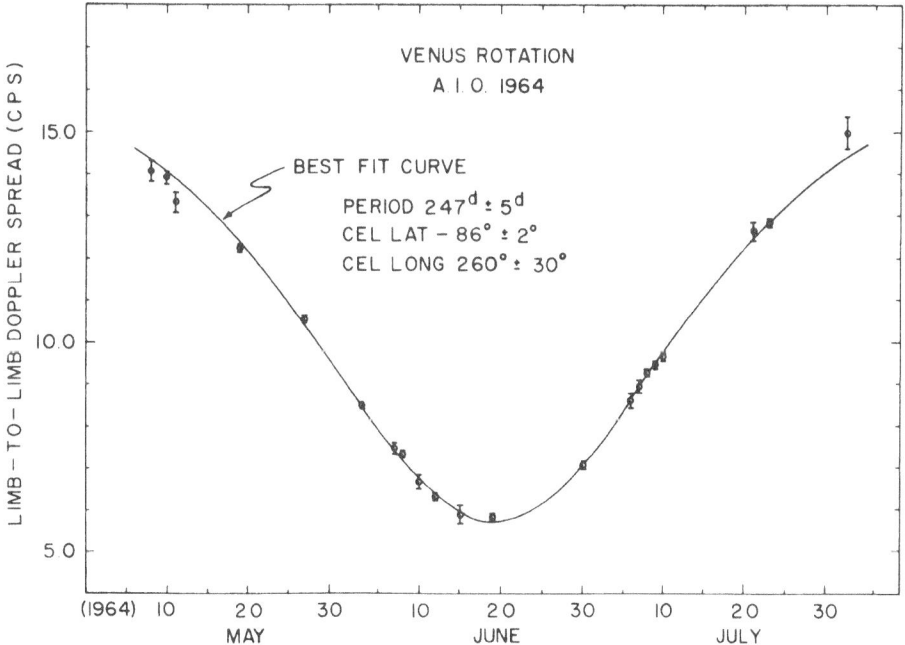

Fig. 3. Plot of limb-to-limb Doppler spread *vs.* date observed for Venus during the 1964 inferior conjunction. The solid curve represents the least-mean-squares fit to the data and corresponds to the rotation axis specified. Data taken at the Arecibo Ionospheric Observatory.

For Mercury, visual observations have indicated that the planet is rotating synchronously in its orbit around the sun, much as the moon does around the earth. To date, insufficient signal strength has been available to confirm these results by radar observation.

In the case of Mars and Jupiter, the rotation has been clearly determined by direct visual observation and it is unlikely that radar observation can add greatly to the presently known accuracy.

SUMMARY

Radar has been used with great success to improve our knowledge of the planetary orbits. So far, its most important result is an improved value for the astronomical unit, with a value of 149,598,000 ±500 km widely accepted. Further improvement will require adjustments in the elements of the planetary orbits, which are now being calculated. Current delay measurement accuracy is approaching 10 μseconds or a few parts in 10^8 of the total delay. Information should also be forthcoming on planetary radii.

From the strength of the returned echoes it is possible to exclude the presence of substantial amounts of liquid water on Mercury, Venus, and Mars. Venus appears to have the densest surface, approximately the same as for solid terrestrial rocks. Mercury appears to reflect in most respects like the moon, but Mars has a surface reflectivity which is highly variable and appears to be correlated with surface position. The surface slopes of Venus appear to be less steep than in the case of the moon, while for Mars they are smoother yet. The presence of an ionosphere has not yet been established from any of the radar measurements.

In the case of Venus, rotation has very reliably been established to be 247 ±5 days retrograde sidereal with an axis oriented very nearly perpendicular to the orbital plane of the planet. For the other planets, the radar echoes have been too weak to permit an analysis leading to a value for rotation.

REFERENCES

1. R. L. Carpenter, *Astron. J.* 69, 2 (1964).
2. R. B. Dyce, G. H. Pettengill, and A. D. Sanchez, Paper presented at URSI (Washington, D.C., April 1964).
3. J. V. Evans and G. N. Taylor, *Nature* 184, 1358 (1959).
4. R. M. Goldstein, *Astron. J.* 69, 12 (1964).
5. R. M. Goldstein and W. F. Gillmore, *Science* 141, 1171 (1963).
6. J. C. James and R. P. Ingalls, *Astron. J.* 69, 19 (1964).
7. F. J. Kerr, *Proc. IRE* 40, 660 (1952).
8. W. K. Klemperer, G. R. Ochs, and K. L. Bowles, *Astron. J.* 69, 22 (1964).
9. V. A. Kotelnikov *et al.*, *Radiotekhn. i Elektron.* 7, 1860 (1962).
10. V. A. Kotelnikov *et al.*, *Dokl. Akad. Nauk* 148, 1320 (1962).
11. V. A. Kotelnikov *et al.*, *Dokl. Akad. Nauk* 151, 532 (1963).
12. V. A. Kotelnikov *et al.*, *Dokl. Akad. Nauk* 151, 811 (1963).
13. I. Maron, G. Luchak, and W. Blitzstein, *Science* 134, 1419 (1961).
14. D. O. Muhleman, Paper presented at Intern. Astron. Union Symp., 21st (Paris, 1963).
15. D. O. Muhleman, *Astron. J.* 69, 34 (1964).
16. D. O. Muhleman, D. B. Holdridge, and N. Block, *Astron. J.* 67, 191 (1962).

17. G. H. Pettengill *et al.*, *Astron. J.* **67**, 181 (1962).
18. F. E. B. Ponsonby, J. H. Thomson, and K. S. Imrie, To appear in *Monthly Notices, Roy. Astron. Soc.* (1963).
19. R. Price, *et al.*, *Science* **129**, 751 (1959).
20. R. Price and G. H. Pettengill, *Planetary Space Sci.* **5**, 71 (1961).
21. W. B. Smith, *Astron J.* **68**, 15 (1963).
22. J. H. Thomson, *et al.*, *Nature* **190**, 519 (1961).
23. W. K. Victor and R. Stevens, *Science* **134**, 46 (1961).
24. W. K. Victor, R. Stevens, and S. W. Golomb, Jet Propulsion Lab. Tech. Rept. No. 32-132 (1961).

Distribution of the Ionizing Radiation
on the Solar Disc During the
Solar Eclipse of Feb. 15, 1961

Dr. D. Ilias

Ionospheric Institute of the National Observatory of Athens
Athens, Greece

Two inactive regions were localized on the solar disc during the eclipse by the comparing of observations of E layer critical frequencies at Athens with other ionospheric and solar radio noise observations. These regions were found to correspond to the intersection of arcs representing the position of the moon's limb on the solar disc when sharp changes of gradient occurred in the electron density or solar radio noise data at the three stations (Athens, Florence, and Garchy).

This method was also applied for the low F region variations using true height profile data from six well-spaced stations (Athens, Ebro, Garchy, Graz, Lindau, and Rostov). When changes of the slope of the electron density variation curves are considered with reference to the phase of the eclipse at each station, an approximate model of the Feb. 15, 1961, solar map was obtained. In this map, which correlated well with the 9.1 cm spectroheliogram published by Stanford, both hot and cold regions were observed. The possibilities of existence of such cold regions for the ionizing sun are discussed.

Index